T0293845

Robust Response Surfaces, Regression, and Positive Data Analyses

Robust Response Surfaces, Regression, and Positive Data Analyses

Rabindra Nath Das

CRC Press
Taylor & Francis Group
Boca Raton London New York

CRC Press is an imprint of the
Taylor & Francis Group, an **informa** business

A CHAPMAN & HALL BOOK

CRC Press
Taylor & Francis Group
6000 Broken Sound Parkway NW, Suite 300
Boca Raton, FL 33487-2742

© 2014 by Taylor & Francis Group, LLC
CRC Press is an imprint of Taylor & Francis Group, an Informa business

No claim to original U.S. Government works

Printed on acid-free paper
Version Date: 20140404

International Standard Book Number-13: 978-1-4665-0677-0 (Hardback)

Visit the Taylor & Francis Web site at
http://www.taylorandfrancis.com

and the CRC Press Web site at
http://www.crcpress.com

Dedicated to
my beloved mother
and to the memory of my father

Contents

List of Figures

List of Tables

Preface

The present book initiates the concept of robust response surface designs, along with the relevant regression and positive data analysis techniques. Response surface methodology (RSM), well-known in literature, is widely used in every field of science and technology such as biology, natural (physical/chemical), environmental, medical, agricultural, quality engineering, etc. RSM is the most popular experimental data generating, modeling, and optimization technique in every field of science. It is a particular case of robust response surface methodology (RRSM). RSM has many limitations, and RRSM aims to overcome many of such limitations. Thus, RRSM will be much better than RSM. It is intended for anyone who knows basic concepts of experimental designs and regression analysis.

This is the first unique book on RRSM. Every chapter is unique regarding its contents, presentation, and organization. Problems on robust response surface designs such as rotatability, slope-rotatability, weak rotatability, optimality, and along with the method of estimation of model parameters, positive data analysis techniques are considered in this book. Some real examples on lifetime responses, resistivity, replicated measures, medical, demography, hydrogeology data, etc., are analyzed. Some examples (considered in this book) on design of experiments do not satisfy the classical assumptions of response surface methodology.

This book is intended as an introductory book on robust response surface designs. The related topics of RRSM, such as regression analysis for correlated errors and positive data analysis, have been included in this book. It is aimed primarily at theoretical and practical statisticians. The range of topics and applications gives this book broad appeal both to theoreticians and practicing professionals in a variety of fields of science. This book will be for a second course in design of experiments along with the regression and data analysis. It is intended to allow a broad group of quality engineers, scientists (in any field), medical practitioners, demographers, economists and statisticians to learn about RRSM, regression and positive data analysis, including both the theory and how to apply it.

The main emphasis of this book is on examining the concept of rotatability, weak rotatability, D-optimality, slope-rotatability, weak slope-rotatability, D-optimal slope rotatability, regression analysis with correlated errors, and positive data analysis. All the contents of this book are taken from the research articles of the author and his collaborators. Thus, the book is unique in every sense. The contents included in this book are described below chapter wise.

The problems included in this book are described in Chapter 1. Short literature reviews related to these problems is presented herein. A detailed summary of each chapter is given. Concepts of robust response surface designs are also clearly explained in this chapter.

Chapter 2 starts with correlated first-order regression designs. The concept of robust first-order rotatable and D-optimal designs are described. The rotatability and D-optimality conditions of first-order regression designs are described for correlated errors. These conditions are further simplified for different error variance-covariance structures (i.e., intra-class, inter-class, compound symmetry, tri-diagonal, and autocorrelated), which are commonly encountered in practical situations. Some construction methods of *robust rotatable* and *D-optimal* designs are described for each error structure.

Chapter 3 describes second-order regression designs for correlated errors. Robust second-order rotatability conditions are examined for a general error variance-covariance structure. Second-order rotatability conditions are simplified for intra-class, inter-class, compound symmetry, tri-diagonal, and autocorrelated error structures. Robust second-order rotatable designs (RSORDs) are constructed for each of the above error structures.

Chapter 4 includes location-scale lifetime models which are derived based on *constant* variance assumption from a lifetime distribution, such as log-normal or exponential. Assuming a first- or second-order location-scale model with correlated errors, first or second-order rotatability and D-optimality conditions are derived for a general error variance-covariance structure. It is observed that the designs are *invariant* of these two non-normal distributions, but they depend on the error variance-covariance structure. Similar results as in Chapters 2 and 3 are presented for both the non-normal distributions, and for some error structures as mentioned in Chapter 2.

In Chapter 5, we present some measures for the degree of closeness to *robust rotatability* of first- and second-order regression designs, which are *not exactly robust rotatable* for a fixed pattern of error correlation structure. The measures are based on the (a) dispersion matrix of regression coefficients and (b) moment matrix. A rule for comparison between robust rotatable and non-rotatable designs are described based on variance and cost function. All the results of closeness to robust rotatability (i.e., weak rotatability) are critically studied with examples.

In Chapter 6, we consider second-order correlated model as in Chapter 3. Robust second-order slope-rotatability along axial directions are described. Conditions of robust second-order slope-rotatability and modified slope-rotatability along axial directions are derived for a general error variance-covariance structure. These conditions are further simplified for intra-class error structure, and are compared with the usual and modified second-order slope-rotatability conditions.

Chapter 7 describes some concepts of optimal robust second-order slope-rotatable designs. Robust second-order slope-rotatability over all directions, with equal maximum directional variance and D-optimal slope are examined in Chapter 7. It is examined that a robust second-order rotatable design is also a robust second-order slope-rotatable design over all directions (for any factors) and with equal maximum directional variance (only for two factors). It is also established that within the robust second-order symmetric balanced designs, robust rotatable designs are also robust D-optimal slope and slope-rotatable with equal maximum directional variance for more than two factors. A class of robust second-order slope-rotatable designs over all directions, with equal maximum directional variance and D-optimal slope are derived for some special error correlation structures.

Chapter 8 considers weakly robust second-order slope-rotatable designs. Some measures are described for robust second-order slope-rotatability along axial directions, over all directions and with equal maximum directional variance. These measures are illustrated with examples.

In Chapter 9, we consider the regression analysis with correlated error models that are described in Chapters 2, 3, and 4. In general, the form of the correlation structure is known for a given situation of the data set, but the parameter that is involved in the correlation structure is always unknown. Some robust methods of estimating the best linear unbiased estimators of all the regression parameters, *except* the intercept are described. In this connection, we also describe some testing procedures for any set of linear hypotheses regarding the unknown regression parameters. Confidence ellipsoid of a set of estimable functions of regression parameters is described. Index of fit for the fitted regression model is presented. Two examples with simulated data illustrate the methods of estimation. Two applications of correlated regression analyses in block designs are described. Randomized block design under compound symmetry structure and a reinforced randomized block design under compound autocorrelated error structure are analysed using correlated regression analysis. In the process, confidence ellipsoid, confidence interval and multiple comparison techniques are studied.

In Chapter 10, log-normal and gamma models are described for constant and non-constant variances. In general, experimental observations are positive, which are generally analysed by log-normal and gamma models. Discrepancy of regression estimates and also in fitting between the log-normal and gamma models for constant and non-constant variances are explained and illustrated. Some applications of log-normal and gamma models analysis with real data are illustrated in quality engineering, medical science, demography, hydrogeology etc.

For smooth readability, we attempted to present each chapter more or less independent of other chapters. As a result, it has not been possible to avoid a few repetitions in some places. Hope that readers will enjoy such repetitions. I have tried my best level to present all the above concepts clearly. It may contain some typographical errors. I would like to request the kind hearted readers to send their comments, suggestions and corrections to my e-mail address: rabin.bwn@gmail.com for further improvement.

I would like to acknowledge the contribution of many people to the conception and completion of this book. With the greatest pleasure I take this opportunity to record my sincerest gratitude to Prof. A.C. Mukhopadhyay for his helpful directions all through the investigation, reviewing and numerous instructive suggestions that have improved the presentation of the book. I also express my deepest gratitude to Prof. B.K. Sinha for his help, direction and valuable suggestions during the work. I am particularly grateful to my teacher Prof. R.N. Panda for kindly introducing the problem to me. I am very much indebted to the three referees who have provided valuable comments and suggestions to improve this book. For collection of several variance-covariance structures, I am also grateful to the late Prof. B.N. Mukherjee. Prof. S.H. Park and Prof. Y. Lee have greatly influenced my career and research work. I have been motivated to carry on research on slope-rotatability by Prof. S.H. Park and on data analysis by Prof. Y. Lee. I would like to thank my co-researchers Prof. A.C. Mukhopadhyay, Prof. R.N. Panda, Prof. A. Sengupta, Dr. S. Dihidar, Prof. R.R. Verdugo, Prof. S.H. Park, Prof. Y. Lee, Prof. M.L. Aggarwal, Prof. D.K.J. Lin, Prof. S. Huda, Prof. J.S. Park, Prof. J. Paul, and Dr. J. Kim for their support of this work. I would like to thank my friend and co-researcher Dr. Kim for his constant inspiration and kind help in type setting. Finally, I would like to thank my family members: my mother Mrs. L.L. Das, my Mom Mrs. Ju-hwa Lee, wife Dr. K.K. Das, and my beloved daughter Mahashweta Das for their love and support.

Rabindra Nath Das

Burdwn, West Bengal, India

Author

Rabindra Nath Das is an Associate Professor in the Department of Statistics, The University of Burdwan, Burdwan, W.B., India. He holds Ph. D. in Statistics from Burdwan University and Post-Doc from Seoul National University, Seoul, Korea. He has authored several articles on design of experiments, quality engineering, regression analysis, epidemiology, medical science, and demography. His special area of interest is on design of experiments, regression analysis, quality engineering, and epidemiology.

Chapter 1

INTRODUCTION

Response surface methodology is a collection of mathematical and statistical techniques useful for analyzing problems where several independent variables influence a dependent variable. The independent variables are often called the input or explanatory variables and the dependent variable is often called the response variable. In response surface methodology, a natural and desirable property is that of *rotatability*, which requires that the variance of the predicted response at a point remains constant at all such points that are equidistant from the design center. The present research monograph confines to first- and second-order regression models with correlated errors. Robust rotatability, slope-rotatability and optimality are described for the first- and second-order designs. Weakly robust rotatable, slope-rotatable designs, and their different measures are discussed. Robust first- and second-order optimal and rotatable designs are discussed for different lifetime distributions in quality improvement experiments. Regression analyses are described for correlated observations. Generally, experimental observations are positive, and they are analysed by log-normal and gamma models. These two models are described for constant and non-constant variances. The discrepancy of regression estimates and the model fittings between these two models are described. Some applications of correlated regression analyses in block designs, and positive data analyses (with real data) are illustrated in the different fields of science.

1.1 THE PROBLEM AND PERSPECTIVE

The basic problem of response surface designs can be stated like this: In an experiment there are k quantitative factors $(x_1, x_2, ..., x_k)$ whose levels can be accurately controlled. For any choice of these levels there would be a response which can only be observed subject to a random error. The true response is functionally dependent on the levels; however, the functional relation between the true response and the levels (the true 'response function' or 'response surface') is not completely known. The object is to obtain information about this relation on the basis of the responses observed at some chosen combinations of the levels of $(x_1, x_2, ..., x_k)$. Specifically, the aim is to obtain an approximate model that is good enough for the region of interest. The success of response surface methodology (RSM) lies in the fact that a simple second-order model can serve as a very useful approximation of the true response surface. The problem of designing is that of choosing the level combinations or the 'design points' in the specified domain of $x_1, x_2, ..., x_k$ (the experimental region) so that one can explore the response surface adequately and efficiently.

Within the above general framework the type of the problem would be determined by various considerations such as the extent of knowledge assumed for the response function (whether a polynomial, and if so, of what degree), the nature of experimentation (whether a single experiment or sequential experiments), the nature of experimental errors (whether uncorrelated or correlated), the aspect of the response surface which is of interest, etc.

The present book concerns with the response surface designs for correlated errors. It concerns with the following specific problems: (i) Let us assume that the true response function is a first-order polynomial and errors are correlated with a general correlated error structure given by the variance-covariance matrix of errors (W). We are concerned with the (a) conditions of first-order rotatability and D-optimality for a general W, (b) determination of first-order rotatable and D-optimal designs which remain valid for all possible values of correlation parameter or parameters belonging to a well-defined class $W_0 = \{W$ positive definite: $W_{N \times N}$ defined by a particular correlation structure neatly specified$\}$, (c) construction of rotatable and D-optimal first-order regression designs for different correlation structures.

(ii) Let us assume that the true response function is a second-order polynomial and errors are correlated with a general variance-covariance structure W. We are concerned with the (a) second-order rotatability conditions for a general W, (b) determination and construction of second-order rotatable designs for all possible values of correlation parameter or parameters belonging to a well defined class $W_0 = \{W$ positive definite: $W_{N \times N}$ defined by a particular correlation structure neatly specified$\}$, (c) determination of general conditions for second-order optimality of rotatable designs as specified in (ii)(b), (d) construction of second-order rotatable designs for different correlation structures.

(iii) Let us assume that the true response function is a first-order polynomial and errors are correlated (with a general correlation structure W). Also assume that the correlated responses are lifetimes having log-normal and exponential distributions. We are concerned with the (a) conditions of first-order rotatability and D-optimality for a general W, (b) determination of first-order rotatable and D-optimal designs which remain valid for all possible values of correlation parameter or parameters belonging to a well defined class $W_0 = \{W$ positive definite: $W_{N \times N}$ defined by a particular correlation structure neatly specified$\}$, (c) construction of rotatable and D-optimal first-order regression designs under these two lifetime distributions and for different correlation structures.

(iv) Let us assume that the true response function is a second-order polynomial and errors are correlated (with a general correlation structure W). Also assume that the correlated responses are lifetimes having log-normal and exponential distributions. We are concerned with the (a) second-order rotatability conditions for a general W, (b) determination of second-order rotatable designs which remain valid for all possible values of correlation parameter or parameters belonging to a well defined class $W_0 = \{W$ positive definite: $W_{N \times N}$ defined by a particular correlation structure neatly specified$\}$, (c) construction of second-order rotatable designs under these two lifetime distributions and for different correlation structures.

(v) Let us assume a design is not first-order rotatable as specified in (i)(b). We are concerned with the (a) determination of the measure of degree of closeness to first-order rotatability, (b) comparison of a non-rotatable design with a first-order rotatable design based on cost and variance functions.

Let us assume a design is not second-order rotatable as specified in (ii)(b). We are concerned with the (c) determination of the measure of degree of closeness to second-order rotatability, (d) comparison of a non-rotatable design with a second-order rotatable design based on cost and variance functions.

(vi) Let us assume that the true response function is a second-order polynomial and errors are correlated (with a general correlation structure W). We are concerned with the (a) second-order slope-rotatability and modified slope-rotatability conditions along axial directions for a general correlation structure W, and (b) verification of these conditions with the usual slope-rotatability and modified slope-rotatability conditions.

(vii) Let us assume that the true response function is a second-order polynomial and errors are correlated (with a general correlation structure W). We are concerned with the (a) second-order slope-rotatability conditions over all directions for a general correlation structure W, and (b) determination of second-order slope-rotatable designs over all directions for some well-known correlated error structures.

(viii) Let us assume that the true response function is a second-order polynomial and errors are correlated (with a general correlation structure W). For a general correlation structure W, we are concerned with the (a) second-order slope-rotatability conditions with equal maximum directional variance, (b) second-order D-optimal slope-rotatability conditions, and (c) determination of second-order slope-rotatable designs with equal maximum directional variance and D-optimal for some well-known correlated error structures.

(ix) Let us assume that the true response function is a second-order polynomial and errors are correlated (with a general correlation structure W). We are concerned with the determination of the measures of degree of closeness to second-order slope-rotatability (a) along axial directions, (b) over all directions and (c) with equal maximum directional variance.

(x) Correlated regression models are introduced in the list of problems (i), (ii), (iii), (iv) and (v). For correlated regression analysis, we are concerned with the (a) estimation of unknown regression parameters and correlation parameter (or parameters), (b) determination of testing procedure for testing any set of linear hypotheses regarding unknown regression parameters, (c) determination of confidence interval, confidence ellipsoid and index of fit, (d) applications of correlated regression analysis in block designs, (e) estimation of regression and correlation parameters for the location-scale regression models for lifetime distributions: log-normal and exponential.

(xi) Generally, experimental responses are positive which are analysed by log-normal and gamma models. We are concerned with the (a) derivation of log-normal and gamma models for constant and non-constant variances, (b) examining the discrepancy of the regression estimates and the model fittings between these two models for constant and non-constant variances, (c) analysis of some real examples based on design of experiments (for lifetime distributions, resistivity, multiple responses, dual response surface designs) which do not satisfy the classical assumptions of response surface designs, (d) applications of these two modeling techniques in quality engineering, medical science, demography, hydrogeology, etc.

1.2   A BRIEF REVIEW OF THE LITERATURE

The existing literature on response surface designs and optimality of designs is already so vast that it will not be possible to give a comprehensive resume within a short span. Instead of attempting that we make here a selective review concentrating on those works which have direct relevance to the problems mentioned in Section 1.1.

Although the idea of response surface models is contained in the classical theory of regression and some early references to the design problem can be found in the literature (see Smith, 1918; Wald, 1943; Placket and Burman, 1946), a full-scale study of the problem can be said to have started with the pioneering work of Box and Wilson (1951). In that paper, Box and Wilson described a sequential procedure for determining the optimal factor combination in a multifactor experiment with a small error variance. They first outlined how to reach the neighbourhood of the stationary point along a 'line of steepest ascent' based on first-order designs. After the neighbourhood was reached, they recommended the inclusion of additional design points for exploring the surface by fitting a second-degree polynomial. The next important step in the development of response surface designs was the introduction of rotatable designs, by Box and Hunter (1957). These designs generate information in all directions symmetrically and are the natural choice of the experimenter in studies where the orientation of the response surface with respect to the co-ordinate axes is not known in advance. Box and Draper (1959) considered the minimization of the integrated mean squared error over the region of interest and generally came to the conclusion that the design which minimizes the bias factor alone is very close to the design which minimizes the integrated mean squared error. Among other contributions to response surface methodology we may refer to Herzberg (1966, 1967), Das and Dey (1967), Dey and Nigam (1968), Mukhopadhyay (1969), Adhikary and Sinha (1976), Adhikary and Panda (1992).

Box and Hunter (1957) introduced the concept of rotatable designs (for uncorrelated and homoscedastic errors) which generate information symmetrically in all directions and stressed the desirability of a design used for exploring the nature of the response surface to be *rotatable*. Emphasis was placed on judging a design on the basis of *prediction variance*, $\text{Var}(\hat{y}/\sigma^2)$. This was a vital contribution, since it underscored very early that a single quantity may not provide enough information when comparing designs. The distribution of $\text{Var}(\hat{y}/\sigma^2)$ over the space of the design variables was explored. A natural and easily attainable property is that of *rotatability* which requires that the variance of the predicted value remains constant at points that are equidistant from the design center. It is rare when a heavy price is to be paid for achieving near rotatability. Rotatability in the case of first-order models is attainable with standard orthogonal arrays of strength two that have many other important properties too. In the case of second-order model, central composite designs defined by Box and Hunter (1957) and other designs can be made to be rotatable very easily.

The introduction of rotatable designs by Box and Hunter (1957) led to the development of a large volume of literature on the construction of such designs. One can refer, in particular, to Bose and Draper (1959), Bose and Carter (1959), Debaun (1959), Gardiner, Grandage and Hader (1959), Hartley (1959), Box and Draper (1959, 1963), Box and Behnken (1960a, 1960b), Draper (1960a, 1960b, 1960c, 1961, 1962), Das and Narasimham (1962), Das (1963), Herzberg (1964), Draper and Herzberg (1971, 1973, 1979). Further references to work prior to 1966 can be obtained in a review article on response surface methodology by Hill and Hunter (1966), and to later work in review articles by Mood and Pike (1975); Myers, Khuri and Carter, Jr. (1989) (Response surface methodology: 1966-1988), and also by Myers, Montgomery, Vining, Borrow, and Kowalski (2004).

The historical importance of the property evolved naturally from the need to achieve *stability in prediction variance*. Too many designs have very unstable prediction variance even in the region of experimentation. Rotatability was a first step towards stabilization and, unlike many other criteria, it could be put to use immediately with standard designs. Obviously, in experiments with mixtures in which a simplex region is important, rotatability is less appealing. Moreover, in early exploratory phases of response surface methodology (RSM) when the learning process about the design variables is in its infancy, the *rotatability* property is certainly *not* important and may even be compromised infavour of other desirable design features. Nevertheless, it is always a good idea to *preserve* some degree of rotatability wherever possible. Recently, some measures of degree of rotatability have been introduced. The first one, introduced by Khuri (1988), is a quantitative measure expressible as a percentage with the value 100 attained when the design is rotatable. The second measure, introduced by Draper and Guttman (1988), provides information about the overall shape of the variance contours for symmetrical second-order designs. Recently, Draper and Pukelsheim (1990), have generalized the measures of *near* rotatability. It provides a criterion that is easy to compute and is invariant under design rotation. It also easily extends to higher degree models. Park, Lim, and Baba (1993) also developed a measure of second-order rotatability based on polar transformation.

The need of the practical user of RSM had been more or less satisfied by the establishment of families of useful experimental designs for first- and second-order models. In the first-order case, the need for orthogonal designs was motivated by Box and Wilson (1951), Box (1952, 1954), and in an excellent book edited by Davies (1954). Specific rotatable designs from two-level factorial and fractional factorial experiments were made available by Plackett and Burman (1946) and Box and Hunter (1961a, 1961b). As far as second-order models are concerned, many subject matter scientists and engineers have a working knowledge of the family of central composite designs (CCDs) given by Box and Hunter (1957) and a class of special three-level designs by Box and Behnken (1960a). Another important contribution came from Hartley (1959), who made an effort to create a more economical or small composite designs. Subsequently a large volume of literature dealing with this problem has evolved with significant contribution from Bose and Draper (1959), Draper (1960a), Box and Behnken (1960a), Das (1961), Das and Narasimhan (1962), Raghavarao (1963), Das (1963), Nigam and Das (1966), Herzberg (1967) and many other workers. Nevertheless, the requirement of rotatability, while it reduces the set of our choices, does not by itself go far enough to determine the optimum design in a particular situation.

In view of optimality of regression designs, rotatable designs play an *important* role. A large volume of significant contributions with respect to optimality are from Elfving (1952, 1955, 1959), Chernoff (1953), Kiefer (1958, 1959, 1960, 1962a, 1962b, 1974, 1975), Karlin and Studden (1966), Kiefer and Wolfowitz (1959, 1960), Galil and Kiefer (1977, 1980). Kiefer, in particular, was instrumental in providing the mathematical groundwork for optimal design theory. Actually, the origin of this theory can be traced back to an article by Smith (1918) who considered different design configurations for polynomial models up to sixth degree. Interestingly enough, Smith (1918) examined cases in which the error variance was not constant within the experimental region. Some early work in the area was done by Wald (1943) and Mood (1946).

Optimal design theory has clearly become an important component in the general development of experimental designs for the case of regression models. For general references in the area one is referred to an excellent book by Pukelsheim (1993) on optimal regression design of experiments. Kiefer and Wolfowitz (1959, 1960) examined the criterion for the regression design and established its equivalence with the minimax criterion proposed earlier by Smith (1918). The work of Kiefer and Wolfowitz was followed up by significant contributions from Karlin and Studden (1966), who established the *admissibility* of rotatable designs in a given context. Later on Galil and Kiefer (1977) derived the designs that are optimum among the rotatable designs with respect to the D-, A- and E-optimality criteria and designs were compared in their performance relative to these and other criteria. Kiefer (1961, 1962a, 1962b) in a series of papers extended the D-optimality criterion to that of Ds-optimality applicable to the case when one is interested in a few but not all of the parameters of the model. The most prominent design optimality criterion happens to be D-optimality, the minimization of the generalized variance of the model coefficients. This criterion has received a great deal of attention, as evidenced by the numerous articles written about it.

The review articles by St. John and Draper (1975), Ash and Hedayat (1978), and Atkinson (1982a) contain many references on D-optimality. Another design criterion that is closely related to D-optimality is G-optimality, in which the maximum prediction variance over the experimental region is minimized. Kiefer and Wolfowitz (1959, 1960) proved the equivalence of G- and D-optimality criterion for a compact experimental region. Other variance related criterion include A-optimality and E-optimality and their detailed descriptions are given in Silvey (1978, 1980). Bandemer (1980) and Atkinson (1982a) provided surveys of optimal experimental designs. A review of algorithms for constructing D_N-optimal designs was given by Cook and Nachtsheim (1980), Johnson and Nachtsheim (1983). Galil and Kiefer (1980) and many other authors developed a family of computer search methods for finding D-optimum or G-optimum designs. For a recent book on optimal regression designs one is referred to the book by Pukelshiem (1993).

In view of slope-rotatability of response surface methodology, a good estimation of the derivatives of the response function may be as important or perhaps more important than the estimation of mean response. Atkinson (1970a, 1972) considered designs for estimation of the slope at a fixed point with the response function being of order 1. Murty and Studden (1972) considered polynomial-regression models with the criterion being the variance of an estimated slope at a fixed point and averaged over an interval. Hader and Park (1978) extended the notion of rotatability to cover the slope for second-order models. They catalogued designs that result in *slope-rotatability*, that is, the variance of the estimated derivatives is constant for all points equidistant from the design center. Myers and Lahoda (1975) extended the Box-Draper integrated mean squared error criterion under model misspecification to cover sets of parametric functions, with the slope being the area of primary application. Mukherjee and Huda (1985) developed designs associated with minimum variance of the estimated slope, maximized over all points in the factor space for second- and third-order polynomial models over a spherical region. Huda (2006) and Park (2006) reviewed the literature on experimental designs for the estimation of the differences between responses, and most of the relevant references are cited there. Huda (2006) discussed various concepts of optimality for estimating the slopes of a response surface in the context of uncorrelated and homoscedastic errors. In recent years, interests in RSM have been increased, and books on this subject have been written by some authors such as Box and Draper (1987, 2007), Pukelsheim (1993), Khuri and Cornell (1996), Park (1996), Myers and Montgomery (1995), Khuri (2006), etc.

1.3 EXISTING LITERATURE IN THE DIRECTION OF PRESENT RESEARCH MONOGRAPH

All the literature for response surface methodology (RSM) cited in Section 1.2 is studied only for uncorrelated and homoscedastic errors. It is not uncommon to come across practical situations where the errors are correlated violating the usual assumptions (Vide article 8.7, pages 273-275, Table 8.3 of Chatterjee and Hadi, 1988). Gennings *et al.* (1989) also discuss practical situations warranting correlated errors and provide estimation of nonlinear model parameters, accommodating both within-unit and between-unit variabilities in fitting a response surface. Bischoff (1992, 1995a, 1995b 1996) studied D-optimal designs for regression models with correlated observations. Myers and Montgomery (1995) (also reproduced in Myers *et al.*, 2002, Example 6.3, page 220) presented an experimental data in a semiconductor plant of a lamination process, and the chamber measurement is made four times on the same device procedure. The chamber measurement is known to be non-normal with a heavy right-tailed distribution. In addition, it is clear that the measurement taken on the same device may well be correlated. This process is a 'repeated measures', which represents the source of the correlation. Myers *et al.* (2002, p.128) also noticed that in *industrial applications* experimental units are not independent at times *by design*. At times this leads to correlation among observations via a *repeated measures* scenario as in *split-plot* design. Several causes of correlation that may occur in the observations are given in Palta (2003). Designing problems for regression with correlated observations have, so far as the author knows, only been solved in an asymptotic-optimal way (Sacks and Ylvisaker, 1966, 1968, 1970; Bickel and Herzberg, 1979; Bickel *et al.*, 1981).

So far all the authors studied rotatability assuming errors are uncorrelated and homoscedastic. But rotatability is a very important and highly desirable property in the response surface methodology. Therefore, in the present monograph we are motivated to study rotatability and the related concepts such as optimality, near rotatability, slope-rotatability, near slope-rotatability, regression analysis with correlated errors, and also positive data analysis.

1.4 ROBUST REGRESSION DESIGNS

In response surface methodology, regression design robustness first appeared in the work of Box and Draper (1959, 1963), dealing with protection against model under specification. When formalized, in an RSM setting, it boiled down to the notion of proper placing of the design points in the region of interest, when under-specification is a concern. Although specifics in the Box and Draper work dealt with spherical regions, Draper and Lawrence (1965) applied the Box and Draper approach to generate designs that are robust to model inadequacies in case of cuboidal regions.

Apart from model misspecification, RSM design robustness which has attracted various authors' attention includes the following categories:
(a) robustness to outliers in the data,
(b) robustness to errors in the design levels,
(c) designs for extrapolation under conditions of model misspecification.
To these, the present monograph adds the fourth category, viz.,
(d) robustness to correlation parameter or parameters of the variance-covariance matrix of correlated errors (in observations) (Panda and Das, 1994; Das 1997, 1999, 2003a, 2003b, 2004, 2009, 2010; Das and Park 2006, 2007, 2008a, 2008b, 2008c, 2009, 2010; Das and Huda, 2011; Das and Lin 2011; Das and Park, 2013; Das, Park and Aggarwal 2010a, 2010b; Das, Pal and Park, 2013; Kim, Das, Sengupta and Paul, 2009; Mukhopadhyay, Bagchi and Das, 2002; Park, Jung and Das, 2009).

Box and Draper (1975) linked the awareness of *outliers* to notions of RSM experimental designs. Design properties were sought that resulted in "minimal impact" of outliers in the resulting data. Box and Draper (1975) argued that in many cases an experiment is designed for many different purposes. There is an implication that optimal designs should not be sought nearly as often as designs that are 'good' in many areas, that is, we seek *compromise designs*. The same opinion is expressed by many authors (e.g., Atkinson, 1982a; Kiefer, 1975).

Herzberg and Andrews (1976) dealt with optimal designs under non-optimal conditions such as missing observations and the presence of outliers. This article considered protection against model misspecification and outliers simultaneously. Draper and Herzberg (1979) dealt with an integrated mean squared error criterion much like that of Box and Draper (1959, 1963). A related notion of robustness deals with errors in the factor levels. Vuchkov and Boyadjieva (1983) considered this problem and attempted to determine design families that are robust. Box (1963) considered the effect of errors in factor levels in both first- and second-order models, although no specific design criteria were considered in Box's article. Another type of design robustness is concerned with designs that are *resistant to errors in extrapolation*. This is particularly important in RSM work, since a response surface is often used by necessity for extrapolation purposes. Draper and Herzberg (1971, 1973, 1979) discussed designs that are robust in this sense under the settings of first- and second-order fitted models. It is assumed that one's ability to extrapolate is influenced by the existence of higher order terms in the true structure.

The last category cited deals with the effect of correlated structure of errors on rotatability and optimality of a design. The purpose of study is to examine the robustness of rotatability, slope-rotatability, and optimality property for different variations in the correlated parameter or parameters of the correlation structure which the observation errors are supposed to satisfy. Linking rotatability, slope-rotatability and optimality with the correlation structures of errors is a topic which, to the best of our knowledge is considered for the first time by the present author with and without the collaboration of his seniors. The present monograph is actually based on the work done in this direction. Panda and Das (1994) initiated a study of first-order rotatable designs with correlated errors. Various correlated error structures are adopted and rotatability of the usual (for uncorrelated and homoscedastic errors) first-order rotatable designs (FORDs) are investigated under such correlated error structures. Robust second-order rotatable designs (with correlated errors) have been initiated by Das (1997). Second-order rotatability conditions are derived for a general correlated error structure. First- and second-order weak rotatability and their respective measures have been derived by Panda and Das (1994), Das (1999) and Das and Park (2007). Robust second-order rotatable designs are examined by Das (2003a, 2004) for several well known correlated error structures. First-order D-optimality has been studied by Das and Park (2008a, 2008b, 2010). D-optimal robust first-order designs for lifetime responses have been developed by Das and Lin (2011), Das and Huda (2011). Robust second-order rotatable designs for lifetime responses have been studied by the author. Robust second-order slope-rotatability and modified slope-rotatability along axial directions have been introduced by Das (2003b) and Das, Pal and Park (2013). Robust slope-rotatability over all directions has been introduced by Das and Park (2006). Park, Jung and Das (2009) studied robust slope-rotatability over all directions, with equal maximum directional variance, and along with their measures. Das and Park (2009) derived measures of robust slope-rotatability along axial directions. Das, Park and Aggarwal (2010a, 2010b) studied robust second-order slope D-optimality and slope-rotatability with equal maximum directional variance.

The above problems deal with correlated first- and second-order regression models. Correlated regression analyses have been studied by Mukhopadhyay, Bagchi and Das (2002); Kim, Das, Sengupta and Paul (2009), Das (2009, 2010). Applications of correlated regression analyses in block designs have been considered by Das and Park (2008c; 2014).

Generally, experimental observations are positive and so they are analysed by log-normal and gamma models (Firth, 1988; McCullagh and Nelder, 1989; Myers *et al.*, 2002). The discrepancy of regression estimates between the log-normal and gamma models are studied by Das and Park (2012) (for constant variance) and Das and Lee (2009) (for non-constant variance). The discrepancy in the model fittings between these two models is studied by Das (2012). Real data (in quality engineering) analyses related to design of experiments (which do not satisfy the usual assumptions of classical design of experiments) are presented by Das (2011a, 2011d, 2013a) and Das and Lee (2008, 2009, 2010). Applications of log-normal and gamma models analyses are given by Das (2011b, 2011c, 2013b) (Medical science), Das and Mukhopadhya (2009)(Medical science), Das, Dihidar and Verdugo (2011), Das and Dihidar (2013) (Demography), Das and Kim (2012) (Hydrogeology).

1.5 SUMMARY OF THE RESEARCH MONOGRAPH

Before presenting a detail account of the work done in the present monograph, a few remarks about the general approach (used here) and their consequences are given below. Regarding the first step in the solution of the problems stated at the end of Section 1.1, we fit the response functions (first- and second-order) to the data by the least squares method and find the estimate of the parameters of interest in a straight forward manner, assuming correlation parameter or parameters in the correlation structure are *known*. In general, the covariance matrix W is unknown, but for all the calculations W is assumed to be known. In practice, however, W includes a number of unknown parameters, and in the calculations which follow, the expressions for W and W^{-1} are replaced by those obtained after altering the unknown parameters by suitable estimates or some assumed values. Actually, the correlation parameter or parameters in the specified correlation structure are unknown and as such we want the design to be rotatable (or optimal) for all possible values of the correlation parameter or parameters in the structure. At least the property should hold for a reasonable range of the correlation parameter or parameters within which the experimenter is supposed to work.

The main emphasis of this book is on examining the concept of rotatability, weak rotatability, D-optimality, slope-rotatability—along axial, over all directions, and with equal maximum directional variance, weak slope-rotatability, D-optimal slope-rotatability, regression analysis with correlated errors. However, in practice the iterative least squares method similar to the one outlined in Chapter 9 can be adopted to estimate the correlation parameters in the model. In addition, this book also includes some real experimental data analysis based on log-normal and gamma models with constant and non-constant variances. Below the contents of the book are described chapter wise.

Chapter 2 starts with first-order correlated regression designs. The first-order rotatability and D-optimal conditions are given for different variance-covariance structures (i.e., intra-class, inter-class, compound symmetry, tri-diagonal and autocorrelated structures) of errors which are commonly come across in practical situations. The concept of robust first-order rotatable and D-optimal designs are described. D-optimal robust first-order designs are examined for different well-known error variance-covariance structures. Different methods of construction of *robust first-order rotatable* and *D-optimal* designs are described for each error variance-covariance structure. For autocorrelated structure, *nearly* optimum robust first-order designs are described, and their *efficiency ratios* are computed.

It is readily observed that a first-order rotatable design with correlated errors under the intra-class structure, is also a first-order rotatable design (FORD) with uncorrelated errors and vice-versa. This holds irrespective of the value of intra-class correlation coefficient, ρ. So, the FORDs with uncorrelated errors are also robust first-order rotatable designs (RFORDs) under the intra-class structure of errors. It is further seen that the variance of the estimated response of a RFORD under the intra-class correlation model is more or less than that of a FORD under uncorrelated model, according as the intra-class correlation coefficient is positive or negative. It is also observed that a design is D-optimum robust first-order design (D-ORFOD) under the intra-class structure of errors, whatever be the value of intra-class correlation coefficient ρ, if it is a D-optimum first-order design (D-OFOD) in the usual model, i.e., when $\rho = 0$.

Other major findings include the following: If the design points are divided into some disjoint sets and if the points have an orthogonal structure within each set, then the over-all design is rotatable under the inter-class covariance structure, whatever be the value of intra-class correlation coefficient ρ, but the converse is *not* always true. The same design satisfies the rotatability property for all possible values of the correlation coefficients in case of compound symmetry structure. In the same chapter, we also undertake a detail study of RFORDs and D-ORFODs including their constructions under the following correlation structures: intra-class, inter-class, compound symmetry, autocorrelated and tri-diagonal structures. The structures considered here are the ones commonly confronted in practice and are the well-known structures cited in textbooks on multivariate statistics.

Chapter 3 describes second-order regression designs with correlated errors. Robust second-order rotatability conditions are examined for a general variance-covariance structure of errors. These conditions are further studied for each specific variance-covariance structure of errors. Second-order rotatability conditions are simplified for intra-class, inter-class, compound symmetry, tri-diagonal and autocorrelated error structures. Robust second-order rotatable designs (RSORDs) are constructed for each of the above well-known correlation structures.

It is readily observed that a second-order rotatable design (SORD) in the usual model preserves the property of rotatability under the intra-class structure of errors. This holds irrespective of the value of the intra-class correlation coefficient, ρ. It is further seen that the variance of the estimated response of a robust second-order rotatable design under the intra-class correlation structure is more or less (under some mild conditions) than that of a SORD under uncorrelated model, according as the intra-class correlation coefficient is positive or negative.

Again, if the design points are divided into some disjoint sets and if the design points of each set satisfy the second-order rotatability conditions in the usual sense, then the overall design is a second-order rotatable under the inter-class structure of errors but the converse is *not* true. Two methods of construction of RSORDs are described under the inter-class structure of errors. The designs described under the inter-class structure also happen to be rotatable (i.e., RSORDs) for all possible values of correlation coefficients under the compound symmetry structure of errors. Finally, a detail study of second-order rotatability for tri-diagonal and autocorrelated structure of errors are undertaken. In the process, one / two construction methods of RSORDs are described under autocorrelated and tri-diagonal error structures.

Chapter 4 includes location-scale lifetime models which are derived based on *constant* variance assumption from a lifetime distribution such as log-normal and exponential. Assuming a first-order location-scale model with correlated errors, first-order rotatability and D-optimality conditions are described for a general variance-covariance structure. It is observed that the designs are *invariant* for the above two lifetime distributions, but they depend on the error variance-covariance structures. Similar results as in Chapter 2 are presented for each lifetime distribution, and for each correlation structure as mentioned in Chapter 2. Second-order location-scale lifetime models are also considered for the same lifetime distributions. Assuming correlated errors, second-order rotatability conditions are derived for a general variance-covariance structure of errors. Robust second-order rotatability conditions are examined for each lifetime distribution and for each correlation structure as mentioned in Chapter 2. Similar results as in Chapter 3 are presented for each lifetime distribution and for each stated correlation structure.

Chapter 5 presents some measures of degree of closeness to *robust rotatability* of first- and second-order regression designs which are *not exactly robust rotatable* for a fixed pattern of correlation structure of errors. The degree of closeness to robust first-order rotatability are described in the following two ways: The measure may be based on the (a) dispersion matrix of regression coefficients, and (b) moment matrix, following Draper and Pukelsheim (1990).

For second-order correlated regression model, the degree of closeness to robust second-order rotatability is derived, based on the moment matrix, following Draper and Pukelsheim (1990), and Park, Lim and Baba (1993). For the first- and second-order designs, a comparison rule is described for comparing between the robust rotatable and non-rotatable designs, based on cost and variance functions. All the results of closeness to robust rotatability (i.e., weak rotatability) are critically studied with examples, separately for first- and second-order regression models.

In Chapter 6, second-order correlated models are considered as in Chapter 3. Robust second-order slope-rotatability along axial directions are described. Conditions of robust second-order slope-rotatability and modified slope-rotatability along axial directions are derived for a general variance-covariance structure of errors. These conditions are further simplified for intra-class structure of errors, and are compared with the usual and modified second-order slope-rotatability conditions.

Chapter 7 describes some optimality criteria of robust second-order slope-rotatable designs. Robust second-order slope-rotatability over all directions, with equal maximum directional variance and D-optimal slope-rotatable designs are described. Conditions of robust second-order slope-rotatability over all directions are derived for a general variance-covariance structure of errors. These conditions are examined for each correlation structure of errors as mentioned in Chapter 2. It is examined that robust second-order rotatable designs are also robust second-order slope-rotatable designs over all directions. Robust second-order slope-rotatable designs over all directions are derived for each correlation structure of errors.

The concept of robust second-order slope-rotatable designs with equal maximum directional variance is described. This requires that the maximum variance of the estimated slope over all possible directions to be only a function of the distance of the point from the design origin, and independent of correlation parameter or parameters involved in the variance-covariance matrix of errors. It is derived that robust second-order rotatable designs of two factors are also robust slope-rotatable designs with equal maximum directional variance. It is also established that within the robust second-order symmetric balanced designs, robust rotatable designs are also robust slope-rotatable with equal maximum directional variance for more than two factors.

The concept of robust second-order D-optimal slope-rotatable designs is described. It is established that robust second-order rotatable designs are also D-optimal robust slope-rotatable designs. A class of robust second-order slope-rotatable designs with equal maximum directional variance and D-optimal slope-rotatable designs are derived for some special correlated error structures.

Chapter 8 considers weakly robust second-order slope-rotatable designs. Some measures are described for robust second-order slope-rotatability along axial directions, over all directions and with equal maximum directional variance. These measures are illustrated with examples.

Chapter 9 considers the problems of correlated regression analyses which are developed in Chapters 2, 3 and 4. In general, the form of the correlation structure is known for a given situation of data set but the parameter (or parameters) that is involved in the correlation structure is always unknown. Here some robust methods are described for estimating the best linear unbiased estimators of all the regression parameters *except* the intercept, which is often unimportant in practice. Some testing procedures are described for testing any set of linear hypotheses regarding the unknown regression parameters. Confidence ellipsoid of a set of estimable functions of regression parameters is described. Index of fit is described for a fitted regression equation. Two examples with simulated data illustrate the estimation methods.

This chapter also describes a few applications of correlated regression analyses in block designs. A randomized block design under compound symmetry structure and a reinforced randomized block design under compound autocorrelated structure are analyzed through correlated regression analysis. In the process, confidence ellipsoid, confidence interval and multiple comparison techniques are studied with correlated errors.

Chapter 10 presents positive data analyses. In general, experimental observations are positive, which are generally analysed by log-normal and gamma models. Log-normal and gamma models are described for constant and non-constant variances. The discrepancy of regression estimates and also the model fittings between the log-normal and gamma models are explained and illustrated for constant and non-constant variances.

This chapter also presents analyses of some real positive demography, hydrogeology, medical data, and experimental designs data related to quality engineering, lifetime, natural science etc. It is shown that observations related to some experimental designs data do not satisfy the assumptions of classical experimental designs. Some applications of log-normal and gamma model analyses (for real positive data) are presented in quality engineering, medical science, demography, hydrogeology, natural science etc. Experimental designs real data related to resistivity, lifetime, multiple responses are analysed using log-normal and gamma models with constant and non-constant variances. Medical data on human biochemical parameters are analysed. Demography data on Indian infant and child mortality are analysed. Hydrogeology data on drinking ground water are analysed.

In Chapter 9, the robust rotatable designs, i.e., those which are developed in Chapters 2, 3 and 4 are used for estimating the regression coefficients in a regression problem arising in the area of reliability. The mean performance of a system depends on k operating factors and its logarithm is expressed by a regression of first- or second-order response surface model. The performance under measurement is actually the life of a system. The performance variable (y) is assumed to follow log-normal and exponential distributions with k operating factors viz., $x_1, x_2,...,x_k$ which determine the operating conditions of the system. The mean performance level is an exponential function whose exponent is assumed to represent a first- or second-degree polynomial in the k operating factors. Treating the logarithm of the performance measure as the response variable, the assumption implies that the response is a first- or second-degree polynomial in the k operating variables plus an error which follows the well-known extreme value or standard normal distribution.

To determine the constants in the model, some observations are required, which may be obtained by conducting an experiment. The experiment when actually executed introduces an extra error (which may be called the experimental error) to the observations obtained. So, what is actually observed is the response already described plus an experimental error. Thus, the logarithm of performance variable y, i.e., response is a linear or quadratic function of the operating variables plus an extreme value (or standard normal) variable, plus a random error (e). The experimental conditions are such that the errors introduced by them may follow some structured pattern as dealt in the Chapters 2, 3 and 4.

The problem considered is one which can commonly arise in practice. The estimation problem of the parameters in the regression model is actually taken up when the experimental errors introduced in the observations follow an intra-class, inter-class, compound symmetry and compound autocorrelated correlation structure, as an illustration. The standard methods available have to be manipulated and altered significantly to make the estimation procedure meaningful in the present context. The developed methods are illustrated with the help of some examples consisting of simulated observational data. Similar methods can be developed for experimental errors following other correlation structures included in the book. It may be pointed out that the investigation is carried out for two correlation structures viz., compound symmetry and compound autocorrelated, and no attempts are made to extend the same to other correlation structures.

1.6 CONCLUDING REMARKS

A brief survey of the work undertaken in the present research monograph is stated in the above. The problems considered have their origin in practice, as in reality the usual assumptions related to random errors have to be relaxed and generalized to include some specific pattern in the correlation structure of errors. It is assumed that the response function can be represented by a polynomial of degree one or two. All the described results related to robust rotatability, slope-rotatability, optimality etc., are only for first- and second-order response models. Similar study can be made for higher-order polynomials also.

For first-order models, the concepts such as robust rotatability, optimality, weak robust rotatability and the corresponding designs are investigated in details for several well-known correlated error structures. For second-order models, the concept of rotatability is studied similar to first-order models. Regarding second-order optimality, details are not worked out for all the different correlation structures, as the algebraic expressions dealt with do not simplify enough to produce really useful results in general. For second-order model, robust slope-rotatability, modified slope-rotatability along axial directions, over all directions, slope-rotatable designs with equal maximum directional variance and D-optimal slope-rotatability are studied. Different measures of degree of closeness to robust rotatability and slope-rotatability are described. Correlated regression models are considered in Chapters 2, 3 and 4 and the technique of correlated regression analyses are described. Applications of correlated regression analyses in block designs are illustrated.

Positive experimental observations are generally analysed by log-normal and gamma models (Firth, 1988). Log-normal and gamma models for constant and non-constant variances are described. The discrepancy of regression estimates and the model fitting between these two models are described. Applications of these two modeling techniques are described in quality engineering, medical science, demography, hydrogeology, natural science etc.

As has been already sufficiently indicated, some of the material of the present research monograph has already been published. A part of the material of Chapter 2 is contained in Panda and Das (1994) and the remaining part is in (Das and Park, 2008a, 2008b, 2010). A part of the material in Chapter 3 is contained in Das (1997, 2003a, 2004). The material in Chapter 4 is partly contained in Panda and Das (1994), Das and Huda (2011) and Das and Lin (2011). The material in Chapter 5 is contained in Panda and Das (1994), Das (1999), and Das and Park (2007). Content of Chapter 6 is included in Das (2003b) and Das, Pal and Park (2014). The material from Chapter 7 is contained in Das and Park (2006); Park, Jung and Das (2009); Das, Park and Aggarwal (2010a, 2010b). Material from Chapter 8 is included in Das and Park (2009) and Park, Jung and Das (2009). The material in Chapter 9 is contained in Kim, Das, Sengupta and Paul (2009) and Das (2009, 2010). Material from Chapter 10 is included in Das (2011a, 2011b, 2011c, 2011d, 2012); Das and Lee (2008, 2009, 2010); Das and Park (2012); Das, Dihidar and Verdugo (2011); Das and Dihidar (2013).

Chapter 2

ROBUST FIRST-ORDER DESIGNS

Optimality or at least rotatability is a highly desirable criterion for fitting a response surface design. This chapter confines to first-order linear regression models with correlated errors. The concept of robust first-order rotatability and optimality are examined for different well-known error variance-covariance structures. Generally, it is very difficult to derive optimal or at least efficient designs for even linear models with correlated errors, and for some correlation structures, D-optimal design does not exist. It is shown that D-optimal robust first-order designs are always robust first-order rotatable *but* the converse is not true. For compound symmetry, inter-class, intra-class and tri-diagonal correlation error structures, some construction methods of D-optimal robust first-order designs are discussed. For autocorrelated error structure, efficient robust rotatable designs are presented.

2.1   INTRODUCTION AND OVERVIEW

Rotatability is one of the desirable criterion of response surface designs. As mentioned in Chapter 1, so far all the authors have studied rotatability assuming errors in observations are uncorrelated and homoscedastic. However, it is *not* uncommon to come across practical situations where the errors are correlated (Gennings *et al.*, 1989; Bischoff, 1992, 1995b, 1996; Myers *et al.*, 2002), violating the usual assumptions. This chapter studies first-order *D-optimality* and *rotatability* for correlated errors.

First-order conditions of rotatability and *D*-optimality are determined for a general correlated error structure. Various correlated error structures of neat forms which are usually encountered in practice, have been adopted and rotatability, *D*-optimality of the usual first-order response surface designs with these generalized classes of covariance structures have been investigated. The present monograph examines the conditions under which first-order regression designs which are known to be rotatable for uncorrelated errors, satisfy the rotatability conditions for a meaningfully enlarged class of correlation structures i.e., the variance of the estimated response at a point as a function of distance square from the origin of the design space, holds for a wide variety of dispersion matrices of errors, assuming a *fixed pattern* of error correlation structure. Designs having this property are called *robust rotatable* designs under the given *fixed pattern* of error correlation structure. This also examines the conditions of first-order designs for which the variance of the estimated response at any point remains *optimum*, for all the dispersion matrices of errors within the class of a *fixed pattern* of correlation structure. Designs having this property are called *optimum robust* first-order designs.

Different construction methods of robust first-order rotatable and optimum designs are discussed. Efficiency ratios are also computed for some designs which are *nearly optimum robust* first-order rotatable designs. The present chapter undertakes a detail study of robust first-order rotatable designs (RFORDs) and optimum robust first-order designs (ORFODs) under the following correlation structures: intra-class, inter-class, compound symmetry, autocorrelated and tri-diagonal structures of errors (Panda and Das, 1994; Das and Park, 2008a, 2008b, 2010). The structures considered here are the ones commonly confronted in practice and are the well-known structures which have found their place in standard textbooks dealing with the theoretical and practical aspects of multivariate analysis.

2.2   FIRST-ORDER CORRELATED MODEL

2.2.1   Model

Suppose there are k factors denoted by $x_1, x_2, ..., x_k$, and a design point $(x_{u1}, x_{u2}, ..., x_{uk})$, $1 \leq u \leq N$, yields a response y_u on the study variable y. Assuming that the response surface is of first-order, we adopt the model

$$y_u = \beta_0 + \sum_{i=1}^{k} \beta_i x_{ui} + e_u;\ 1 \leq u \leq N,$$

or,

$$\mathbf{Y} = X\boldsymbol{\beta} + \mathbf{e}, \qquad\qquad (2.1)$$

where $\mathbf{Y} = (y_1, y_2, ..., y_N)'$ is the vector of recorded observations on the study variable y, $\boldsymbol{\beta} = (\beta_0, \beta_1, ..., \beta_k)'$ is the vector of regression coefficients, $X = (\mathbf{1} : (x_{ui}); 1 \leq u \leq N; 1 \leq i \leq k)$ is the model (or design) matrix, $\mathbf{e}$ is the vector of errors which are assumed to be normally distributed with expectation, i.e., $E(\mathbf{e}) = \mathbf{0}$, $Dis(\mathbf{e}) = W$ and rank $(W) = N$. The matrix W may represent any variance-covariance matrix of correlated errors. For example, $W = [\sigma^2 \{\rho^{|i-j|}\}_{1 \leq i, j \leq N}]$ is well known as an autocorrelated error structure. Note that the matrix W is unknown but for all the calculations W is assumed to be known. In practice, however, W includes a number of unknown parameters, and in the calculations which follow, the expressions for W and W^{-1} are replaced by those obtained after replacing the unknown parameters with suitable estimates or some assumed values based on experience.

2.2.2 Analysis and rotatability

The best linear unbiased estimator of $\boldsymbol{\beta}$ (for known W and $(X'W^{-1}X)$ positive definite) is

$$\hat{\boldsymbol{\beta}} = (X'W^{-1}X)^{-1}(X'W^{-1}\mathbf{Y})$$

with $Dis(\hat{\boldsymbol{\beta}}) = (X'W^{-1}X)^{-1}$. Thus, the moment matrix of any first-order regression design ξ (under the model (2.1)) is denoted by $M(\xi) = (X'W^{-1}X)$. The necessary and sufficient conditions of robust first-order rotatability are given in the following theorem.

Theorem 2.1 (Panda and Das, 1994) *The necessary and sufficient conditions of a robust first-order rotatable design (for all values of ρ in W and for all $1 \leq i, j \leq k$) in the model (2.1) are*

(i) $v_{0j} = \mathbf{1}'W^{-1}\mathbf{x}_j = 0,$

(ii) $v_{ij} = \mathbf{x}_i'W^{-1}\mathbf{x}_j = 0; \; i \neq j,$

(iii) $v_{ii} = \mathbf{x}_i'W^{-1}\mathbf{x}_i = constant = \lambda > 0.$ (2.2)

Proof: The variance of the estimated response $\hat{y}_\mathbf{x}$ at $\mathbf{x} = (x_1, x_2, ..., x_k)'$ is given by

$$Var(\hat{y}_\mathbf{x}) = Var(\hat{\beta}_0) + 2\sum_{i=1}^{k} x_i Cov(\hat{\beta}_0, \hat{\beta}_i) + \sum_{i=1}^{k} x_i^2 Var(\hat{\beta}_i) + 2\sum\sum_{i<j}^{k} x_i x_j Cov(\hat{\beta}_i, \hat{\beta}_j)$$

$$= v^{00} + 2\sum_{i=1}^{k} x_i v^{0i} + \sum_{i=1}^{k} x_i^2 v^{ii} + 2\sum\sum_{i<j}^{k} x_i x_j v^{ij},$$ (2.3)

where v^{ij}'s are the elements of the matrix $Dis(\hat{\boldsymbol{\beta}}) = (X'W^{-1}X)^{-1} = ((v_{ij}))^{-1} = ((v^{ij}))$ say, $0 \leq i, j \leq k$; $v_{00} = \mathbf{1}'W^{-1}\mathbf{1}$, $v_{0j} = \mathbf{1}'W^{-1}\mathbf{x}_j$, $v_{ij} = \mathbf{x}_i'W^{-1}\mathbf{x}_j$ and $\mathbf{x}_i = (x_{1i}, ..., x_{Ni})'$; $1 \leq i, j \leq k$. The design is called a robust first-order rotatable one when the variance function in (2.3) is a function of $r^2 = \sum_{i=1}^{k} x_i^2$ only. This happens for *all* $\mathbf{x}$ in the factor region of interest *if and only if* $v^{0i} = 0$, $v^{ij} = 0$ and $v^{ii} =$ constant, for $1 \leq i \neq j \leq k$, for *all* values of ρ's in W, where ρ's are the correlation parameters in W. Equivalent conditions are in (2.2). This completes the proof. $\square$

Following (2.2) and (2.3), the variance of the estimated response at $\mathbf{x}$ of a first-order rotatable design with the general variance-covariance matrix W of errors reduces to

$$Var(\hat{y}_\mathbf{x}) = \frac{1}{v_{00}} + \frac{1}{\lambda}\sum_{i=1}^{k} x_i^2 = f(r^2),$$ (2.4)

where $r^2 = \sum_{i=1}^{k} x_i^2$, $v_{00} = \mathbf{1}'W^{-1}\mathbf{1}$ and λ is as in (2.2).

2.2.3   Robust rotatable and optimum designs

This section presents regression designs under the correlated regression model (2.1), such that variance of the estimated response at $\mathbf{x} = (x_1, x_2, ..., x_k)'$ is a function of the distance square from the origin of the design space, for a well defined wide class of the dispersion matrices W. For different optimality criteria including G-optimality and D-optimality one is referred to Kiefer and Wolfowitz (1960), Silvey (1980) and Pukelsheim (1993). Some relevant definitions are given below.

Definition 2.1 Rotatable design: A design is said to be *rotatable* if the variance of the estimated response at a point is a function of only the distance from the design center (i.e., center of the coordinate axes, or at $(0, 0, ..., 0)$) to that point.

Definition 2.2 Robust first-order rotatable design: A design D of k factors under the correlated model (2.1) which remains first-order rotatable for *all* the variance-covariance matrices belonging to a well defined class $W_0 = \{W$ positive definite: $W_{N \times N}$ defined by a particular correlation structure possessing a definite pattern$\}$ is called a *Robust First-Order Rotatable Design* (RFORD), with reference to the variance-covariance class W_0.

Definition 2.3 Optimum robust first-order design: Assuming $\mathcal{X}$, the design space under consideration to be circumscribing the unit cube, viz., the unit ball $\mathcal{X} = \{\mathbf{x}: \mathbf{x}'\mathbf{x} \leq k\}$, a design d which is G-optimal or D-optimal within the set of first-order regression designs for a class of variance-covariance matrices for the errors W_0 (determined by a well defined correlation structure) has been termed an *Optimum Robust First-Order Design* (ORFOD) with reference to the variance-covariance class W_0.

Definition 2.4 G-optimal design: Suppose the primary interest lies in predicting the expected value of the estimated response $\hat{y}_{\mathbf{x}}$ i.e., $\mathrm{E}(\hat{y}_{\mathbf{x}})$ for $\mathbf{x} \in \mathcal{X}$, some region in R^k, where $\mathcal{X}$ is as in Definition 2.3 and the model for $y_{\mathbf{x}}$ is as in (2.1). The design matrix X as in 2.1 is so determined that $Max_{\mathbf{x} \in \mathcal{X}}(\mathrm{Var}\,(\hat{y}_{\mathbf{x}}))$ is minimized. Such a design, when it exists is called G-optimal in the design space $\mathcal{X}$ for the model in 2.1. The G-optimal design defined above, when it exists is also known to be equivalent to the D-optimal design in the same context, i.e., the design matrix corresponding to it minimizes $\mathrm{Dis}(\hat{\beta})$, the expression for which is given in Section 2.2.2 for the whole class of non-singular design matrices with the design points belonging to $\mathcal{X}$ (Kiefer and Wolfowitz, 1960).

The definition of G-optimal or equivalently D-optimal design is dependent on the polynomial model assumed for $\mathrm{E}(y_{\mathbf{x}})$. Irrespective of whether $\mathrm{E}(y_{\mathbf{x}})$ is a linear function of x_i's as is assumed in the present chapter or $\mathrm{E}(y_{\mathbf{x}})$ is a quadratic function of x_i's as considered in the next chapter, the substance in the definition remains same.

Definition 2.5 D-optimal robust first-order design: A first-order regression design ξ of k factors in the correlated model (2.1) is said to be *D-Optimal Robust First-Order Design* (*D*-ORFOD) if the determinant $\mid X'W^{-1}X \mid$ is uniformly maximum (over the design space $\mathcal{X} = \{\mid x_{ui} \mid \leq 1; 1 \leq u \leq N,\ 1 \leq i \leq k\}$) for *all* the variance-covariance matrices belonging to a well-defined class W_0.

Theorem 2.2 (Das and Park, 2008a; Das and Lin, 2011) *The necessary and sufficient conditions for a D-optimal robust first-order design in the model (2.1) are (for all $1 \leq i, j \leq k$)*

(i) $v_{0j} = \mathbf{1}'W^{-1}\mathbf{x}_j = 0$,

(ii) $v_{ij} = \mathbf{x}_i'W^{-1}\mathbf{x}_j = 0,\ i \neq j$;

(iii) $v_{ii} = \mathbf{x}_i'W^{-1}\mathbf{x}_i = \mu$,

for the design matrix $X \in \mathcal{X}$, where μ is the maximum possible value (a positive constant) over the design space $\mathcal{X}$ as in Definition 2.5.

Proof: For a first-order design ξ, the moment matrix is $M(\xi) = X'W^{-1}X = (\{v_{ij}\})$ as given in (2.3). Let $M_{i,i}$ be the $i \times i$ leading principal submatrix of $M(\xi)$, $1 \le i \le k+1$. Since $M(\xi)$ is a positive definite matrix, with $v_{ij} = v_{ji}$, for $0 \le i, j \le k$; $v_{ii} > 0$, for $0 \le i \le k$; and $M_{i,i}$ is positive definite, for $1 \le i \le k + 1$; then,

$$M(\xi) = M = M_{k+1,k+1} = \begin{pmatrix} M_{k,k} & \mathbf{v} \\ \mathbf{v}' & v_{kk} \end{pmatrix} = \begin{pmatrix} M_{k,k} & \mathbf{0} \\ \mathbf{v}' & 1 \end{pmatrix} \times \begin{pmatrix} I_k & M_{k,k}^{-1}\mathbf{v} \\ \mathbf{0}' & v_{kk} - \mathbf{v}'M_{k,k}^{-1}\mathbf{v} \end{pmatrix},$$

where $\mathbf{v} = (v_{0k}, ..., v_{(k-1)k})'$. Therefore

$$det(M) = det(M_{k,k})det(v_{kk} - \mathbf{v}'M_{k,k}^{-1}\mathbf{v}).$$

Since both $det(M_{k,k})$ and v_{kk} are positive, both $M_{k,k}$ and $M_{k,k}^{-1}$ are positive definite and $\mathbf{v}'M_{k,k}^{-1}\mathbf{v} \ge 0$, this implies

$$det(M) \le det(M_{k,k})v_{kk}.$$

with the equality holds *if and only if* $\mathbf{v} = \mathbf{0}$ (equivalently, $\mathbf{v}'M_{k,k}^{-1}\mathbf{v} = \mathbf{0}$), that is $v_{ik} = 0$, for all $0 \le i \le k - 1$.

Similarly, expanding $M_{k,k}$, we can show that $det(M) \le det(M_{k-1,k-1})v_{(k-1)(k-1)}v_{kk}$, with equality holds if and only if $v_{i(k-1)} = 0, 0 \le i \le k - 2$. By applying induction on $M_{k-1,k-1}, M_{k-2,k-2}, ..., M_{2,2}$, we have

$$det(M) \le \prod_{i=0}^{k} v_{ii},$$

with equality holds if and only if $v_{ij} = 0, 0 \le i < j, 1 \le j \le k$. This inequality is also known as the Hadamard's inequality (Anderson, 1984, p.54).

Since M is symmetric matrix, this implies that $v_{ij} = 0, 0 \le i \ne j \le k$. Thus, the maximum value of $det(M)$ is $v_{00}v_{11} \ldots v_{kk}$ if and only if $v_{ij} = 0, 0 \le i \ne j \le k$. Based on the design matrix X over the design space $\mathcal{X}$ given in Definition 2.5, the maximum value possible for each v_{ii} is the same (equal to μ, say). Therefore, from the above result $det(M)$ will attain maximum possible value ($= v_{00}\mu^k$) if and only if $v_{ij} = 0, 0 \le i \ne j \le k$; and $v_{ii} = \mu, 1 \le i \le k$. This completes the proof. $\square$

The variance of the estimated response $\hat{y}_{\mathbf{x}}$ at $\mathbf{x}$ of a first-order D-optimal design with correlated errors is given by $\text{Var}(\hat{y}_{\mathbf{x}}) = \frac{1}{v_{00}} + \frac{1}{\mu}\sum_{i=1}^{k} x_i^2 = f(r^2)$, where $r^2 = \sum_{i=1}^{k} x_i^2$, $v_{00} = \mathbf{1}'W^{-1}\mathbf{1}$ and μ as in Theorem 2.2. If v_{ii} is constant $= \lambda$ ($< \mu$), $1 \le i \le k$, then the design is a robust first-order rotatable as the variance of the estimated response $\hat{y}_{\mathbf{x}}$ is a function of only the distance from the design center, but not a D-optimal (as $\lambda < \mu$) robust first-order design. This can be stated as the following theorem.

Theorem 2.3 (Das and Park, 2008a) *A D-optimal robust first-order regression design is always a robust first-order rotatable design but the converse is not true.*

Let $\text{V}(\hat{y}_{\mathbf{x}})_0$ be the variance of the estimated response at $\mathbf{x}$ of an optimum robust first-order design (ORFOD) d_0, as defined in Definition 2.3 and $V(\hat{y}_{\mathbf{x}})$ be the variance of the estimated response at $\mathbf{x}$ of any other first-order design d (possibly near optimum), under the model (2.1). Note that $\text{V}(\hat{y}_{\mathbf{x}})_0$ and $V(\hat{y}_{\mathbf{x}})$ both are functions of $r^2 = \sum_{i=1}^{k} x_i^2$ and correlation parameter or parameters dictated by the correlation structure class W_0 in Definition 2.2. Efficiency ratio of the given design d is denoted by ER and is naturally given by

$$ER\% = \frac{Max_{0 \leq r^2 \leq k}\{V(\hat{y}_\mathbf{x})_0\}}{Max_{0 \leq r^2 \leq k}\{V(\hat{y}_\mathbf{x})\}} \times 100, \qquad (2.5)$$

and ER is computed for the class of all the covariance matrices in Definition 2.2, i.e., in the well-defined range of correlation parameters as dictated by the class. Here $V(\hat{y}_\mathbf{x})_0$ may denote the variance of $\hat{y}_\mathbf{x}$ for the *hypothetical optimum* design which may or may not be possible to arrive at explicitly.

Errors in observations may have some simple variance-covariance structures i.e., some *simplified pattern* of *regular* variance-covariance structures. Or, they may have some *irregular pattern* of variance-covariance structure. In Section 2.2, the conditions of robust first-order rotatability, D-optimality, variance function are derived, and the corresponding rotatable and optimum designs are defined for a general class of variance-covariance matrices as in Definition 2.2 dictated by some regular pattern or structure usually considered in literature.

Following sections focus on robust first-order rotatable designs (RFORDs) and optimum robust first-order designs (ORFODs) for the following error variance-covariance structures: Intra-class, Inter-class, Compound symmetry, Autocorrelated and Tri-diagonal (Press, 1972; Morrison, 1976; Seber, 1984; Lindsay, 1993; Palta, 2003; Panda and Das, 1994; Das and Park, 2008a, 2008b, 2010). The conditions of robust first-order rotatability, D-optimality, the variance function and the corresponding rotatable, D-optimal and nearly optimum designs are discussed for these correlation structures.

2.3   ROBUST FIRST-ORDER DESIGNS FOR INTRA-CLASS STRUCTURE

Intra-class structure is the simplest variance-covariance structure which arises when any two errors have the same correlation and each has the same variance. It is also known as of uniform correlation structure. This can happen easily in a situation when all the observations studied are from the same batch or from the same run in a furnace. Let ρ be the correlation between any two errors and each has the same variance σ^2. Then intra-class error variance-covariance structure is given by the class

$$\text{Dis}(\mathbf{e}) = \sigma^2[(1-\rho)I_N + \rho E_{N \times N}] = W_{N \times N}(\rho) \text{ (say)} : \sigma > 0, \ \tfrac{-1}{N-1} < \rho < 1. \qquad (2.6)$$

Here, I_N indicates an identity matrix of order N and $E_{N \times N}$ is an $N \times N$ matrix of all elements 1.

Observe that,

$$W_{N \times N}^{-1}(\rho) = \sigma^{-2}[(\delta_0 - \gamma_0)I_N + \gamma_0 E_{N \times N}], \qquad (2.7)$$

where $\delta_0 = \frac{1+(N-2)\rho}{(1-\rho)\{1+(N-1)\rho\}}$, $\gamma_0 = -\frac{\rho}{(1-\rho)\{1+(N-1)\rho\}}$ and $\rho > -(N-1)^{-1}$.

The following theorem is immediate from (2.2) under the structure (2.6).

Theorem 2.4 (Panda and Das, 1994) *The necessary and sufficient conditions of a robust first-order rotatable design for all possible values of ρ in $W_{N \times N}(\rho)$ under the intra-class variance-covariance structure (2.6) in the model (2.1) are (for all $1 \leq i, j \leq k$)*

(1) $\displaystyle\sum_{u=1}^{N} x_{uj} = 0,$

(2) $\sum_{u=1}^{N} x_{ui}x_{uj} = 0; i \neq j,$

(3) $\sum_{u=1}^{N} x_{ui}^2 = constant = \lambda.$ \hfill (2.8)

These are the well-known conditions of a first-order rotatable (orthogonal) design (FORD) in the usual model (i.e., for uncorrelated and homoscedastic errors). The following theorems are now immediate.

Theorem 2.5 (Panda and Das, 1994) *A design is robust first-order rotatable with intra-class error structure if and only if it is a FORD in the usual model, whatever be the value of intra-class correlation coefficient* ρ.

Theorem 2.6 (Das and Park, 2010) *A design is a D-optimal robust first-order under the intra-class structure (2.6) if and only if it is a D-optimal first-order design in the usual model, whatever be the value of intra-class correlation coefficient* ρ *in* $W_{N \times N}(\rho)$.

2.3.1 Comparison between RFORD and FORD

Under the intra-class structure in (2.6), $v_{00} = \mathbf{1}'W^{-1}\mathbf{1} = \frac{N}{\{1+(N-1)\rho\}\sigma^2}$, $\lambda = \mathbf{x}_i'W^{-1}\mathbf{x}_i = v_{ii} = \frac{N}{(1-\rho)\sigma^2}$ and following (2.4), the variance of the estimated response at $\mathbf{x}$ of a RFORD under the intra-class structure is given by

$$\mathrm{Var}(\hat{y}_{\mathbf{x}}|W_{N\times N}(\rho)) = \frac{\sigma^2}{N}[a(N,\rho) + b(N,\rho)r^2] = V_\rho(r^2), \text{say} \tag{2.9}$$

provided $N \geq (k+1)$, where $a(N,\rho) = \{1+(N-1)\rho\}$ and $b(N,\rho) = (1-\rho)$, and $r^2 = \displaystyle\sum_{i=1}^{k} x_i^2$.

Let $V_0(r^2)$ be the variance function of a FORD in the uncorrelated case. This corresponds to $\rho = 0$ and is given by

$$\mathrm{Var}(\hat{y}_{\mathbf{x}}) = \frac{\sigma^2}{N}(1+r^2) = V_0(r^2). \tag{2.10}$$

It is interesting to observe that, in general, $V_0(r^2) \overset{<}{>} V_\rho(r^2)$ for all $0 \leq r^2 \leq k$, according as $\rho \overset{>}{<} 0$, provided $(N-1) > r^2$.

2.4 ROBUST FIRST-ORDER DESIGNS FOR INTER-CLASS STRUCTURE

It is an extension of of intra-class structure. This situation is observed if the errors are grouped into some groups such that within each group there is the same intra-class structure and between groups there is no correlation. Such a situation may arise when the observations correspond to objects from a distinct number of batches, say furnace runs, pairs of observations within a batch being all equally correlated, whereas the correlation is zero for each pair from two distinct batches. Inter-class structure is obtained from the correlation structure (2.6) when it is altered to

$$\text{Dis}(\mathbf{e}) = I_m \otimes W_{n \times n}(\rho) = W_1 \text{ say,} \tag{2.11}$$

where $\otimes$ denotes Kronecker product, $N = mn$, and $W_{n \times n}(\rho)$ as in (2.6). Variance-covariance structure (2.11) is known as inter-class correlation structure. By virtue of (2.9), whenever $n \geq (k+1)$, since the between groups correlations are all zero,

$$\text{Var}(\hat{y}_{\mathbf{x}} | I_m \otimes W_{n \times n}(\rho)) = \frac{\sigma^2}{nm}[a(n, \rho) + b(n, \rho)r^2] = \frac{\sigma^2}{N}[a(n, \rho) + b(n, \rho)r^2], \tag{2.12}$$

where $N = mn$, $n \geq (k+1)$ and $a(n, \rho), b(n, \rho)$ are as given in (2.9). Thus, we have the following corollary.

Corollary 2.1 (Panda and Das, 1994) *If the observations are divided into* m *sets of* n *each and the corresponding design matrix has an orthogonal structure (i.e., FORD for uncorrelated and homoscedastic errors) within each set, then the overall design is rotatable under the inter-class variance-covariance structure (2.11), whatever be the value of intra-class correlation coefficient* ρ.

Remark 2.1 The observational vector $\mathbf{Y}_{N \times 1}$ ($N = mn$) is partitioned as $\mathbf{Y}_{N \times 1} = (\mathbf{Y}_1', \mathbf{Y}_2', ..., \mathbf{Y}_m')'$, where each $\mathbf{Y}_i$ is a vector of order $n \times 1$ and the corresponding design matrix X is accordingly partitioned as $X = (X_1', X_2', .., X_m')'$ where each X_i is of order $n \times (k+1)$. If every X_i is orthogonal i.e., FORD, the design matrix X will be robust FORD under the inter-class structure (2.11), whatever be the value of correlation coefficient ρ.

It may, however, be pointed out in connection with the Remark 2.1 that it is not necessary that the components in each X_i should satisfy all the orthogonality conditions, say, for instance $\sum x_{uj}^{(i)} = 0$, where $x_{uj}^{(i)}$ is the uth run of jth factor in X_i for the design to be RFORD. This aspect is examined below. Towards this, one needs to examine the first-order rotatability conditions (following (2.2)) under the inter-class structure (2.11), the following theorem states these conditions.

Theorem 2.7 (Panda and Das, 1994) *The necessary and sufficient conditions of a robust first-order rotatable design for all values of ρ in W_1 under the inter-class structure (2.11) in the model (2.1) are (N=mn and for all $1 \leq i, j \leq k$)*

(1) $v_{0j} = \displaystyle\sum_{u=1}^{N} x_{uj} = 0,$

(2) $v_{ij} = \frac{1}{(1-\rho)} \displaystyle\sum_{u=1}^{N} x_{ui}x_{uj} - \frac{\rho}{(1-\rho)\{1+(n-1)\rho\}}\{(\displaystyle\sum_{u=1}^{n} x_{ui})(\displaystyle\sum_{u=1}^{n} x_{uj}) + ... +$

$(\displaystyle\sum_{u=n(m-1)+1}^{mn} x_{ui})(\displaystyle\sum_{u=n(m-1)+1}^{mn} x_{uj})\} = 0; i \neq j,$

(3) $v_{ii} = \{\sigma^2\}^{-1}[\frac{1}{(1-\rho)} \displaystyle\sum_{u=1}^{N} x_{ui}^2 - \frac{\rho}{(1-\rho)\{1+(n-1)\rho\}}\{(\displaystyle\sum_{u=1}^{n} x_{ui})^2 + ... + (\displaystyle\sum_{u=n(m-1)+1}^{mn} x_{ui})^2\}] =$ *constant* $= \lambda.$

$$\tag{2.13}$$

It is obviously possible to construct designs satisfying ((1)–(3)) in (2.13) where some of the necessary and sufficient conditions given in (2.2) relating to the individual group of observations are violated. The following example illustrates it.

Example 2.1 Consider an experimental design consisting of three factors, four independent groups, each of seven observations, i.e., $k = 3$, $m = 4$, $n = 7$ and $\text{Dis}(\mathbf{e}) = I_4 \otimes W_{7 \times 7}(\rho)$, $N = mn = 28$.

$GroupI$:	1	2	3	4	5	6	7
$\mathbf{x}_1$	1	-1	1	-1	1	0	0
$\mathbf{x}_2$	1	1	-1	-1	0	1	0
$\mathbf{x}_3$	1	-1	-1	1	0	0	1

$GroupII$:	1	2	3	4	5	6	7
$\mathbf{x}_1$	1	-1	1	-1	1	0	0
$\mathbf{x}_2$	1	1	-1	-1	0	-1	0
$\mathbf{x}_3$	1	-1	-1	1	0	0	-1

$GroupIII$:	1	2	3	4	5	6	7
$\mathbf{x}_1$	1	-1	1	-1	-1	0	0
$\mathbf{x}_2$	1	1	-1	-1	0	1	0
$\mathbf{x}_3$	1	-1	-1	1	0	0	-1

$GroupIV$:	1	2	3	4	5	6	7
$\mathbf{x}_1$	1	-1	1	-1	-1	0	0
$\mathbf{x}_2$	1	1	-1	-1	0	-1	0
$\mathbf{x}_3$	1	-1	-1	1	0	0	1

Group totals for $\mathbf{x}_i$'s are given in the following table.

$Group$:	I	II	III	IV
$\mathbf{x}_1$	1	1	-1	-1
$\mathbf{x}_2$	1	-1	1	-1
$\mathbf{x}_3$	1	-1	-1	1

The conditions (1) and (2) in (2.13) are readily satisfied. The expression (3) in (2.13) simplifies to $\frac{4(5+29\rho)}{(1-\rho)(1+6\rho)}$ and it is the same for $i = 1,2,3$. Variance of the estimated response at $\mathbf{x}$ of the above design is given by

$$\frac{\text{Var}(\hat{y}_{\mathbf{x}})}{\sigma^2} = \frac{1+6\rho}{8} + \frac{(1-\rho)(1+6\rho)r^2}{4(5+29\rho)} \tag{2.14}$$

where $r^2 = \sum_{i=1}^{k} x_i^2$. But the design clearly, does not satisfy the condition: $\sum_u x_{uj}^{(i)} = 0$ for all $j = 1,2,3$ and $i = 1,2,3,4$.

2.4.1   Optimum robust first-order designs under inter-class structure

This section determines the optimum robust first-order designs under the inter-class structure (2.11). The following theorem gives the necessary and sufficient conditions of a D-ORFOD as given in Definition 2.5 (from Theorem 2.2).

Theorem 2.8 (Das and Park, 2010) *The necessary and sufficient conditions of a D-optimal robust first-order design for all ρ in W_1 under the inter-class structure (2.11), in the model (2.1) are (for all $1 \le i, j \le k$)*

(i) $v_{0j} = \sum_{u=1}^{N} x_{uj} = 0,$

(ii) $v_{ij} = \frac{1}{(1-\rho)} \sum_{u=1}^{N} x_{ui}x_{uj} - \frac{\rho}{(1-\rho)\{1+(n-1)\rho\}}\{(\sum_{u=1}^{n} x_{ui})(\sum_{u=1}^{n} x_{uj}) + ...$

$+ (\sum_{u=n(m-1)+1}^{mn} x_{ui})(\sum_{u=n(m-1)+1}^{mn} x_{uj})\} = 0; i \neq j,$

(iii) $\sum_{u=1}^{N} x_{ui}^2 = N; \sum_{u=1}^{n} x_{ui} = \sum_{u=(n+1)}^{2n} x_{ui} = ... = \sum_{u=n(m-1)+1}^{mn} x_{ui} = 0; if \rho > 0,$

(iv) $\sum_{u=1}^{N} x_{ui}^2 = N; \sum_{u=1}^{n} x_{ui} = \sum_{u=(n+1)}^{2n} x_{ui} = ... = \sum_{u=n(m-1)+1}^{mn} x_{ui} = \pm n; \ if \ \rho < 0.$ (2.15)

The variance of the estimated response at $\mathbf{x}$ of a D-ORFOD under the inter-class structure (2.11), using (2.15) over the design space $\mathcal{X}$ as in Definition 2.3 is

$$\text{Var}(\hat{y}_{\mathbf{x}})_{01} = \frac{\sigma^2}{N}[\{1+(n-1)\rho\} + (1-\rho)r^2], \ \ if \ \ \rho > 0,$$ (2.16)

and

$$\text{Var}(\hat{y}_{\mathbf{x}})_{02} = \frac{\sigma^2}{N}[\{1+(n-1)\rho\}(1+r^2)], \ \ if \ \ \rho < 0,$$ (2.17)

where $r^2 = \sum_{i=1}^{k} x_i^2.$

2.4.2   RFORD and D-ORFOD under inter-class structure

It is clear from (2.15), that there are two sets of D-ORFODs under the inter-class structure (2.11). One set is related to $\rho > 0$ and the other set is related to $\rho < 0$. The following first two methods give D-ORFODs when $\rho > 0$ and Method IV gives D-ORFODs when $\rho < 0$. All the four methods give RFORDs.

Method I: If the design points are divided into m sets of n each, where $n \equiv 0$ (Mod 4) and an H_n (Hadamard) matrix exists and $k < n$, then from H_n, an orthogonal structure within each set can be constructed such that $x_{ui} = 1$ or -1; $1 \leq u \leq N = mn$; $1 \leq i \leq k$. The overall design so constructed is a D-ORFOD under the inter-class structure (2.11), whatever be the value of intra-class correlation coefficient $\rho > 0$.

Remark 2.2 If n is not a multiple of 4, we can add a suitable number of central points i.e., $\mathbf{0} = (0,0,....,0)$ to each set of the design points of m sets obtained by Method I. The resulting designs thus obtained are not exactly optimum, but are nearly so when $\rho > 0$.

Method II: If $k \leq m \equiv 0$ (Mod 4) and n is even, a D-optimal robust first-order design of k factors can be constructed under the inter-class structure (2.11), whatever be the value of intra-class correlation coefficient $\rho > 0$, by selecting any k columns from $H_m \otimes U_n$, where $\otimes$ denotes Kronecker product, assuming H_m a Hadamard matrix of order m exists. U_n is a column vector of order $n \times 1$ with elements $+1$ or -1 such that sum of elements of U_n is zero. For the existence of Hadamard matrices and their construction procedures one is referred to Geramita and Seberry(1979), Hedayat and Wallis (1978), Wallis, Street and Wallis (1972).

Method III: If $k \leq m$, an H_m exists and n is odd i.e., $n = 2t + 1$, one can construct a robust first-order rotatable (*not* exactly optimum) design by following Method II, and by replacing U_n by $U_n^* = (U_{n-1}', 0)'$, where U_{n-1} is given as in Method II. The variance of the estimated response at $\mathbf{x}$ of a robust first-order rotatable design obtained by Method III is

$$\text{Var}(\hat{y}_{\mathbf{x}}) = \sigma^2 \left[\frac{\{1+(n-1)\rho\}}{N} + \frac{(1-\rho)}{m(n-1)} r^2 \right] \qquad (2.18)$$

where $N = mn$ and $r^2 = \sum_{i=1}^{k} x_i^2$.

Remark 2.3 For $\rho > 0$, Method I and Method II give D-optimal robust first-order designs and Method III gives *nearly* optimal robust rotatable designs, as the designs obtained by Method III do not satisfy the condition (iii) in (2.15) i.e., $\sum_{u=1}^{N} x_{ui}^2 < N$.

Following (2.5), efficiency ratio of a design constructed by Method III is

$$ER\% = \frac{Max_{0 \leq r^2 \leq k} \{ \frac{1+(n-1)\rho}{n} + \frac{(1-\rho)r^2}{n} \}}{Max_{0 \leq r^2 \leq k} \{ \frac{1+(n-1)\rho}{n} + \frac{(1-\rho)r^2}{n-1} \}} \times 100, \text{ if } \rho > 0. \qquad (2.19)$$

Note that ER is an increasing function of n for a fixed value of ρ.

Method IV: If $k < m$, where H_m, a Hadamard matrix of order m exists, a D-optimal robust first-order design of k factors can be constructed under the inter-class structure (2.11), whatever be the value of intra-class correlation coefficient $\rho < 0$, by selecting any k columns barring the first column from $H_m \otimes J_n$, where $\otimes$ denotes Kronecker product, H_m is written as a Standard Hadamard matrix of order m, having all the elements of the first row and the first column as 1's, and J_n is a column vector of order $n \times 1$ of all 1's.

Variance function of a design obtained by this method satisfies (2.17). Note that it can be readily verified that (2.13) holds for these designs obtained from the above four methods for all possible values of ρ. Therefore, the designs thus obtained from the above four methods are all robust first-order rotatable designs under the inter-class structure (2.11), for all ρ, positive or negative.

Example 2.2 Following the above four methods, four examples (one from each method) of D-ORFOD (for Method I, II, and IV) and RFORD (Method III) with 2 factors, 4 groups, each of 4 (or 5) observations, i.e., $k = 2$, $m = 4$, $n = 4$ (or 5), $N = mn = 16$ (or 20), under the inter-class structure, using Hadamard matrix of order 4, are given below. The designs (denoted by d_i from Method i; $i = $ I, II, III, IV) are displayed below (rows being factors and column being runs). We construct D-ORFODs and RFORD from

$$H_4 = \begin{pmatrix} 1 & 1 & 1 & 1 \\ 1 & -1 & 1 & -1 \\ 1 & 1 & -1 & -1 \\ 1 & -1 & -1 & 1 \end{pmatrix}, H_4 \otimes U_4 = \begin{pmatrix} U_4 & U_4 & U_4 & U_4 \\ U_4 & -U_4 & U_4 & -U_4 \\ U_4 & U_4 & -U_4 & -U_4 \\ U_4 & -U_4 & -U_4 & U_4 \end{pmatrix},$$

$$H_4 \otimes J_4 = \begin{pmatrix} J_4 & J_4 & J_4 & J_4 \\ J_4 & -J_4 & J_4 & -J_4 \\ J_4 & J_4 & -J_4 & -J_4 \\ J_4 & -J_4 & -J_4 & J_4 \end{pmatrix},$$

where $U_4 = (1, 1, -1, -1)'$ and $J_4 = (1, 1, 1, 1)'$.

d_I :	1	2	3	4	5	6	7	8	9	10	11	12	13	14	15	16
$\mathbf{x}_1$	1	-1	1	-1	1	1	-1	-1	1	-1	1	-1	1	1	-1	-1
$\mathbf{x}_2$	1	1	-1	-1	1	-1	-1	1	1	1	-1	-1	1	-1	-1	1

d_{II} :	1	2	3	4	5	6	7	8	9	10	11	12	13	14	15	16
$\mathbf{x}_1$	1	1	-1	-1	1	1	-1	-1	1	1	-1	-1	1	1	-1	-1
$\mathbf{x}_2$	1	1	-1	-1	-1	-1	1	1	1	1	-1	-1	-1	-1	1	1

d_{III} :	1	2	3	4	5	6	7	8	9	10	11	12	13	14	15	16	17	18	19	20
$\mathbf{x}_1$	1	1	-1	-1	0	1	1	-1	-1	0	1	1	-1	-1	0	1	1	-1	-1	0
$\mathbf{x}_2$	1	1	-1	-1	0	-1	-1	1	1	0	1	1	-1	-1	0	-1	-1	1	1	0

d_{IV} :	1	2	3	4	5	6	7	8	9	10	11	12	13	14	15	16
$\mathbf{x}_1$	1	1	1	1	-1	-1	-1	-1	1	1	1	1	-1	-1	-1	-1
$\mathbf{x}_2$	1	1	1	1	1	1	1	1	-1	-1	-1	-1	-1	-1	-1	-1

Note that the designs d_I, d_{II}, and d_{IV} are constructed, respectively, by using last three columns of H_4, first two columns of $H_4 \otimes U_4$, and second and third columns of $H_4 \otimes J_4$. Also note that the first 4 runs are associated to first group, second 4 runs are associated to second group, and so on.

2.5   ROBUST FIRST-ORDER DESIGNS FOR GENERALIZED INTER-CLASS STRUCTURE

The population variance-covariance matrix will have the generalized inter-class structure if the errors are divided into m sets of different sizes such that each error has the *same* variance, and within each set, there is an intra-class structure and the intra-class correlation parameter may vary from set to set. Also, between any two different sets, the covariances are *all* equal to zero. Generalized inter-class variance-covariance structure of errors is given by

$$\text{Dis}(\mathbf{e}) = \text{Diag}(W_1(\rho_1), W_2(\rho_2), ..., W_m(\rho_m)) = W_2 \text{ say}, \qquad (2.20)$$

where $W_i(\rho_i)$ is a matrix of order $n_i \times n_i$ and is given by

$$W_i(\rho_i) = \sigma^2[(1 - \rho_i)I_{n_i} + \rho_i E_{n_i \times n_i}]; \ \ 1 \le i \le m.$$

Note that,

$$W_i^{-1}(\rho_i) = \sigma^{-2}[(\zeta_i - \psi_i)I_{n_i} + \psi_i E_{n_i \times n_i}]$$

where $\zeta_i = \frac{1+(n_i-2)\rho_i}{(1-\rho_i)\{1+(n_i-1)\rho_i\}}$, $\psi_i = -\frac{\rho_i}{(1-\rho_i)\{1+(n_i-1)\rho_i\}}$ and $\rho_i > \frac{-1}{n_i-1}$; $1 \le i \le m$. (2.21)

The first-order rotatability conditions under the generalized inter-class structure (2.20), following (2.2), are given in the following theorem.

Theorem 2.9 *The necessary and sufficient conditions of a robust first-order rotatable design for all possible values of ρ_i under the generalized inter-class variance-covariance structure (2.20) in the model (2.1) are (ζ_i, ψ_i, are given in (2.21), and for all $1 \le i, j \le k$)*

$$\textbf{(1)} \ \ v_{0j} = \{\zeta_1 + (n_1 - 1)\psi_1\} \sum_{u=1}^{n_1} x_{uj} + \{\zeta_2 + (n_2 - 1)\psi_2\} \sum_{u=n_1+1}^{n_2} x_{uj} + ... + \{\zeta_m + (n_m - $$

$$1)\psi_m\} \sum_{u=n_1+n_2+...+n_{m-1}+1}^{n_m} x_{uj} = 0;$$

$$\textbf{(2)}\ v_{ij} = (\zeta_1 - \psi_1) \sum_{u=1}^{n_1} x_{ui} x_{uj} + \psi_1 (\sum_{u=1}^{n_1} x_{ui})(\sum_{u=1}^{n_1} x_{uj}) + (\zeta_2 - \psi_2) \sum_{u=n_1+1}^{n_2} x_{ui} x_{uj}$$

$$+ \psi_2 (\sum_{u=n_1+1}^{n_2} x_{ui})(\sum_{u=n_1+1}^{n_2} x_{uj}) + ... + (\zeta_m - \psi_m)(\sum_{u=n_1+...+n_{m-1}+1}^{n_m} x_{ui} x_{uj})$$

$$+ \psi_m (\sum_{u=n_1+...+n_{m-1}+1}^{n_m} x_{ui})(\sum_{u=n_1+...+n_{m-1}+1}^{n_m} x_{uj}) = 0;\ i \neq j,$$

$$\textbf{(3)}\ v_{ii} = \{\sigma^2\}^{-1} [(\zeta_1 - \psi_1) \sum_{u=1}^{n_1} x_{ui}^2 + \psi_1 (\sum_{u=1}^{n_1} x_{ui})^2 + (\zeta_2 - \psi_2) \sum_{u=n_1+1}^{n_2} x_{ui}^2 + \psi_2 (\sum_{u=n_1+1}^{n_2} x_{ui})^2 +$$

$$... + (\zeta_m - \psi_m)(\sum_{u=n_1+...+n_{m-1}+1}^{n_m} x_{ui}^2) + \psi_m (\sum_{u=n_1+...+n_{m-1}+1}^{n_m} x_{ui})^2 = constant = \lambda. \quad (2.22)$$

RFORD under the generalized inter-class structure: If the design points are divided into m sets of sizes $n_1, n_2,...,n_m$, respectively, and each set has an orthogonal structure, i.e., each set forms a FORD in the usual model (for uncorrelated and homoscedastic errors), then the overall design is a robust first-order rotatable design under the generalized inter-class structure (2.20), whatever be the values of $\rho_1, \rho_2,...,\rho_m$. Note that Method II of Section 2.4.2 actually gives a design which is RFORD in this general situation, if $n_1 = n_2 = ... = n_m = n$ is even.

2.6 ROBUST FIRST-ORDER DESIGNS FOR COMPOUND SYMMETRY STRUCTURE

This section focuses to an error variance-covariance structure which is an extension of inter-class structure. According to the hypothesis of compound symmetry, as formulated by Vataw (1948), the population variance-covariance matrix will have such a pattern when the errors are divided into m sets of size n each such that each error has the *same* variance and within each set of errors, the covariances are equal (i.e., every set has the *same* intra-class structure) and the covariance between any two errors from two distinct sets is a constant which may be different from the constant intra-class covariance within each set. Essentially, compound symmetry hypothesis is an extension of the intra-class (or inter-class) model and is often known to be used in psychometry as elucidated by Mukherjee (1981) in relation to problems dealing with reliability and validity of tests. With the additional assumption of normality the same model has been characterized as representing bipolar normal distributions by Roy (1954), who also has provided test procedures for some hypotheses of interest connected to the model.

Compound symmetry structure is given by the variance-covariance matrix of errors as

$$\text{Dis}(\mathbf{e}) = \sigma^2 [I_m \otimes (A - B) + E_{m \times m} \otimes B] = W_{N \times N}(\rho, \rho_1) \text{ say}, \quad (2.23)$$

where $A = (1 - \rho)I_n + \rho E_{n \times n}$, $B = \rho_1 E_{n \times n}$, $N = mn$ and $\otimes$ denotes Kronecker product.

Note that,

$$W_{N \times N}^{-1}(\rho, \rho_1) = (\sigma^2)^{-1} [I_m \otimes (A_1 - B_1) + E_{m \times m} \otimes B_1],$$

where $A_1 = (\delta_2 - \gamma_2)I_n + \gamma_2 E_{n \times n}$, $B_1 = \delta_3 E_{n \times n}$, $\delta_2 = \gamma_2 + 1/(1 - \rho)$, $\gamma_2 = [(m-1)n\rho_1^2 - (m-2)n\rho\rho_1 - \{1 + (n-1)\rho\}\rho]/R$, $\delta_3 = \{1 - \delta_2 - (n-1)\rho\gamma_2\}/(m-1)n\rho_1$, and $R = (1 - \rho)[(m-2)n\rho_1\{1 + (n-1)\rho\} + \{1 + (n-1)\rho\}^2 - (m-1)n^2\rho_1^2]$. \quad (2.24)

The first-order rotatability conditions (following (2.2)) under the compound symmetry structure (2.23) are given in the following theorem.

Theorem 2.10 (Panda and Das, 1994) *The necessary and sufficient conditions of a robust first-order rotatable design, for all possible values of ρ, ρ_1 in $W_{N \times N}(\rho, \rho_1)$ under the compound symmetry variance-covariance structure (2.23) in the model (2.1) are (δ_2, γ_2, δ_3 are as in (2.24), and for all $1 \leq i,\ j \leq k$)*

(1) $v_{0j} = \sum_{u=1}^{N} x_{uj} = 0$;

(2) $v_{ij} = (\delta_2 - \gamma_2) \sum_{u=1}^{N} x_{ui} x_{uj} + (\gamma_2 - \delta_3)\{(\sum_{u=1}^{n} x_{ui})(\sum_{u=1}^{n} x_{uj}) + (\sum_{u=(n+1)}^{2n} x_{ui})(\sum_{u=(n+1)}^{2n} x_{uj}) +$

.... $+ (\sum_{u=n(m-1)+1}^{mn} x_{ui})(\sum_{u=n(m-1)+1}^{mn} x_{uj})\} + \delta_3(\sum_{u=1}^{N} x_{ui})(\sum_{u=1}^{N} x_{uj}) = 0;\ i \neq j,$

(3) $v_{ii} = \{\sigma^2\}^{-1}[(\delta_2 - \gamma_2) \sum_{u=1}^{N} x_{ui}^2 + (\gamma_2 - \delta_3)\{(\sum_{u=1}^{n} x_{ui})^2 + ... + (\sum_{u=n(m-1)+1}^{mn} x_{ui})^2\} +$

$\delta_3(\sum_{u=1}^{N} x_{ui})^2] = constant = \lambda.$ (2.25)

2.6.1 Optimum robust first-order designs under compound symmetry structure

This section examines optimum robust first-order designs under the compound symmetry structure (2.23). The following theorem gives the necessary and sufficient conditions of a *D-ORFOD*.

Theorem 2.11 (Das and Park, 2010) *The necessary and sufficient conditions of a D-optimal robust first-order design, for all ρ, ρ_1 in $W_{N \times N}(\rho, \rho_1)$ under the compound symmetry structure (2.23), in the model (2.1) are (for all $1 \leq i,\ j \leq k$)*

(i) $v_{0j} = \sum_{u=1}^{N} x_{uj} = 0$;

(ii) $v_{ij} = (\delta_2 - \gamma_2) \sum_{u=1}^{N} x_{ui} x_{uj} + (\gamma_2 - \delta_3)\{(\sum_{u=1}^{n} x_{ui})(\sum_{u=1}^{n} x_{uj}) + (\sum_{u=(n+1)}^{2n} x_{ui})(\sum_{u=(n+1)}^{2n} x_{uj}) +$

... $+ (\sum_{u=n(m-1)+1}^{mn} x_{ui})(\sum_{u=n(m-1)+1}^{mn} x_{uj})\} + \delta_3(\sum_{u=1}^{N} x_{ui})(\sum_{u=1}^{N} x_{uj}) = 0;\ i \neq j,$

(iii) $\sum_{u=1}^{N} x_{ui}^2 = N; \sum_{u=1}^{n} x_{ui} = \sum_{u=(n+1)}^{2n} x_{ui} = ... = \sum_{u=n(m-1)+1}^{mn} x_{ui} = 0;\ if\ (\gamma_2 - \delta_3) < 0,$

(iv) $\sum_{u=1}^{N} x_{ui}^2 = N; \sum_{u=1}^{n} x_{ui} = \sum_{u=(n+1)}^{2n} x_{ui} = ... = \sum_{u=n(m-1)+1}^{mn} x_{ui} = \pm n; if (\gamma_2 - \delta_3) >$
0.   (2.26)

Note that $(\gamma_2 - \delta_3) > 0$ holds in the following two cases: Case I: $\rho_1 > \rho > 0$, $0 < \rho_1 \leq 0.5$; m, $n < 6$; and Case II: $0 < \rho_1 < \rho$; m, $n < 12$. The basic parameters characterizing the correlation structure (2.23) being ρ and ρ_1, the interpretation of $(\gamma_2 - \delta_3) > 0$ can be obtained from the above two cases only, and $(\gamma_2 - \delta_3) < 0$ everywhere else. Below only the conditions either $(\gamma_2 - \delta_3) < 0$ or $(\gamma_2 - \delta_3) > 0$ will be considered.

Following (2.26), the variance of the estimated response at $\mathbf{x}$ of a D-ORFOD under the compound symmetry structure (2.23) is

$$\text{Var}(\hat{y}_\mathbf{x})_{01} = \frac{\sigma^2}{N}[a(\delta_2,\ \gamma_2,\ \delta_3) + \frac{1}{\delta_2 - \gamma_2}r^2], \text{ if } (\gamma_2 - \delta_3) < 0, \tag{2.27}$$

$$\text{Var}(\hat{y}_\mathbf{x})_{02} = \frac{\sigma^2}{N}[a(\delta_2,\ \gamma_2,\ \delta_3) + \frac{1}{(\delta_2 - \gamma_2) + (\gamma_2 - \delta_3)n}r^2], \text{ if } (\gamma_2 - \delta_3) > 0, \tag{2.28}$$

where $a(\delta_2,\ \gamma_2,\ \delta_3) = \frac{1}{\{\delta_2 + (n-1)\gamma_2 + n(m-1)\delta_3\}}$ and $r^2 = \sum_{i=1}^{k} x_i^2$.

2.6.2 RFORD and D-ORFOD under compound symmetry structure

From (2.26), it follows that there are two sets of D-ORFODs under the compound symmetry structure (2.23) for the design space as given in Definition 2.3. One set is related to $(\gamma_2 - \delta_3) < 0$ and the other set is related to $(\gamma_2 - \delta_3) > 0$.

Method I: If the design points are divided into m sets of n each, where $n \equiv 0$ (Mod 4) and an H_n (Hadamard) matrix exists and $k < n$, then from H_n, an orthogonal structure within each set can be constructed such that $x_{ui} = 1$ or -1; $1 \le u \le N = mn$; $1 \le i \le k$. Then the overall design so constructed is a D-ORFOD under the compound symmetry structure (2.23), whatever be the values of correlation coefficient ρ and ρ_1 satisfying $(\gamma_2 - \delta_3) < 0$. The resulting designs satisfy (i), (ii) and (iii) of (2.26), and its variance function is as (2.27).

Method II: For the values of correlation coefficient ρ and ρ_1 satisfying $(\gamma_2 - \delta_3) < 0$, a D-ORFOD of k factors under the compound symmetry structure as in (2.23) can be constructed by selecting any k columns from $H_m \otimes U_n$, where $\otimes$ denotes Kronecker product, H_m is a Hadamard matrix of order $m \times m$, and U_n is a column vector of order $n \times 1$ with elements $+1$ or -1 such that sum of elements of U_n is zero. Note that the resulting designs satisfy (i), (ii) and (iii) of (2.26), and its variance function is as (2.27).

Method III: A D-ORFOD of k factors under the compound symmetry structure as in (2.23), whatever be the values of correlation coefficient ρ and ρ_1 satisfying $(\gamma_2 - \delta_3) > 0$, can be constructed by selecting any k columns barring the first column from $H_m \otimes J_n$, where $\otimes$ denotes Kronecker product, H_m is the Standard Hadamard matrix of order $m \times m$ (having all the elements of the first row and the first column as 1's), and J_n is a column vector of order $n \times 1$ of all 1's. It can be readily verified that (i), (ii) and (iv) of (2.26) hold for these designs and its variance function is as (2.28).

Method III as in Section 2.4.2 gives *nearly* optimum robust first-order rotatable designs under the compound symmetry structure (2.23), for the values of correlation coefficient ρ and ρ_1 satisfying $(\gamma_2 - \delta_3) < 0$. The variance of the estimated response at $\mathbf{x}$ of a RFORD under the structure (2.23), obtained by Method III as in Section 2.4.2 is

$$\text{Var}(\hat{y}_\mathbf{x}) = \sigma^2 \left[a(\delta_2,\ \gamma_2,\ \delta_3) + \frac{1}{m(n-1)(\delta_2 - \gamma_2)}r^2 \right], \tag{2.29}$$

where $a(\delta_2,\ \gamma_2,\ \delta_3)$ as in (2.28) and $r^2 = \sum_{i=1}^{k} x_i^2$.

Following (2.5), efficiency ratio of a design obtained by Method III as in Section 2.4.2, under the compound symmetry structure (2.23) is

$$ER\% = \frac{Max_{0 \leq r^2 \leq k}\left[\frac{1}{n\{\delta_2+(n-1)\gamma_2+n(m-1)\delta_3\}} + \frac{1}{n(\delta_2-\gamma_2)}r^2\right]}{Max_{0 \leq r^2 \leq k}\left[\frac{1}{n\{\delta_2+(n-1)\gamma_2+n(m-1)\delta_3\}} + \frac{1}{(n-1)(\delta_2-\gamma_2)}r^2\right]} \times 100, \text{ if } (\gamma_2 - \delta_3) < 0.$$

Note that d_I, d_{II}, d_{IV} (in Example 2.2) are, respectively, examples of D-ORFODs for the methods I, II and III under the compound symmetry structure.

2.7   ROBUST FIRST-ORDER DESIGNS FOR TRI-DIAGONAL STRUCTURE

Tri-diagonal error variance-covariance structure relaxes the intra-class structure in the form of a *lag-model* error variance-covariance structure. The symmetric matrix of the uniform tri-diagonal variance-covariance structure which arises in distributed lag-models in regression as illustrated in Press (1972) and its spectral decomposition is given by O' Hagan (1975). Also maximum likelihood (ML) estimates of unknown parameters of this structure under normality assumption has been provided by Mukherjee (1981). Tri-diagonal structure of the variance-covariance matrix arising from the circular stochastic model (Anderson, 1969, 1970) where the elements in the principal diagonal are identically equal to $(1 + \beta^2)$ with the elements in the first sub- and super diagonals all equal to $(-\beta)$. The remaining off-diagonal elements are null.

This section examines RFORD and D-ORFOD under the tri-diagonal structure. The tri-diagonal structure of errors arises when the variance of each error is *same* say, σ^2 and the correlation between any two errors having *lag n* is ρ, and 0 (zero) otherwise. The tri-diagonal error structure with $2n$ observations is given below.

$$\text{Dis}(\mathbf{e}) = \sigma^2 \left[\begin{pmatrix} I_n & I_n \\ I_n & I_n \end{pmatrix} \frac{1+\rho}{2} + \begin{pmatrix} I_n & -I_n \\ -I_n & I_n \end{pmatrix} \frac{1-\rho}{2} \right] = W_3 \text{ say.} \qquad (2.30)$$

Observe that,

$$W_3^{-1} = (\sigma^2)^{-1} \left[\begin{pmatrix} I_n & I_n \\ I_n & I_n \end{pmatrix} \frac{1}{2(1+\rho)} + \begin{pmatrix} I_n & -I_n \\ -I_n & I_n \end{pmatrix} \frac{1}{2(1-\rho)} \right].$$

The following theorem states the necessary and sufficient conditions (from (2.2)) of a robust first-order rotatable design under the tri-diagonal structure (2.30).

Theorem 2.12 (Panda and Das, 1994) *The necessary and sufficient conditions of a robust first-order rotatable design, for all possible values of ρ in W_3 under the tri-diagonal variance-covariance structure (2.30) in the model (2.1) are (for $1 \leq i$, $j \leq k$)*

(1) $v_{0j} = \sum_{u=1}^{2n} x_{uj} = 0,$

(2) $v_{ij} = \frac{1}{1-\rho^2} \sum_{u=1}^{2n} x_{ui}x_{uj} - \frac{\rho}{1-\rho^2} \{\sum_{u=1}^{n} x_{(n+u)i}x_{uj} + \sum_{u=1}^{n} x_{ui}x_{(n+u)j}\} = 0; i \neq j,$

(3) $v_{ii} = (\sigma^2)^{-1}\left[\frac{1}{(1-\rho^2)} \sum_{u=1}^{2n} x_{ui}^2 - \frac{2\rho}{1-\rho^2} \sum_{u=1}^{n} x_{ui}x_{(n+u)i}\right] = constant = \lambda.$ (2.31)

A design of k factors which satisfies (1) through (3) in (2.31) for all ρ, is obviously called robust first-order rotatable design under the structure (2.30).

2.7.1 Optimum robust first-order designs under tri-diagonal structure

This section describes optimum robust first-order designs under the tri-diagonal structure (2.30). The following theorem presents the necessary and sufficient conditions of a D-ORFOD under the tri-diagonal structure (2.30).

Theorem 2.13 (Das and Park, 2008b) *The necessary and sufficient conditions of a D-optimal robust first-order design, for all ρ in W_3 under the tri-diagonal structure (2.30), in the model (2.1) are (for all $1 \le i,\ j \le k$)*

(i) $v_{0j} = \displaystyle\sum_{u=1}^{2n} x_{uj} = 0,$

(ii) $v_{ij} = \dfrac{1}{1-\rho^2} \displaystyle\sum_{u=1}^{2n} x_{ui}x_{uj} - \dfrac{\rho}{1-\rho^2}\{\displaystyle\sum_{u=1}^{n} x_{(n+u)i}x_{uj} + \displaystyle\sum_{u=1}^{n} x_{ui}x_{(n+u)j}\} = 0;\ i \ne j,$

(iii) $\displaystyle\sum_{u=1}^{2n} x_{ui}^2 = 2n,\ and\ \displaystyle\sum_{u=1}^{n} x_{ui}x_{(n+u)i} = -n;\ if\ \rho > 0,$

(iv) $\displaystyle\sum_{u=1}^{2n} x_{ui}^2 = 2n,\ and\ \displaystyle\sum_{u=1}^{n} x_{ui}x_{(n+u)i} = n;\ if\ \rho < 0.$ (2.32)

The variance of the estimated response at $\mathbf{x}$ of a D-ORFOD under the tri-diagonal structure (2.30), over the design space as in Definition 2.3 is (from (2.4) and using (2.32))

$$\text{Var}(\hat{y}_{\mathbf{x}})_{01} = \frac{\sigma^2}{2n}[(1+\rho) + (1-\rho)r^2],\ \text{if}\ \rho > 0,$$ (2.33)

$$\text{and}\ \ \text{Var}(\hat{y}_{\mathbf{x}})_{02} = \frac{\sigma^2}{2n}[(1+\rho)(1+r^2)],\ \text{if}\ \rho < 0,$$ (2.34)

where $r^2 = \displaystyle\sum_{i=1}^{k} x_i^2.$

2.7.2 RFORD and D-ORFOD under tri-diagonal structure

One can easily construct two (clear from (2.32)) sets of D-ORFODs under the tri-diagonal structure (2.30). One set is related to $\rho > 0$ and the other is related to $\rho < 0$. The following Method I and Method II give, respectively, D-ORFODs for $\rho > 0$ and $\rho < 0$, and Method III, Method IV give nearly D-ORFODs for all ρ, whereas all the designs developed by these four methods are RFORDs (Das and Park, 2008b).

Method I: If $k \le n$, D-optimum robust first-order designs of k factors can be constructed under the tri-diagonal structure (2.30), whatever be the value of correlation coefficient $\rho > 0$, by selecting any k columns from $(H_n' : - H_n')'$, where H_n is a Hadamard matrix of order n assuming it exists for n. It can be readily verified that (i), (ii) and (iii) in (2.32) hold for these designs. Variance function of a design, constructed by this method, is the same as in (2.33).

Method II: Let us start with a usual FORD of k factors and n design points with levels $+1$ or -1. With these k factors, a D-ORFOD of k factors with $2n$ design points can be constructed under the tri-diagonal structure (2.30), whatever be the value of correlation coefficient $\rho < 0$, by repeating ith factor column of the usual FORD just below itself, $1 \le i \le k$. It can be readily verified that (i), (ii) and (iv) in (2.32) hold for these designs. Variance function of a design, constructed by this method, is the same as in (2.34).

Method III: Let us start with a usual FORD of k factors and n design points. Let $\mathbf{x}_1$, $\mathbf{x}_2,...,\mathbf{x}_k$ be the k factors of the selected FORD (with design levels $x_{ui} = 1$ or -1; $1 \leq u \leq n$, $1 \leq i \leq k$). The set of n design points can be extended to $2n$ points by adding n central points or **0**-points (**0**-point $= (0, 0, ..., 0)$) together just below or above the n design points of the selected FORD. It can be readily verified that (2.31) holds (for all possible values of ρ) for these designs.

Note that $\sum_{u=1}^{2n} x_{ui}^2 = n$ and $\sum_{u=1}^{n} x_{ui} x_{(n+u)i} = 0$; $1 \leq i \leq k$. The variance of the estimated response at $\mathbf{x}$ of a design constructed by this method under the structure (2.30), is given by

$$\text{Var}(\hat{y}_{\mathbf{x}}) = \frac{\sigma^2}{n} \left[\frac{1+\rho}{2} + (1-\rho^2)r^2 \right] \tag{2.35}$$

where $r^2 = \sum_{i=1}^{k} x_i^2$.

This above method gives RFORDs under the tri-diagonal structure (2.30) for all ρ. Following (2.5), efficiency ratio of a design developed by Method III is

$$ER\% = \frac{Max_{0 \leq r^2 \leq k}\{(1+\rho) + (1-\rho)r^2\}}{Max_{0 \leq r^2 \leq k} 2\{(1+\rho)/2 + (1-\rho^2)r^2\}} \times 100, \quad \text{if} \ \ \rho > 0, \tag{2.36}$$

$$ER\% = \frac{Max_{0 \leq r^2 \leq k}\{(1+\rho)(1+r^2)\}}{Max_{0 \leq r^2 \leq k} 2\{(1+\rho)/2 + (1-\rho^2)r^2\}} \times 100, \quad \text{if} \ \ \rho < 0. \tag{2.37}$$

Method IV: We start with a usual FORD of $2k$ factors and n design points with levels $+1$ or -1. With these $2k$ factors we form a k factors design under the tri-diagonal structure (2.30) with $2n$ design points by arranging any k factors of usual FORD just below the other set of remaining k factors. It can be readily verified that (2.31) holds (for all possible values of ρ) for these designs.

Note that $\sum_{u=1}^{2n} x_{ui}^2 = 2n$ and $\sum_{u=1}^{n} x_{ui} x_{(n+u)i} = 0$; $1 \leq i \leq k$. Variance of the estimated response at $\mathbf{x}$ of a design developed by Method IV is given by

$$\text{Var}(\hat{y}_{\mathbf{x}}) = \frac{\sigma^2}{2n}[(1+\rho) + (1-\rho^2)r^2].$$

This method gives RFORDs under the correlation structure (2.30) for *all* ρ. Following (2.5), efficiency ratio of a design developed by Method IV is

$$ER\% = \frac{Max_{0 \leq r^2 \leq k}\{(1+\rho) + (1-\rho)r^2\}}{Max_{0 \leq r^2 \leq k}\{(1+\rho) + (1-\rho^2)r^2\}} \times 100, \quad \text{if} \ \ \rho > 0, \tag{2.38}$$

$$ER\% = \frac{Max_{0 \leq r^2 \leq k}\{(1+\rho)(1+r^2)\}}{Max_{0 \leq r^2 \leq k}\{(1+\rho) + (1-\rho^2)r^2\}} \times 100, \quad \text{if} \ \ \rho < 0. \tag{2.39}$$

For the tri-diagonal error structure (2.30), below we provide two examples of D-ORFODs (d_1 and d_2) for $\rho > 0$ and one (d_3) for $\rho < 0$, following Method I and Method II, respectively. These designs are constructed from H_2 and H_4 only, where the Hadamard matrices are formed by Kronecker product method as given below:

$$H_2 = \begin{pmatrix} 1 & 1 \\ 1 & -1 \end{pmatrix}, \quad H_4 = H_2 \otimes H_2 = H_4 = \begin{pmatrix} 1 & 1 & 1 & 1 \\ 1 & -1 & 1 & -1 \\ 1 & 1 & -1 & -1 \\ 1 & -1 & -1 & 1 \end{pmatrix}.$$

Example 2.3 A D-ORFOD, denoted by d_1 with 2 factors $(\mathbf{x}_1, \mathbf{x}_2)$ and N (number of runs) = 4 for $\rho > 0$, is given below (following Method I and from H_2) (rows being factors and columns being runs)

d_1 :	1	2	3	4
$\mathbf{x}_1$	1	1	-1	-1
$\mathbf{x}_2$	1	-1	-1	1

Example 2.4 A D-ORFOD, d_2 with 4 factors $(\mathbf{x}_1, \mathbf{x}_2, \mathbf{x}_3, \mathbf{x}_4)$ and N (number of runs) = 8 for $\rho > 0$, is given below (following Method I and from H_4) (rows being factors and columns being runs)

d_2 :	1	2	3	4	5	6	7	8
$\mathbf{x}_1$	1	1	1	1	-1	-1	-1	-1
$\mathbf{x}_2$	1	-1	1	-1	-1	1	-1	1
$\mathbf{x}_3$	1	1	-1	-1	-1	-1	1	1
$\mathbf{x}_4$	1	-1	-1	1	-1	1	1	-1

Example 2.5 A D-ORFOD, d_3 with 3 factors $(\mathbf{x}_1, \mathbf{x}_2, \mathbf{x}_3)$ and N (number of runs) = 8 for $\rho < 0$, is given below (following Method II and from H_4) (rows being factors and columns being runs)

d_3 :	1	2	3	4	5	6	7	8
$\mathbf{x}_1$	1	-1	1	-1	1	-1	1	-1
$\mathbf{x}_2$	1	1	-1	-1	1	1	-1	-1
$\mathbf{x}_3$	1	-1	-1	1	1	-1	-1	1

2.8 ROBUST FIRST-ORDER DESIGNS FOR AUTOCORRELATED STRUCTURE

The present section studies RFORDs and D-ORFODs for autocorrelated error structure given by

$$\text{Dis}(\mathbf{e}) = [\sigma^2 \{\rho^{|i-j|}\}_{1 \leq i,\, j \leq N}] = W_4 \text{ say.} \qquad (2.40)$$

This structure is very well known and profusely used in Econometrics. Fields of application where autocorrelated disturbances or errors constitute the right kind of assumption are briefly described in Johnston (1984, p. 309) and Press (1972), to name only two of the many such books available. First-order autoregressive AR(1) process in errors gives rise to an autocorrelated structure as described below.

The AR(1) process is defined as $e_t = \rho\, e_{t-1} + \epsilon_t$ where $E(\epsilon) = \mathbf{0}$ and $E(\epsilon\epsilon') = \sigma_\epsilon^2 I$, i.e., ϵ_t's are uncorrelated and have equal variances. The $\{\epsilon_t\}$ forms an autoregressive process with the autocorrelated structure for the variance-covariance matrix. Use of autoregressive AR(1) process for errors in regression problems in the usual normal set up is very widely known. In this connection it may be mentioned that autocorrelated structure for the variance-covariance matrix although generally studied under normality assumptions may also hold good for other distributions having useful applications in practice. For instance Kotz and Adams (1964) have investigated the distribution of the sum of autocorrelated Gamma variables which has many interesting applications in engineering, insurance and meteorology and has generated a lot of interest among scientists working in those areas.

Observe that,

$$W_4^{-1} = \{\sigma^2(1-\rho^2)\}^{-1}[(1+\rho^2)I_N - \rho^2 A_0 - \rho B_0]$$

where I_N is the $N \times N$ identity matrix, A_0 is the $N \times N$ matrix with elements $a_{11} = a_{NN}$ = 1 and all other elements 0 (zeros), and B_0 is the $N \times N$ matrix with $b_{ij} = 1$ for $|\,i - j\,|$ = 1 and all other elements 0 (zeros).

First-order rotatability conditions for the above autocorrelated error structure is given in the following theorem.

Theorem 2.14 (Panda and Das, 1994) *The necessary and sufficient conditions of a robust first-order rotatable design, for all possible values of ρ in W_4 under the autocorrelated error variance-covariance structure (2.40) in the model (2.1) are (for all $1 \leq i,\ j \leq k$)*

(1) $v_{0j} = \sum_{u=1}^{N} x_{uj} - \rho \sum_{u=2}^{N-1} x_{uj} = 0,$

(2) $v_{ij} = \sum_{u=1}^{N} x_{ui}x_{uj} + \rho^2 \sum_{u=2}^{N-1} x_{ui}x_{uj} - \rho\{\sum_{u=1}^{N-1} x_{ui}x_{(u+1)j} + \sum_{u=1}^{N-1} x_{(u+1)i}x_{uj}\} = 0;\ i \neq j,$

(3) $v_{ii} = \{\sigma^2(1-\rho^2)\}^{-1}[\sum_{u=1}^{N} x_{ui}^2 + \rho^2 \sum_{u=2}^{N-1} x_{ui}^2 - 2\rho \sum_{u=1}^{N-1} x_{ui}x_{(u+1)i}] = constant = \lambda.$ (2.41)

A design of k factors $(\mathbf{x}_1, \mathbf{x}_2, ..., \mathbf{x}_k)$ which satisfies (1) through (3) in (2.41) for *all* ρ, is called robust first-order rotatable design under the autocorrelated error structure (2.40).

2.8.1   Optimum robust designs under autocorrelated structure

This section examines optimum first-order designs under autocorrelated errors. The following theorem gives the necessary and sufficient conditions of a *D*-ORFOD under the autocorrelated error structure (2.40).

Theorem 2.15 (Das and Park, 2008a) *The necessary and sufficient conditions of a D-optimal robust first-order design, for all possible values of ρ in W_4 under the autocorrelated error structure (2.40), in the model (2.1) are (for all $1 \leq i,\ j \leq k$)*

(i) $v_{0j} = \sum_{u=1}^{N} x_{uj} - \rho \sum_{u=2}^{N-1} x_{uj} = 0,$

(ii) $v_{ij} = \sum_{u=1}^{N} x_{ui}x_{uj} + \rho^2 \sum_{u=2}^{N-1} x_{ui}x_{uj} - \rho\{\sum_{u=1}^{N-1} x_{ui}x_{(u+1)j} + \sum_{u=1}^{N-1} x_{(u+1)i}x_{uj}\} = 0;\ i \neq j,$

(iii) $\sum_{u=1}^{N} x_{ui}^2 = N;\ \sum_{u=2}^{N-1} x_{ui}^2 = N - 2;\ and\ \sum_{u=1}^{N-1} x_{ui}x_{(u+1)i} = -(N-1);\ if\ \rho > 0,$

(iv) $\sum_{u=1}^{N} x_{ui}^2 = N;\ \sum_{u=2}^{N-1} x_{ui}^2 = N - 2;\ and\ \sum_{u=1}^{N-1} x_{ui}x_{(u+1)i} = (N-1);\ if\ \rho < 0.$ (2.42)

Note that, ρ is positive in most of the practical situations for the autocorrelated structure (2.40). We examine here *D*-ORFODs for positive and negative values of ρ. Following (2.4) and using (2.42), the variance of the estimated response at $\mathbf{x}$ of a *D*-ORFOD under the autocorrelated structure (2.40), over the factor space as in Definition 2.3 is

$$\mathrm{Var}(\hat{y}_\mathbf{x})_{01} = \sigma^2 \left[\frac{1+\rho}{N-(N-2)\rho} + \frac{(1-\rho^2)r^2}{N+(N-2)\rho^2 + 2\rho(N-1)} \right]\ if\ \rho > 0,$$ (2.43)

$$\text{and Var}(\hat{y}_{\mathbf{x}})_{02} = \sigma^2 \left[\frac{1+\rho}{N - (N-2)\rho} + \frac{(1-\rho^2)r^2}{N + (N-2)\rho^2 - 2\rho(N-1)} \right] \text{ if } \rho < 0, \quad (2.44)$$

where $r^2 = \sum_{i=1}^{k} x_i^2$.

2.8.2 RFORD and nearly *D*-ORFOD under autocorrelated structure

This section describes three construction methods of robust first-order rotatable designs under the autocorrelated error structure (2.40). Following (2.42), it is clear that one should construct two sets of *D*-ORFODs under the autocorrelated structure (2.40). One set is related to $\rho > 0$ and the other is related to $\rho < 0$.

Panda and Das (1994) have developed a construction method of robust first-order rotatable designs under the autocorrelated error structure (2.40) and the designs thus obtained are *not* optimal. In addition, **Das and Park (2008a)** have developed **two more construction** methods of RFORDs under the autocorrelated structure (2.40), using augmentation of columns of Hadamard matrices, and the designs thus obtained are *not* optimal but *nearly* so and have better efficiency than the designs described by Panda and Das (1994). For the sake of simplicity, as examples, explicit constructions of RFORDs are described using augmentation of columns of Hadamard matrices H_4, H_8 and H_{16} only. Obviously, similar methods can be applied to other Hadamard matrices to get useful designs. Method II involves the augmentation of columns of the one Hadamard matrix, and Method III involves the augmentation of columns of two or more Hadamard matrices of different orders (Das and Park, 2008a).

Method I: Following Panda and Das (1994), let us start with an usual FORD, say d_1 with n design points and k factors. The set of n design points can be extended to $(2n+1)$ points by incorporating $(n+1)$ central points (or **0** point $= (0,0,...,0)$) in the following way. One central point is placed in between a pair of points in d_1. The other two central points are placed one at the beginning and one at the end.

It can be readily verified that (2.41) holds for this design for *all* ρ. Thus the designs which are obtained by the above method are RFORDs under the autocorrelated structure (2.40) for all positive or negative ρ. Further, the variance of the estimated response at **x** of a RFORD under the autocorrelated structure (2.40) constructed by Method I is given by

$$\text{Var}(\hat{y}_{\mathbf{x}}) = \sigma^2 \left[\frac{1+\rho}{N - (N-2)\rho} + \frac{(1-\rho^2)r^2}{n(1+\rho^2)} \right], \quad (2.45)$$

where $N = 2n + 1$ and $r^2 = \sum_{i=1}^{k} x_i^2$.

Remark 2.4 The RFORDs under the autocorrelated structure (2.40) thus constructed as in Method I are *not* invariant with respect to rotatability under the permutation of design points.

Following (2.5), efficiency ratio of a design obtained by Method I is

$$ER\% = \frac{Max_{0 \leq r^2 \leq k} \left[\frac{1+\rho}{N-(N-2)\rho} + \frac{(1-\rho^2)r^2}{N+(N-2)\rho^2+2\rho(N-1)} \right]}{Max_{0 \leq r^2 \leq k} \left[\frac{1+\rho}{N-(N-2)\rho} + \frac{2(1-\rho^2)r^2}{(N-1)(1+\rho^2)} \right]} \times 100, \text{ if } \rho > 0,$$

and

$$ER\% = \frac{Max_{0 \leq r^2 \leq k} \left[\frac{1+\rho}{N-(N-2)\rho} + \frac{(1-\rho^2)r^2}{N+(N-2)\rho^2-2\rho(N-1)} \right]}{Max_{0 \leq r^2 \leq k} \left[\frac{1+\rho}{N-(N-2)\rho} + \frac{2(1-\rho^2)r^2}{(N-1)(1+\rho^2)} \right]} \times 100, \text{ if } \rho < 0.$$

The following *two* methods (Das and Park, 2008a) describe RFORDs, which are *not* optimal but *nearly* so and better than the designs described by Panda and Das (1994). Let $h_j^{(i)}$ be the jth column of the Hadamard matrix of order $i = 4, 8, 16$ where the Hadamard matrices are formed by Kronecker product method as given below:

$$H_2 = \begin{pmatrix} 1 & 1 \\ 1 & -1 \end{pmatrix}, \; H_4 = H_2 \otimes H_2, H_8 = H_2 \otimes H_4, H_{16} = H_2 \otimes H_8,$$

where $\otimes$ denotes Kronecker product.

Method II(a): (i) Robust FORD of 2 factors (from H_4) with $N = 11$ and $\sum_{u=1}^{N-1} x_{ui}x_{(u+1)i} = 0$ is $(\mathbf{x}_1, \mathbf{x}_2)$, where

$$\mathbf{x}_1 = (0, h_3^{(4)\prime}, 0, h_4^{(4)\prime}, 0)', \quad \mathbf{x}_2 = (0, h_4^{(4)\prime}, 0, h_3^{(4)\prime}, 0)'.$$

(ii) Robust FORD of 6 factors (from H_8) with $N = 19$ and $\sum_{u=1}^{N-1} x_{ui}x_{(u+1)i} = 0$ is $(\mathbf{x}_1, \mathbf{x}_2, \mathbf{x}_3, \mathbf{x}_4, \mathbf{x}_5, \mathbf{x}_6)$, where

$$\mathbf{x}_1 = (0, h_3^{(8)\prime}, 0, h_4^{(8)\prime}, 0)', \; \mathbf{x}_2 = (0, h_4^{(8)\prime}, 0, h_3^{(8)\prime}, 0)', \; \mathbf{x}_3 = (0, h_5^{(8)\prime}, 0, h_6^{(8)\prime}, 0)',$$

$$\mathbf{x}_4 = (0, h_6^{(8)\prime}, 0, h_5^{(8)\prime}, 0)', \; \mathbf{x}_5 = (0, h_7^{(8)\prime}, 0, h_8^{(8)\prime}, 0)', \; \mathbf{x}_6 = (0, h_8^{(8)\prime}, 0, h_7^{(8)\prime}, 0)'.$$

(iii) Robust FORD of 14 factors (from H_{16}) with $N = 35$ and $\sum_{u=1}^{N-1} x_{ui}x_{(u+1)i} = 0$ is $(\mathbf{x}_1, \mathbf{x}_2,...,\mathbf{x}_{14})$, where,

$$\mathbf{x}_1 = (0, h_3^{(16)\prime}, 0, h_4^{(16)\prime}, 0)', \; \mathbf{x}_2 = (0, h_4^{(16)\prime}, 0, h_3^{(16)\prime}, 0)', \; \mathbf{x}_3 = (0, h_5^{(16)\prime}, 0, h_6^{(16)\prime}, 0)',$$

$$\mathbf{x}_4 = (0, h_6^{(16)\prime}, 0, h_5^{(16)\prime}, 0)', \; \mathbf{x}_5 = (0, h_7^{(16)\prime}, 0, h_8^{(16)\prime}, 0)', \; \mathbf{x}_6 = (0, h_8^{(16)\prime}, 0, h_7^{(16)\prime}, 0)',$$

$$\mathbf{x}_7 = (0, h_9^{(16)\prime}, 0, h_{10}^{(16)\prime}, 0)', \; \mathbf{x}_8 = (0, h_{10}^{(16)\prime}, 0, h_9^{(16)\prime}, 0)', \; \mathbf{x}_9 = (0, h_{11}^{(16)\prime}, 0, h_{12}^{(16)\prime}, 0)',$$

$$\mathbf{x}_{10} = (0, h_{12}^{(16)\prime}, 0, h_{11}^{(16)\prime}, 0)', \; \mathbf{x}_{11} = (0, h_{13}^{(16)\prime}, 0, h_{14}^{(16)\prime}, 0)', \; \mathbf{x}_{12} = (0, h_{14}^{(16)\prime}, 0, h_{13}^{(16)\prime}, 0)',$$

$$\mathbf{x}_{13} = (0, h_{15}^{(16)\prime}, 0, h_{16}^{(16)\prime}, 0)', \; \mathbf{x}_{14} = (0, h_{16}^{(16)\prime}, 0, h_{15}^{(16)\prime}, 0)'.$$

It can be readily verified that (2.41) holds for each of the above designs for *all* ρ and $\sum_{u=1}^{N-1} x_{ui}x_{(u+1)i} = 0; 1 \le i \le k$. The above designs are RFORDs for $\rho > 0$ and $\rho < 0$. Variance of the estimated response at $\mathbf{x}$ of a RFORD as given above in this section, under Method II(a) is

$$\text{Var}(\hat{y}_\mathbf{x}) = \sigma^2 \left[\frac{1+\rho}{N - (N-2)\rho} + \frac{(1-\rho^2)r^2}{(N-3)(1+\rho^2)} \right]$$

where $r^2 = \sum_{i=1}^k x_i^2$ and N is the number of design points.

Following (2.5), the efficiency ratio of a design obtained by above Method II(a)is

$$ER\% = \frac{Max_{0 \le r^2 \le k} \left[\frac{1+\rho}{N-(N-2)\rho} + \frac{(1-\rho^2)r^2}{N+(N-2)\rho^2+2\rho(N-1)} \right]}{Max_{0 \le r^2 \le k} \left[\frac{1+\rho}{N-(N-2)\rho} + \frac{(1-\rho^2)r^2}{(N-3)(1+\rho^2)} \right]} \times 100, \text{ if } \rho > 0,$$

$$ER\% = \frac{Max_{0 \le r^2 \le k}\left[\frac{1+\rho}{N-(N-2)\rho} + \frac{(1-\rho^2)r^2}{N+(N-2)\rho^2-2\rho(N-1)}\right]}{Max_{0 \le r^2 \le k}\left[\frac{1+\rho}{N-(N-2)\rho} + \frac{(1-\rho^2)r^2}{(N-3)(1+\rho^2)}\right]} \times 100, \text{ if } \rho < 0.$$

Note that the efficiency ratio (ER) of designs developed by Method II(a) are greater than ER of designs developed by Method I.

Method II(b): The present method displays RFORDs under the autocorrelated structure (2.40) which satisfy (2.41) and $\sum_{u=1}^{N-1} x_{ui}x_{(u+1)i}$ is a negative constant for all i, $1 \le i \le k$.

(i) Robust FORD of 4 factors (from H_8) with $N = 19$ and $\sum_{u=1}^{N-1} x_{ui}x_{(u+1)i} = -6$ is $(\mathbf{x}_1, \mathbf{x}_2, \mathbf{x}_3, \mathbf{x}_4)$, where

$$\mathbf{x}_1 = (0, h_2^{(8)\prime}, 0, h_3^{(8)\prime}, 0)', \quad \mathbf{x}_2 = (0, h_4^{(8)\prime}, 0, h_6^{(8)\prime}, 0)',$$

$$\mathbf{x}_3 = (0, h_3^{(8)\prime}, 0, h_2^{(8)\prime}, 0)', \quad \mathbf{x}_4 = (0, h_6^{(8)\prime}, 0, h_4^{(8)\prime}, 0)'.$$

The variance of the estimated response at $\mathbf{x}$ of a RFORD as given above is

$$\text{Var}(\hat{y}_{\mathbf{x}}) = \sigma^2\left[\frac{1+\rho}{19-17\rho} + \frac{(1-\rho^2)r^2}{16+16\rho^2+12\rho)}\right].$$

Following (2.5), the efficiency ratio of the above design is

$$ER\% = \frac{Max_{0 \le r^2 \le k}\left[\frac{1+\rho}{19-17\rho} + \frac{(1-\rho^2)r^2}{19+17\rho^2+36\rho}\right]}{Max_{0 \le r^2 \le k}\left[\frac{1+\rho}{19-17\rho} + \frac{(1-\rho^2)r^2}{16+16\rho^2+12\rho}\right]} \times 100, \text{ if } \rho > 0.$$

(ii) Robust FORD of 4 factors (from H_8) with $N = 19$ and $\sum_{u=1}^{N-1} x_{ui}x_{(u+1)i} = -4$ is $(\mathbf{x}_1, \mathbf{x}_2, \mathbf{x}_3, \mathbf{x}_4)$, where

$$\mathbf{x}_1 = (0, h_3^{(8)\prime}, 0, h_6^{(8)\prime}, 0)', \quad \mathbf{x}_2 = (0, h_4^{(8)\prime}, 0, h_8^{(8)\prime}, 0)',$$

$$\mathbf{x}_3 = (0, h_6^{(8)\prime}, 0, h_3^{(8)\prime}, 0)', \quad \mathbf{x}_4 = (0, h_8^{(8)\prime}, 0, h_4^{(8)\prime}, 0)'.$$

It can be readily verified that (2.41) holds for these above designs for *all* ρ.

The variance of the estimated response at $\mathbf{x}$ of a RFORD as given above is

$$\text{Var}(\hat{y}_{\mathbf{x}}) = \sigma^2\left[\frac{1+\rho}{19-17\rho} + \frac{(1-\rho^2)r^2}{(16+16\rho^2+8\rho)}\right].$$

The above designs are RFORDs under the autocorrelated structure (2.40) for $\rho > 0$ and $\rho < 0$.

Following (2.5), the efficiency ratio of the above design is

$$ER\% = \frac{Max_{0 \le r^2 \le k}\left[\frac{1+\rho}{19-17\rho} + \frac{(1-\rho^2)r^2}{19+17\rho^2+36\rho}\right]}{Max_{0 \le r^2 \le k}\left[\frac{1+\rho}{19-17\rho} + \frac{(1-\rho^2)r^2}{16+16\rho^2+8\rho}\right]} \times 100, \text{ if } \rho > 0.$$

Method III (a): Further constructions by following methods similar to II(a) and

II(b) are given here. (i) Robust FORD of 3 factors (from H_4 and H_8) with $N = 15$ and $\sum_{u=1}^{N-1} x_{ui} x_{(u+1)i} = 0$ is $(\mathbf{x}_1, \mathbf{x}_2, \mathbf{x}_3)$, where

$$\mathbf{x}_1 = (0, h_2^{(4)\prime}, 0, h_7^{(8)\prime}, 0)', \quad \mathbf{x}_2 = (0, h_3^{(4)\prime}, 0, h_4^{(8)\prime}, 0)', \quad \mathbf{x}_3 = (0, h_4^{(4)\prime}, 0, h_3^{(8)\prime}, 0)'.$$

(ii) Robust FORD of 7 factors (from H_8 and H_{16}) with $N = 27$ and $\sum_{u=1}^{N-1} x_{ui} x_{(u+1)i} = 0$ is $(\mathbf{x}_1, \mathbf{x}_2, \mathbf{x}_3,...,\mathbf{x}_7)$, where

$$\mathbf{x}_1 = (0, h_2^{(8)\prime}, 0, h_7^{(16)\prime}, 0)', \quad \mathbf{x}_2 = (0, h_5^{(8)\prime}, 0, h_{16}^{(16)\prime}, 0)', \quad \mathbf{x}_3 = (0, h_6^{(8)\prime}, 0, h_{15}^{(16)\prime}, 0)',$$

$$\mathbf{x}_4 = (0, h_7^{(8)\prime}, 0, h_{12}^{(16)\prime}, 0)', \quad \mathbf{x}_5 = (0, h_8^{(8)\prime}, 0, h_{11}^{(16)\prime}, 0)', \quad \mathbf{x}_6 = (0, h_3^{(8)\prime}, 0, h_4^{(16)\prime}, 0)',$$

$$\mathbf{x}_7 = (0, h_4^{(8)\prime}, 0, h_3^{(16)\prime}, 0)'.$$

It can be readily verified that (2.41) holds for these above designs for *all* ρ. Variance of the estimated response at $\mathbf{x}$ and efficiency ratio of a design as given above under the Method III(a) are same as in Method II(a). Note that the number of design points in Method III(a) is less than the number of design points as in Method II(a).

Method III(b): This method describes RFORDs under the autocorrelated structure (2.40), which satisfy (2.41) and $\sum_{u=1}^{N-1} x_{ui} x_{(u+1)i}$ is a positive constant for all i, $1 \leq i \leq k$. (i) Robust FORD of 2 factors (from H_4 and H_8) with $N = 15$ and $\sum_{u=1}^{N-1} x_{ui} x_{(u+1)i} = 2$ is $d_1 = (\mathbf{x}_1, \mathbf{x}_2)$ or $d_2 = (\mathbf{x}_3, \mathbf{x}_4)$, where

$$d_1: \quad \mathbf{x}_1 = (0, h_3^{(4)\prime}, 0, h_3^{(8)\prime}, 0)', \quad \mathbf{x}_2 = (0, h_4^{(4)\prime}, 0, h_7^{(8)\prime}, 0)',$$

$$d_2: \quad \mathbf{x}_3 = (0, h_2^{(4)\prime}, 0, h_5^{(8)\prime}, 0)', \quad \mathbf{x}_4 = (0, h_4^{(4)\prime}, 0, h_7^{(8)\prime}, 0)',$$

where d_1 and d_2 are two designs.

It can be readily verified that (2.41) holds for these above designs for *all* ρ. Following (2.4), the variance of the estimated response at $\mathbf{x}$ of a RFORD as given above is

$$\text{Var}(\hat{y}_\mathbf{x}) = \sigma^2 \left[\frac{1+\rho}{15 - 13\rho} + \frac{(1 - \rho^2)r^2}{(12 + 12\rho^2 - 4\rho)} \right].$$

Following (2.5), efficiency ratio of a design obtained by the above Method III(b) is

$$ER\% = \frac{Max_{0 \leq r^2 \leq k}\left[\frac{1+\rho}{15-13\rho} + \frac{(1-\rho^2)r^2}{15+13\rho^2-28\rho}\right]}{Max_{0 \leq r^2 \leq k}\left[\frac{1+\rho}{15-13\rho} + \frac{(1-\rho^2)r^2}{12+12\rho^2-4\rho}\right]} \times 100, \text{ if } \rho < 0.$$

Note that the designs given by Method I, II(a), II(b), III(a) are nearly optimum RFORDs for $\rho > 0$ and designs given by Method I, II(a), III(a), III(b) are nearly optimum RFORDs when $\rho < 0$. Some RFORDs (constructed by the above methods) are displayed in Table 2.3 to Table 2.6, and their efficiency ratios are presented in Table 2.7 and Table 2.8. Graphical presentations of their efficiency ratios are displayed in Figure 2.1 and Figure 2.2. From Table 2.7 and Table 2.8, and also from Figure 2.1 and Figure 2.2, it is observed that some designs (D_8, D_9 and D_{18}) have very high efficiency around 0.90 for all values of ρ under autocorrelated structure.

TABLE 2.1: Nearly D-optimal design D'_1 for $k = 2$ factors and N=11. Das, R.N. and Park, S.H. (2008a). "On efficient robust rotatable designs with autocorrelated errors," Journal of the Korean Statistical Society, 37(2), pp. 97-106.

$RunNo.$:	1	2	3	4	5	6	7	8	9	10	11
$\mathbf{x}_1$	0	1	-1	1	-1	0	1	-1	-1	1	0
$\mathbf{x}_2$	0	1	-1	-1	1	0	1	-1	1	-1	0

TABLE 2.2: Nearly D-optimal design D'_2 for $k = 2$ factors and N=15. Das, R.N. and Park, S.H. (2008a). "On efficient robust rotatable designs with autocorrelated errors," Journal of the Korean Statistical Society, 37(2), pp. 97-106.

$RunNo.$:	1	2	3	4	5	6	7	8	9	10	11	12	13	14	15
$\mathbf{x}_1$	0	1	-1	-1	1	0	1	-1	1	-1	1	-1	1	-1	0
$\mathbf{x}_2$	0	1	-1	1	-1	0	1	-1	1	-1	-1	1	-1	1	0

TABLE 2.3: Nearly D-optimal designs for $k = 2$ factors when $\rho \geq 0$. Das, R.N. and Park, S.H. (2008a). "On efficient robust rotatable designs with autocorrelated errors," Journal of the Korean Statistical Society, 37(2), pp. 97-106.

Design	no. of runs	$\sum_{u=1}^{N} x_{ui}^2$	$\sum_{u=2}^{N-1} x_{ui}^2$	$\sum_{u=1}^{N-1} x_{ui}x_{(u+1)i}$	$ER\%$
D_1 : $\mathbf{x}_1 = (0, h_2^{(4)\prime}, 0, h_4^{(4)\prime}, 0)'$, $\mathbf{x}_2 = (0, h_4^{(4)\prime}, 0, h_2^{(4)\prime}, 0)'$	11	8 ($\neq 11$)	8 ($\neq 9$)	-4 ($\neq -10$)	$\dfrac{[\frac{1+\rho}{11-9\rho} + \frac{2(1-\rho^2)}{11+9\rho^2+20\rho}]}{[\frac{1+\rho}{11-9\rho} + \frac{2(1-\rho^2)}{8+8\rho^2+8\rho}]}$
D_2 : $\mathbf{x}_1 = (0, h_4^{(4)\prime}, 0, h_2^{(8)\prime}, 0)'$, $\mathbf{x}_2 = (0, h_2^{(4)\prime}, 0, h_6^{(8)\prime}, 0)'$	15	12 ($\neq 15$)	12 ($\neq 13$)	-8 ($\neq -14$)	$\dfrac{[\frac{1+\rho}{15-13\rho} + \frac{2(1-\rho^2)}{15+13\rho^2+28\rho}]}{[\frac{1+\rho}{15-13\rho} + \frac{2(1-\rho^2)}{12+12\rho^2+16\rho}]}$
D_3 : $\mathbf{x}_1 = (0, h_2^{(8)\prime}, 0, h_6^{(8)\prime}, 0)'$, $\mathbf{x}_2 = (0, h_6^{(8)\prime}, 0, h_2^{(8)\prime}, 0)'$	19	16 ($\neq 19$)	16 ($\neq 17$)	-12 ($\neq -18$)	$\dfrac{[\frac{1+\rho}{19-17\rho} + \frac{2(1-\rho^2)}{19+17\rho^2+36\rho}]}{[\frac{1+\rho}{19-17\rho} + \frac{2(1-\rho^2)}{16+16\rho^2+24\rho}]}$
D_4 : $\mathbf{x}_1 = (0, h_2^{(8)\prime}, 0, h_{10}^{(16)\prime}, 0)'$, $\mathbf{x}_2 = (0, h_6^{(8)\prime}, 0, h_2^{(16)\prime}, 0)'$	27	24 ($\neq 27$)	24 ($\neq 25$)	-20 ($\neq -26$)	$\dfrac{[\frac{1+\rho}{27-25\rho} + \frac{2(1-\rho^2)}{27+25\rho^2+52\rho}]}{[\frac{1+\rho}{27-25\rho} + \frac{2(1-\rho^2)}{24+24\rho^2+40\rho}]}$
D_5 : $\mathbf{x}_1 = (0, h_2^{(16)\prime}, 0, h_{10}^{(16)\prime}, 0)'$, $\mathbf{x}_2 = (0, h_{10}^{(16)\prime}, 0, h_2^{(16)\prime}, 0)'$	35	32 ($\neq 35$)	32 ($\neq 33$)	-28 ($\neq -34$)	$\dfrac{[\frac{1+\rho}{35-33\rho} + \frac{2(1-\rho^2)}{35+33\rho^2+68\rho}]}{[\frac{1+\rho}{35-33\rho} + \frac{2(1-\rho^2)}{32+32\rho^2+56\rho}]}$

2.9   CONCLUDING REMARKS

This chapter focuses on first-order optimality and rotatability conditions of response surface designs for correlated errors. The pattern of the correlation structure is known, and of some simple type, usually encountered in practice. First-order optimality and rotatability have been examined for intra-class, inter-class, compound symmetry, tri-diagonal and auto-correlated error structures. Several construction methods of robust first-order rotatable and optimum designs are discussed under the above error variance-covariance structures. Only a few straight forward construction methods of RFORDs and ORFODs are discussed as examples. Of course there may be many more construction methods of RFORDs and ORFODs which will possibly be developed in future. In case of autocorrelated structure, RFORDs have been discussed. It has not been possible to develop a single method of ORFODs under autocorrelated structure (Das and Park, 2008a).

This chapter has only considered the mentioned five variance-covariance structures of errors for their usefulness in practice and because of their wide prevalence in existing literature. There are many other variance-covariance structures, say for instance, the Guttman circumplex structure, Democratic structure, Quasi-persymmetric structure, Equipredictability structure, etc., (Mukherjee, 1981), which exist in literature, though found with much less frequency in comparison with the structures considered in this chapter. But they too have their uses, and similar problems as provided in this chapter can be investigated under those structures of errors too.

TABLE 2.4: Nearly D-optimal designs for $k = 3$ and 4 factors when $\rho \geq 0$. Das, R.N. and Park, S.H. (2008a). "On efficient robust rotatable designs with autocorrelated errors," Journal of the Korean Statistical Society, 37(2), pp. 97-106.

Design	no. of factors	no. of runs	$\sum\limits_{u=1}^{N} x_{ui}^2$	$\sum\limits_{u=2}^{N-1} x_{ui}^2$	$\sum\limits_{u=1}^{N-1} x_{ui}x_{(u+1)i}$
D_6 : $\mathbf{x}_1 = (0, h_2^{(4)\prime}, 0, h_4^{(8)\prime}, 0)'$, $\mathbf{x}_2 = (0, h_3^{(4)\prime}, 0, h_6^{(8)\prime}, 0)'$, $\mathbf{x}_3 = (0, h_4^{(4)\prime}, 0, h_8^{(8)\prime}, 0)'$	3	15	12 $(\neq 15)$	12 $(\neq 13)$	-4 $(\neq -14)$
D_7 : $\mathbf{x}_1 = (0, h_2^{(8)\prime}, 0, h_4^{(8)\prime}, 0)'$, $\mathbf{x}_2 = (0, h_4^{(8)\prime}, 0, h_2^{(8)\prime}, 0)'$, $\mathbf{x}_3 = (0, h_6^{(8)\prime}, 0, h_8^{(8)\prime}, 0)'$, $\mathbf{x}_4 = (0, h_8^{(8)\prime}, 0, h_6^{(8)\prime}, 0)'$	4	19	16 $(\neq 19)$	16 $(\neq 17)$	-8 $(\neq -18)$
D_8 : $\mathbf{x}_1 = (0, h_2^{(8)\prime}, 0, h_6^{(16)\prime}, 0)'$, $\mathbf{x}_2 = (0, h_4^{(8)\prime}, 0, h_2^{(16)\prime}, 0)'$, $\mathbf{x}_3 = (0, h_6^{(8)\prime}, 0, h_{14}^{(16)\prime}, 0)'$, $\mathbf{x}_4 = (0, h_8^{(8)\prime}, 0, h_{10}^{(16)\prime}, 0)'$	4	27	24 $(\neq 27)$	24 $(\neq 25)$	-16 $(\neq -26)$
D_9 : $\mathbf{x}_1 = (0, h_2^{(16)\prime}, 0, h_6^{(16)\prime}, 0)'$, $\mathbf{x}_2 = (0, h_6^{(16)\prime}, 0, h_2^{(16)\prime}, 0)'$, $\mathbf{x}_3 = (0, h_{10}^{(16)\prime}, 0, h_{14}^{(16)\prime}, 0)'$, $\mathbf{x}_4 = (0, h_{14}^{(16)\prime}, 0, h_{10}^{(16)\prime}, 0)'$	4	35	32 $(\neq 35)$	32 $(\neq 33)$	-24 $(\neq -34)$

TABLE 2.5: Nearly D-optimal designs for $k = 2$ factors when $\rho \leq 0$. Das, R.N. and Park, S.H. (2008a). "On efficient robust rotatable designs with autocorrelated errors," Journal of the Korean Statistical Society, 37(2), pp. 97-106.

Design	no. of runs	$\sum_{u=1}^{N} x_{ui}^2$	$\sum_{u=2}^{N-1} x_{ui}^2$	$\sum_{u=1}^{N-1} x_{ui}x_{(u+1)i}$	ER
D_{10} : $\mathbf{x}_1 = (0, h_3^{(4)\prime}, 0, h_4^{(4)\prime}, 0)'$, $\mathbf{x}_2 = (0, h_4^{(4)\prime}, 0, h_3^{(4)\prime}, 0)'$	11	8 $(\neq 11)$	8 $(\neq 9)$	0 $(\neq 10)$	$\dfrac{[\frac{1+\rho}{11-9\rho} + \frac{2(1-\rho^2)}{11+9\rho^2-20\rho}]}{[\frac{1+\rho}{11-9\rho} + \frac{2(1-\rho^2)}{8+8\rho^2}]}$
D_{11} : $\mathbf{x}_1 = (0, h_3^{(4)\prime}, 0, h_7^{(8)\prime}, 0)'$, $\mathbf{x}_2 = (0, h_4^{(4)\prime}, 0, h_5^{(8)\prime}, 0)'$	15	12 $(\neq 15)$	12 $(\neq 13)$	4 $(\neq 14)$	$\dfrac{[\frac{1+\rho}{15-13\rho} + \frac{2(1-\rho^2)}{15+13\rho^2-28\rho}]}{[\frac{1+\rho}{15-13\rho} + \frac{2(1-\rho^2)}{12+12\rho^2-8\rho}]}$
D_{12} : $\mathbf{x}_1 = (0, h_5^{(8)\prime}, 0, h_8^{(8)\prime}, 0)'$, $\mathbf{x}_2 = (0, h_7^{(8)\prime}, 0, h_5^{(8)\prime}, 0)'$	19	16 $(\neq 19)$	16 $(\neq 17)$	8 $(\neq 18)$	$\dfrac{[\frac{1+\rho}{19-17\rho} + \frac{2(1-\rho^2)}{19+17\rho^2-36\rho}]}{[\frac{1+\rho}{19-17\rho} + \frac{2(1-\rho^2)}{16+16\rho^2-16\rho}]}$
D_{13} : $\mathbf{x}_1 = (0, h_5^{(8)\prime}, 0, h_{13}^{(16)\prime}, 0)'$, $\mathbf{x}_2 = (0, h_7^{(8)\prime}, 0, h_9^{(16)\prime}, 0)'$	27	24 $(\neq 27)$	24 $(\neq 25)$	16 $(\neq 26)$	$\dfrac{[\frac{1+\rho}{27-25\rho} + \frac{2(1-\rho^2)}{27+25\rho^2-52\rho}]}{[\frac{1+\rho}{27-25\rho} + \frac{2(1-\rho^2)}{24+24\rho^2-32\rho}]}$
D_{14} : $\mathbf{x}_1 = (0, h_9^{(16)\prime}, 0, h_{13}^{(16)\prime}, 0)'$, $\mathbf{x}_2 = (0, h_{13}^{(16)\prime}, 0, h_9^{(16)\prime}, 0)'$	35	32 $(\neq 35)$	32 $(\neq 33)$	24 $(\neq 34)$	$\dfrac{[\frac{1+\rho}{35-33\rho} + \frac{2(1-\rho^2)}{35+33\rho^2-68\rho}]}{[\frac{1+\rho}{35-33\rho} + \frac{2(1-\rho^2)}{32+32\rho^2-48\rho}]}$

TABLE 2.6: Nearly D-optimal designs for $k = 3$ and 4 factors when $\rho \leq 0$. Das, R.N. and Park, S.H. (2008a). "On efficient robust rotatable designs with autocorrelated errors," Journal of the Korean Statistical Society, 37(2), pp. 97-106.

Design	no. of factors	no. of runs	$\sum_{u=1}^{N} x_{ui}^2$	$\sum_{u=2}^{N-1} x_{ui}^2$	$\sum_{u=1}^{N-1} x_{ui}x_{(u+1)i}$
$D_{15}:$ $\mathbf{x}_1 = (0, h_2^{(4)\prime}, 0, h_7^{(8)\prime}, 0)'$, $\mathbf{x}_2 = (0, h_3^{(4)\prime}, 0, h_4^{(8)\prime}, 0)'$, $\mathbf{x}_3 = (0, h_4^{(4)\prime}, 0, h_3^{(8)\prime}, 0)'$	3	15	12 $(\neq 15)$	12 $(\neq 13)$	0 $(\neq 14)$
$D_{16}:$ $\mathbf{x}_1 = (0, h_3^{(8)\prime}, 0, h_7^{(8)\prime}, 0)'$, $\mathbf{x}_2 = (0, h_7^{(8)\prime}, 0, h_3^{(8)\prime}, 0)'$, $\mathbf{x}_3 = (0, h_4^{(8)\prime}, 0, h_5^{(8)\prime}, 0)'$, $\mathbf{x}_4 = (0, h_5^{(8)\prime}, 0, h_4^{(8)\prime}, 0)'$	4	19	16 $(\neq 19)$	16 $(\neq 17)$	4 $(\neq 18)$
$D_{17}:$ $\mathbf{x}_1 = (0, h_3^{(8)\prime}, 0, h_{13}^{(16)\prime}, 0)'$, $\mathbf{x}_2 = (0, h_4^{(8)\prime}, 0, h_9^{(16)\prime}, 0)'$, $\mathbf{x}_3 = (0, h_5^{(8)\prime}, 0, h_7^{(16)\prime}, 0)'$, $\mathbf{x}_4 = (0, h_7^{(8)\prime}, 0, h_5^{(16)\prime}, 0)'$	4	27	24 $(\neq 27)$	24 $(\neq 25)$	12 $(\neq 26)$
$D_{18}:$ $\mathbf{x}_1 = (0, h_5^{(16)\prime}, 0, h_{13}^{(16)\prime}, 0)'$, $\mathbf{x}_2 = (0, h_{13}^{(16)\prime}, 0, h_5^{(16)\prime}, 0)'$, $\mathbf{x}_3 = (0, h_7^{(16)\prime}, 0, h_9^{(16)\prime}, 0)'$, $\mathbf{x}_4 = (0, h_9^{(16)\prime}, 0, h_7^{(16)\prime}, 0)'$	4	35	32 $(\neq 35)$	32 $(\neq 33)$	20 $(\neq 34)$

TABLE 2.7: Computation of efficiency ratio of the designs from D_1 to D_9. Das, R.N. and Park, S.H. (2008a). "On efficient robust rotatable designs with autocorrelated errors," Journal of the Korean Statistical Society, 37(2), pp. 97-106.

ρ	Efficiency of								
	D_1	D_2	D_3	D_4	D_5	D_6	D_7	D_8	D_9
0.0	80.000	85.714	88.889	92.308	94.118	84.21	86.95	90.91	93.02
0.1	78.392	84.913	88.415	92.091	93.996	79.29	82.87	88.22	91.02
0.2	78.842	85.612	89.113	92.684	94.493	77.37	81.26	87.32	90.42
0.3	80.805	87.339	90.577	93.780	95.361	77.76	81.54	87.75	90.84
0.4	83.810	89.681	92.562	95.120	96.398	79.97	83.32	89.21	92.02
0.5	87.385	92.260	94.460	96.490	97.440	83.50	86.23	91.35	93.72
0.6	91.056	94.736	96.317	97.722	98.360	87.78	89.81	93.82	95.59
0.7	94.394	96.850	97.852	98.710	99.085	92.13	93.49	96.21	97.35
0.8	97.080	98.447	98.976	99.409	99.591	92.13	96.67	98.16	98.75
0.9	98.951	99.483	99.677	99.826	99.885	98.58	98.90	99.44	99.64

TABLE 2.8: Computation of efficiency ratio of the designs from D_{10} to D_{18}. Das, R.N. and Park, S.H. (2008a). "On efficient robust rotatable designs with autocorrelated errors," Journal of the Korean Statistical Society, 37(2), pp. 97-106.

ρ	Efficiency of								
	D_{10}	D_{11}	D_{12}	D_{13}	D_{14}	D_{15}	D_{16}	D_{17}	D_{18}
0.0	80.00	85.71	88.89	92.31	94.12	84.21	86.96	90.91	93.02
−0.1	70.75	79.46	84.17	89.14	91.74	73.55	78.49	85.14	88.65
−0.2	63.93	75.03	80.87	86.98	90.13	65.82	72.52	81.15	85.66
−0.3	58.99	71.87	78.59	85.51	89.05	60.28	68.34	78.40	83.61
−0.4	55.50	69.71	77.04	84.53	88.33	56.38	65.45	76.52	82.22
−0.5	53.10	68.26	76.02	83.90	87.87	53.69	63.49	75.25	81.30
−0.6	51.52	67.35	75.38	83.51	87.60	51.91	62.22	74.47	80.72
−0.7	50.55	66.82	75.03	83.30	87.45	50.80	61.45	73.99	80.38
−0.8	50.05	66.58	74.88	83.23	87.41	50.19	61.04	73.76	80.21
−0.9	49.90	66.54	74.78	83.24	87.43	49.96	60.91	73.70	80.18

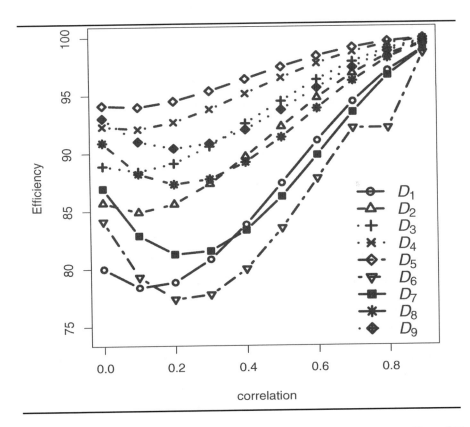

FIGURE 2.1: The efficiency of the designs D_1 through D_9 for $\rho \geq 0$. Das, R.N. and Park, S.H. (2008a). "On efficient robust rotatable designs with autocorrelated errors," Journal of the Korean Statistical Society, 37(2), pp. 97-106.

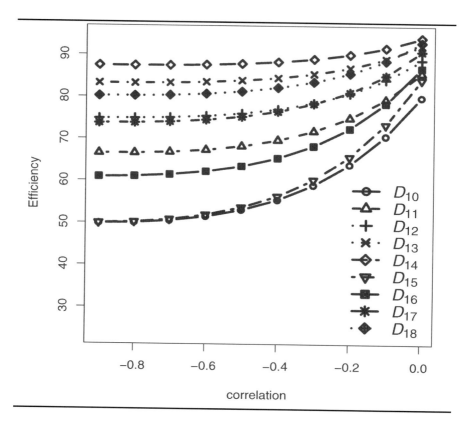

FIGURE 2.2: The efficiency of the designs D_{10} through D_{18} for $\rho \leq 0$. Das, R.N. and Park, S.H. (2008a). "On efficient robust rotatable designs with autocorrelated errors," Journal of the Korean Statistical Society, 37(2), pp. 97-106.

Chapter 3

ROBUST SECOND-ORDER DESIGNS

This chapter offers a detailed study of robust second-order rotatable designs for correlated errors. Second-order rotatability conditions are derived for a general variance-covariance structure of errors. General second-order rotatability conditions are examined for different well-known error variance-covariance structures. Some construction methods of robust second-order rotatable designs are developed under different correlated error structures.

3.1   INTRODUCTION AND OVERVIEW

As it has been pointed out in Chapter 1, rotatability of different orders in connection with regression designs has been studied extensively by a host of authors beginning with Box and Wilson (1951) and Box and Hunter (1957), of course, in the context of uncorrelated and homoscedastic random errors. But our main concern in the current research monograph is to explore the situation of correlated errors, particularly when errors follow a well-known structured pattern of correlatedness usually encountered in practice and justified by their inclusion for consideration in standard textbooks by previous authors working in the area of multivariate regression in particular, and multivariate analysis in general.

After dealing with the RFORDs and ORFODs in Chapter 2, in the present chapter, it is planned to consider the problem of determining second-order rotatable designs for correlated errors. Assuming a general correlated error structure, the conditions of second-order rotatability are determined for a second-degree response surface design. The particular variance-covariance structures of errors explored in the present chapter are exactly the same as those included in Chapter 2, viz., intra-class, inter-class, compound symmetry, tri-diagonal and autocorrelated.

Just like the robust first-order rotatable designs (RFORDs) and optimal robust first-order designs (ORFODs) given in Definition 2.2 and Definition 2.3, their counter parts are considered in the context of second-order rotatability viz., robust second-order rotatable designs (RSORDs) and optimal robust second-order rotatable designs (ORSORDs). All the error variance-covariance structures are investigated intensively to arrive at the necessary and sufficient conditions for a design to be RSORD. Consideration of robust optimality in general, in second-order designs seems to be quite complicated. Explicit conditions of ORSORD are obtained only in the case of intra-class variance-covariance structure. In the other cases of variance-covariance structures considered, only some general features of robust optimality are touched upon and no detailed investigations of robust optimality are undertaken. The construction of RSORDs are also taken up in the different situations included.

3.2   SECOND-ORDER CORRELATED MODEL

3.2.1   Model

Suppose there are k factors denoted by $x_1, x_2, ...,x_k$, and a design point $(x_{u1}, x_{u2},..., x_{uk})$, $1 \leq u \leq N$ yields a response y_u on the study variable y. Assuming that the response surface is of second-order, we adopt the model:

$$y_u = \beta_0 + \sum_{i=1}^{k} \beta_i x_{ui} + \sum_{i \leq j=1}^{k} \beta_{ij} x_{ui} x_{uj} + e_u; \; 1 \leq u \leq N,$$

or, $\mathbf{Y} = X\boldsymbol{\beta} + \mathbf{e}$,   (3.1)

where $\mathbf{Y}$ is the vector of recorded observations on the study variable y; $\boldsymbol{\beta} = (\beta_0, \beta_{11},...,\beta_{kk}, \beta_1,...,\beta_k, \beta_{12}, \beta_{13},...,\beta_{(k-1)k})'$ is the vector of regression coefficients of order $\binom{k+2}{2} \times 1$; $X = (\mathbf{1} : Z)$ is the model (or design) matrix, $Z = (\mathbf{x}_1 \otimes_1 \mathbf{x}_1,...,\mathbf{x}_k \otimes_1 \mathbf{x}_k, \; \mathbf{x}_1,...,\mathbf{x}_k,$

$\mathbf{x}_1 \otimes_1 \mathbf{x}_2, ..., \mathbf{x}_{k-1} \otimes_1 \mathbf{x}_k)$; $\mathbf{x}_i = (x_{1i}, ..., x_{Ni})'$; $1 \leq i \leq k$; $\mathbf{x}_i \otimes_1 \mathbf{x}_j = (x_{1i} x_{1j}, ..., x_{Ni} x_{Nj})'$; $1 \leq i, j \leq k$.

Here $\otimes_1$ denotes the Hadamard product of two matrices of the *same* order and it is defined as follows: Let $L = ((a_{ij}))$ and $M = ((b_{ij}))$ be two matrices of the same type, say $p \times q$. The Hadamard product $L \otimes_1 M$ of L and M is a matrix H, of order $p \times q$, where H $= ((a_{ij} b_{ij}))$. Further, $\mathbf{e}$ is the vector of errors (of order $N \times 1$) which are as assumed to be normally distributed with expectation $E(\mathbf{e}) = \mathbf{0}$ and variance-covariance matrix $\mathrm{Dis}(\mathbf{e}) = W$ with rank $(W) = N$.

3.2.2   Analysis

Assuming $(X'W^{-1}X)$ is positive definite for known W, the best linear unbiased estimator of $\boldsymbol{\beta}$ is $\hat{\boldsymbol{\beta}} = (X'W^{-1}X)^{-1}(X'W^{-1}\mathbf{Y})$ with

$$\mathrm{Dis}(\hat{\boldsymbol{\beta}}) = (X'W^{-1}X)^{-1} = \begin{pmatrix} A & B & C \\ B' & P & Q \\ C' & Q' & R \end{pmatrix}_{\binom{k+2}{2} \times \binom{k+2}{2}},$$

where A, P and R are all symmetric matrices given by

$$A_{(k+1) \times (k+1)} = \begin{pmatrix} v_{00} & v_{0.11} & v_{0.22}\cdots & v_{0.kk} \\ v_{11.0} & & (v_{ii.jj}); \ 1 \leq i, j \leq k & \\ \cdots & \cdots & \cdots & \cdots \\ v_{kk.0} & \cdots & \cdots & \cdots \end{pmatrix}$$

$v_{00} = \mathbf{1}'W^{-1}\mathbf{1}$, $v_{0.jj} = \mathbf{1}'W^{-1}(\mathbf{x}_j \otimes_1 \mathbf{x}_j) = (\mathbf{x}_j \otimes_1 \mathbf{x}_j)'W^{-1}\mathbf{1} = v_{jj.0}$; $1 \leq j \leq k$, $v_{ii.jj} = (\mathbf{x}_i \otimes_1 \mathbf{x}_i)'W^{-1}(\mathbf{x}_j \otimes_1 \mathbf{x}_j)$; $1 \leq i, j \leq k$,

$$B_{(k+1) \times k} = \begin{pmatrix} v_{0.1} & v_{0.2}\cdots & v_{0.k} \\ & (v_{ii.j})_{k \times k}; \ 1 \leq i, j \leq k & \end{pmatrix},$$

$v_{0.j} = \mathbf{1}'W^{-1}\mathbf{x}_j$; $v_{ii.j} = (\mathbf{x}_i \otimes_1 \mathbf{x}_i)'W^{-1}\mathbf{x}_j$; $1 \leq i, j \leq k$,

$$C_{(k+1) \times \binom{k}{2}} = \begin{pmatrix} v_{0.12} & v_{0.13}\cdots & v_{0.(k-1)k} \\ & (v_{ii.jl})_{k \times \binom{k}{2}}; \ 1 \leq i, j < l \leq k & \end{pmatrix},$$

$v_{0.jl} = \mathbf{1}'W^{-1}(\mathbf{x}_j \otimes_1 \mathbf{x}_l)$; $v_{ii.jl} = (\mathbf{x}_i \otimes_1 \mathbf{x}_i)'W^{-1}(\mathbf{x}_j \otimes_1 \mathbf{x}_l)$; $1 \leq i \leq k$, $1 \leq j < l \leq k$,

$P_{k \times k} = ((v_{i.j}))$, $v_{i.j} = \mathbf{x}_i'W^{-1}\mathbf{x}_j$; $1 \leq i, j \leq k$,

$Q_{k \times \binom{k}{2}} = ((v_{i.jl}))$, $v_{i.jl} = \mathbf{x}_i'W^{-1}(\mathbf{x}_j \otimes_1 \mathbf{x}_l)$; $1 \leq i, j < l \leq k$,

$R_{\binom{k}{2} \times \binom{k}{2}} = ((v_{ij.lt}))$, $v_{ij.lt} = (\mathbf{x}_i \otimes_1 \mathbf{x}_j)'W^{-1}(\mathbf{x}_l \otimes_1 \mathbf{x}_t)$;
$\qquad 1 \leq i < j \leq k$, $1 \leq l < t \leq k$. $\qquad\qquad$ (3.2)

Note that $v_{0.j} = v_{j.0}$, $v_{0.jj} = v_{jj.0}$, $v_{0.jl} = v_{jl.0}$, $v_{i.j} = v_{j.i}$, $v_{ii.j} = v_{j.ii}$, $v_{i.jl} = v_{jl.i}$, $v_{ii.jj} = v_{jj.ii}$, $v_{ij.lt} = v_{lt.ij}$. In the inverse matrix $(X'W^{-1}X)^{-1}$, the element corresponding to v_m in $(X'W^{-1}X)$ is denoted by v^m for *all* m included in the preceding expressions.

3.3   ROBUST SECOND-ORDER ROTATABILITY

3.3.1   Robust second-order rotatability conditions

The necessary and sufficient conditions of robust second-order rotatability for a general correlated error structure are given in the following theorem.

Theorem 3.1 (Das, 1997) *The necessary and sufficient conditions of a robust second-order rotatable design under the general variance-covariance structure W in the model (3.1) are (for all values of ρ in W, and v's are as in (3.2))*

(I): (i) $v_{0.j} = v_{0.jl} = 0$; $1 \le j < l \le k$,

(ii) $v_{i.j} = 0$; $1 \le i, j \le k$, $i \ne j$,

(iii) : (1) $v_{ii.j} = 0$; $1 \le i \le k$, $1 \le j \le k$,

(2) $v_{i.jl} = 0$; $1 \le i \le k$, $1 \le j < l \le k$,

(3) $v_{ii.jl} = 0$; $1 \le i \le k$, $1 \le j < l \le k$,

(4) $v_{ij.lt} = 0$; $1 \le i < j \le k$, $1 \le l < t \le k$, $(i,j) \ne (l,t)$,

(II): (i) $v_{0.jj} = \text{constant} = a_0(>0)$, say; $1 \le j \le k$,

(ii) $v_{i.i} = \text{constant} = \frac{1}{e}(>0)$, say; $1 \le i \le k$,

(iii) $v_{ii.ii} = \text{constant} = (\frac{2}{c} + d)(>0)$, say; $1 \le i \le k$,

(III): (i) $v_{ii.jj} = \text{constant} = d(>0)$; $1 \le i, j \le k$, $i \ne j$,

(ii) $v_{ij.ij} = \text{constant} = \frac{1}{c}(>0)$; $1 \le i < j \le k$,

(IV) :$v_{ii.ii} = 2v_{ij.ij} + v_{ii.jj}$; $1 \le i < j \le k$. $\qquad\qquad$ (3.3)

Proof: The estimated response at $\mathbf{x}$ is given by

$$\hat{y}_{\mathbf{x}} = \hat{\beta}_0 + \sum_{i=1}^{k}\hat{\beta}_{ii}x_i^2 + \sum_{i=1}^{k}\hat{\beta}_i x_i + \sum_{i<j=1}^{k}\hat{\beta}_{ij}x_i x_j.$$

According to (3.2.), the variance of the estimated response $\hat{y}_{\mathbf{x}}$ is given by

$$\text{Var}(\hat{y}_x) = \text{Var}(\hat{\beta}_0) + \sum_{i=1}^{k}x_i^4\text{Var}(\hat{\beta}_{ii}) + 2\sum_{i=1}^{k}x_i^2\text{Cov}(\hat{\beta}_0,\hat{\beta}_{ii}) + 2\sum_{i<j=1}^{k}x_i^2 x_j^2\text{Cov}(\hat{\beta}_{ii},\hat{\beta}_{jj})$$

$$+ \sum_{i=1}^{k}x_i^2\text{Var}(\hat{\beta}_i) + 2\sum_{i=1}^{k}x_i\text{Cov}(\hat{\beta}_0,\hat{\beta}_i) + 2\sum_{i<j=1}^{k}x_i x_j\text{Cov}(\hat{\beta}_i,\hat{\beta}_j) + 2\sum_{i=1}^{k}x_i^3\text{Cov}(\hat{\beta}_i,\hat{\beta}_{ii})$$

$$+2\sum_{i<j=1}^{k}\sum x_i^2 x_j \text{Cov}(\hat{\beta}_{ii}, \hat{\beta}_j) + \sum_{i<j=1}^{k}\sum x_i^2 x_j^2 \text{Var}(\hat{\beta}_{ij}) + 2\sum_{i<j=1}^{k}\sum x_i x_j \text{Cov}(\hat{\beta}_0, \hat{\beta}_{ij})$$

$$+2\sum_{s=1i<j=1}^{k}\sum\sum x_s^2 x_i x_j \text{Cov}(\hat{\beta}_{ss}, \hat{\beta}_{ij}) + 2\sum_{s=1i<j=1}^{k}\sum\sum x_s x_i x_j \text{Cov}(\hat{\beta}_s, \hat{\beta}_{ij})$$

$$+2\sum_{i<j=1l<t=1(i,j)\neq(l,t)}^{k}\sum\sum\sum x_i x_j x_l x_t \text{Cov}(\hat{\beta}_{ij}, \hat{\beta}_{lt})$$

$$= v^{00} + \sum_{i=1}^{k} x_i^4 v^{ii.ii} + 2\sum_{i=1}^{k} x_i^2 v^{0.ii} + 2\sum_{i<j=1}^{k}\sum x_i^2 x_j^2 v^{ii.jj} + \sum_{i=1}^{k} x_i^2 v^{i.i}$$

$$+ 2\sum_{i=1}^{k} x_i v^{0.i} + 2\sum_{i<j=1}^{k}\sum x_i x_j v^{i.j} + 2\sum_{i=1}^{k} x_i^3 v^{i.ii} + 2\sum_{i<j=1}^{k}\sum x_i^2 x_j v^{ii.j}$$

$$+ \sum_{i<j=1}^{k}\sum x_i^2 x_j^2 v^{ij.ij} + 2\sum_{i<j=1}^{k}\sum x_i x_j v^{0.ij} + 2\sum_{s=1i<j=1}^{k}\sum\sum x_s^2 x_i x_j v^{ss.ij}$$

$$+ 2\sum_{s=1i<j=1}^{k}\sum\sum x_s x_i x_j v^{s.ij} + 2\sum_{i<j=1l<t=1(i,j)\neq(l,t)}^{k}\sum\sum\sum x_i x_j x_l x_t v^{ij.lt}. \qquad (3.4)$$

The variance function in (3.4) will be a function of $\sum_{i=1}^{k} x_i^2$ for second-order rotatability for *all* $x_1, x_2,...,x_k$, and for *all* values of ρ in W, where ρ is the correlation parameter in W, *if and only if*

(i) $v^{0.i} = v^{0.ij} = v^{i.j} = v^{ii.j} = 0;\ 1 \leq i, j \leq k,\ i \neq j$,

(ii) $v^{s.ij} = v^{ss.ij} = 0;\ 1 \leq s \leq k,\ 1 \leq i < j \leq k$;

(iii) $v^{ij.lt} = 0;\ 1 \leq i < j \leq k,\ 1 \leq l < t \leq k,\ (i,j) \neq (l,t)$;

(iv) $v^{0.ii} = $ constant $= a_1$, say; $1 \leq i \leq k$;

(v) $v^{i.i} = $ constant $= e$, say; $1 \leq i \leq k$;

(vi) $v^{ii.jj} = $ constant $= d_1$, say; $1 \leq i, j \leq k,\ i \neq j$;

(vii) $v^{ij.ij} = $ constant $= c$, say; $1 \leq i < j \leq k$;

(viii) $v^{ii.ii} = $ constant $= (\frac{c}{2} + d_1);\ 1 \leq i \leq k$.

Clearly, the necessary and sufficient conditions (of the above equivalent conditions) of robust second-order rotatability (for the correlated error model (3.1)) in terms of the elements of the moment matrix are in (3.3). This completes the proof. $\square$

Under second-order robust rotatability as in (3.3), the dispersion matrix of $\hat{\beta}$, i.e., $\text{Dis}(\hat{\beta})$ reduces to the following form:

$$\text{Dis}(\hat{\boldsymbol{\beta}}) = \begin{pmatrix} A_1 & O & O \\ O & D_1 & O \\ O & O & D_2 \end{pmatrix}^{-1} = \begin{pmatrix} A_1^{-1} & O & O \\ O & D_1^{-1} & O \\ O & O & D_2^{-1} \end{pmatrix}.$$

where

$$A_{1_{(k+1)\times(k+1)}} = \begin{pmatrix} v_{00} & a_0 E_{1k} \\ a_0 E_{k1} & \frac{2}{c} I_k + d E_{kk} \end{pmatrix}, \quad D_{1_{k\times k}} = \text{Diag}(\frac{1}{e}, \frac{1}{e}, ..., \frac{1}{e}),$$

$$D_{2_{\binom{k}{2}\times\binom{k}{2}}} = \text{Diag}(\frac{1}{c}, \frac{1}{c}, ..., \frac{1}{c}), \quad A_{1_{(k+1)\times(k+1)}}^{-1} = \begin{pmatrix} v^{00} & a_1 E_{1k} \\ a_1 E_{k1} & \frac{c}{2} I_k + d_1 E_{kk} \end{pmatrix},$$

where E_{mn} is an $m \times n$ matrix with all elements 1,

$$v^{00} = \frac{(\frac{2}{c} + kd)}{\{v_{00}(\frac{2}{c} + kd) - ka_0^2\}}, \quad a_1 = \frac{-a_0}{\{v_{00}(\frac{2}{c} + kd) - ka_0^2\}}, \quad d_1 = \frac{c(a_0^2 - dv_{00})}{2\{v_{00}(\frac{2}{c} + kd) - ka_0^2\}},$$

(3.5)

and v_{00}, a_0, $\frac{1}{c}$, d, $\frac{1}{e}$ and are as in (3.2) and (3.3).

3.3.2   Robust second-order rotatable non-singularity condition

A robust second-order rotatable design is said to be a non-singular if the matrix

$$\begin{pmatrix} A_1 & O & O \\ O & D_1 & O \\ O & O & D_2 \end{pmatrix}$$

is non-singular, i.e., A_1 is non-singular, which means $|A_1| > 0$.

Now, $|A_1| = |\begin{pmatrix} v_{00} & a_0 E_{1k} \\ a_0 E_{k1} & \frac{2}{c} I_k + d E_{kk} \end{pmatrix}|$

$$= v_{00}\{(\frac{2}{c} + d - \frac{a_0^2}{v_{00}}) + (k-1)(d - \frac{a_0^2}{v_{00}})\} \times \{(\frac{2}{c} + d - \frac{a_0^2}{v_{00}}) - (d - \frac{a_0^2}{v_{00}})\}^{k-1}$$

$$= v_{00}\{\frac{2}{c} + k(d - \frac{a_0^2}{v_{00}})\}(\frac{2}{c})^{k-1}.$$

Hence the non-singularity condition of a robust second-order rotatable design is

$$(\mathbf{V}): \frac{2}{c} + k(d - \frac{a_0^2}{v_{00}}) > 0,$$

(3.6)

where v_{00}, a_0, $\frac{1}{c}$, and d are as in (3.2) and (3.3).

For a robust second-order rotatable non-singular design

$$\text{Dis}(\hat{\boldsymbol{\beta}}) = \begin{pmatrix} A_1 & O & O \\ O & D_1 & O \\ O & O & D_2 \end{pmatrix}^{-1} = \begin{pmatrix} A_1^{-1} & O & O \\ O & D_1^{-1} & O \\ O & O & D_2^{-1} \end{pmatrix}.$$

Following (3.3) and (3.5), variance of the estimated response at $\mathbf{x}$ of a robust second-order rotatable design is given by

$$\text{Var}(\hat{y}_\mathbf{x}) = \text{Var}(\hat{\beta}_0) + 2\sum_{i=1}^{k} x_i^2 \text{Cov}(\hat{\beta}_0, \hat{\beta}_{ii}) + \sum_{i=1}^{k} x_i^4 \text{Var}(\hat{\beta}_{ii}) + 2\sum_{i<j=1}^{k} x_i^2 x_j^2 \text{Cov}(\hat{\beta}_{ii}, \hat{\beta}_{jj})$$

$$+ \sum_{i=1}^{k} x_i^2 \mathrm{Var}(\hat{\beta}_i) + \sum_{i<j=1}^{k} \sum x_i^2 x_j^2 \mathrm{Var}(\hat{\beta}_{ij})$$

$$= v^{00} + 2a_1 \sum_{i=1}^{k} x_i^2 + (\frac{c}{2} + d_1) \sum_{i=1}^{k} x_i^4 + 2d_1 \sum_{i<j=1}^{k} \sum x_i^2 x_j^2 + e \sum_{i=1}^{k} x_i^2 + c \sum_{i<j=1}^{k} \sum x_i^2 x_j^2$$

$$= v^{00} + (2a_1 + e) \sum_{i=1}^{k} x_i^2 + \frac{c}{2}(\sum_{i=1}^{k} x_i^2)^2 + d_1(\sum_{i=1}^{k} x_i^2)^2$$

$$= v^{00} + (2a_1 + e)r^2 + (\frac{c}{2} + d_1)r^4 = A_0 + A_1 r^2 + A_2 r^4, \tag{3.7}$$

which is a function of $r^2 = \sum_{i=1}^{k} x_i^2$, where $A_0 = v^{00}$, $A_1 = (2a_1 + e)$, $A_2 = (\frac{c}{2} + d_1)$ and c, e are as in (3.4) and v^{00}, a_1, d_1 are as in (3.5).

Remark 3.1 If errors are uncorrelated and homoscedastic, (3.3), (3.6) and (3.7) reduce, respectively, to the well-known expressions of second-order rotatability conditions, non-singularity condition and variance function in that situation.

3.3.3 Robust second-order rotatable and optimum designs

Chapter 2 defines robust first-order and optimum robust first-order designs. Such designs in view of their robustness to correlation parameter or parameters of correlated errors, but follow a structured pattern are presented to be quite useful in practice. But as first-order designs are found inadequate, particularly when the design space of experimentation includes the optimal design point as an interior point, it becomes absolutely necessary to consider designs which are used for fitting a second- or higher-degree response surface function. In the present chapter, the same principle of robustness and optimality is utilized as in Chapter 2, but only with the difference that the appropriate degree of the response surface function to be fitted is a second-degree polynomial instead of a first-degree polynomial. Below are given some relevant definitions.

Definition 3.1 Robust second-order rotatable design: A design D on k factors which remains second-order rotatable for *all* the variance covariance matrices of errors belonging to a well defined class $W_0 = \{W$ positive definite: $W_{N \times N}$ defined by a particular correlation structure neatly specified$\}$ is called a Robust Second-Order Rotatable Design (RSORD), with reference to the variance covariance class W_0.

Definition 3.2 Optimum robust second-order rotatable design: Assuming $\mathcal{X}$, the design space under consideration to be the unit ball i.e., $\mathcal{X} = \{\mathbf{x}: \mathbf{x}'\mathbf{x} \leq k\}$, a design D which is G-optimal within the set of robust second-order rotatable designs, defined for a class of variance-covariance matrices of errors W_0 (determined by a well defined correlation structure) of the model $y_{\mathbf{x}}$ as in (3.1) has been termed an Optimum Robust Second-Order Rotatable Design (ORSORD) with reference to the variance-covariance class W_0.

Following (3.7), the variance function of a robust second-order rotatable design is

$$\mathrm{Var}(\hat{y}_{\mathbf{x}}) = A_0 + A_1 r^2 + A_2 r^4$$

where A_0, A_1 and A_2 are as in (3.7). Here $A_0 > 0$ and $A_2 > 0$ and $\mathrm{Var}(\hat{y}_{\mathbf{x}})$ being a second-degree function of r^2 represents a parabola in r^2. Hence there can be at most one minimum between 0 and k, the limits of values of r^2 within which we are interested. If $\mathrm{Var}(\hat{y}_{\mathbf{x}})$ does not possess a minimum in between values 0 and k of r^2, it is either monotonically increasing or monotonically decreasing in $[0, k]$. Thus the maximum of $Var(\hat{y}_\mathbf{x})$ is obtained under all circumstances as

$$Max.\{A_0, \quad A_0 + A_1 k + A_2 k^2\}. \tag{3.8}$$

For any correlation structure considered in the chapter, (3.8) can be evaluated and for the given values of the correlations or structure parameters, the design which minimizes (3.8) can be recognized as the optimum design. In case of the intra-class correlated structure, the G-optimal design so found in the situation where intra-class correlation coefficient = 0 works out to be the optimum robust second-order rotatable design for all feasible values of the intra-class correlation coefficient. In cases of other correlation structures considered, the problem is quite complicated as the designs which may be obtained as optimum for some particular values of the correlation parameters do not usually satisfy the optimality property in the extended situation where the values of the correlation parameters are altered. It is quite possible that for a certain range of values of the correlation coefficients, there may exist a design which happens to be optimal robust for the whole range of values of parameters specified. But exploration of such conditions seem to be quite cumbersome and has been avoided in the present monograph. Thus, for all the correlation structures besides intra-class correlation structure, no attempts have been made to investigate the existence of optimal robust second-order rotatable designs.

The remaining part of this chapter deals with the interesting correlated structures, already introduced in Chapter 2, viz., the intra-class, inter-class, compound symmetry, tri-diagonal and autocorrelated structures. Mainly the existence and construction of RSORDs are studied under the above five correlated error structures.

3.4   ROBUST SECOND-ORDER DESIGNS FOR INTRA-CLASS STRUCTURE

In Chapter 2 we have already mentioned that errors may have some simple and patterned variance-covariance structures which are of interest to us and form the scope of the area of our investigation. Similar to what has been done in connection with first-order rotatable designs, in Section 3.3, the conditions of second-order rotatability, non-singularity, variance function, robust second-order rotatable designs (RSORDs) are derived for the general classes of variance-covariance matrices as in Definition 3.1.

The different regular patterns of variance-covariance matrices considered here are same as those considered for first-order rotatable designs. In this section we study robust second-order rotatable designs (RSORDs) when errors have the following variance covariance structures: intra-class, inter-class, compound symmetry, tri-diagonal and autocorrelated. Some arguments are provided for practical usefulness of these structures in connection with response surface designs in Chapter 2.

Intra-class correlation structure of errors is described in Chapter 2, and is given in (2.6). Let ρ be the intra-class correlation coefficient between any two errors and each has the *same* variance σ^2.

3.4.1   Second-order rotatability conditions under intra-class structure

The necessary and sufficient conditions of second-order rotatability under the intra-class structure (2.6) can be easily obtained from (3.3), which are given in the following theorem.

Theorem 3.2 (Das, 2003a) *The necessary and sufficient conditions of a robust second-order rotatable design under the intra-class variance-covariance structure, for all values of ρ in $W_{N \times N}(\rho)$ as in (2.6), in the model (3.1) are (for all $1 \le i, j \le k$)*

(I) $\sum_{u=1}^{N} x_{ui_1}^{\alpha_1} x_{ui_2}^{\alpha_2} x_{ui_3}^{\alpha_3} x_{ui_4}^{\alpha_4} = 0$; *for α_i's non-negative integers, any α_i odd and $\sum_{i=1}^{4} \alpha_i \le 4$,*

(II) (i) $\sum_{u=1}^{N} x_{ui}^2 = constant$,

(ii) $\sum_{u=1}^{N} x_{ui}^4 = constant$,

(III) $\sum_{u=1}^{N} x_{ui}^2 x_{uj}^2 = constant$; $i \ne j$,

(IV) $\sum_{u=1}^{N} x_{ui}^4 = 3 \sum_{u=1}^{N} x_{ui}^2 x_{uj}^2$, $i \ne j$. $\qquad\qquad$ (3.9)

Note that the conditions (I)–(IV) as in (3.9) are independent of intra-class correlation coefficient ρ and are *same* as the necessary and sufficient conditions of a usual SORD.

3.4.2 Non-singularity condition under intra-class structure

Using standard representation, the following are the moment relations: $\sum_{u=1}^{N} x_{ui}^2 = N\lambda_2$; $\sum_{u=1}^{N} x_{ui}^4 = 3N\lambda_4$; $\sum_{u=1}^{N} x_{ui}^2 x_{uj}^2 = N\lambda_4$; $1 \le i, j \le k$, $i \ne j$. Using the above, and following (3.2) and (3.3), the design parameters of a RSORD under the intra-class structure (2.6) are the following: $a_0 = \frac{N\lambda_2}{\sigma^2 \{1+(N-1)\rho\}}$, $d = \frac{\{1+(N-1)\rho\}N\lambda_4 - \rho N^2 \lambda_2^2}{\sigma^2(1-\rho)\{1+(N-1)\rho\}}$, $\frac{1}{e} = \frac{N\lambda_2}{\sigma^2(1-\rho)}$, $\frac{1}{c} = \frac{N\lambda_4}{\sigma^2(1-\rho)}$, $v_{00} = \frac{N}{\sigma^2\{1+(N-1)\rho\}}$.

As regards non-singularity, we check (3.6). The expression $\frac{2}{c} + k(d - \frac{a_0^2}{v_{00}})$ simplifies to

$$\frac{N}{\sigma^2(1-\rho)}[(k+2)\lambda_4 - k\lambda_2^2].$$

Hence the non-singularity condition of the above design is

$$\frac{\lambda_4}{\lambda_2^2} > \frac{k}{k+2} \qquad\qquad (3.10)$$

which is the *same* as that of a usual SORD. Thus the following theorem is immediate.

Theorem 3.3 (Das, 2003a) *A design is a robust second-order rotatable under the intra-class structure of errors if and only if it is a usual SORD, whatever be the value of intra-class correlation coefficient ρ.*

3.4.3 Estimated response variance under intra-class structure

The variance of the estimated response (variance function) at $\mathbf{x}$ of a RSORD under the intra-class structure is obtained by following (3.7) and using the above results. The components v^{00}, $(2a_1 + e)$, and $(c/2 + d_1)$ in (3.5) under the intra-class structure (2.6) simplify to

$$v^{00} = \frac{\sigma^2[(k+2)\{1+(N-1)\rho\}\lambda_4 - kN\rho\lambda_2^2]}{T} = A(N, k, \rho); \text{ say,} \quad a_1 = -\frac{\sigma^2(1-\rho)\lambda_2}{T}, \quad d_1 = \frac{\sigma^2(1-\rho)[\lambda_2^2 - \lambda_4]}{2\lambda_4 T}, \quad B(N, k, \rho) = (2a_1 + e) = \frac{\sigma^2(1-\rho)(k+2)[\lambda_4 - \lambda_2^2]}{\lambda_2 T}, \quad C(N, k, \rho) = (c/2 + d_1) = \frac{\sigma^2(1-\rho)[(k+1)\lambda_4 - \lambda_2^2(k-1)]}{2\lambda_4 T}, \text{ and } T = N\{(k+2)\lambda_4 - k\lambda_2^2\}.$$
$$\qquad\qquad (3.11)$$

Hence the variance function under intra-class structure at $\mathbf{x}$ is given by,

$$\text{Var}(\hat{y}_\mathbf{x}|W_{N \times N}(\rho)) = A(N, k, \rho) + B(N, k, \rho)r^2 + C(N, k, \rho)r^4 = V_\rho(r^2), \qquad (3.12)$$

where $r^2 = \sum_{i=1}^{k} x_i^2$ and $A(N, k, \rho)$, $B(N, , k, \rho)$, $C(N, k, \rho)$ are as in (3.12).

The non-singularity condition is assumed to hold as otherwise invertibility condition would not have been satisfied. Let $V_0(r^2)$ be the variance function of a SORD in the uncorrelated case. This corresponds to $\rho = 0$ and is given by

$$\text{Var}(\hat{y}_\mathbf{x}) = \frac{\sigma^2}{T}[(k+2)\lambda_4 + \frac{(k+2)(\lambda_4 - \lambda_2^2)r^2}{\lambda_2} + \frac{\{(k+1)\lambda_4 - (k-1)\lambda_2^2\}r^4}{2\lambda_4}] = V_0(r^2),$$

where T as in (3.11).

It is interesting to observe that, in general, $V_0(r^2) \overset{<}{\underset{>}{>}} V_\rho(r^2)$ for all $0 \leq r^2 \leq k$, according as $\rho \overset{>}{\underset{<}{<}} 0$, provided

$$N[(k+2)\lambda_4 - k\lambda_2^2] > [\frac{(k+2)}{\lambda_2}\{\lambda_4(\lambda_2 + r^2) - \lambda_2^2 r^2\} + \frac{\{(k+1)\lambda_4 - (k-1)\lambda_2^2\}r^4}{2\lambda_4}].$$

3.4.4   Optimum RSORD under intra-class structure

Optimality of robust second-order rotatable designs as defined in Definition 3.2 has been explicitly explored only in case of the simplest correlation structure, viz., the intra-class correlation structure. The same arguments which establish the SORD property in the simple structure of uncorrelated errors extend the robust rotatability property of second-order to the enlarged class of intra-class correlated structure. Theorem 3.3 establishes the robust optimality in the latter situation of the SORDs which are found to be optimal in the former situation, i.e., the designs which are optimal robust second-order rotatable in the case of intra-class correlated structure are precisely the second-order rotatable designs which satisfy the conditions of Theorem 3.3.2 of Kiefer (1960). According to Kiefer (1960), these designs besides being SORD should satisfy the following conditions: There are exactly two sets of points. The first set consists of points on the surface of the hypersphere defined by $\sum_{i=1}^{k} x_i^2 = k$ and the second consists of only the center points. Moreover, cardinality of the two sets should be in the ratio $1 - p : p$ where $p = \frac{2}{(k+1)(k+2)}$, where $\frac{(k+1)(k+2)}{2}$ is actually the total number of parameters in the second-order response surface regression model.

3.5   ROBUST SECOND-ORDER DESIGNS FOR INTER-CLASS STRUCTURE

The situations where the inter-class error structure in observations could arise is described in Chapter 2, and is given in (2.11).

3.5.1 Second-order rotatability conditions under inter-class structure

The second-order rotatability conditions under the inter-class structure (2.11) can be easily derived from (3.3). The following theorem presents the necessary and sufficient conditions of second-order rotatability under the inter-class structure.

Theorem 3.4 (Das, 2003a) *The necessary and sufficient conditions of a robust second-order rotatable design under the inter-class variance-covariance structure, for all values of ρ in W_1 as in (2.11), in the model (3.1) are*

(I): (i) $v_{0.j} = 0 \Leftrightarrow \sum_{u=1}^{N} x_{uj} = 0; \ 1 \leq j \leq k,$

$v_{0.jl} = 0 \Leftrightarrow \sum_{u=1}^{N} x_{uj} x_{ul} = 0; \ 1 \leq j < l \leq k,$

(ii) $v_{i.j} = 0 \Leftrightarrow \frac{1}{(1-\rho)} \sum_{u=1}^{N} x_{ui} x_{uj} - \frac{\rho}{(1-\rho)\{1+(n-1)\rho\}} \{(\sum_{u=1}^{n} x_{ui})(\sum_{u=1}^{n} x_{uj}) + ... + (\sum_{u=n(m-1)+1}^{mn} x_{ui})(\sum_{u=n(m-1)+1}^{mn} x_{uj})\} = 0; \ 1 \leq i, j \leq k, \ i \neq j,$

(iii): (1)$v_{ii.j} = 0 \Leftrightarrow \frac{1}{(1-\rho)} \sum_{u=1}^{N} x_{ui}^2 x_{uj} - \frac{\rho}{(1-\rho)\{1+(n-1)\rho\}} \{(\sum_{u=1}^{n} x_{ui}^2)(\sum_{u=1}^{n} x_{uj}) + ... + (\sum_{u=n(m-1)+1}^{mn} x_{ui}^2)(\sum_{u=n(m-1)+1}^{mn} x_{uj})\} = 0; \ 1 \leq i \leq k, \ 1 \leq j \leq k,$

(2) $v_{ij.l} = 0 \Leftrightarrow \frac{1}{(1-\rho)} \sum_{u=1}^{N} x_{ui} x_{uj} x_{ul} - \frac{\rho}{(1-\rho)\{1+(n-1)\rho\}} \{(\sum_{u=1}^{n} x_{ui} x_{uj})(\sum_{u=1}^{n} x_{ul}) + ... + (\sum_{u=n(m-1)+1}^{mn} x_{ui} x_{uj})(\sum_{u=n(m-1)+1}^{mn} x_{ul})\} = 0; \ 1 \leq i < j \leq k, \ 1 \leq l \leq k,$

(3) $v_{ii.jl} = 0 \Leftrightarrow \frac{1}{(1-\rho)} \sum_{u=1}^{N} x_{ui}^2 x_{uj} x_{ul} - \frac{\rho}{(1-\rho)\{1+(n-1)\rho\}} \{(\sum_{u=1}^{n} x_{ui}^2)(\sum_{u=1}^{n} x_{uj} x_{ul}) + ... + (\sum_{u=n(m-1)+1}^{mn} x_{ui}^2)(\sum_{u=n(m-1)+1}^{mn} x_{uj} x_{ul})\} = 0; 1 \leq i \leq k, 1 \leq j < l \leq k,$

(4) $v_{ij.lt} = 0 \Leftrightarrow \frac{1}{(1-\rho)} \sum_{u=1}^{N} x_{ui} x_{uj} x_{ul} x_{ut} - \frac{\rho}{(1-\rho)\{1+(n-1)\rho\}} \{(\sum_{u=1}^{n} x_{ui} x_{uj})(\sum_{u=1}^{n} x_{ul} x_{ut}) + ... + (\sum_{u=n(m-1)+1}^{mn} x_{ui} x_{uj})(\sum_{u=n(m-1)+1}^{mn} x_{ul} x_{ut})\} = 0; 1 \leq i < j \leq k, 1 \leq l < t \leq k, \ (i,j) \neq (l,t)$

(II): (i) $v_{0.jj} = constant \Leftrightarrow \{\sigma^2(1 + (n-1)\rho)\}^{-1} \sum_{u=1}^{N} x_{uj}^2 = a_0(> 0); \ 1 \leq j \leq k,$

(ii) $v_{i.i} = constant \Leftrightarrow \{\sigma^2\}^{-1}[\frac{1}{(1-\rho)} \sum_{u=1}^{N} x_{ui}^2 - \frac{\rho}{(1-\rho)\{1+(n-1)\rho\}} \{(\sum_{u=1}^{n} x_{ui})^2 + ... + (\sum_{u=n(m-1)+1}^{mn} x_{ui})^2\}] = \frac{1}{e}(> 0); \ 1 \leq i \leq k,$

(iii) $v_{ii.ii} = constant \Leftrightarrow \{\sigma^2\}^{-1}[\frac{1}{(1-\rho)} \sum_{u=1}^{N} x_{ui}^4 - \frac{\rho}{(1-\rho)\{1+(n-1)\rho\}} \{(\sum_{u=1}^{n} x_{ui}^2)^2 + ... + (\sum_{u=n(m-1)+1}^{mn} x_{ui}^2)^2\}] = (\frac{2}{c} + d)(> 0); \ 1 \leq i \leq k,$

(III): (i) $v_{ii.jj} = constant \Leftrightarrow \{\sigma^2\}^{-1}[\frac{1}{(1-\rho)} \sum_{u=1}^{N} x_{ui}^2 x_{uj}^2 - \frac{\rho}{(1-\rho)\{1+(n-1)\rho\}} \{(\sum_{u=1}^{n} x_{ui}^2)(\sum_{u=1}^{n} x_{uj}^2) + ... + (\sum_{u=n(m-1)+1}^{mn} x_{ui}^2)(\sum_{u=n(m-1)+1}^{mn} x_{uj}^2)\}] = d \ (> 0); \ 1 \leq i, \ j \leq k, \ i \neq j,$

(ii) $v_{ij.ij} = constant \Leftrightarrow \{\sigma^2\}^{-1}[\frac{1}{(1-\rho)} \sum_{u=1}^{N} x_{ui}^2 x_{uj}^2 - \frac{\rho}{(1-\rho)\{1+(n-1)\rho\}} \{(\sum_{u=1}^{n} x_{ui} x_{uj})^2 + ... + (\sum_{u=n(m-1)+1}^{mn} x_{ui} x_{uj})^2\}] = \frac{1}{c}(> 0); \ 1 \leq i < j \leq k,$

(IV) $v_{ii.ii} = 2v_{ij.ij} + v_{ii.jj}; \ 1 \leq i < j \leq k.$ $\qquad(3.13)$
where $v_{ii.ii}, v_{ii.jj}$ and $v_{ij.ij}$ are as in **(II)** *(iii) and* **(III)** *(i), (ii) in* (3.13).
Non-singularity condition is given by

$$\textbf{(V)} \ \{\frac{2}{c} + k(d - \frac{a_0^2}{v_{00}})\} > 0 \qquad (3.14)$$

where $v_{00} = \frac{N}{1+(n-1)\rho}$ and $\frac{1}{c}$, d, a_0 are as in (3.13).

From (3.13), the following corollary is now immediate.

Corollary 3.1 (Das, 2003a) *If the design points are divided into* m *sets of* n *each, having a second-order rotatable design in the usual sense within each set, and* $n > (k+1)$, *then the overall design is a robust second-order rotatable design under the inter-class structure (2.11), whatever be the value of intra-class correlation coefficient* ρ.

Remark 3.2 It is not essential that the design points in each set should be a second-order rotatable design in the usual sense in order that the overall design becomes a robust second-order rotatable design for *all* ρ under the inter-class error structure (2.11).

This aspect is examined below. It is now possible to construct RSORDs satisfying **(I)** through **(IV)** as in (3.13) and non-singularity condition **(V)** as in (3.14) but violating the second-order rotatability conditions within each group of observations as stated in Corollary 3.1. An example is provided below to highlight this point.

Example 3.1 Consider an experimental design consisting of three factors, four independent groups, each of eleven observations, i.e., $k = 3$, $m = 4$, $n = 11$ and Dis(**e**) $= I_4 \otimes W_{11 \times 11}(\rho)$, $N = mn = 44$.

GroupI:	1	2	3	4	5	6	7	8	9	10	11
x_1	1	-1	1	-1	1	-1	1	-1	2	0	0
x_2	1	-1	-1	1	1	-1	-1	1	0	2	0
x_3	1	-1	-1	-1	-1	1	1	1	0	0	2

GroupII:	1	2	3	4	5	6	7	8	9	10	11
x_1	1	-1	1	-1	1	-1	1	-1	2	0	0
x_2	1	-1	-1	1	1	-1	-1	1	0	-2	0
x_3	1	-1	-1	-1	-1	1	1	1	0	0	-2

GroupIII:	1	2	3	4	5	6	7	8	9	10	11
x_1	1	-1	1	-1	1	-1	1	-1	-2	0	0
x_2	1	-1	-1	1	1	-1	-1	1	0	2	0
x_3	1	-1	-1	-1	-1	1	1	1	0	0	-2

GroupIV:	1	2	3	4	5	6	7	8	9	10	11
x_1	1	-1	1	-1	1	-1	1	-1	-2	0	0
x_2	1	-1	-1	1	1	-1	-1	1	0	-2	0
x_3	1	-1	-1	-1	-1	1	1	1	0	0	2

Group totals for x_i's are given in the following table

Groups:	I	II	III	IV
x_1	2	2	-2	-2
x_2	2	-2	2	-2
x_3	2	-2	-2	2

Obviously, the design points within a group do not satisfy the usual SORD conditions. The condition (I) in (3.13) is readily satisfied. In case of above example, the expressions (i), (ii) and (iii) of (II) in (3.13) simplify respectively to

$$\frac{48}{(1+10\rho)} = a_0, \quad \frac{48}{(1-\rho)} - \frac{16\rho}{(1-\rho)(1+10\rho)} = \frac{1}{e}, \quad \frac{96}{(1-\rho)} - \frac{576\rho}{(1-\rho)(1+10\rho)} = \left(\frac{2}{c}+d\right),$$

and they are the *same* for $i = 1,2,3$. Again, the conditions (i) and (ii) of (III) in (3.13) reduce respectively to $\frac{32}{(1-\rho)} - \frac{576\rho}{(1-\rho)(1+10\rho)} = d$ and $\frac{32}{(1-\rho)} = \frac{1}{c}$, and they are the *same* for $1 \le i < j \le 3$. Therefore, (II) and (III) of (3.13) are satisfied. Thus, it is readily verified that (IV) of (3.13) is also satisfied. The other rotatability conditions enumerated naturally hold.

Finally, non-singularity condition is verified by checking (3.14). For the above example, the expression

$$\frac{2}{c} + k(d - \frac{a_0^2}{v_{00}})$$

simplifies to

$$\frac{64}{(1-\rho)} + 3\left\{ \frac{32}{(1-\rho)} - \frac{576\rho}{(1-\rho)(1+10\rho)} - \frac{(48)^2}{44(1+10\rho)} \right\}$$

$$= \frac{160}{(1-\rho)} - \frac{(12)^3\rho}{(1-\rho)(1+10\rho)} - \frac{1728}{11(1+10\rho)} = \frac{32}{11(1-\rho)}$$

which is positive. Hence the non-singularity condition is also satisfied. Thus the overall design satisfies SORD conditions for *all* ρ for the inter-class correlation structure (2.11). Therefore, the design given in Example 3.1 is a RSORD under the inter-class structure.

3.5.2   RSORD construction methods under inter-class structure

Method I (Description of the method): From Corollary 3.1, RSORD can be constructed under the inter-class structure (2.11) with the help of usual SORDs by arranging the design points into m sets of n each such that a second-order rotatable structure in the usual sense is satisfied for design points within each set. Thus, all the different SORDs known in the literature in the usual circumstances of uncorrelated errors can be repeatedly used to construct a RSORD under the inter-class structure (Das, 2003a).

Features of the method:
(a) It can be readily verified that (3.13) holds for these designs.
(b) As regards non-singularity, we note the following design parameters

$$a_0 = \frac{mn\lambda_2}{\sigma^2\{1+(n-1)\rho\}} = \frac{N\lambda_2}{\sigma^2\{1+(n-1)\rho\}}; \quad \frac{1}{e} = \frac{N\lambda_2}{\sigma^2(1-\rho)}, \quad d = \frac{1}{\sigma^2}\left[\frac{N\lambda_4}{(1-\rho)} - \frac{m\rho(n\lambda_2)^2}{(1-\rho)\{1+(n-1)\rho\}}\right],$$
$$\frac{1}{c} = \frac{1}{\sigma^2}\left[\frac{N\lambda_4}{(1-\rho)}\right], \quad v_{00} = \frac{N}{\sigma^2\{1+(n-1)\rho\}}.$$

To examine non-singularity, we examine the expression $\frac{2}{c} + k(d - \frac{a_0^2}{v_{00}})$ using the above expressions. This expression simplifies to

$$\frac{\lambda_4}{\lambda_2^2} > \frac{k}{k+2}, \tag{3.15}$$

which is the non-singularity condition of each set of observations out of m sets. Therefore, the designs described by Method I satisfy non-singularity condition for all values of ρ.

Variance Function: The variance function of a RSORD under the inter-class structure constructed by the above method is obtained by using (3.7), (3.13), the above results, and noting the following:

$$v_{00} = \frac{N}{\sigma^2\{1+(n-1)\rho\}}, \quad v^{00} = \frac{2/c+kd}{T_1} = \frac{\sigma^2}{T}[(k+2)\lambda_4\{1+(n-1)\rho\} - kn\rho\lambda_2^2] = A(n,k,\rho)$$

say; $a_1 = -\frac{a_0}{T_1} = -\frac{\sigma^2(1-\rho)\lambda_2}{T}$, $d_1 = \frac{c(a_0^2-dv_{00})}{2T_1} = \frac{\sigma^2(1-\rho)(\lambda_2^2-\lambda_4)}{2\lambda_4 T}$, where $T_1 = \{v_{00}(2/c +$

$kd) - ka_0^2\} = \frac{N^2\{(k+2)\lambda_4-k\lambda_2^2\}}{\sigma^4(1-\rho)\{1+(n-1)\rho\}}$ and $(2a_1 + e) = \frac{\sigma^2(1-\rho)(k+2)[\lambda_4-\lambda_2^2]}{\lambda_2 T} = B(n,k,\rho)$ say;

$(c/2 + d_1) = \frac{\sigma^2(1-\rho)[(k+1)\lambda_4-\lambda_2^2(k-1)]}{2\lambda_4 T} = C(n,k,\rho)$ say; $T = N\{(k+2)\lambda_4 - k\lambda_2^2\}$.

Therefore, the variance function is given by (following (3.7))

$$V(\hat{y}_\mathbf{x}) = v^{00} + (2a_1 + e)r^2 + (c/2 + d_1)r^4,$$

$$= A(n,k,\rho) + B(n,k,\rho)r^2 + C(n,k,\rho)r^4, \tag{3.16}$$

where $r^2 = \sum_{i=1}^{k} x_i^2$ and $A(n,k,\rho)$, $B(n,k,\rho)$, $C(n,k,\rho)$ are as in above.

Method II (description of the method): (Das, 2004) The compound design constructed by Das (2004) is a particular SORD, known in the literature as a central composite design (CCD), used widely in the context of uncorrelated errors, with the following modifications with reference to the different sets of observations:

Suppose that we have a design matrix D_0: $k \times 2^p$ obtained from $\frac{1}{2^{k-p}}$ fraction of 2^k-experiment with levels $+1$ and -1, and the defining equation in obtaining the fraction does not involve any main effect or interactions involving four or fewer factors (Resolution V). Let us write the array of treatment combinations $L_{(k\times G)} = D_0$ with $G = 2^p$. Assume L can be partitioned as $L_{(k\times G)} = [L_{1(k\times G_0)} : L_{2(k\times G_0)} : \ldots : L_{s(k\times G_0)}]$, where $sG_0 = G$, and the inner product of any two distinct rows of L_i is zero, $i = 1, 2,...,s$.

Suppose that there exists a matrix H_1 of order $(k+1) \times s$ written with elements $+1$ and -1 such that the inner product of every pair of rows of the matrix is 0 (zero). From the given condition one can assume arbitrarily that the first row of H_1 consists of all $+1$'s. Now deleting this first row from H_1, we get, say

$$H_{10} = \begin{pmatrix} h_{11} & h_{12} & h_{13}..... & h_{1s} \\ h_{21} & h_{22} & h_{23}......... & h_{2s} \\ ..... & ...... & .... & ... \\ .... & ...... & .... & .... \\ h_{k1} & h_{k2} & h_{k3}........... & h_{ks} \end{pmatrix}.$$

Let $D_i = \varsigma$ Diag $(h_{1i}, h_{2i}, ..., h_{ki})$, $i = 1, 2,...,s$. We can write $F_i = [L_i : D_i]$, $i = 1,2,...,s$. Then $F = (F_1 : F_2 : ... : F_s)$. Here, the k rows of matrix F form a RSORD under the inter-class structure (2.11), of k factors with $m = s$ groups, of size $n = (G_0 + k)$ and the number of design points $N = s(G_0 + k) = sG_0 + sk = G + sk = 2^p + mk$, $s = m$ being the number of sets in which the observations are divided. Note that ς is determined from the condition (IV) of (3.13). It is to be noted that the design points obtained from the part L_i of F_i are from the appropriate fractional factorial 2^k experiment. The design points given by D_i, $i = 1, 2,...,m$ are the so called star points. The overall design thus constructed constitute a kind of a central composite design usually constructed.

In the design constructed by Method II above, obviously a factor point (obtained either from the appropriate fraction of a 2^k experiment or the star points) may have to be repeated. In certain situations, the repetition has to be accepted so that different parameters of the design match in order to give an overall rotatable design in the present context. The part of the design belonging to the different sets do not give a SORD.

Features of the method:

(a) The condition (I) of (3.13) is readily satisfied for these designs. The expressions (i), (ii) and (iii) in (II) of (3.13) simplify respectively to

$$a_0 = \frac{(G_0 + \varsigma^2)m}{\sigma^2\{1 + (n-1)\rho\}}, \quad \frac{1}{e} = \frac{1}{\sigma^2}[\frac{(G_0 + \varsigma^2)m}{1 - \rho} - \frac{\rho\varsigma^2 m}{(1 - \rho)\{1 + (n-1)\rho\}}],$$

$$(\frac{2}{c} + d) = \frac{1}{\sigma^2}[\frac{(G_0 + \varsigma^4)m}{1 - \rho} - \frac{\rho(G_0 + \varsigma^2)^2 m}{(1 - \rho)\{1 + (n-1)\rho\}}]$$

and they are the *same* for $i = 1, 2, ..., k$.

Note that $n = (G_0 + k)$, $N = mn = (G_0 + k)m = G + mk = 2^p + mk$, $s = m$. Again, the expressions (i) and (ii) in (III) of (3.13) reduce respectively to

$$d = \frac{1}{\sigma^2}[\frac{G_0 m}{1 - \rho} - \frac{\rho(G_0 + \varsigma^2)^2 m}{(1 - \rho)\{1 + (n-1)\rho\}}], \quad \frac{1}{c} = \frac{1}{\sigma^2}(\frac{G_0 m}{1 - \rho})$$

and they are the *same* for $1 \leq i < j \leq k$.

Using the above results, ς is determined from (IV) of (3.13), considering the following equation

$$2\frac{G_0 m}{1 - \rho} + [\frac{G_0 m}{1 - \rho} - \frac{\rho(G_0 + \varsigma^2)^2 m}{(1 - \rho)\{1 + (n-1)\rho\}}] = \frac{(G_0 + \varsigma^4)m}{1 - \rho} - \frac{\rho(G_0 + \varsigma^2)^2 m}{(1 - \rho)\{1 + (n-1)\rho\}}$$

or, $3G_0 = G_0 + \varsigma^4$
or, $\varsigma = (2G_0)^{1/4}$.

Therefore, with $\varsigma = (2G_0)^{1/4}$, rotatability conditions (3.13) are satisfied by the design points described by Method II for *all* ρ.

(b) To ensure non-singularity, we examine the expression

$$\frac{2}{c} + k(d - \frac{a_0^2}{v_{00}})$$

i.e., the condition (V) of (3.14), with respect to design parameters as in above and noting $v_{00} = \frac{N}{\sigma^2\{1 + (n-1)\rho\}}$, the expression

$$\frac{2}{c} + k(d - \frac{a_0^2}{v_{00}})$$

simplifies to

$$\frac{m}{\sigma^2(1 - \rho)}[(k + 2)G_0 - k\frac{(G_0 + \varsigma^2)^2}{n}].$$

Hence the non-singularity condition boils down to

$$(k + 2)G_0 - k\frac{(G_0 + \varsigma^2)^2}{n} > 0$$

$$\Rightarrow \frac{nG_0}{(G_0 + \varsigma^2)^2} > \frac{k}{k + 2}$$

$$\Rightarrow \frac{G_0 + k}{G_0 + 2\{1 + (2G_0)^{1/2}\}} > \frac{k}{k + 2}. \qquad (3.17)$$

Thus the designs described in Method II satisfies the non-singularity condition (3.17) for *all* values of ρ as G_0 is always greater than k. Therefore, the designs obtained by following Method II are RSORD, under the inter-class structure (2.11).

RSORDs construction Method II given above is found to be quite powerful in the sense that the original design D_o can also be obtained from some source other than factorial experiments as such, say, e.g., BIB designs which have been utilized extensively by various authors in the construction of SORDs in the context of uncorrelated errors. In the following lines is described the construction of RSORD with $k = 7$ factors with the help of the BIB design with parameters $v = b = 7, r = k = 3, \lambda = 1$ in such a manner that the design can be made to satisfy Kiefer's (1960) optimality conditions for the uncorrelated error case. Here the design space is $\mathcal{X} = \{x : x'x \leq 3\}$. Let the symmetrical BIB design with parameters $v = b = 7, r = k = 3, \lambda = 1$ be given as (0, 1, 3), (1, 2, 4), (2, 3, 5), (3, 4, 6), (4, 5, 0), (5, 6, 1), (6, 0, 2).

Let there be 7 factors corresponding to the 7 treatments in the BIB design. Construct a 7×8 matrix D_i from the ith block as follows, $i = 1,2,...,7$. Suppose the treatments in the ith block of the design be (i_1, i_2, i_3). Then the three rows of the 7×8 matrix D_i corresponding to the factors represented by the treatments i_1, i_2 and i_3 of the block give a full factorial experiment with 3 factors each having two levels $+1$ and -1. Let us write 0 as the level of every other factor in each of the columns. The 7×8 matrix of $-1, +1$ and 0's thus obtained is called D_i. Let the same D_i be justaposed one by one 35 times and then augment the columns of the matrix by writing at the end $O_{7 \times 8}$ which represents a matrix of type 7×8 with all elements 0.

We construct $L_i = [D_i : D_i : D_i :...: D_i : O_{7 \times 8}]$ (such an L_i) for each one of the 7 blocks of the BIB design, $i = 1, 2,...,7$. The complete design thus constructed gives a RSORD with $k = 7$ factors, $m = 7$ and $n = 35 \times 8 + 8 = 288$. This design besides being an RSORD satisfies Kiefer's (1960) optimality condition of an optimal second-order regression design in the uncorrelated error case. The points are partitioned in two sets. The first set consists of points on the surface of the hypersphere given by $x'x = 3$. The second set consists of only center points. The cardinality of the second set constitute a proportion $\frac{2}{(k+1)(k+2)} = 2/72 = 1/36$ of the total set of all points in the design.

The design thus constructed is exactly optimal according to Kiefer (1960) in the context of uncorrelated errors. Of course, it includes a very large number of design points and will not be of much use in practice. This is done only as an exercise to demonstrate that exact optimality can be attained in some cases. If we are satisfied with approximate optimality, the number of design points can be drastically reduced. Say, for instance we can write an L_i as $[D_i : D_i : D_i : D_i : O_{7 \times 8}]$, $i = 1,2,3,...,7$.

Then the design will have $k = 7$, $m = 7$ and $n = 33$. In this design the center points constitute a $1/33$ proportion of all points in the design. The expression of the variance function of estimated response for particular designs obtained by Method II can be easily found out by following the methods used in case of Method I. This exercise is not carried out to avoid repetitions.

Example 3.2 Consider an experimental design consisting of three factors, four independent groups, each of seven observations i.e., $k = 3, m = 4, n = 7$ and $\text{Dis}(\mathbf{e}) = I_4 \otimes W_{7 \times 7}(\rho)$, $N = mn = 28$ (using factorial experiment) (Method II).

Gr. 1:	1	2	3	4	5	6	7	Gr. 2:	1	2	3	4	5	6	7
x_1	1	-1	-1	1	a	0	0	x_1	1	-1	-1	1	a	0	0
x_2	-1	1	-1	1	0	a	0	x_2	-1	1	-1	1	0	-a	0
x_3	-1	-1	1	1	0	0	a	x_3	-1	-1	1	1	0	0	-a

Gr. 3:	1	2	3	4	5	6	7	Gr. 4:	1	2	3	4	5	6	7
x_1	-1	1	1	-1	-a	0	0	x_1	-1	1	1	-1	-a	0	0
x_2	-1	1	-1	1	0	a	0	x_2	-1	1	-1	1	0	-a	0
x_3	-1	-1	1	1	0	0	-a	x_3	-1	-1	1	1	0	0	a

Group totals for x_i's are given in the following table.

Group :	1	2	3	4
x_1	a	a	-a	-a
x_2	a	-a	a	-a
x_3	a	-a	-a	a

where $a=(2)^{3/4}$. So the design points within a group do not satisfy the usual SORD conditions, but they form a RSORD under the inter-class structure (2.11).

Example 3.3 Consider an experimental design consisting of seven factors, eight independent groups, each of fifteen observations i.e., $k = 7$, $m = 8$, $n = 15$ and $\text{Dis}(\mathbf{e}) = I_8 \otimes W_{15\times15}(\rho)$, $N = mn = 120$ (using BIB design; Method II).

Gr. 1:	1	2	3	4	5	6	7	8	9	10	11	12	13	14	15
x_1	0	0	0	0	0	0	0	0	2	0	0	0	0	0	0
x_2	-1	-1	1	1	-1	-1	1	1	0	2	0	0	0	0	0
x_3	-1	-1	-1	-1	1	1	1	1	0	0	2	0	0	0	0
x_4	0	0	0	0	0	0	0	0	0	0	0	2	0	0	0
x_5	1	-1	1	-1	-1	1	-1	1	0	0	0	0	2	0	0
x_6	0	0	0	0	0	0	0	0	0	0	0	0	0	2	0
x_7	0	0	0	0	0	0	0	0	0	0	0	0	0	0	2

Gr. 2:	1	2	3	4	5	6	7	8	9	10	11	12	13	14	15
x_1	0	0	0	0	0	0	0	0	-2	0	0	0	0	0	0
x_2	0	0	0	0	0	0	0	0	0	2	0	0	0	0	0
x_3	-1	-1	-1	-1	1	1	1	1	0	0	-2	0	0	0	0
x_4	1	-1	-1	1	1	-1	-1	1	0	0	0	2	0	0	0
x_5	0	0	0	0	0	0	0	0	0	0	0	0	-2	0	0
x_6	1	1	-1	-1	-1	-1	1	1	0	0	0	0	0	2	0
x_7	0	0	0	0	0	0	0	0	0	0	0	0	0	0	-2

Gr. 3:	1	2	3	4	5	6	7	8	9	10	11	12	13	14	15
x_1	0	0	0	0	0	0	0	0	2	0	0	0	0	0	0
x_2	0	0	0	0	0	0	0	0	0	-2	0	0	0	0	0
x_3	0	0	0	0	0	0	0	0	0	0	-2	0	0	0	0
x_4	1	-1	-1	1	1	-1	-1	1	0	0	0	2	0	0	0
x_5	1	-1	1	-1	-1	1	-1	1	0	0	0	0	2	0	0
x_6	0	0	0	0	0	0	0	0	0	0	0	0	0	-2	0
x_7	-1	1	1	-1	1	-1	-1	1	0	0	0	0	0	0	-2

Gr. 4:	1	2	3	4	5	6	7	8	9	10	11	12	13	14	15
x_1	-1	1	-1	1	-1	1	-1	1	-2	0	0	0	0	0	0
x_2	0	0	0	0	0	0	0	0	0	-2	0	0	0	0	0
x_3	0	0	0	0	0	0	0	0	0	0	2	0	0	0	0
x_4	0	0	0	0	0	0	0	0	0	0	0	2	0	0	0
x_5	1	-1	1	-1	-1	1	-1	1	0	0	0	0	-2	0	0
x_6	1	1	-1	-1	-1	-1	1	1	0	0	0	0	0	-2	0
x_7	0	0	0	0	0	0	0	0	0	0	0	0	0	0	2

Gr. 5:	1	2	3	4	5	6	7	8	9	10	11	12	13	14	15
x_1	0	0	0	0	0	0	0	0	2	0	0	0	0	0	0
x_2	-1	-1	1	1	-1	-1	1	1	0	2	0	0	0	0	0
x_3	0	0	0	0	0	0	0	0	0	0	2	0	0	0	0
x_4	0	0	0	0	0	0	0	0	0	0	0	-2	0	0	0
x_5	0	0	0	0	0	0	0	0	0	0	0	0	-2	0	0
x_6	1	1	-1	-1	-1	-1	1	1	0	0	0	0	0	-2	0
x_7	-1	1	1	-1	1	-1	-1	1	0	0	0	0	0	0	-2

Gr. 6:	1	2	3	4	5	6	7	8	9	10	11	12	13	14	15
x_1	-1	1	-1	1	-1	1	-1	1	-2	0	0	0	0	0	0
x_2	0	0	0	0	0	0	0	0	0	2	0	0	0	0	0
x_3	-1	-1	-1	-1	1	1	1	1	0	0	-2	0	0	0	0
x_4	0	0	0	0	0	0	0	0	0	0	0	-2	0	0	0
x_5	0	0	0	0	0	0	0	0	0	0	0	0	2	0	0
x_6	0	0	0	0	0	0	0	0	0	0	0	0	0	-2	0
x_7	-1	1	1	-1	1	-1	-1	1	0	0	0	0	0	0	2

Gr. 7:	1	2	3	4	5	6	7	8	9	10	11	12	13	14	15
x_1	-1	1	-1	1	-1	1	-1	1	2	0	0	0	0	0	0
x_2	-1	-1	1	1	-1	-1	1	1	0	-2	0	0	0	0	0
x_3	0	0	0	0	0	0	0	0	0	0	-2	0	0	0	0
x_4	-1	-1	-1	-1	1	1	1	1	0	0	0	-2	0	0	0
x_5	0	0	0	0	0	0	0	0	0	0	0	0	-2	0	0
x_6	0	0	0	0	0	0	0	0	0	0	0	0	0	2	0
x_7	0	0	0	0	0	0	0	0	0	0	0	0	0	0	2

Gr. 8:	1	2	3	4	5	6	7	8	9	10	11	12	13	14	15
x_1	0	0	0	0	0	0	0	0	-2	0	0	0	0	0	0
x_2	-1	-1	1	1	-1	-1	1	1	0	-2	0	0	0	0	0
x_3	-1	-1	-1	-1	1	1	1	1	0	0	2	0	0	0	0
x_4	0	0	0	0	0	0	0	0	0	0	0	-2	0	0	0
x_5	1	-1	1	-1	-1	1	-1	1	0	0	0	0	2	0	0
x_6	0	0	0	0	0	0	0	0	0	0	0	0	0	2	0
x_7	0	0	0	0	0	0	0	0	0	0	0	0	0	0	-2

Group totals for $\mathbf{x}_i$'s are given in the following table.

Group :	1	2	3	4	5	6	7	8
$\mathbf{x}_1$	2	-2	2	-2	2	-2	2	-2
$\mathbf{x}_2$	2	2	-2	-2	2	2	-2	-2
$\mathbf{x}_3$	2	-2	-2	2	2	-2	-2	2
$\mathbf{x}_4$	2	2	2	2	-2	-2	-2	-2
$\mathbf{x}_5$	2	-2	2	-2	-2	2	-2	2
$\mathbf{x}_6$	2	2	-2	-2	-2	-2	2	2
$\mathbf{x}_7$	2	-2	-2	2	-2	2	2	-2

So the design points within a group do not satisfy the usual SORD conditions, but they form a RSORD under the inter-class structure (2.11).

A Table is given below to compare the number of design points between Method I and Method II, where designs are constructed from factorial or fractional factorial experiment (using Central Composite Designs), for some useful values of k and m.

The present section develops RSORDs (or design matrices X's) under the inter-class structure (2.11), whatever be the value of correlation parameter ρ in W_1 in (2.11). Note that W_1 is unknown as the correlation parameter ρ involved in W_1 is unknown. As W_1 is unknown, it is a very difficult problem to estimate the regression parameters and error variance in the model using the design matrix as developed in this section. In this connection, a method of estimation of unknown regression parameters and analysis of the design, is given in Chapter 9.

3.6 ROBUST SECOND-ORDER DESIGNS FOR COMPOUND SYMMETRY STRUCTURE

The present section deals with a variance-covariance structure of errors which is an extension of inter-class structure. It is termed as compound symmetry structure and is given in (2.23).

3.6.1 Second-order rotatability conditions under compound symmetry structure

The following theorem presents the second-order rotatability conditions under the compound symmetry structure (2.23), which can be easily obtained from (3.3).

TABLE 3.1: Design points between Method I and Method II

k, m :	Method I	Method II
$k = 2,\ m = 4$	52	28
$k = 3,\ m = 4$	80	28
$k = 3,\ m = 4$	80	44
$k = 4,\ m = 8$	248	92
$k = 5,\ m = 8$	256	92
$k = 6,\ m = 8$	424	156

Theorem 3.5 (Das, 2003a) *The necessary and sufficient conditions of a robust second-order rotatable design under the compound symmetry variance-covariance structure, for all values of ρ, ρ_1 in $W_{N \times N}(\rho, \rho_1)$ as in (2.23), in the model (3.1) are*

(I) : (i) $v_{0.j} = 0 \Leftrightarrow \sum_{u=1}^{N} x_{uj} = 0; 1 \leq j \leq k,$

$v_{0.jl} = 0 \Leftrightarrow \sum_{u=1}^{N} x_{uj} x_{ul} = 0; 1 \leq j < l \leq k,$

(ii) $v_{i.j} = 0 \Leftrightarrow (\delta_2 - \gamma_2) \sum_{u=1}^{N} x_{ui} x_{uj} + (\gamma_2 - \delta_3)\{(\sum_{u=1}^{n} x_{ui})(\sum_{u=1}^{n} x_{uj}) + ... + (\sum_{u=n(m-1)+1}^{mn} x_{ui})(\sum_{u=n(m-1)+1}^{mn} x_{uj})\} + \delta_3(\sum_{u=1}^{N} x_{ui})(\sum_{u=1}^{N} x_{uj}) = 0; 1 \leq i, j \leq k, \ i \neq j,$

(iii): (1) $v_{ii.j} = 0 \Leftrightarrow (\delta_2 - \gamma_2) \sum_{u=1}^{N} x_{ui}^2 x_{uj} + (\gamma_2 - \delta_3)\{(\sum_{u=1}^{n} x_{ui}^2)(\sum_{u=1}^{n} x_{uj}) + ... + (\sum_{u=n(m-1)+1}^{mn} x_{ui}^2)(\sum_{u=n(m-1)+1}^{mn} x_{uj})\} + \delta_3(\sum_{u=1}^{N} x_{ui}^2)(\sum_{u=1}^{N} x_{uj}) = 0; 1 \leq i \leq k, \ 1 \leq j \leq k,$

(2) $v_{ij.l} = 0 \Leftrightarrow (\delta_2 - \gamma_2) \sum_{u=1}^{N} x_{ui} x_{uj} x_{ul} + (\gamma_2 - \delta_3)\{(\sum_{u=1}^{n} x_{ui} x_{uj})(\sum_{u=1}^{n} x_{ul}) + ... + (\sum_{u=n(m-1)+1}^{mn} x_{ui} x_{uj})(\sum_{u=n(m-1)+1}^{mn} x_{ul})\} + \delta_3(\sum_{u=1}^{N} x_{ui} x_{uj})(\sum_{u=1}^{N} x_{ul}) = 0; 1 \leq i < j \leq k, \ 1 \leq l \leq k,$

(3) $v_{ii.jl} = 0 \Leftrightarrow (\delta_2 - \gamma_2) \sum_{u=1}^{N} x_{ui}^2 x_{uj} x_{ul} + (\gamma_2 - \delta_3)\{(\sum_{u=1}^{n} x_{ui}^2)(\sum_{u=1}^{n} x_{uj} x_{ul}) + ... + (\sum_{u=n(m-1)+1}^{mn} x_{ui}^2)(\sum_{u=n(m-1)+1}^{mn} x_{uj} x_{ul})\} + \delta_3(\sum_{u=1}^{N} x_{ui}^2)(\sum_{u=1}^{N} x_{uj} x_{ul}) = 0; 1 \leq i \leq k, \ 1 \leq j < l \leq k,$

(4) $v_{ij.lt} = 0 \Leftrightarrow (\delta_2 - \gamma_2) \sum_{u=1}^{N} x_{ui} x_{uj} x_{ul} x_{ut} + (\gamma_2 - \delta_3)\{(\sum_{u=1}^{n} x_{ui} x_{uj})(\sum_{u=1}^{n} x_{ul} x_{ut}) + ... + (\sum_{u=n(m-1)+1}^{mn} x_{ui} x_{uj})(\sum_{u=n(m-1)+1}^{mn} x_{ul} x_{ut})\} + \delta_3(\sum_{u=1}^{N} x_{ui} x_{uj})(\sum_{u=1}^{N} x_{ul} x_{ut}) = 0; 1 \leq i < j \leq k, 1 \leq l < t \leq k, (i,j) \neq (l,t),$

(II): (i) $v_{0.jj} = constant \Leftrightarrow \{\sigma^2\}^{-1}[\delta_2 + (n-1)\gamma_2 + n(m-1)\delta_3] \sum_{u=1}^{N} x_{uj}^2 = a_0 \ (> 0); 1 \leq j \leq k,$

(ii) $v_{i.i} = constant \Leftrightarrow \{\sigma^2\}^{-1}[(\delta_2 - \gamma_2) \sum_{u=1}^{N} x_{ui}^2 + (\gamma_2 - \delta_3)\{(\sum_{u=1}^{n} x_{ui})^2 + ... + (\sum_{u=n(m-1)+1}^{mn} x_{ui})^2\} + \delta_3(\sum_{u=1}^{N} x_{ui})^2] = \frac{1}{e} \ (> 0); 1 \leq i \leq k,$

(iii) $v_{ii.ii} = constant \Leftrightarrow \{\sigma^2\}^{-1}[(\delta_2 - \gamma_2) \sum_{u=1}^{N} x_{ui}^4 + (\gamma_2 - \delta_3)\{(\sum_{u=1}^{n} x_{ui}^2)^2 + ... + (\sum_{u=n(m-1)+1}^{mn} x_{ui}^2)^2\} + \delta_3(\sum_{u=1}^{N} x_{ui}^2)^2] = (\frac{2}{c} + d)(> 0); 1 \leq i \leq k,$

(III): (i) $v_{ii.jj} = constant \Leftrightarrow \{\sigma^2\}^{-1}[(\delta_2 - \gamma_2) \sum_{u=1}^{N} x_{ui}^2 x_{uj}^2 + (\gamma_2 - \delta_3)\{(\sum_{u=1}^{n} x_{ui}^2)(\sum_{u=1}^{n} x_{uj}^2) + ... + (\sum_{u=n(m-1)+1}^{mn} x_{ui}^2)(\sum_{u=n(m-1)+1}^{mn} x_{uj}^2)\} + \delta_3(\sum_{u=1}^{N} x_{ui}^2)(\sum_{u=1}^{N} x_{uj}^2)] = d \ (> 0); 1 \leq i, \ j \leq k, \ i \neq j,$

(ii) $v_{ij.ij} = constant \Leftrightarrow \{\sigma^2\}^{-1}[(\delta_2 - \gamma_2) \sum_{u=1}^{N} x_{ui}^2 x_{uj}^2 + (\gamma_2 - \delta_3)\{(\sum_{u=1}^{n} x_{ui} x_{uj})^2 + ... + (\sum_{u=n(m-1)+1}^{mn} x_{ui} x_{uj})^2\} + \delta_3(\sum_{u=1}^{N} x_{ui} x_{uj})^2] = \frac{1}{c} \ (> 0); \ 1 \leq i < j \leq k,$

(IV): $v_{ii.ii} = 2v_{ij.ij} + v_{ii.jj}; 1 \leq i < j \leq k,$ (3.18)

where $v_{ii.ii}, v_{ii.jj}$ and $v_{ij.ij}$ are as in **(II)** *(iii) and* **(III)** *(i), (ii) in (3.18) and δ_2, γ_2 and δ_3 are as in (2.24).*

Non-singularity condition is given by

(V) : $\frac{2}{c} + k(d - \frac{a_0^2}{v_{00}}) > 0,$ (3.19) where $v_{00} = N[(\delta_2 + (n-1)\gamma_2 + n(m-1)\delta_3]$ and $\frac{1}{c}$, d, a_0 are as in (3.18).

The following theorem is now immediate for the compound symmetry structure.

Theorem 3.6 (Das, 2003a) *If the design points are divided into* m *sets of* n *each, having a second-order rotatable structure in the usual sense within each set, then the overall design is a robust second-order rotatable under the compound symmetry structure (2.23), whatever be the values of intra-class and inter-class correlation coefficients ρ, and ρ_1, respectively.*

Proof. It is clear from the specifications of the given design that n design points in each of the m sets satisfy the following moment relations:

(1) $\displaystyle\sum_{u=(i-1)n+1}^{in} x_{uj_1}^{\alpha_1} x_{uj_2}^{\alpha_2} x_{uj_3}^{\alpha_3} x_{uj_4}^{\alpha_4} = 0;\ 1 \le i \le m;$ for α_j's non-negative integers, any α_j

odd and $\displaystyle\sum_{j=1}^{4} \alpha_j \le 4,$

(2) : (i) $\displaystyle\sum_{u=(i-1)n+1}^{in} x_{uj}^2 = constant = n\lambda_2^2; 1 \le j \le k,\ 1 \le i \le m,$

(ii) $\displaystyle\sum_{u=(i-1)n+1}^{in} x_{uj}^4 = constant = 3n\lambda_4; 1 \le j \le k,\ 1 \le i \le m,$

(3) $\displaystyle\sum_{u=(i-1)n+1}^{in} x_{uj_1}^2 x_{uj_2}^2 = constant = n\lambda_4; 1 \le j_1,\ j_2 \le k,\ j_1 \ne j_2, 1 \le i \le m,$

(4) $\displaystyle\sum_{u=(i-1)n+1}^{in} x_{uj}^4 = 3 \sum_{u=(i-1)n+1}^{in} x_{uj_1}^2 x_{uj_2}^2; 1 \le j,\ j_1,\ j_2 \le k,\ j_1 \ne j_2,\ 1 \le i \le m.$

(5) $\frac{\lambda_4}{\lambda_2^2} > \frac{k}{k+2}.$ (3.20)

Using (3.20), it can be readily verified that (3.18) holds for all ρ and ρ_1. As regards non-singularity, we check (3.19), using (3.20) and the following

$$a_0 = mn\lambda_2[\delta_2 + (n-1)\gamma_2 + n(m-1)\delta_3]/\sigma^2,\ 1/c = mn\lambda_4(\delta_2 - \gamma_2)/\sigma^2,$$

$$d = [mn\lambda_4(\delta_2 - \gamma_2) + mn^2\lambda_2^2(\gamma_2 - \delta_3) + \delta_3(mn\lambda_2)^2]/\sigma^2,\ 1/e = [(\delta_2 - \gamma_2)mn\lambda_2]/\sigma^2$$

$$v_{00} = N[\delta_2 + (n-1)\gamma_2 + n(m-1)\delta_3]/\sigma^2,\quad N = mn.$$

The expression $\frac{2}{c} + k(d - \frac{a_0^2}{v_{00}})$ simplifies to $(\delta_2 - \gamma_2)[(k+2)\lambda_4 - k\lambda_2^2]$. Hence, the non-singularity condition boils down to

$\frac{\lambda_4}{\lambda_2^2} > \frac{k}{k+2}$ as $(\delta_2 - \gamma_2) = \frac{1}{(1-\rho)} > 0,$

and this is also the non-singularity condition of each set of n observations out of the m sets. Therefore, the theorem is verified.

Converse part of Theorem 3.6 does not hold as a RSORD for the total design under the correlation structure (2.23) does not necessarily imply the holding of SORD conditions for each of the m sets of observations separately (for uncorrelated and homoscedastic errors). This is shown to be obviously true, by considering Example 3.1, in Section 3.5 which satisfies all conditions of a RSORD in the present situation of compound symmetry error structure (2.23). For the example provided, all the conditions (3.18) hold and the non-singularity condition (3.19) after simplification reduces to $\frac{32(\delta_2 - \gamma_2)}{11}$, which is always positive. But the usual SORD conditions do not hold for each one of the sets under consideration, as pointed out before.

3.6.2   RSORD construction for compound symmetry structure

The design given by Method I in Section 3.5.2 is also a RSORD under the compound symmetry structure (2.23) for all values of ρ and ρ_1. Verification of the statement follows from the proof of the Theorem 3.6.

Variance Function: The variance function of a RSORD under the compound symmetry structure (2.23) as constructed by the Method I as in Section 3.5.2, is obtained from following (3.7) and using the above design parameters, and considering the following

$$v^{00} = \frac{2/c + kd}{T_1} = \frac{\sigma^2[(k+2)\lambda_4(\delta_2 - \gamma_2) + k\lambda_2^2\{n\gamma_2 + n(m-1)\delta_3\}]}{\{\delta_2 + (n-1)\gamma_2 + n(m-1)\delta_3\}T_2} = B_0, \text{ say,}$$

$$a_1 = -\frac{a_0}{T_1} = -\frac{\sigma^2\lambda_2}{T_2}, \quad d_1 = \frac{c(a_0^2 - dv_{00})}{2T_1} = \frac{\sigma^2(\lambda_2^2 - \lambda_4)}{2\lambda_4 T_2},$$

where $T_1 = \{v_{00}(2/c+kd) - ka_0^2\} = \frac{N^2}{\sigma^4}[\{\delta_2 + (n-1)\gamma_2 + n(m-1)\delta_3\}(\delta_2 - \gamma_2)\{(k+2)\lambda_4 - k\lambda_2^2\}]$ and

$$(2a_1 + e) = \frac{\sigma^2(k+2)(\lambda_4 - \lambda_2^2)}{\lambda_2 T_2} = B_1, \text{ say,}$$

$$(c/2 + d_1) = \frac{\sigma^2[(k+1)\lambda_4 - \lambda_2^2(k-1)]}{2\lambda_4 T_2} = B_2, \text{ say,}$$

$$T_2 = N(\delta_2 - \gamma_2)\{(k+2)\lambda_4 - k\lambda_2^2\}.$$

Therefore, the variance function is given by (using the above)

$$V(\hat{y}_{\mathbf{x}}) = v^{00} + (2a_1 + e)r^2 + (c/2 + d_1)r^4$$

$$= B_0 + B_1 r^2 + B_2 r^4 \tag{3.21}$$

where $r^2 = \sum_{i=1}^{k} x_i^2$ and B_0, B_1, B_2 are as in above.

The designs obtained by Method II in Section 3.5.2 is also a RSORD in the present situation of compound symmetry error structure for all feasible values of the correlations coefficients ρ and ρ_1. Explicit expression of variance function for a design obtained by Method II follows similar arguments as in case of designs by Method I and is not undertaken for the sake of precision.

3.7   ROBUST SECOND-ORDER DESIGNS FOR TRI-DIAGONAL STRUCTURE

Tri-diagonal error variance-covariance structure is a relaxation of intra-class structure or *lag-model* variance-covariance structure of errors and is given in (2.30).

3.7.1   Second-order rotatability conditions under tri-diagonal structure

The following theorem gives the second-order rotatability conditions (from (3.3)) under the tri-diagonal structure (2.30).

Theorem 3.7 *The necessary and sufficient conditions of a robust second-order rotatable design under the tri-diagonal variance-covariance structure, for all values of ρ in $W_3(\rho)$ as in (2.30), in the model (3.1) are*

(I): (i) $v_{0.j} = 0 \Leftrightarrow \sum_{u=1}^{N} x_{uj} = 0; 1 \le j \le k,$

$v_{0.jl} = 0 \Leftrightarrow \sum_{u=1}^{N} x_{uj} x_{ul} = 0; 1 \le j < l \le k,$

(ii) $v_{i.j} = 0 \Leftrightarrow \sum_{u=1}^{N} x_{ui} x_{uj} - \rho\{\sum_{u=1}^{n} x_{(n+u)i} x_{uj} + \sum_{u=1}^{n} x_{ui} x_{(n+u)j}\} = 0; 1 \le i, j \le k, \ i \ne j,$

(iii): (1)$v_{ii.j} = 0 \Leftrightarrow \sum_{u=1}^{N} x_{ui}^2 x_{uj} - \rho\{\sum_{u=1}^{n} x_{(n+u)i}^2 x_{uj} + \sum_{u=1}^{n} x_{ui}^2 x_{(n+u)j}\} = 0; 1 \le i \le k, \ 1 \le j \le k,$

(2) $v_{ij.l} = 0 \Leftrightarrow \sum_{u=1}^{N} x_{ui} x_{uj} x_{ul} - \rho(\sum_{u=1}^{n} x_{(n+u)i} x_{(n+u)j} x_{ul} + \sum_{u=1}^{n} x_{ui} x_{uj} x_{(n+u)l}) = 0; 1 \le i < j \le k, \ 1 \le l \le k,$

(3) $v_{ii.jl} = 0 \Leftrightarrow \sum_{u=1}^{N} x_{ui}^2 x_{uj} x_{ul} - \rho(\sum_{u=1}^{n} x_{(n+u)i}^2 x_{uj} x_{ul} + \sum_{u=1}^{n} x_{ui}^2 x_{(n+u)j} x_{(n+u)l}) = 0; 1 \le i \le k, \ 1 \le j < l \le k,$

(4) $v_{ij.lt} = 0 \Leftrightarrow \sum_{u=1}^{N} x_{ui} x_{uj} x_{ul} x_{ut} - \rho(\sum_{u=1}^{n} x_{(n+u)i} x_{(n+u)j} x_{ul} x_{ut} + \sum_{u=1}^{n} x_{ui} x_{uj} x_{(n+u)l} x_{(n+u)t}) = 0; 1 \le i < j \le k, \ 1 \le l < t \le k, \ (i,j) \ne (l,t),$

(II): (i) $v_{0.jj} = constant \Leftrightarrow (1-\rho)\{\sigma^2(1-\rho^2)\}^{-1} \sum_{u=1}^{N} x_{uj}^2 = a_0(> 0); 1 \le j \le k,$

(ii) $v_{i.i} = constant \Leftrightarrow \{\sigma^2(1-\rho^2)\}^{-1}[\sum_{u=1}^{N} x_{ui}^2 - 2\rho \sum_{u=1}^{n} x_{ui} x_{(n+u)i}] = \frac{1}{e}(> 0); 1 \le i \le k,$

(iii) $v_{ii.ii} = constant \Leftrightarrow \{\sigma^2(1-\rho^2)\}^{-1}[\sum_{u=1}^{N} x_{ui}^4 - 2\rho \sum_{u=1}^{n} x_{ui}^2 x_{(n+u)i}^2] = (\frac{2}{c} + d) (> 0); 1 \le i \le k,$

(III): (i) $v_{ii.jj} = constant \Leftrightarrow \{\sigma^2(1-\rho^2)\}^{-1}[\sum_{u=1}^{N} x_{ui}^2 x_{uj}^2 - \rho(\sum_{u=1}^{n} x_{(n+u)i}^2 x_{uj}^2 + \sum_{u=1}^{n} x_{ui}^2 x_{(n+u)j}^2)] = d(> 0); 1 \le i, \ j \le k, \ i \ne j,$

(ii) $v_{ij.ij} = constant \Leftrightarrow \{\sigma^2(1-\rho^2)\}^{-1}[\sum_{u=1}^{N} x_{ui}^2 x_{uj}^2 - 2\rho \sum_{u=1}^{n} x_{ui} x_{uj} x_{(n+u)i} x_{(n+u)j}] = \frac{1}{c}(> 0); 1 \le i < j \le k,$

(IV): $v_{ii.ii} = 2 v_{ij.ij} + v_{ii.jj}; 1 \le i < j \le k,$ \hfill (3.22)
where $N = 2n$, $v_{ii.ii}$, $v_{ii.jj}$ and $v_{ij.ij}$ are as in **(II)** *(iii) and* **(III)** *(i), (ii) in (3.22).*
Non-singularity condition is given by \hfill (3.23)
(V) : $\frac{2}{c} + k(d - \frac{a_0^2}{v_{00}}) > 0,$
where $v_{00} = \frac{N}{\sigma^2(1+\rho)}$, $N = 2n$ and $\frac{1}{c}, d, a_0$ are as in (3.22).

3.7.2 RSORD construction for tri-diagonal structure

The present section develops *two* methods of construction of RSORDs under the tri-diagonal error variance-covariance structure (2.30). The designs which are obtained by these methods satisfy the moment conditions given in (3.22) for *all* values of ρ.

Method I (description of the method) (Das, 2003a): We start with a usual second-order rotatable design (SORD) having n design points involving k factors. The set of n

design points can be extended to $2n$ points by incorporating n central points together just below or above the n design points of the original design with which we started.

Features of the method:

(a) It can be readily verified that (3.22) holds for these designs.

(b) To ensure non-singularity, we first examine the original design and the newly constructed design. For the original design with which we started, the following are the moment relations:

$$\sum_{u=1}^{n} x_{ui}^2 = n\lambda_2; \quad \sum_{u=1}^{n} x_{ui}^4 = 3n\lambda_4; \quad \sum_{u=1}^{n} x_{ui}^2 x_{uj}^2 = n\lambda_4; \quad 1 \le i, j \le k, \quad i \ne j.$$

Using the above results for the newly constructed design, we get the following:

$$\sum_{u=1}^{2n} x_{ui}^2 = n\lambda_2; \quad \sum_{u=1}^{2n} x_{ui}^4 = 3n\lambda_4; \quad \sum_{u=1}^{2n} x_{ui}^2 x_{uj}^2 = n\lambda_4; 1 \le i, j \le k, \ i \ne j.$$

Using the above results, the design parameters of the newly constructed design are the following:

$$a_0 = \frac{(1-\rho)n\lambda_2}{\sigma^2(1-\rho^2)}; \quad d = \frac{n\lambda_4}{\sigma^2(1-\rho^2)}; \quad \frac{1}{e} = \frac{n\lambda_2}{\sigma^2(1-\rho^2)}; \quad \frac{1}{c} = \frac{n\lambda_4}{\sigma^2(1-\rho^2)}.$$

Using the above design parameters and noting $v_{00} = \frac{2n(1-\rho)}{\sigma^2(1-\rho^2)}$, the expression $\frac{2}{c} + k(d - \frac{a_0^2}{v_{00}})$ simplifies to

$$\frac{n}{\sigma^2(1-\rho^2)}[(k+2)\lambda_4 - \frac{k\lambda_2^2(1-\rho)}{2}].$$

Hence the non-singularity condition of the above design is

$$\frac{\lambda_4}{\lambda_2^2} > \frac{k}{k+2}\frac{(1-\rho)}{2}. \tag{3.24}$$

It is readily seen that $0 \le \frac{(1-\rho)}{2} \le 1$, $-1 \le \rho \le 1$. The design with n points, to start with, is a SORD in the usual sense so that $\frac{\lambda_4}{\lambda_2^2} > \frac{k}{k+2}$. This condition indeed does imply the revised condition (3.24) derived here for *all* values of ρ. Thus, the designs described in Method I satisfies the non-singularity condition for all values of ρ. Therefore, the designs obtained by following Method I are RSORD, under the tri-diagonal structure (2.30).

(c) **Variance Function:** The variance function of a RSORD under the tri-diagonal structure constructed by the above method is obtained by using (3.7), and the above design parameters and noting the following:

$$v_{00} = \frac{2n(1-\rho)}{\sigma^2(1-\rho^2)}, \quad v^{00} = \frac{2/c + kd}{T_1} = \frac{\sigma^2}{T_3}[(k+2)\lambda_4(1+\rho)] = C_0, \text{ say,}$$

$$a_1 = -\frac{a_0}{T_1} = -\frac{\sigma^2(1-\rho^2)\lambda_2}{T_3}, \quad d_1 = \frac{c(a_0^2 - dv_{00})}{2T_1} = \frac{\sigma^2(1-\rho^2)[(1-\rho)\lambda_2^2 - 2\lambda_4]}{2\lambda_4 T_3},$$

$$\text{where } T_1 = \{v_{00}(2/c + kd) - ka_0^2\} = \frac{n^2(1-\rho)\{2(k+2)\lambda_4 - k(1-\rho)\lambda_2^2\}}{\{\sigma^2(1-\rho^2)\}^2}$$

$$\text{and } (2a_1 + e) = \frac{\sigma^2(1-\rho^2)[2(k+2)\lambda_4 - \{k(1-\rho)+2\}\lambda_2^2]}{\lambda_2 T_3} = C_1, \text{ say,}$$

$$(c/2 + d_1) = \frac{\sigma^2(1 - \rho^2)[2(k+1)\lambda_4 - (1 - \rho)(k-1)\lambda_2^2]}{2\lambda_4 T_3} = C_2, \text{ say,}$$

$$T_3 = n\{2(k+2)\lambda_4 - k(1 - \rho)\lambda_2^2\}.$$

Therefore, from (3.7), the variance function is given by

$$\text{Var}(\hat{y}_{\mathbf{x}}) = v^{00} + (2a_1 + e)r^2 + (c/2 + d_1)r^4$$

$$= C_0 + C_1 r^2 + C_2 r^4 \tag{3.25}$$

where $r^2 = \sum_{i=1}^{k} x_i^2$ and C_0, C_1, C_2 are as in above.

Example 3.4 A RSORD with 2 groups under tri-diagonal structure: An example of a RSORD with 2 factors, 2 groups, each of 9 observations under the tri-diagonal structure, is given below (following Method I). The RSORD (denoted by 'd_4') is displayed below (rows being factors and columns being runs).

d_4 :	1	2	3	4	5	6	7	8	9
$\mathbf{x}_1$	-1	1	-1	1	-1.414	1.414	0	0	0
$\mathbf{x}_2$	-1	-1	1	1	0	0	-1.414	1.414	0
d_4 :	10	11	12	13	14	15	16	17	18
$\mathbf{x}_1$	0	0	0	0	0	0	0	0	0
$\mathbf{x}_2$	0	0	0	0	0	0	0	0	0

Method II (Description of the method) (Das, 2003a): We start with a usual SORD having n design points involving k factors. Let $(\mathbf{z}_1, \mathbf{z}_2,...,\mathbf{z}_k)$ be a usual SORD of k factors with n design points, where $\mathbf{z}_i$ is a vector of order $n \times 1$, $1 \leq i \leq k$ representing the levels of the ith explanatory variable or factor in different design points. The set of n design points can be extended to $2n$ points by repeating ith factor column just below itself, i.e., $(\mathbf{z}_i' : \mathbf{z}_i')'$ $= \mathbf{x}_i$, say; $1 \leq i \leq k$. Now $\mathbf{x}_i$ is a vector of order $2n \times 1$ and $(\mathbf{x}_1, \mathbf{x}_2,...,\mathbf{x}_k)$ form a RSORD under the tri-diagonal variance-covariance structure (2.30).

Features of the method:

(i) It can be readily verified that (3.22) holds for these designs.

(ii) To ensure non-singularity condition, we first examine the original design and the newly constructed design. For the original design with which we started, the following are the moment relations:

$$\sum_{u=1}^{n} z_{ui}^2 = n\lambda_2; \ \sum_{u=1}^{n} z_{ui}^4 = 3n\lambda_4; \ \sum_{u=1}^{n} z_{ui}^2 z_{uj}^2 = n\lambda_4; 1 \leq i, j \leq k, \ i \neq j.$$

Using the above results for the newly constructed design, we get the following:

$$\sum_{u=1}^{2n} x_{ui}^2 = 2n\lambda_2; \ \sum_{u=1}^{2n} x_{ui}^4 = 6n\lambda_4; \ \sum_{u=1}^{2n} x_{ui}^2 x_{ju}^2 = 2n\lambda_4; 1 \leq i, j \leq k, \ i \neq j.$$

Using these above results, the design parameters of the newly constructed designs are the following:

$$a_0 = \frac{(1 - \rho)2n\lambda_2}{\sigma^2(1 - \rho^2)}; \ d = \frac{2n\lambda_4}{\sigma^2(1 - \rho^2)}; \ \frac{1}{e} = \frac{2n\lambda_2}{\sigma^2(1 - \rho^2)}; \ \frac{1}{c} = \frac{2n\lambda_4}{\sigma^2(1 - \rho^2)}.$$

Using the above design parameters and noting $v_{00} = \frac{2n(1-\rho)}{\sigma^2(1-\rho^2)}$, the expression $\frac{2}{c} + k(d - \frac{a_0^2}{v_{00}})$ simplifies to

$$\frac{2n}{\sigma^2(1-\rho^2)}[(k+2)\lambda_4 - k\lambda_2^2].$$

Hence the non-singularity condition of the above design is

$$\frac{\lambda_4}{\lambda_2^2} > \frac{k}{k+2}, \tag{3.26}$$

which is the non-singularity condition of the usual SORD we start with.

Thus, the designs described in Method II satisfies the non-singularity condition for *all* values of ρ. Therefore, the designs obtained by following Method II are RSORD, under the tri-diagonal error variance-covariance structure (2.30).

(c) Variance Function: The variance function of a RSORD under the tri-diagonal structure constructed by the above Method II is obtained by using (3.7), and using the above design parameters and noting the following:

$$v_{00} = \frac{2n(1-\rho)}{\sigma^2(1-\rho^2)}, \quad v^{00} = \frac{2/c+kd}{T_1} = \frac{\sigma^2(1-\rho^2)(k+2)\lambda_4}{T_4(1-\rho)} = C_{01}, \text{ say,}$$

$$a_1 = -\frac{a_0}{T_1} = -\frac{\sigma^2(1-\rho^2)\lambda_2}{T_4}, \quad d_1 = \frac{c(a_0^2 - dv_{00})}{2T_1} = \frac{\sigma^2(1-\rho^2)[(1-\rho)\lambda_2^2 - \lambda_4]}{2\lambda_4 T_4},$$

$$\text{where } T_1 = \{v_{00}(2/c+kd) - ka_0^2\} = \frac{4n^2(1-\rho)\{(k+2)\lambda_4 - k(1-\rho)\lambda_2^2\}}{\{\sigma^2(1-\rho^2)\}^2}$$

$$\text{and } (2a_1 + e) = \frac{\sigma^2(1-\rho^2)[(k+2)\lambda_4 - \{k(1-\rho)+2\}\lambda_2^2]}{\lambda_2 T_4} = C_{11}, \text{ say,}$$

$$(c/2 + d_1) = \frac{\sigma^2(1-\rho^2)[(k+1)\lambda_4 - (1-\rho)(k-1)\lambda_2^2]}{2\lambda_4 T_4} = C_{21}, \text{ say,}$$

$$T_4 = 2n\{(k+2)\lambda_4 - k(1-\rho)\lambda_2^2\}.$$

Therefore, the variance function is given by (following (3.7))

$$V(\hat{y}_\mathbf{x}) = v^{00} + (2a_1 + e)r^2 + (c/2 + d_1)r^4,$$

$$= C_{01} + C_{11}r^2 + C_{21}r^4, \tag{3.27}$$

where $r^2 = \sum_{i=1}^{k} x_i^2$ and C_{01}, C_{11}, C_{21} are as in above.

Example 3.5 A RSORD with 2 groups under tri-diagonal structure: An example of a RSORD with 2 factors, 2 groups, each of 9 observations under the tri-diagonal structure, is given below (following Method II). The RSORD (denoted by 'd_5') is displayed below (rows being factors and columns being runs).

d_5 :	1	2	3	4	5	6	7	8	9
$\mathbf{x}_1$	-1	1	-1	1	-1.414	1.414	0	0	0
$\mathbf{x}_2$	-1	-1	1	1	0	0	-1.414	1.414	0
d_5 :	10	11	12	13	14	15	16	17	18
$\mathbf{x}_1$	-1	1	-1	1	-1.414	1.414	0	0	0
$\mathbf{x}_2$	-1	-1	1	1	0	0	-1.414	1.414	0

3.8   ROBUST SECOND-ORDER DESIGNS FOR AUTOCORRELATED STRUCTURE

The present section studies second-order rotatable designs when errors have the autocorrelated structure given in (2.40).

3.8.1   Second-order rotatability conditions for autocorrelated structure

The second-order rotatability conditions under the autocorrelated structure (2.40) can be derived from (3.3). These are given in the following theorem.

Theorem 3.8 *The necessary and sufficient conditions of a robust second-order rotatable design under the autocorrelated variance-covariance structure, for all values of ρ in W_4 as in (2.40), in the model (3.1) are*

(I): (i) $v_{0.j} = 0 \Leftrightarrow \sum_{u=1}^{N} x_{uj} - \rho \sum_{u=2}^{N-1} x_{uj} = 0; 1 \leq j \leq k,$

$v_{0.jl} = 0 \Leftrightarrow \sum_{u=1}^{N} x_{uj} x_{ul} - \rho \sum_{u=2}^{N-1} x_{uj} x_{ul} = 0; 1 \leq j < l \leq k,$

(ii) $v_{i.j} = 0 \Leftrightarrow \sum_{u=1}^{N} x_{ui} x_{uj} + \rho^2 \sum_{u=2}^{N-1} x_{ui} x_{uj} - \rho \{ \sum_{u=1}^{N-1} x_{ui} x_{(u+1)j} + \sum_{u=1}^{N-1} x_{(u+1)i} x_{uj} \} = 0; 1 \leq i, j \leq k, \ i \neq j,$

(iii): (1) $v_{ii.j} = 0 \Leftrightarrow \sum_{u=1}^{N} x_{ui}^2 x_{uj} + \rho^2 \sum_{u=2}^{N-1} x_{ui}^2 x_{uj} - \rho \{ \sum_{u=1}^{N-1} x_{ui}^2 x_{(u+1)j} + \sum_{u=1}^{N-1} x_{(u+1)i}^2 x_{uj} \} = 0; 1 \leq i \leq k, \ 1 \leq j \leq k,$

(2) $v_{ij.l} = 0 \Leftrightarrow \sum_{u=1}^{N} x_{ui} x_{uj} x_{ul} + \rho^2 \sum_{u=2}^{N-1} x_{ui} x_{uj} x_{ul} - \rho (\sum_{u=1}^{N-1} x_{ui} x_{uj} x_{(u+1)l} +$
$\sum_{u=1}^{N-1} x_{(u+1)i} x_{(u+1)j} x_{ul}) = 0; 1 \leq i < j \leq k, 1 \leq l \leq k,$

(3) $v_{ii.jl} = 0 \Leftrightarrow \sum_{u=1}^{N} x_{ui}^2 x_{uj} x_{ul} + \rho^2 \sum_{u=2}^{N-1} x_{ui}^2 x_{uj} x_{ul} - \rho (\sum_{u=1}^{N-1} x_{ui}^2 x_{(u+1)j} x_{(u+1)l} +$
$\sum_{u=1}^{N-1} x_{(u+1)i}^2 x_{uj} x_{ul}) = 0; 1 \leq i \leq k, \ 1 \leq j < l \leq k,$

(4) $v_{ij.lt} = 0 \Leftrightarrow \sum_{u=1}^{N} x_{ui} x_{uj} x_{ul} x_{ut} + \rho^2 \sum_{u=2}^{N-1} x_{ui} x_{uj} x_{ul} x_{ut} - \rho (\sum_{u=1}^{N-1} x_{ui} x_{uj} x_{(u+1)l} x_{(u+1)t} +$
$\sum_{u=1}^{N-1} x_{(u+1)i} x_{(u+1)j} x_{ul} x_{ut}) = 0; 1 \leq i < j \leq k, \ 1 \leq l < t \leq k, \ (i,j) \neq (l,t)$

(II): (i) $v_{0.jj} = constant \Leftrightarrow \{\sigma^2(1 - \rho^2)\}^{-1}(1 - \rho)[\sum_{u=1}^{N} x_{uj}^2 - \rho \sum_{u=2}^{N-1} x_{uj}^2] = a_0(> 0); 1 \leq j \leq k,$

(ii) $v_{i.i} = constant \Leftrightarrow \{\sigma^2(1-\rho^2)\}^{-1}[\sum_{u=1}^{N} x_{ui}^2 + \rho^2 \sum_{u=2}^{N-1} x_{ui}^2 - 2\rho \sum_{u=1}^{N-1} x_{ui}x_{(u+1)i}] = \frac{1}{e}(> 0); 1 \le i \le k,$

(iii) $v_{ii.ii} = constant \Leftrightarrow \{\sigma^2(1-\rho^2)\}^{-1}[\sum_{u=1}^{N} x_{ui}^4 + \rho^2 \sum_{u=2}^{N-1} x_{ui}^4 - 2\rho \sum_{u=1}^{N-1} x_{ui}^2 x_{(u+1)i}^2] = (\frac{2}{c} + d)(> 0); 1 \le i \le k,$

(III): (i) $v_{ii.jj} = constant \Leftrightarrow \{\sigma^2(1-\rho^2)\}^{-1}[\sum_{u=1}^{N} x_{ui}^2 x_{uj}^2 + \rho^2 \sum_{u=2}^{N-1} x_{ui}^2 x_{uj}^2 - \rho(\sum_{u=1}^{N-1} x_{ui}^2 x_{(u+1)j}^2 + \sum_{u=1}^{N-1} x_{(u+1)i}^2 x_{uj}^2)] = d(> 0); 1 \le i, j \le k,\ i \ne j,$

(ii) $v_{ij.ij} = constant \Leftrightarrow \{\sigma^2(1-\rho^2)\}^{-1}[\sum_{u=1}^{N} x_{ui}^2 x_{uj}^2 + \rho^2 \sum_{u=2}^{N-1} x_{ui}^2 x_{uj}^2 - 2\rho \sum_{u=1}^{N-1} x_{ui}x_{uj} x_{(u+1)i}x_{(u+1)j}] = \frac{1}{c}(> 0); 1 \le i < j \le k,$

(IV): $v_{ii.ii} = 2v_{ij.ij} + v_{ii.jj}; 1 \le i < j \le k,$
where $v_{ii.ii}$, $v_{ii.jj}$ and $v_{ij.ij}$ are as in **(II)** *(iii) and* **(III)** *(i), (ii) in (3.28).*
The condition for non-singularity is given by (3.28)

(V): $\frac{2}{c} + k(d - \frac{a_0^2}{v_{00}}) > 0,$ (3.29)

where $v_{00} = \frac{\{N-(N-2)\rho\}}{\sigma^2(1+\rho)}$ and $\frac{1}{c}$, d, a_0 are as in (3.28).

Note that the conditions for rotatability, as stipulated above, are hard to satisfy in practice since ρ is unknown unless special efforts are made regarding choice of the underlying *robust* designs. Towards this, a remarkable simplification obtains whenever *all* odd moments of various types vanish. The moments which should all equal to 0 (zero) are listed below for ready reference (the divisors are easy to sort out and are omitted for brevity):

$$\sum_{u=1}^{N} x_{uj}, \sum_{u=2}^{N-1} x_{uj}, \sum_{u=1}^{N} x_{ui}x_{uj}, \sum_{u=2}^{N-1} x_{ui}x_{uj}, \sum_{u=1}^{N-1} x_{ui}x_{(u+1)j}, \sum_{u=1}^{N-1} x_{(u+1)i}x_{uj},$$

$$\sum_{u=1}^{N} x_{ui}^2 x_{uj}, \sum_{u=2}^{N-1} x_{ui}^2 x_{uj}, \sum_{u=1}^{N-1} x_{ui}^2 x_{(u+1)j}, \sum_{u=1}^{N-1} x_{(u+1)i}^2 x_{uj}, \sum_{u=1}^{N} x_{ui}x_{uj}x_{ul}, \sum_{u=2}^{N-1} x_{ui}x_{uj}x_{ul},$$

$$\sum_{u=1}^{N-1} x_{ui}x_{uj}x_{(u+1)l}, \sum_{u=1}^{N-1} x_{(u+1)i}x_{(u+1)j}x_{ul}, \sum_{u=1}^{N} x_{ui}^2 x_{uj}x_{ul},$$

$$\sum_{u=2}^{N-1} x_{ui}^2 x_{uj}x_{ul}, \sum_{u=1}^{N-1} x_{ui}^2 x_{(u+1)j}x_{(u+1)l}, \sum_{u=1}^{N-1} x_{(u+1)i}^2 x_{uj}x_{ul}, \sum_{u=1}^{N} x_{ui}x_{uj}x_{ul}x_{ut},$$

$$\sum_{u=2}^{N-1} x_{ui}x_{uj}x_{ul}x_{ut}, \sum_{u=1}^{N-1} x_{ui}x_{uj}x_{(u+1)l}x_{(u+1)t},$$

$$\sum_{u=1}^{N-1} x_{(u+1)i}x_{(u+1)j}x_{ul}x_{ut}; 1 \le i < j < l \le k;\ 1 \le l < t \le k,\ (i,j) \ne (l,t).$$

As to the even order moments (up to order 4) mere constancy of such moments (of various types) suffices to ensure exact rotatability. Specially, we desire to have each of the following terms independent of i and j:

$$\sum_{u=1}^{N} x_{ui}^2, \; \sum_{u=2}^{N-1} x_{ui}^2, \; \sum_{u=1}^{N} x_{ui}^4, \; \sum_{u=2}^{N-1} x_{ui}^4, \; \sum_{u=1}^{N-1} x_{ui}x_{(u+1)i}, \; \sum_{u=1}^{N-1} x_{ui}^2 x_{(u+1)i}^2, \; \sum_{u=1, i\neq j}^{N} x_{ui}^2 x_{uj}^2,$$

$$\sum_{u=2, i\neq j}^{N-1} x_{ui}^2 x_{uj}^2, \; \sum_{u=1, i\neq j}^{N-1} x_{ui}^2 x_{(u+1)j}^2, \; \sum_{u=1, i\neq j}^{N-1} x_{(u+1)i}^2 x_{uj}^2, \; \sum_{u=1, i\neq j}^{N-1} x_{ui}x_{uj}x_{(u+1)i}x_{(u+1)j};$$

$1 \leq i, j \leq k, \; i \neq j.$

Thus, a design satisfying the conditions mentioned above may be successfully utilized to get a rotatable design when observations violate uncorrelatedness assumption of errors subject to the stipulation that the covariance structure is of the type (2.40). Such a design will be *robust* in the sense described.

3.8.2   RSORD construction for autocorrelated structure

The present section discusses a construction method of a RSORD under the autocorrelated error variance-covariance structure as in (2.40). The designs obtainable by this method satisfy the moment conditions stipulated above along with other conditions required for rotatability, viz., (3.28) and (3.29).

Description of the method (Das, 1997): We start with a usual SORD having n non-central design points involving k factors. The set of n design points can be extended to $(2n+1)$ points by incorporating $(n+1)$ central points in the following way. One central point is placed in between each pair of non-central design points in the sequence, utilizing thus $(n-1)$ such central points. Of the other two central points one is placed at the beginning and one at the end. The number of central points of the usual SORD with which we started may be different from the number of central points required in the design so constructed in the autocorrelated error variance-covariance structure situation.

Features of the method:

(a) It can be readily verified that the above simplified moment conditions hold for these designs. Also conditions (3.28) hold as the original design is a usual SORD.

(b) To ensure non-singularity, we first examine the original design and the newly constructed design. n is the number of non-central points in the original design which is usual SORD. After the addition of $(n+1)$ central points the constructed design has, say, N_1 design points, where $N_1 = 2n + 1$. Assuming the original design with which we started contains N points in all with $(N - n)$ central points obviously, the moments of the constructed design will be as follows.

$$\sum_{u=1}^{N_1} x_{ui}^2 = N\lambda_2, \; \sum_{u=1}^{N_1} x_{ui}^4 = 3N\lambda_4, \; \sum_{u=1}^{N_1} x_{ui}^2 x_{uj}^2 = N\lambda_4, \; \sum_{u=2}^{N_1-1} x_{ui}^2 = N\lambda_2, \; \sum_{u=2}^{N_1-1} x_{ui}^4 = 3N\lambda_4,$$

$$\sum_{u=2}^{N_1-1} x_{ui}^2 x_{uj}^2 = N\lambda_4; \; 1 \leq i, j \leq k, \; i \neq j.$$

Here, N is the number of design points of the original design and λ_2 and λ_4 are respectively, its second and fourth moments. Using (3.28) and the above results, the design parameters of the newly constructed designs are the following:

$$a_0 = \frac{(1-\rho)^2 N\lambda_2}{\sigma^2(1-\rho^2)}; \; d = \frac{(1+\rho^2)N\lambda_4}{\sigma^2(1-\rho^2)}; \; \frac{1}{e} = \frac{N\lambda_2(1+\rho^2)}{\sigma^2(1-\rho^2)}; \; \frac{1}{c} = \frac{N\lambda_4(1+\rho^2)}{\sigma^2(1-\rho^2)}.$$

Using the above design parameters and noting $v_{00} = \frac{\{N_1 - (N_1 - 2)\rho\}(1-\rho)}{\sigma^2(1-\rho^2)}$, the expression $\frac{2}{c} + k(d - \frac{a_0^2}{v_{00}})$ simplifies to

$$\frac{1}{\sigma^2(1-\rho^2)}[(k+2)(1+\rho^2)N\lambda_4 - \frac{kN^2\lambda_2^2(1-\rho)^3}{\{N_1 - (N_1 - 2)\rho\}}].$$

Hence the non-singularity condition of the above design is

$$\frac{\lambda_4}{\lambda_2^2} > \frac{k}{k+2}[\frac{N(1-\rho)^3}{\{N_1 - (N_1 - 2)\rho\}(1+\rho^2)}]. \tag{3.30}$$

Let $f(N, N_1, \rho) = \frac{N(1-\rho)^3}{\{N_1 - (N_1 - 2)\rho\}(1+\rho^2)}$. It is readily seen that $0 \le f(N, N_1, \rho) \le \frac{2N}{N_1 - 1} = \frac{N}{n} \le 2$, $-1 \le \rho \le 1$. The design, to start with, is a SORD in the usual sense so that $\frac{\lambda_4}{\lambda_2^2} > \frac{k}{k+2}$. The constructed design is non-singular, if the number of central points in the original SORD is less than n as is the case in all possible situations.

(c) Variance Function: The variance function of a RSORD under the autocorrelated structure constructed by the above method is obtained by following (3.7) and the above design parameters, and noting the following:

$$v_{00} = \frac{\{N_1 - (N_1 - 2)\rho\}(1-\rho)}{\sigma^2(1-\rho^2)}, (2/c + kd) = \frac{(k+2)(1+\rho^2)N\lambda_4}{\sigma^2(1-\rho^2)T}, a_1 = -\frac{(1-\rho)^2 N\lambda_2}{\sigma^2(1-\rho^2)T},$$

$$d_1 = \frac{(1-\rho)[N\lambda_2^2(1-\rho)^3 - \lambda_4(1+\rho^2)\{N_1 - (N_1 - 2)\rho\}]}{2\lambda_4(1+\rho^2)\{\sigma^2(1-\rho^2)\}T},$$

$$T = \frac{(1-\rho)N}{\{\sigma^2(1-\rho^2)\}^2}[\lambda_4(k+2)\{N_1 - (N_1 - 2)\rho\}(1+\rho^2) - kN\lambda_2^2(1-\rho)^3],$$

$$v^{00} = (2/c + kd)/T = \frac{(k+2)(1+\rho^2)(1-\rho^2)\lambda_4\sigma^2}{(1-\rho)T_1} = A(N, N_1, k, \rho), \text{say}$$

$$(2a_1 + e) = B(N, N_1, k, \rho), \text{say}$$

$$= \frac{\sigma^2(1-\rho^2)[\lambda_4(k+2)(1+\rho^2)\{N_1 - (N_1 - 2)\rho\} - N\lambda_2^2(1-\rho)\{k(1-\rho)^2 + 2(1+\rho^2)\}]}{N\lambda_2(1+\rho^2)T_1},$$

$$(c/2 + d_1) = C(N, N_1, k, \rho), \text{say}$$

$$= \frac{\sigma^2(1-\rho^2)[\lambda_4(k+1)(1+\rho^2)\{N_1 - (N_1 - 2)\rho\} - N\lambda_2^2(k-1)(1-\rho)^3]}{2N\lambda_4(1+\rho^2)T_1},$$

$$T_1 = [\lambda_4(k+2)(1+\rho^2)\{N_1 - (N_1 - 2)\rho)\} - kN\lambda_2^2(1-\rho)^3].$$

Therefore, the variance function is given by (following (3.7))

$$\text{Var}(\hat{y}_\mathbf{x}) = v^{00} + (2a_1 + e)r^2 + (c/2 + d_1)r^4$$

$$= A(N, N_1, k, \rho) + B(N, N_1, k, \rho)r^2 + C(N, N_1, k, \rho)r^4,$$

where $r^2 = \sum_{i=1}^{k} x_i^2$ and $A(N, N_1, k, \rho)$, $B(N, N_1, k, \rho)$, $C(N, N_1, k, \rho)$ are as in above.

An example is given below of a design obtained by the above method.

Example 3.6 We start with a central composite SORD with 3 factors and 20 design points consisting of 14 non-central and 6 central design points. Following the above method in Section 3.8.2, we construct a RSORD under the autocorrelated structure (2.40) with 3 factors and 29 design points. The design (denoted by d_o) is displayed below for ready reference (rows being factors and column being runs).

d_0 :	1	2	3	4	5	6	7	8	9	10	11	12	13	14	15	16	17
x_1	0	-1	0	1	0	-1	0	1	0	-1	0	1	0	-1	0	1	0
x_2	0	-1	0	-1	0	1	0	1	0	-1	0	-1	0	1	0	1	0
x_3	0	-1	0	-1	0	-1	0	-1	0	1	0	1	0	1	0	1	0

d_0 :	18	19	20	21	22	23	24	25	26	27	28	29
x_1	-1.682	0	1.682	0	0	0	0	0	0	0	0	0
x_2	0	0	0	0	-1.682	0	1.682	0	0	0	0	0
x_3	0	0	0	0	0	0	0	0	-1.682	0	1.682	0

3.9   CONCLUDING REMARKS

The present chapter has determined the second-order rotatability conditions for correlated errors, but the correlation structure is known and of some simple type usually encountered in practice. Second-order rotatability has been examined for intra-class, inter-class, compound symmetry structure, tri-diagonal and autocorrelated error structures. Several construction methods of robust second-order rotatable designs are developed under the above correlation structures. Only a few construction methods of RSORDs are provided as examples. Of course there may be many more construction methods of RSORDs which will possibly be developed in future. Nonspecific general results are stated for optimum robust second-order rotatable designs. For intra-class correlation structure only, specific conditions optimum robust second-order rotatable designs are examined, but for other correlation structures, optimum robust second-order rotatable designs (ORSORDs) are not studied in details in the present monograph. The concluding remarks made in the second paragraph of Section 2.9 hold good for this chapter too.

Chapter 4

ROBUST REGRESSION DESIGNS FOR NON-NORMAL DISTRIBUTIONS

In Chapters 2 and 3, robust first- and second-order designs are examined assuming the response (y) follows Normal distribution. In the present chapter, robust first- and second-order regression designs are considered for two non-normal distributions (i.e., log-normal and exponential). For example, in quality engineering, the most commonly used lifetime response distributions are log-normal and exponential. For lifetime improvement experiments, generally, design of experiments are mostly used in locating the optimal operating conditions of the process parameters. In the present chapter, D-optimal robust first-order and robust second-order rotatable designs are described for correlated lifetime responses having log-normal and exponential distributions. Robust designs are described for some well-known error variance covariance structures as mentioned in Chapter 2. It is shown that robust first-order D-optimal designs for these two lifetime distributions are always robust rotatable, but the converse is *not* true.

4.1   INTRODUCTION AND OVERVIEW

It is known to be difficult to derive optimal designs for linear models with correlated errors. For some cases, the D-optimal design may not exist. Bischoff (1995a) has proved two determinant formulas which are useful for deriving optimal or efficient designs with respect to the D-criterion. Bischoff (1996) pointed that D-optimal design does not exist for linear models with some correlated error structures, and for the tri-diagonal structure he derived *maximin designs*. Kiefer and Wynn (1981) suggested seeking optimal designs only in the class of uncorrelated and homoscedastic errors. For special factorial linear models with correlated errors, optimal designs are known (see, for example, Kiefer and Wynn, 1981, 1984). Gennings *et al.* (1989) studied optimality under non-linear models with correlated errors. Little is known on optimal designs of linear regression models for correlated errors (Bischoff, 1992, 1995b, 1996). Recently, Das and Park (2008a) derived *efficient robust rotatable* designs for autocorrelated error structure instead of D-optimal designs, and they (2008b) developed D-optimal robust first-order designs under tri-diagonal correlation structure with lag n. Improvement of performance of a system plays an important role in reliability theory. Statistical design of experiments have been popularly used in reliability theory in order to improve the quality of a system (see, for example, Taguchi, 1986, 1987; Condra, 1993).

In response surface methodology (RSM), when we are remote from the optimum operating conditions for the system, we usually assume that a first-order model is an adequate approximation to the true surface in a small region of the explanatory variables x's. If there is a curvature in the system, then a polynomial of higher degree, such as the second-order model must be used. The method of steepest ascent helps us for moving sequentially along the path of steepest ascent, that is, the direction of maximum increase in response through sequentially selecting an appropriate response function. A practitioner is experimenting with a system in which the goal is not to find a *point of optimum response*, but to search for a new *region* in which the process or product is improved (see, for example, Draper and Lin, 1996).

In reliability theory, the random variable under consideration is the lifetime of a system. A question of fundamental importance is how to improve the mean lifetime or reliability of the system for a given mission time. Mean lifetime is conveniently written as a function of the exploratory variables. For exponential lifetime distribution a similar study was done by Mukhopadhyay *et al.* (2002). Aitkin and Clayton (1980) fitted complex censored survival data to exponential, Weibull and extreme value distributions. Design of experiments and concepts of optimality have been abundantly used in reliability theory, in order to improve the mean lifetime. The conventional response surface designs are *not* appropriate in lifetime distributions, because they do not satisfy the assumptions of lifetime distributions (i.e., non-linearity, dependence of errors, non-normality, etc.).

In the present chapter, the necessary and sufficient conditions of first-order D-optimality and second-order rotatability are described for log-normal and exponential lifetime distributions with a general correlated error structure; specifically, compound symmetry, inter-class, intra-class and tri-diagonal correlation structures. For each lifetime distribution (log-normal or exponential) and for each correlation structure, construction of first-order D-optimal and second-order rotatable designs are described.

4.2   CORRELATED ERROR MODELS FOR NON-NORMAL DISTRIBUTIONS

In the usual response surface models, error is assumed to be normally distributed, consequently, the response follows Normal distribution. Recently, Das and Lin (2011) and Das and Huda (2011) studied robust first-order designs for log-normal and exponential distributions. In the present chapter, it is assumed that the response distribution is log-normal or exponential, which is observed in lifetime distribution. Assuming the response follows log-normal or exponential distribution with correlated errors, two linear models are displayed below.

4.2.1   Correlated error models for log-normal distribution

Let T be the lifetime of a component or a system, measured in some unit, which follows log-normal or exponential distribution. Let $x_1, x_2,...,x_k$ be k controllable explanatory variables in the system which are highly related with lifetime T. The log-normal probability density function (p.d.f.) of lifetime T given the vector $\mathbf{x} = (x_1, x_2, \ldots, x_k)'$, is of the form

$$f(t) = \frac{1}{t\,\delta\,(2\pi)^{1/2}}\ \exp[-\ \frac{(\ln t - \ln h(\mathbf{x}))^2}{2\delta^2}];\ t \geq 0,\ \delta \geq 0. \tag{4.1}$$

Therefore, $\ln T$ is assumed to follow Normal distribution with mean $\ln h(\mathbf{x})$ and variance 'δ^2' for a given $\mathbf{x}$. In fact that δ is independent of $\mathbf{x}$ implies proportional hazards for lifetimes and constant variance for log lifetimes of individuals. For more details of such a popular model and its applications, see, Lieblein and Zelen (1956), Kao (1959), Pike (1966), Nelson (1972), Peto and Lee (1973), Lawless (1982), and Meeker and Escobar (1998).

The probability density function of the log lifetime Y^* $(= \ln T)$, given $\mathbf{x}$, is a Normal distribution with mean $\ln h(\mathbf{x})$ and variance δ^2. Take $\ln h(\mathbf{x}) = g(\mathbf{x}, \boldsymbol{\beta})$, thus, $E(T) = h(\mathbf{x})\ \exp(\frac{\delta^2}{2}) = e^{g(\mathbf{x},\boldsymbol{\beta})}\ \exp(\frac{\delta^2}{2}) > 0$, where $\boldsymbol{\beta}$ is a vector of unknown coefficients. In general, $g(\mathbf{x}, \boldsymbol{\beta})$ is usually taken to represent a polynomial in some well defined factors, the number of which may be denoted by p. We take $g(\mathbf{x}, \boldsymbol{\beta})$ to be a linear $(p = k + 1)$ or a second-order $(p = \binom{k+2}{2})$ function in k factors. A well laid out experimental plan which in the present case is a D-optimal (for first-order model) or rotatable (for second-order model) robust design may be utilized for realizing the data on T for N given values of $\mathbf{x}$ which indicate N different operating conditions under the plan.

Assuming that the response surface $g(\mathbf{x}, \boldsymbol{\beta})$ is of first-order, we adopt the model:

$$\ln T = \beta_0 + \beta_1 x_1 + \beta_2 x_2 + \ldots\ldots + \beta_k x_k + \delta\tau,$$

where τ follows the Standard Normal distribution with

$$E(\tau) = 0;\quad \text{Var}(\tau) = 1.$$

Take $(\ln T)$ as the response variable. For estimating the parameters $\boldsymbol{\beta} = (\beta_0, \beta_1, \ldots, \beta_k)'$ we have to conduct an experiment to collect the data on the basis of which the estimation to be done. Once an experiment is conducted, the experiment introduces some noise factors which may be numerous, some may be unidentifiable. The total impact on $(\ln T)$ of all these noise factors represented by the experimental condition is denoted by e. Consider the response surface $g(\mathbf{x}, \boldsymbol{\beta})$ is of the first-order. We write $y = \ln T + e = \beta_0 + \beta_1 x_1 + \ldots + \beta_k x_k + \delta\tau + e$, or

$$y_u = \beta_0 + \beta_1 x_{u1} + \beta_2 x_{u2} + \ldots + \beta_k x_{uk} + \delta\tau_u + e_u;\ 1 \leq u \leq N, \tag{4.2}$$

where β_i's are unknown regression parameters and x_{ui}'s are experimental levels which are non-stochastic, τ_u follows Standard Normal distribution, and e_u's represent experimental error given rise to by the noise factors. Note that there are two random variables 'τ' and e in the model (4.2). Some authors (for example, Aalen, 1988, 1989, 1992, 1994) studied the *random effects* e in the frailty model in survival analysis. The use of random effects in the survival analysis models is completely different issues. All the earlier authors used response surface designs for improving the quality of a unit without considering the random term $\delta\tau$ in the regression model (4.2), which arises from the distribution of lifetime T. Thus, the model (4.2) is completely different from the usual response surface model.

It is assumed that τ_u's are all uncorrelated and τ_u and e_u are uncorrelated. e_u's represent the experimental errors with expectation, $E(\mathbf{e}) = \mathbf{0}$ and the dispersion matrix, $\mathrm{Dis}(\mathbf{e}) = \sigma_1^2 W_1$, where $\mathbf{e}$ is the vector of errors e's. No mention has so far been made about the distributional assumption of $\mathbf{e}$. The most convenient distributional choice of $\mathbf{e}$, the vector of errors e_u's, is a Multivariate Normal distribution with $E(\mathbf{e}) = \mathbf{0}$, $\mathrm{Dis}(\mathbf{e}) = \sigma_1^2 W_1$, and rank $(W_1) = N$, where W_1 is any unknown general variance covariance structure of errors. The choice of the experimental levels x_{ui}'s, i.e., $X = ((x_{ui}); 1 \leq u \leq N, \ 1 \leq i \leq k)$, the design matrix is the most important part for getting maximum information regarding the unknown regression parameters. The prescription of the proper design matrix is a problem of regression design of experiments (See Box and Hunter, 1957; Pukelsheim 1993; Khuri and Cornell, 1996; Box and Draper, 2007; Panda and Das, 1994; Das, 1997, 2003a, 2004; Das and Park, 2008a, 2008b).

4.2.2 Correlated error models for exponential distribution

The exponential probability density function (p.d.f.) of lifetime T given the vector $\mathbf{x} = (x_1, x_2, \ldots, x_k)'$, is of the form

$$f(t) \ = \ \frac{1}{\delta(\mathbf{x})} \ \exp[-\frac{t}{\delta(\mathbf{x})}]; \ t \geq 0, \ \delta(\mathbf{x}) \geq 0. \tag{4.3}$$

Note that $E(T) = \delta(\mathbf{x})$ and $\mathrm{Var}(T) = \{\delta(\mathbf{x})\}^2$. Here $\delta(\mathbf{x})$ depends on $\mathbf{x}$. Thus, both the mean and variance are functions of $\mathbf{x}$. Let us take $\delta(\mathbf{x}) = \exp[g(\mathbf{x}, \boldsymbol{\beta})] = e^R$ say, where $R = g(\mathbf{x}, \boldsymbol{\beta})$ (as in Section 4.2.1) and $\boldsymbol{\beta}$ is a vector of unknown regression coefficients.

The probability density function of the log lifetime $Y_1 = \ln T$, given $\mathbf{x}$ is

$$f_1(y_1) = \exp(y_1 - R) \ \exp[-\exp(y_1 - R)]; \ -\infty \ < y_1 < \ \infty, \tag{4.3}$$

which is an Extreme value distribution.

Let us transform $h = Y_1 - R$. Therefore, the probability density function of h, given $\mathbf{x}$ is

$$f_2(h) = \exp(h) \ \exp[-\exp(h)]; \ -\infty \ < h < \ \infty, \tag{4.4}$$

which is the standard extreme value distribution. Note that $E(h) = -\nu$, where $\nu = 0.5772...$, known as Euler's constant and $\mathrm{Var}(h) = \pi^2/6$.

Now it is assumed that given $\mathbf{x}$, T follows an exponential distribution in an *ideal situation* with mean $e^{g(\mathbf{x}, \boldsymbol{\beta})}$. This is the same as assuming $\ln T = g(\mathbf{x}, \boldsymbol{\beta}) + h$, where h follows the standard extreme value distribution. Assuming that the response surface $g(\mathbf{x}, \boldsymbol{\beta})$ is of first-order, we adopt the model:

$$\ln T = \beta_0 \ + \beta_1 x_1 + \beta_2 x_2 + \ldots \ldots + \beta_k x_k + h,$$

where h follows the standard extreme value distribution whose mean and variance are given above. Take $(\ln T)$ as the response variable. Similarly, for the model (4.2), the above model can be written as

$$y^* \ = \ln T \ + e = \ \beta_0 + \beta_1 x_1 + \beta_2 x_2 + \ldots + \beta_k x_k + h + e,$$

$$\text{or,} \ \ y_u^* \ = \ \beta_0 + \beta_1 x_{u1} + \beta_2 x_{u2} + \ldots + \beta_k x_{uk} + h_u + e_u; \ 1 \leq u \leq N, \tag{4.5}$$

where β_i's are unknown regression parameters and x_{ui}'s are experimental levels which are non-stochastic, h follows standard extreme value distribution, and e_u's represent experimental error given rise to by the noise factors. We assume that h_u's are all uncorrelated and h_u and e_u are uncorrelated. Similarly, for the model (4.2), $\mathbf{e}$ follows Multivariate Normal distribution with $E(\mathbf{e}) = \mathbf{0}$, $\text{Dis}(\mathbf{e}) = \sigma_1^2 W_1$, and rank $(W_1) = N$, where W_1 is any unknown general error variance covariance structure. In all calculations, W_1 needs to be known.

There are no such optimal or nearly optimal designs that exist in literature as in the situation mentioned above. So we need to describe such designs which are more appropriate in this situation. Little is known in literature for the specific situation of log-normal (Das and Lin, 2011) and exponential (Das and Huda, 2011) lifetime distributions. In Section 4.4 we describe some D-optimal robust first-order designs for some special error correlation structures W_1 for the situation as mentioned herein.

4.3   ROBUST FIRST-ORDER DESIGNS FOR LOG-NORMAL AND EXPONENTIAL DISTRIBUTIONS

For log-normal lifetime distribution, the model as explained in (4.2) can be written as

$$y_u = \beta_0 + \beta_1 x_{u1} + \ldots + \beta_k x_{uk} + \epsilon_u; \ 1 \le u \le N, \quad \text{say,}$$

$$\text{or,} \quad \mathbf{Y} = X\boldsymbol{\beta} + \boldsymbol{\epsilon}, \tag{4.6}$$

where $\epsilon_u = \delta \tau_u + e_u$, $\mathbf{Y} = (y_1, y_2, \ldots, y_N)'$, is the vector of recorded observations, $X = (1 : (x_{ui}); 1 \le u \le N, 1 \le i \le k)$ is the design (or model) matrix; $\boldsymbol{\beta} = (\beta_0, \beta_1, \ldots, \beta_k)'$ is the vector of regression coefficients and $\boldsymbol{\epsilon} = (\epsilon_1, \epsilon_2, \ldots, \epsilon_N)'$. Then $E(\epsilon_u) = 0$, $\text{Var}(\epsilon_u) = (\delta^2 + \sigma_1^2) = \sigma^2$ say, $\text{Cov}(\epsilon_u, \epsilon_{u'}) = \text{Cov}(e_u, e_{u'})$; $E(\boldsymbol{\epsilon}) = \mathbf{0}$, $\text{Dis}(\boldsymbol{\epsilon}) = \delta^2 I_N + \sigma_1{}^2 W_1 = \sigma^2 W$ say, and $\boldsymbol{\epsilon} \sim N(\mathbf{0}, \ \sigma^2 W)$, where W is an $N \times N$ matrix whose $(i,j)th$ element $(i \ne j)$ is $\text{Cov}(e_i, e_j)/\sigma^2$ and all diagonal elements are unity. Note that if $\text{Cov}(e_i, e_j) = 0$ for all $i \ne j$, then $\text{Dis}(\boldsymbol{\epsilon}) = \sigma^2 I_N$.

For exponential lifetime distribution, the model as explained in (4.5) is

$$y_u^* = \beta_0 + \beta_1 x_{u1} + \beta_2 x_{u2} + \ldots + \beta_k x_{uk} + h_u + e_u; \ 1 \le u \le N.$$

Let us consider

$$y_u = y_u^* + \nu = \beta_0 + \beta_1 x_{u1} + \beta_2 x_{u2} + \ldots + \beta_k x_{uk} + \epsilon_u; \ 1 \le u \le N,$$

$$\text{or,} \quad \mathbf{Y} = X\boldsymbol{\beta} + \boldsymbol{\epsilon}, \tag{4.7}$$

where $\epsilon_u = \nu + h_u + e_u$, $y_u = y_u^* + \nu$ $(-\nu = E(h)$ and ν is known as Euler's constant); $\mathbf{Y} = (y_1, y_2, \ldots, y_N)'$, is the vector of recorded observations, $X = (1 : (x_{ui}); 1 \le u \le N; 1 \le i \le k)$ is the design (or model) matrix; $\boldsymbol{\beta} = (\beta_0, \beta_1, \ldots, \beta_k)'$ is the vector of regression coefficients and $\boldsymbol{\epsilon} = (\epsilon_1, \epsilon_2, \ldots, \epsilon_N)'$. Then $E(\epsilon_u) = 0$, $\text{Var}(\epsilon_u) = (\pi^2/6 + \sigma_1^2) = \sigma^2$ say, $\text{Cov}(\epsilon_u, \epsilon_{u'}) = \text{Cov}(e_u, e_{u'})$; $E(\boldsymbol{\epsilon}) = \mathbf{0}$, $\text{Dis}(\boldsymbol{\epsilon}) = (\pi^2/6) I_N + \sigma_1{}^2 W_1 = \sigma^2 W$ say, where

$$W = \begin{pmatrix} 1 & \frac{\text{Cov}(e_1, e_2)}{\sigma^2} & \frac{\text{Cov}(e_1, e_3)}{\sigma^2} & \cdots & \frac{\text{Cov}(e_1, e_N)}{\sigma^2} \\ \frac{\text{Cov}(e_2, e_1)}{\sigma^2} & 1 & \frac{\text{Cov}(e_2, e_3)}{\sigma^2} & \cdots & \frac{\text{Cov}(e_2, e_N)}{\sigma^2} \\ \cdots & \cdots & \cdots & & \cdots \\ \frac{\text{Cov}(e_N, e_1)}{\sigma^2} & \frac{\text{Cov}(e_N, e_2)}{\sigma^2} & \frac{\text{Cov}(e_N, e_3)}{\sigma^2} & \cdots & 1 \end{pmatrix}.$$

Note that if $\text{Cov}(e_i, e_j) = 0$ for all $i \ne j$, $\text{Dis}(\boldsymbol{\epsilon}) = \sigma^2 I_N$, i.e., $W = I_N$. In general, the matrix W is unknown, but for all the calculations W is assumed to be known. In practice, however, W includes a number of unknown parameters, and in the calculations which follow, the expressions for W and W^{-1} are replaced by those obtained after replacing the unknown parameters with suitable estimates or some assumed values.

For both the models (4.6) and (4.7), the best linear unbiased estimator (BLUE) of β, for known W is $\hat{\beta} = (X'W^{-1}X)^{-1} (X'W^{-1}\mathbf{Y})$, with

$$\text{Dis}(\hat{\beta}) = \sigma^2(X'W^{-1}X)^{-1} = \sigma^2(\{v_{ij}\})^{-1} \text{ say, } 0 \leq i, j \leq k, \tag{4.8}$$

where $v_{00} = \mathbf{1}'W^{-1}\mathbf{1}$, $v_{0j} = \mathbf{1}'W^{-1}\mathbf{x}_j$, $v_{ij} = \mathbf{x}_i'W^{-1}\mathbf{x}_j$ and $\mathbf{x}_i = (x_{1i}, x_{2i}, ..., x_{Ni})'$.

In this section, we are interested in finding regression designs under the correlated regression models (4.6) and (4.7) such that variance of the estimated response at $\mathbf{x}$ is optimum in certain sense, for a wide variety of the dispersion matrix W. The objective of the D-optimality criteria is to *minimize* $| (X'W^{-1}X)^{-1} |$, or equivalently *maximize* $| X'W^{-1}X |$ where $(X'W^{-1}X)$ is known as the moment matrix. Some relevant definitions (similar as in Chapter 2) for the model (4.6) or (4.7) are given below.

Definition 4.1 Rotatable design: A design is said to be *rotatable* if the variance of the estimated response of the correlated lifetime model (4.6) or (4.7) at a point is a function of only the distance from the design center (i.e., center of the coordinate axes, or at $(0, 0, \ldots, 0)$) to that point.

Definition 4.2 Robust first-order rotatable design: A design D on k factors under the correlated lifetime model (4.6) or (4.7) which remains first-order rotatable for *all* the variance-covariance matrices belonging to a well-defined class $W_0 = \{W$ positive definite: $W_{N\times N}$ defined by a particular correlation structure possessing a definite pattern$\}$ is called a robust first-order rotatable design (RFORD). Note that W involves correlation parameter (ρ) or parameters $(\rho_i\text{'s})$, with reference to the variance-covariance class W_0.

Definition 4.3 D-optimal robust first-order design: A first-order regression design ξ of k factors in the correlated lifetime model (4.6) or (4.7) is said to be D-optimal robust first-order design (D-ORFOD) if the determinant $| M(\xi) | = | X'W^{-1}X |$ is uniformly maximum (over the design space $\mathcal{X} = \{ | x_{ui} | \leq 1; 1 \leq u \leq N, 1 \leq i \leq k\}$) for *all* the variance-covariance matrices belonging to a well defined class W_0 as in Definition 4.2, with reference to the variance-covariance class W_0.

$$\text{Note that } | M(\xi) | \leq \prod_{i=0}^{k} v_{ii}, \tag{4.9}$$

and equality holds if and only if $v_{0i} = 0$ and $v_{ij} = 0$; $1 \leq i \neq j \leq k$, and v_{ij}'s are as in (4.8) (proof is given in Chapter 2). The inequality in (4.9) is known as Hadamard's inequality (Anderson, 1984, p. 54). Hence we have the following theorem (the proof follows directly from (4.9) and Definition 4.3).

Theorem 4.1 *The necessary and sufficient conditions for a D-optimal robust first-order design in the correlated lifetime model (4.6) or (4.7) are (for all $1 \leq j \leq k$)*

(i) $v_{0j} = \mathbf{1}'W^{-1}\mathbf{x}_j = 0$,

(ii) $v_{ij} = \mathbf{x}_i'W^{-1}\mathbf{x}_j = 0, \quad i \neq j$;

(iii) $v_{ii} = \mathbf{x}_i'W^{-1}\mathbf{x}_i = \mu$ *say*, $\tag{4.10}$
for the design matrix $X \in \mathcal{X}$ where μ is the maximum possible value (constant) over the design space $\mathcal{X}$ as in Definition 4.3.

The variance of the estimated response $\hat{y}_\mathbf{x}$ (for the correlated lifetime model (4.6) or (4.7)) at $\mathbf{x}$ of a first-order D-optimal design is given by

$$V(\hat{y}_\mathbf{x}) = \frac{1}{v_{00}} + \frac{1}{\mu} \sum_{i=1}^{k} x_i^2 = f(r^2) \text{ say}, \tag{4.11}$$

where $r^2 = \sum_{i=1}^{k} x_i^2$, $v_{00} = \mathbf{1}'W^{-1}\mathbf{1}$ and μ as in (4.10). Following (4.11), we have the following theorem.

Theorem 4.2 *A D-optimal robust first-order regression design for the correlated error lifetime model (4.6) or (4.7) is always a robust first-order rotatable design but the converse is not true.*

The proof follows directly from (4.10) and (4.11). If v_{ii} is constant $= \lambda$ ($< \mu$), $1 \leq i \leq k$, where μ is as in (4.10), then the design is a robust first-order rotatable but not a D-optimal robust first-order design.

Note that the model (4.6) or (4.7) looks similar to the model (2.1) (Chapter 2) but the differences exist between the observational vectors ($\mathbf{Y}$) and the random errors ($\mathbf{e}$ or ϵ) for these models. These differences can be noted from the models (2.1), (4.6) and (4.7). But the results and the designs of the model (2.1) are *exactly* same to the model (4.6) or (4.7) for some correlated error structures (intra-class, inter-class, compound symmetry and tri-diagonal). Below we examine the necessary and sufficient conditions for first-order D-optimality, variance function and the corresponding D-optimal designs under the compound symmetry, inter-class, intra-class and tri-diagonal error variance-covariance structures for the models (4.6) and (4.7). For ready reference, only the necessary and sufficient conditions are reproduced.

4.4 ROBUST FIRST-ORDER DESIGNS FOR TWO NON-NORMAL DISTRIBUTIONS

In the present section, the necessary and sufficient conditions of robust first-order D-optimality for the correlated lifetime model (4.6) or (4.7) and the variance function are described under the compound symmetry, inter-class, intra-class and tri-diagonal error structures.

4.4.1 Compound symmetry correlation structure

Compound symmetry variance covariance matrix (2.23) (Chapter 2) will have such a pattern when the observations are divided into m sets of size n each such that every observation has the *same* variance, and within each set of observations, the covariances are equal and the covariance between two observations from two distinct sets is a constant which may be different from the constant intra-class covariance in a set. For ready reference, this variance covariance matrix of errors is reproduced below

$$\text{Dis}(\mathbf{e}) = \sigma_1^2[I_m \otimes (A-B) + E_{m \times m} \otimes B] = \sigma_1^2 W_1(\rho^*, \rho_1^*), \tag{4.12}$$

where $A = (1-\rho^*)I_n + \rho^* E_{n \times n}$, $B = \rho_1^* E_{n \times n}$, I_n indicates an identity matrix of order n, $E_{n \times n}$ is an $n \times n$ matrix with all elements 1, $N = mn$ and $\otimes$ denotes Kronecker product.

Under the compound symmetry variance-covariance structure $\sigma_1^2 W_1(\rho^*, \rho_1^*)$, $\sigma^2 W$ (for log-normal distribution model (4.6), $\sigma^2 W = \delta^2 I_N + \sigma_1^2 W_1$) will be reduced to $\sigma^2 W_1(\rho, \rho_1)$, as in (4.12), where $\sigma^2 = \delta^2 + \sigma_1^2$, $\rho = q\rho^*$, $\rho_1 = q\rho_1^*$ and $q = \frac{\sigma_1^2}{\sigma_1^2 + \delta^2}$. Also, for exponential lifetime distribution model (4.7), $\sigma^2 W$ ($= (\pi^2/6)I_N + \sigma_1^2 W_1$) will be reduced to the same form as in (4.12) with $q = \frac{\sigma_1^2}{\sigma_1^2 + \pi^2/6}$ and $\sigma^2 = \pi^2/6 + \sigma_1^2$.

Also note that, $W_1^{-1}(\rho, \rho_1) = [I_m \otimes (A_1 - B_1) + E_{m \times m} \otimes B_1]$, where $A_1 = (\delta_2 - \gamma_2)I_n + \gamma_2 E_{n \times n}$, $B_1 = \delta_3 E_{n \times n}$, $\delta_2 = \gamma_2 + \frac{1}{(1-\rho)}$, $\gamma_2 = [(m-1)n\rho_1^2 - (m-2)n\rho\rho_1 - \{1 + (n-1)\rho\}\rho]/R$, $\delta_3 = \frac{\{1 - \delta_2 - (n-1)\rho\gamma_2\}}{(m-1)n\rho_1}$, and $R = (1-\rho)[(m-2)n\rho_1\{1 + (n-1)\rho\} + \{1 + (n-1)\rho\}^2 - (m-1)n^2\rho_1^2]$.

Following Theorem 4.1, the necessary and sufficient conditions for a D-optimal robust first-order design (D-ORFOD) for the lifetime model (4.6) or (4.7), under the variance-covariance structure $\sigma^2 W(\rho, \rho_1)$ can be summarized in Theorem 4.3 below.

Theorem 4.3 *A set of necessary and sufficient conditions for a D-optimal robust first-order regression design for the lifetime model (4.6) or (4.7) under the compound symmetry variance-covariance structure $\sigma^2 W_1(\rho, \rho_1)$ (as in (4.12)) as over the design space $\mathcal{X}$ as in Definition 4.3 are (for all $1 \leq i, j \leq k$)*

(i) $\displaystyle\sum_{u=1}^{N} x_{uj} = 0;$

(ii) $\displaystyle(\delta_2 - \gamma_2)\sum_{u=1}^{N} x_{ui}x_{uj} + (\gamma_2 - \delta_3)\{(\sum_{u=1}^{n} x_{ui})(\sum_{u=1}^{n} x_{uj}) + (\sum_{u=(n+1)}^{2n} x_{ui})(\sum_{u=(n+1)}^{2n} x_{uj}) +$

$\displaystyle\cdots + (\sum_{u=n(m-1)+1}^{mn} x_{ui})(\sum_{u=n(m-1)+1}^{mn} x_{uj})\} + \delta_3(\sum_{u=1}^{N} x_{ui})(\sum_{u=1}^{N} x_{uj}) = 0, \ i \neq j,$

(iii) if $(\gamma_2 - \delta_3) < 0$,

$\displaystyle\sum_{u=1}^{N} x_{ui}^2 = N; \sum_{u=1}^{n} x_{ui} = \sum_{u=(n+1)}^{2n} x_{ui} = \cdots = \sum_{u=n(m-1)+1}^{mn} x_{ui} = 0;$

(iv) if $(\gamma_2 - \delta_3) > 0$,

$\displaystyle\sum_{u=1}^{N} x_{ui}^2 = N; \sum_{u=1}^{n} x_{ui} = \sum_{u=(n+1)}^{2n} x_{ui} = \cdots = \sum_{u=n(m-1)+1}^{mn} x_{ui} = \pm n. \qquad (4.13)$

Note that $(\gamma_2 - \delta_3) > 0$ holds in the following two cases: Case I: $\rho_1 > \rho > 0$, $0 < \rho_1 \leq 0.5$; m, $n < 6$; and Case II: $0 < \rho_1 < \rho$; m, $n < 12$. The basic parameters characterizing the correlation structure (4.12) being ρ and ρ_1, the interpretation of $(\gamma_2 - \delta_3) > 0$ can be obtained from the above two cases only, and $(\gamma_2 - \delta_3) < 0$ everywhere else. Given below only the conditions either $(\gamma_2 - \delta_3) < 0$ or $(\gamma_2 - \delta_3) > 0$ will be considered.

The variance function of a D-ORFOD under the structure $\sigma^2 W(\rho, \rho_1)$ as in (4.12) is

$$V(\hat{y}_{\mathbf{x}})_{01} = \frac{\sigma^2}{N}[\frac{1}{\{\delta_2 + (n-1)\gamma_2 + n(m-1)\delta_3\}} + \frac{r^2}{\delta_2 - \gamma_2}], \text{ if } (\gamma_2 - \delta_3) < 0, \qquad (4.14)$$

$$V(\hat{y}_{\mathbf{x}})_{02} = \frac{\sigma^2}{N}[\frac{1}{\{\delta_2 + (n-1)\gamma_2 + n(m-1)\delta_3\}} + \frac{r^2}{(\delta_2 - \gamma_2) + (\gamma_2 - \delta_3)n}], \text{ if } (\gamma_2 - \delta_3) > 0, \qquad (4.15)$$

where $r^2 = \displaystyle\sum_{i=1}^{k} x_i^2$.

These above conditions (in (4.13)) are exactly same as in Chapter 2 for the compound symmetry structure. Hence, the same method of constructions of D-optimal robust first-order designs (in Section 2.6, Chapter 2) give the D-optimal designs for the lifetime models as in (4.6) or (4.7).

4.4.2   Inter-class correlation structure

Inter-class variance covariance structure (2.11) is given in Chapter 2. It is an extension of intra-class structure. This situation is observed if the observations are grouped into some groups such that within each group there is the same intra-class structure and between groups there is no correlation. Inter-class structure is actually a particular case of compound symmetry structure which is obtained from $W_1(\rho^*, \rho_1^*)$ as in (4.12), assuming $\rho_1^* = 0$ and is given below

$$\text{Dis}(\mathbf{e}) = \sigma_1^2[I_m \otimes A] = \sigma_1^2 W_1(\rho^*, 0), \qquad (4.16)$$

where $\otimes$ denotes Kronecker product, A is as in (4.12) and $N = mn$. Similarly as the variance-covariance structure $\sigma^2 W_1(\rho, \rho_1)$ (Section 4.4.1), $\sigma^2 W$ under the inter-class variance-covariance structure $\sigma_1^2 W_1(\rho^*, 0)$ as in (4.16) will be reduced to $\sigma^2 W_1(\rho, 0)$, where $\rho = \frac{\sigma_1^2}{\sigma_1^2 + \delta^2}\rho^*$ and $\sigma^2 = \sigma_1^2 + \delta^2$ (for log-normal distribution) or $\rho = \frac{\sigma_1^2}{\sigma_1^2 + \pi^2/6}\rho^*$ and $\sigma^2 = \sigma_1^2 + \pi^2/6$ (for exponential distribution).

The necessary and sufficient conditions for a D-optimal robust first-order regression design for the lifetime model (4.6) or (4.7) under the inter-class variance-covariance structure $\sigma^2 W_1(\rho, 0)$ (4.16) is given below.

Theorem 4.4 *A set of necessary and sufficient conditions for a D-optimal robust first-order regression design for the lifetime model (4.6) or (4.7) under the inter-class variance-covariance structure $\sigma^2 W_1(\rho, 0)$, over the design space $\mathcal{X}$ are (for all $1 \le i, j \le k$ and ρ in $W_1(\rho, 0)$)*

(i) $\displaystyle\sum_{u=1}^{N} x_{uj} = 0;$

(ii) $\displaystyle\alpha_0 \sum_{u=1}^{N} x_{ui}x_{uj} - \alpha_1\{(\sum_{u=1}^{n} x_{ui})(\sum_{u=1}^{n} x_{uj}) + \cdots + (\sum_{u=n(m-1)+1}^{mn} x_{ui})(\sum_{u=n(m-1)+1}^{mn} x_{uj})\} = 0, \; i \ne j;$

(iii) $\displaystyle\sum_{u=1}^{N} x_{ui}^2 = N; \sum_{u=1}^{n} x_{ui} = \sum_{u=(n+1)}^{2n} x_{ui} = \cdots = \sum_{u=n(m-1)+1}^{mn} x_{ui} = 0, \; \text{if } \rho > 0; \;\; and$

(iv) $\displaystyle\sum_{u=1}^{N} x_{ui}^2 = N; \sum_{u=1}^{n} x_{ui} = \sum_{u=(n+1)}^{2n} x_{ui} = \cdots = \sum_{u=n(m-1)+1}^{mn} x_{ui} = \pm n, \; \text{if } \rho < 0,$
   where $\alpha_0 = \frac{1}{(1-\rho)}$ *and* $\alpha_1 = \frac{\rho}{(1-\rho)\{1+(n-1)\rho\}}.$

Note that the robust first-order D-optimal designs in Section 2.4 (Chapter 2) give the D-optimal designs for the model (4.6) or (4.7) under the inter-class error structure.

4.4.3   Intra-class correlation structure

Intra-class structure (2.6) (Chapter 2) is the simplest variance-covariance structure which arises when errors of any two observations have the same correlation and each has the same variance. It is also known as uniform correlation structure. This happens when all the observations studied are from the same batch or from the same run in a furnace. This is actually a special case of inter-class structure when $m = 1$, as given below.

$$\text{Dis}(\mathbf{e}) = \sigma_1^2[(1 - \rho^*)I_N + \rho^* E_{N \times N}] = \sigma_1^2 W_1(\rho^*), \qquad (4.17)$$

where $\sigma_1 > 0$, and $-(N-1)^{-1} < \rho^* < 1$. Similarly as the variance-covariance structure in Sections 4.4.1 and 4.4.2, $\sigma^2 W$ under the intra-class variance-covariance structure $\sigma_1^2 W_1(\rho^*)$ as in (4.17) will be reduced to $\sigma^2 W_1(\rho)$, which is also an intra-class structure as in (4.17), where $\rho = \frac{\sigma_1^2}{\sigma_1^2 + \delta^2}\rho^*$ and $\sigma^2 = \sigma_1^2 + \delta^2$ (for log-normal distribution) or $\rho = \frac{\sigma_1^2}{\sigma_1^2 + \pi^2/6}\rho^*$ and $\sigma^2 = \sigma_1^2 + \pi^2/6$ (for exponential distribution).

Following Theorem 4.1, the necessary and sufficient conditions for D-optimal robust first-order regression design under the intra-class variance-covariance structure $\sigma^2 W_1(\rho)$ can be simplified to $v_{0j} = \sum_{u=1}^{N} x_{uj} = 0$; $v_{ij} = \sum_{u=1}^{N} x_{ui}x_{uj} = 0$, $i \neq j$; and $v_{ii} = \sum_{u=1}^{N} x_{ui}^2 = $ constant $= N = \mu$. Therefore, we have the following theorem.

Theorem 4.5 *A design is a D-optimal robust first-order for the lifetime model (4.6) or (4.7) under the intra-class structure if and only if it is a D-optimal first-order rotatable design in the usual model (i.e., when errors are uncorrelated and homoscedastic), whatever be the value of the intra-class correlation coefficient ρ in $W_1(\rho)$.*

4.4.4   Tri-diagonal correlation structure

Tri-diagonal error variance-covariance structure (2.30) is described in Chapter 2. Bischoff (1996) studied *maximin designs* under the linear model for correlated errors with tri-diagonal structure of lag one. Tri-diagonal structure is a covariance structure of errors which is a relaxation of intra-class structure in the form of a *lag-model* error variance-covariance structure. In the present section, robust first-order D-optimality are described for the lifetime model (4.6) or (4.7), under tri-diagonal error structure. This error structure arises when the variance of each observation is *same* say, σ_1^2 and the correlation between two observations of having *lag* n is ρ_1, and 0 (zero) otherwise. For ready reference, tri-diagonal error variance-covariance structure for $2n$ observations is reproduced below

$$\text{Dis}(\mathbf{e}) = \sigma_1^2 \left[\begin{pmatrix} I_n & I_n \\ I_n & I_n \end{pmatrix} \times \frac{(1 + \rho_1)}{2} + \begin{pmatrix} I_n & -I_n \\ -I_n & I_n \end{pmatrix} \times \frac{(1 - \rho_1)}{2} \right] = \sigma_1^2 W_1(\rho_1), \ \textit{say.} \quad (4.18)$$

Similarly as the variance-covariance structure in Section 4.4.1, $\sigma^2 W$ under the tri-diagonal variance-covariance structure $\sigma_1^2 W_1(\rho_1)$ as in (4.18) will be reduced to $\sigma^2 W_1(\rho)$, which is also a tri-diagonal structure as in (4.18), where $\rho = \frac{\rho_1 \sigma_1^2}{\sigma_1^2 + \delta^2}$ and $\sigma^2 = \sigma_1^2 + \delta^2$ (for log-normal distribution) or $\rho = \frac{\rho_1 \sigma_1^2}{\sigma_1^2 + \pi^2/6}$ and $\sigma^2 = \sigma_1^2 + \pi^2/6$ (for exponential distribution).

From Theorem 4.1, the necessary and sufficient conditions for a D-ORFOD for the lifetime model (4.6) or (4.7), under the tri-diagonal variance-covariance structure $\sigma^2 W_1(\rho)$ as in (4.18) is given below.

Theorem 4.6 *A set of necessary and sufficient conditions for a D-optimal robust first-order regression design for the lifetime model (4.6) or (4.7), under the variance-covariance structure (4.18), over the design space $\mathcal{X}$ as in Definition 4.3 are (for all $1 \leq i, j \leq k$ and ρ in $W_1(\rho)$)*

(i) $\sum_{u=1}^{2n} x_{uj} = 0$,

(ii) $\frac{1}{1-\rho^2} \sum\limits_{u=1}^{2n} x_{ui}x_{uj} - \frac{\rho}{1-\rho^2}\{\sum\limits_{u=1}^{n} x_{(n+u)i}x_{uj} + \sum\limits_{u=1}^{n} x_{ui}x_{(n+u)j}\} = 0, i \neq j;$

(iii) $\sum\limits_{u=1}^{2n} x_{ui}^2 = 2n \ \ and \ \ \sum_{u=1}^{n} x_{ui}x_{(n+u)i} = -n; \ if \ \rho > 0, \ and$

(iv) $\sum\limits_{u=1}^{2n} x_{ui}^2 = 2n \ \ and \ \ \sum_{u=1}^{n} x_{ui}x_{(n+u)i} = n; \ 1 \leq i \leq k; \ if \ \rho < 0.$ (4.19)

The same method of construction as in Section 2.7.2, Chapter 2, give D-optimal designs for the lifetime models (4.6) and (4.7) under the tri-diagonal structure (4.18). Note that the D-optimality conditions and the results as described in Sections 4.4.1 to 4.4.4 are exactly same as in Chapter 2. For ready reference, only these results are reproduced in the present context of non-normal distributions.

4.5   ROBUST SECOND-ORDER DESIGNS FOR TWO NON-NORMAL DISTRIBUTIONS

Robust second-order designs for normal distribution are described in Chapter 3. A detailed discussion of first-order designs for correlated errors with log-normal and exponential distributions are given in above sections. In the present section, robust second-order designs will be described for log-normal and exponential distributions.

Similarly for the model (4.6) under log-normal distribution, assuming that the response surface $g(\mathbf{x}, \boldsymbol{\beta})$ is of second-order, we adopt the model:

$$y_u = \beta_0 + \sum_{i=1}^{k} \beta_i x_{ui} + \sum_{i \leq j=1}^{k} \beta_{ij} x_{ui} x_{uj} + \delta\tau + e; \ 1 \leq u \leq N,$$

$$or, \quad \mathbf{Y} = X\boldsymbol{\beta} + \boldsymbol{\epsilon},$$ (4.20)

where δ, τ, $\boldsymbol{\epsilon}$ are given in (4.6), and the design matrix X and the regression coefficient vector $\boldsymbol{\beta}$ are given in (3.1) (Chapter 3). Same assumptions of $\boldsymbol{\epsilon}$ for the model (4.6) are also valid for the model (4.20).

Similarly for the model (4.7) under exponential distribution, assuming that the response surface $g(\mathbf{x}, \boldsymbol{\beta})$ is of second-order, we adopt the model:

$$y_u = \beta_0 + \sum_{i=1}^{k} \beta_i x_{ui} + \sum_{i \leq j=1}^{k} \beta_{ij} x_{ui} x_{uj} + \nu + h + e; \ 1 \leq u \leq N,$$

$$or, \quad \mathbf{Y} = X\boldsymbol{\beta} + \boldsymbol{\epsilon},$$ (4.21)

where h, ν, $\boldsymbol{\epsilon}$ are given in (4.7), and the design matrix X and the regression coefficient vector $\boldsymbol{\beta}$ are given in (3.1) (Chapter 3). Same assumptions of $\boldsymbol{\epsilon}$ for the model (4.7) are also valid for the model (4.21).

Therefore, for the second-order correlated log-normal model (4.20) or exponential model (4.21), one can derive the rotatability conditions on the same line as in the model (3.1) (Chapter 3). These second-order rotatability conditions will be exactly same as in (3.2)

(Chapter 3). Also, the simplified second-order rotatabilty conditions and the designs in the present situations under intra-class, inter-class, compound symmetry and tri-diagonal error structures will be exactly same as in Chapter 3 for the respective error structure. These are not reproduced here.

4.6   CONCLUDING REMARKS

In this chapter, first-order D-optimal and second-order rotatable designs are examined for log-normal and exponential lifetime distributions with correlated errors. The mean lifetime $g(\mathbf{x}, \boldsymbol{\beta})$ of a component is a reasonable function of the explanatory variables. This chapter derives the model $\ln T = g(\mathbf{x}, \boldsymbol{\beta}) + \delta\tau$, where $\tau \sim N(0,1)$, or, $\ln T = g(\mathbf{x}, \boldsymbol{\beta}) + h$, where h follows standard extreme value distribution. Assuming additive error component e, which arises mainly due to experimentation, the final model $\ln T = g(\mathbf{x}, \boldsymbol{\beta}) + \delta\tau + e$ (or, $\ln T = g(\mathbf{x}, \boldsymbol{\beta}) + h + e$) has been derived. From practical point of view, this model is more appropriate for lifetime improvement experiment.

Recently, Myers *et al.* (2002) analysed "The Worsted Yarn Data" (Myers *et al.*, 2002, Table 2.7, p. 36) using a conventional (errors are uncorrelated and homoscedastic) second-order response surface design, and treating response ($y = T$) as the cycles to failure (T). They noticed that variance is not constant and the analysis is inappropriate. Using the log transformation of the cycles to failure (i.e., $y = \ln T$), they analysed the data, and found that log model, overall, is an improvement on the original quadratic fit. There remains evidences of heterogeneity in variances, however. Myers *et al.* (2002, p.128) also noticed that in *industrial applications* experimental units are not independent at times *by design*. This leads to correlation among observations via a *repeated measures* scenario as in *split plot* design. For the same Worsted data, Das and Lee (2009) showed that simple log transformation is insufficient to reduce the variance constant, and they analysed the data using joint generalized linear models, and found that log-normal or gamma distribution (Chapter 10) gives much more appropriate results. Thus, it is reasonable to consider $y = \ln T$ as the response with correlated errors for the model of a lifetime distribution. If errors are indeed uncorrelated and homoscedastic, the usual response surface designs are appropriate for the response $y = \ln T$. Note that the fitted model of the original response T is non-linear i.e., $\hat{T} = e^{\hat{\beta}_0 + \hat{\beta}_1 x_1 + \ldots + \hat{\beta}_k x_k}$, which has been derived for "The Worsted Yarn Data" by Myers *et al.* (2002, p.38). Evidence (as in above) shows that lifetime observations may be heterogeneous and correlated. Log-normal and exponential lifetime distribution have been widely used for many products, such as various types of manufactured items (Lawless, 1982). For these two distributions (log-normal and exponential), the D-optimal first-order designs and second-order rotatable designs for some well-known correlation structures have been examined. It is crucial to use proper optimal operating conditions of the process parameters in lifetime improvement experiments.

Chapter 5

WEAKLY ROBUST ROTATABLE DESIGNS

Rotatability is one of the most highly desirable property in response surface methodology. For first- and second-order general correlated regression models, the concept of *robust rotatability* is introduced in Chapters 2, 3 and 4. This chapter develops some measures of robust rotatability for first- and second-order response surface designs with correlated errors. These measures are illustrated with examples. In the process, the concept of *weakly robust rotatable design* is introduced.

5.1 INTRODUCTION AND OVERVIEW

It is clear from the discussions in Chapters 2, 3 and 4, one of the major aims of rotatability in response surface functions has been to achieve stability in the variance of the predicted response. Some sort of stability is guaranteed in the case of uncorrelated regression designs, as the desirable rotatability property remains *unaltered* if the *design points* of a rotatable design are *permuted* in any manner.

It is noted that the regression models with correlated errors, rotatability, an important feature of an experimental design, may be *distorted* if we allow some *permutation* of design points of a robust rotatable design developed in the preceding chapters. As for example, this fact can be observed for the designs in Chapters 2, 3 and 4 corresponding to inter-class, compound symmetry, tri-diagonal and autocorrelated error structures. But the reduction of cost and time of experimentation often necessitates a permutation of design points.

In situations, where exact rotatability is unattainable—because of cost and time restriction and more importantly because of technical restrictions like *orthogonal blocking* or model specifications, it is still a good idea to make the design as *rotatable as possible*. Thus, it is important to measure the extent of deviation from rotatability for a design which is not rotatable. The degree or measure of rotatability for a design which is *not* rotatable under usual regression model was assessed to begin with by Draper and Guttman (1988), Khuri (1988), Draper and Pukelsheim (1990) and Park *et al.* (1993).

This chapter presents some measures for the degree of closeness to *robust rotatability* of a first- and second-order regression designs which are *not exactly robust rotatable* under a fixed pattern of correlation structure of observations. Here we introduce the degree of closeness to robust first-order rotatability for correlated first-order regression model (2.1) in the following two ways: (a) The measure may be based on the dispersion matrix of regression coefficients and (b) it may also be based on the moment matrix, following Draper and Pukelsheim (1990).

For the second-order correlated regression model (3.1), the degree of closeness to robust second-order rotatability is measured based on the moment matrix, following Draper and Pukelsheim (1990) and Park *et al.* (1993). For the first- and second-order designs, we introduce a comparison rule between robust rotatable and non-rotatable designs, based on cost and variance function. All the results as mentioned above have been developed in Panda and Das (1994), Das (1999), Das and Park (2007). In Panda and Das (1994), the investigation was first initiated with first-order designs, and it was extended in Das (1999). Recently, Das and Park (2007) have developed a new measure of robust second-order rotatability following Park *et al.* (1993).

5.2   DEVIATION FROM ROTATABILITY

Let us consider the designs constructed by the methods as in Section 2.9.2 and Section 3.8.2, under the autocorrelated error structure. They are respectively robust first- and second-order rotatable designs. It is very important to set the *positions* of central and non-central design points. If one permutes the design points of such a robust rotatable design in such a way that at least two or all the central points are brought together, one after another and/or at least two non-central points (all very similar) occurring in the design are arranged together one after another, the time and cost of the newly obtained (or changed) design will be reduced with respect to the original robust rotatable design with which one started. This is because of the cost component accruable to the "setting up of an experiment by altering levels of factors from one run to the next" and is certainly reduced in the changed design compared to the original design (appropriate cost function is given in Section 5.3.4). But the newly constructed design obtained by a suitable permutation of design points of a robust rotatable design do not satisfy the rotatability conditions in terms of moments. Thus, the changed design is *not* a robust rotatable, which is *deviated* from rotatability. The fact cited above can be observed in connection with almost each of the robust rotatable designs described in Chapters 2, 3 and 4, except those in the intra-class correlation structure, where any permutation of design points of a robust rotatable design will again give a robust rotatable design.

5.3 WEAKLY ROBUST FIRST-ORDER ROTATABLE DESIGNS

The present section introduces some measures of degree of closeness to robust rotatability of a first-order regression design under the correlated model (2.1) in Chapters 2 and 4. These concepts are initiated by Panda and Das (1994). Draper and Pukelsheim (1990) developed the degree of second-order rotatability measures for uncorrelated and homoscedastic errors. The following definitions and terminology are used in measuring the degree of closeness to robust rotatability.

Definition 5.1 Derived design: A design which is obtained by any permutation of design points of a *robust rotatable design* is called a *derived design*.

The permutation of design points is recommended from the practical point of view, with an aim at reducing the cost and time of the experiment. The class of all desirable derived designs obtained by suitable permutation of design points of a robust rotatable design d_0 is denoted by $D(d_0)$. Note that *all* the derived designs are *not* necessarily robust rotatable.

Definition 5.2 Weakly robust first-order rotatable design: When a derived design is not robust first-order rotatable, it may be nearly so in some sense for a certain range of correlation parameter (or parameters) defining the correlation structure of errors with which one may be working. Such a design under a well defined correlation structure of errors may be designated as a weakly robust first-order rotatable design (WRFORD).

In the sections that follow we attempt to investigate the degree of closeness of the weakly robust first-order rotatable designs to those exactly robust rotatable, by considering measure (i) based on dispersion matrix of regression coefficients (Panda and Das, 1994) and those (ii) based on the moment matrix (Panda and Das, 1994).

5.3.1 Robust first-order rotatability measure based on dispersion matrix

It is readily observed that the derived designs obtained by permutation of design points of a robust rotatable design are rotatable if all the correlation coefficients defining the correlated error structure equate to zero. For assessment of weak rotatability (i.e., when one or all the correlation coefficients are not zero in the error structure), one criterion based on dispersion matrix of regression coefficients is given below through the following definitions. The definitions given below are developed by Panda and Das (1994).

Definition 5.3 Weakly robust first-order rotatable design: A derived design $d \in D(d_o)$ is said to be a weakly robust first-order rotatable design (WRFORD) of strength $(100\alpha\%, 100\beta\%, 100\gamma\%)$ where α, β, γ are small, for some $R_{d(\alpha,\beta,\gamma)}(\rho) = \{R_{1d}(\rho) \cap R_{2d}(\rho) \cap R_{3d}(\rho)\}$, if

(1.1) $\text{Max}_i \mid v^{0i} \mid \leq \alpha$ for some $R_{1d}(\rho)$,

(1.2) $\text{Max}_{i \neq j} \mid v^{ij} \mid \leq \beta$ for some $R_{2d}(\rho)$,

and (1.3) $\frac{\text{Min}_i(v^{ii})}{\text{Max}_i(v^{ii})} \geq (1 - \gamma)$ for some $R_{3d}(\rho)$, $\hfill (5.1)$

where $R_{id}(\rho)$ represents the allowable region of correlation coefficients in the correlation structure denoted by ρ satisfying the above condition (1.i), i = 1,2,3 and v^{0i}, v^{ij}, v^{ii} are as in Section 2.2, in Chapter 2. $R_{d(\alpha,\beta,\gamma)}(\rho)$ is called the weak rotatability region of the design d with strength $(100\alpha\%, 100\beta\%, 100\gamma\%)$.

Note that $R_{d(\alpha,\beta,\gamma)}(\rho)$ always includes the point given by all the correlation coefficients in the given structure as zero, whatever be the values of α, β and γ. Depending on the above measure, WRFORDs are classified into the following three types.

Definition 5.4 Type-I weakly robust first-order rotatable design: A WRFORD is said to be Type-I weakly robust first-order rotatable design (T-I WRFORD) of strength $(100\alpha\%, 100\beta\%, 0.0\%)$ if,

(i) $v^{ii} = \text{constant}$; $1 \leq i \leq k$,

and (ii) at least one of v^{ij} or v^{0i} is non zero,

where v^{0i} and v^{ij} satisfy conditions (1.1) and (1.2) of Definition 5.3 respectively, $1 \leq i \neq j \leq k$.

The class of Type-I WRFORD obtained from a robust first-order rotatable design d_0 is denoted by $D_1(d_0)$.

Definition 5.5 Type-II weakly robust first-order rotatable design: A WRFORD is said to be Type-II weakly robust first-order rotatable design (T-II WRFORD) of strength $(0.0\%, 0.0\%, 100\gamma\%)$ if,

(i) $v^{0i} = 0$ and $v^{ij} = 0$; $1 \leq i \neq j \leq k$,

and (ii) at least one of v^{ii} is different from other v^{ii}'s,

where v^{ii}'s satisfy condition (1.3) of Definition 5.3, $1 \leq i \leq k$.

Type-II WRFORD class obtained from a robust first-order rotatable design d_0 is denoted by $D_2(d_0)$.

Definition 5.6 Type-III weakly robust first-order rotatable design: A WRFORD is said to be Type-III weakly robust first-order rotatable design (T-III WRFORD) of strength $(100\alpha\%, 100\beta\%, 100\gamma\%)$ if,

(i) at least one of v^{0i} or v^{ij} is non-zero; $1 \leq i \neq j \leq k$,

and (ii) at least one of v^{ii} is different from other v^{ii}'s, $1 \leq i \leq k$,

where v^{0i}, v^{ij} and v^{ii} satisfy conditions (1.1), (1.2) and (1.3) of Definition 5.3 respectively, $1 \leq i \neq j \leq k$.

Type-III WRFORD class obtained from a robust first-order rotatable design d_0 is denoted by $D_3(d_0)$. Note that $D_1(d_0)$, $D_2(d_0)$ and $D_3(d_0)$ form a partition of $D(d_0)$.

5.3.2 Robust first-order rotatability measure based on moment matrix

This section introduces a measure of robust first-order rotatability based on moment matrix following Draper and Pukelsheim (1990). Draper and Pukelsheim (1990) developed a measure of degree of rotatability for second-order regression designs when errors are uncorrelated and homoscedastic. Panda and Das (1994) developed a measure of robust rotatability for first-order regression designs when errors are correlated in the line of Draper and Pukelsheim (1990). The following lines give the methods developed by Panda and Das (1994). Towards this, we start with *normed* moment matrix, defined as

$$M = \frac{X'W^{-1}X}{1'W^{-1}1}. \tag{5.2}$$

Here X is the model (or design) matrix of the first-order correlated regression model (2.1) and W is any variance-covariance matrix of errors. When $W = \sigma^2 I_N$, (5.2) reduces to the usual moment matrix. The purpose in dividing by $\mathbf{1}'W^{-1}\mathbf{1}$ is to reduce the (1,1)th element to unity. In this sense, M defines *normed* moment matrix.

Suppose $X_o = (\mathbf{1} : (x_{ui}))$ (of order $N \times (k+1)$) is a robust first-order rotatable design matrix under the model (2.1) and M_{d_0} is the corresponding moment matrix of order $(k+1) \times (k+1)$. Then M_{d_0} can be written in the form

$$M_{d_0} = M_0 + h(\rho)k^{1/2}M_1, \tag{5.3}$$

where $h(\rho) = \frac{\mathbf{x}_i'W^{-1}\mathbf{x}_i}{\mathbf{1}'W^{-1}\mathbf{1}}$ is the *same* for all i (in view of rotatability; where $\mathbf{x}_i = (x_{1i}, x_{2i},...,x_{Ni})'$). M_0 is a matrix of order $(k+1) \times (k+1)$ containing 'one' in the (1,1) position and zero everywhere else, where M_1 is a matrix of order $(k+1) \times (k+1)$ whose first row and first column contain zeros only and remaining part of M_1 contains a diagonal matrix $k^{-1/2}I_k$.

Note that M_0 and M_1 are symmetric and orthogonal so that $M_0M_1 = 0$, and also M_i has norm $\| M_i \| = [\text{tr.}(M_iM_i)]^{1/2} = 1$, $i = 0,1$. Let M_d be the moment matrix of a derived design $d \in D(d_0)$. Suppose that we regress M_d on M_0 and M_1 to derive the fitted equation

$$\bar{M}_d = \alpha M_0 + \beta M_1 \tag{5.4}$$

with regression coefficients α and β. These coefficients are determined by multiplying equation (5.4) in turn by M_0, M_1 and taking traces we get,

$$\alpha = tr(M_dM_0) = 1$$

$$\beta = tr(M_dM_1). \tag{5.5}$$

Hence, we obtain the fitted regression equation as

$$\bar{M}_d = M_0 + M_1 tr(M_dM_1), \tag{5.6}$$

and $\bar{M}_d$ is called the rotatable component of M_d.

Two measures based on M_d and $\bar{M}_d$ are defined as follows:
(a) the measure of robust rotatability:

$$Q_d = \| \bar{M}_d - M_0 \|^2 \, / \, \| M_d - M_0 \|^2$$
$$= \{tr(\bar{M}_d - M_0)^2\}/\{tr(M_d - M_0)^2\} \tag{5.7}$$

($Q_d \leq 1$, with equality if and only if M_d is a first-order robust rotatable), and (b) the distance between M_d and $\bar{M}_d$:

$$\delta = \| M_d - \bar{M}_d \| = \{tr(M_d - \bar{M}_d)^2\}^{1/2} \tag{5.8}$$

with smaller δ meaning more robust rotatable.
So,

$$Q_d = \frac{\| \bar{M}_d - M_0 \|^2}{\delta^2 + \| \bar{M}_d - M_0 \|^2}. \tag{5.9}$$

Definition 5.7 Weakly robust first-order rotatable design: A derived design $d \in D(d_0)$ is said to be weakly robust first-order rotatable (WRFOR) of strength ξ if

$$Q_d \geq \xi. \tag{5.10}$$

Here Q_d involves the correlation parameters ρ which define the correlation structure and as such, $Q_d \geq \xi$ for *all* ρ, the set of correlation coefficients is too strong a demand to be met with. On the other hand, for a given ξ, we can possibly find the range of values of the correlation coefficients represented by ρ for which $Q_d \geq \xi$. We will call this range as the *weak rotatability region* (WRR)($R^*_{d(\xi)}(\rho)$) of the design d. Naturally, the desirability of using d will rest on the wide nature of the WRR ($R^*_{d(\xi)}(\rho)$) along with its strength ξ. Generally, we would require ξ to be very high say, around 0.95.

5.3.3 Comparison between robust first-order rotatable and weakly robust rotatable designs

Suppose $C_T(d)$ is the total cost of an experimental design (first-order) d with N design points of k factors. Suppose c_0 is the overhead cost, c_1 is the cost of altering the level of a factor (all changes are assumed to involve same cost) from one run to the next and c_2 is the cost of conducting the experiment and collecting data for *each* level combination of the k factors (i.e., in each run).

For d_0, the RFORD, described by Method I, of Section 2.8.2, in Chapter 2, with N design points and k factors, the cost is obviously

$$C_T(d_0) = c_0 + (N-1)kc_1 + Nc_2. \tag{5.11}$$

It may be noted that for illustration the correlation structure under consideration is the autocorrelation one with correlation coefficient parameter ρ.

Remark 5.1 The components c_0 and Nc_2 of the cost function (5.11) of the robust first-order rotatable design remain unaltered for *all* competing designs. But the component $(N-1)kc_1$ of (5.11) can be reduced to some extent by a suitable sequencing of the design points i.e., permutation of design points of a robust first-order rotatable design d_0.

Let d be a derived first-order design where $d \in D(d_0)$. In order to compare d and d_0, we now introduce the following criteria. Suppose $V_0(d_0)$ is the maximum variance of any robust first-order rotatable design d_0 under correlated regression model (2.1). Clearly the maximum is attained at the point $(\pm 1, \pm 1,...,\pm 1)$. Recall that $C_T(d_0)$ is its total cost. On the other hand, $V_1(d)$ is the maximum variance of the derived design $d \in D(d_0)$ at, say, ith point $(x_{i1}, x_{i2},...,x_{ik})$ and $C_T(d)$ is its total cost.

Per unit of cost efficiency is defined as the reciprocal of the variance divided by the total cost. Thus, we can compare the designs on the basis of this criterion, i.e., per unit of cost efficiency. Hence, if

$$\frac{\frac{1}{V_1(d)}}{C_T(d)} > \frac{\frac{1}{V_0(d_0)}}{C_T(d_0)},$$

$$\text{or,} \quad V_1(d)C_T(d) < V_0(d_0)C_T(d_0), \quad \text{for } \rho \text{ in } D_d(\rho), \tag{5.12}$$

the derived design d may be considered better than d_0 in $D_d(\rho)$, and $D_d(\rho)$ is called *desirable region* of the design d. The above ideas for weakly robust first-order rotatability are critically studied with examples given below.

Example 5.1 Consider a robust first-order rotatable design d_0 constructed by Method I of Section 2.8.2, in Chapter 2, with three factors ($k = 3$) and ($N = 9$) observations corresponding to the autocorrelated structure (2.40). The design (RFORD) is given by d_0 (rows being factors and columns representing runs).

d_0 :	1	2	3	4	5	6	7	8	9
x_1	0	1	0	1	0	-1	0	-1	0
x_2	0	1	0	-1	0	1	0	-1	0
x_3	0	1	0	-1	0	-1	0	1	0

$C_T(d_0) = c_0 + 24c_1 + 9c_2$, $M_{d_0} = M_0 + \frac{4(1+\rho^2)3^{1/2}M_1}{(1-\rho)(9-7\rho)}$.

Let us fix $C_T(d_i) = c_0 + 21c_1 + 9c_2$. We get the following 4 derived designs from d_0 where each $d_i \in D_1(d_0)$; $1 \le i \le 4$.

d_1 :	1	2	3	4	5	6	7	8	9
x_1	1	0	0	1	0	-1	0	-1	0
x_2	1	0	0	-1	0	1	0	-1	0
x_3	1	0	0	-1	0	-1	0	1	0

d_2 :	1	2	3	4	5	6	7	8	9
x_1	1	0	0	1	0	-1	0	-1	0
x_2	-1	0	0	1	0	1	0	-1	0
x_3	-1	0	0	1	0	-1	0	1	0

d_3 :	1	2	3	4	5	6	7	8	9
x_1	-1	0	0	1	0	1	0	-1	0
x_2	1	0	0	1	0	-1	0	-1	0
x_3	-1	0	0	1	0	-1	0	1	0

d_4 :	1	2	3	4	5	6	7	8	9
x_1	-1	0	0	1	0	1	0	-1	0
x_2	-1	0	0	1	0	-1	0	1	0
x_3	1	0	0	1	0	-1	0	-1	0

Moment matrix for kth derived design d_k is M_{d_k}

$$M_{d_k} = \begin{pmatrix} 1 & \frac{b_{1k}\rho}{(9-7\rho)} & \frac{b_{2k}\rho}{(9-7\rho)} & \frac{b_{3k}\rho}{(9-7\rho)} \\ & \frac{(4+3\rho^2)}{\{(1-\rho)(9-7\rho)\}} & \frac{b_{12k}\rho^2}{\{(1-\rho)(9-7\rho)\}} & \frac{b_{13k}\rho^2}{\{(1-\rho)(9-7\rho)\}} \\ Sym. & & \frac{(4+3\rho^2)}{\{(1-\rho)(9-7\rho)\}} & \frac{b_{23k}\rho^2}{\{(1-\rho)(9-7\rho)\}} \\ & & & \frac{(4+3\rho^2)}{\{(1-\rho)(9-7\rho)\}} \end{pmatrix}$$

where in M_{d_k}, $b_{jk} = 1$ or -1 according as the Ist level of the jth factor is $+1$ or -1 of the kth design, $1 \le j \le 3$ and $b_{ijk} = 1$ or -1 according as the product of the two Ist levels of ith and jth factor is -1 or 1 of the kth design, $1 \le i \ne j \le 3$; $1 \le k \le 4$.

If the dispersion matrix of $\hat{\beta}$ for the kth derived design d_k is $\text{Dis}_{d_k}(\hat{\beta}) = z(\rho)((\sigma_{ij})$; $1 \le i, j \le 4)$, then $\sigma_{11} = 2(1 + \rho^2)(4 + \rho^2)$, $\sigma_{12} = \sigma_{21} = c_{1k}2\rho(1 - \rho)(1 + \rho^2)$, $\sigma_{13} = \sigma_{31} = c_{2k}2\rho(1-\rho)(1+\rho^2)$, $\sigma_{14} = \sigma_{41} = c_{3k}2\rho(1-\rho)(1+\rho^2)$, $\sigma_{22} = 2(1-\rho)(9-7\rho+4\rho^2-3\rho^3)$, $\sigma_{23} = \sigma_{32} = c_{12k}\rho^2(1-\rho)(5-4\rho)$, $\sigma_{24} = \sigma_{42} = c_{13k}\rho^2(1-\rho)(5-4\rho)$, $\sigma_{33} = 2(1-\rho)(9-7\rho+4\rho^2-3\rho^3)$, $\sigma_{34} = \sigma_{43} = c_{23k}\rho^2(1 - \rho)(5 - 4\rho)$, $\sigma_{44} = 2(1 - \rho)(9 - 7\rho + 4\rho^2 - 3\rho^3)$, where $c_{jk} = 1$ or -1 according as the Ist level of the jth factor is -1 or $+1$, $1 \le j \le 3$ and $c_{ijk} = 1$ or -1 according as the product of the two Ist levels of ith and jth factor is 1 or -1 of the kth design $1 \le i \ne j \le 3$; $1 \le k \le 4$, $z(\rho) = \sigma^2(1 - \rho^2)/R_1$ and $R_1 = 4(1 - \rho)(1 + \rho^2)[(3 - 2\rho)(4 + \rho^2) + 6(1 - \rho)]$.

$$\text{Max.}_i \mid v^{0ik} \mid /\sigma^2 = -2\rho(1 - \rho)(1 + \rho^2)(1 - \rho^2)/R_1, \text{ if } -1 < \rho < 0$$

$$\text{and Max.}_i \mid v^{0ik} \mid /\sigma^2 = 2\rho(1 - \rho)(1 + \rho^2)(1 - \rho^2)/R_1, \text{ if } 0 < \rho < 1; \ 1 \le k \le 4.$$

$$\text{Max.}_{i \ne j} \mid v^{ijk} \mid /\sigma^2 = \rho^2(1 - \rho)(5 - 4\rho)(1 - \rho^2)/R_1, \text{ if } -1 < \rho < 1; \ 1 \le k \le 4.$$

TABLE 5.1: Robust first-order rotatability measures and efficiency for different values of ρ. Panda, R.N. and Das, R.N. (1994). "First Order Rotatable Designs With Correlated Errors (FORDWCE)," Cal. Statist. Assoc. Bull., 44, pp. 83-101.

ρ	Q_{d_k}	$V_0(d_0)C_T(d_0)$	$V_1(d)C_T(d)$	$Max. \mid v^{0ik} \mid /\sigma^2$	$Max. \mid v^{ijk} \mid /\sigma^2$
-0.9	.8524	15.26	24.19	0.0025	0.0083
-0.8	.8759	16.84	24.88	0.0045	0.0090
-0.7	.9003	18.55	25.12	0.0060	0.0109
-0.6	.9247	20.36	25.50	0.0069	0.0112
-0.5	.9475	22.21	25.83	0.0072	0.0101
-0.4	.9673	24.04	26.11	0.0069	0.0079
-0.3	.9827	25.75	26.32	0.0061	0.0052
-0.2	.9931	27.24	26.48	0.0046	0.0026
-0.1	.9985	28.43	26.60	0.0025	0.0007
0.0	1.0000	29.28	26.69	0.0000	0.0000
0.1	.9998	29.80	27.92	0.0030	0.0007
0.2	.9968	30.11	29.42	0.0063	0.0025
0.3	.9943	30.43	31.34	0.0097	0.0051
0.4	.9918	31.12	33.95	0.0132	0.0077
0.5	.9890	32.76	37.74	0.0163	0.0098
0.6	.9857	36.46	43.71	0.0187	0.0107
0.7	.9814	44.76	54.29	0.0199	0.0103
0.8	.9758	65.55	76.94	0.0187	0.0082
0.9	.9689	140.01	149.84	0.0134	0.0047

$$Q_d = 3\{\frac{(4+3\rho^2)}{(1-\rho)(9-7\rho}\}^2 / [3\{\frac{(4+3\rho^2)}{(1-\rho)(9-7\rho)}\}^2 + 6\frac{\rho^2}{(9-7\rho)^2}(1+\frac{\rho^2}{(1-\rho^2)})]; 1 \le k \le 4.$$

$$Max.\{\frac{V(\hat{y}_{\mathbf{x}})}{\sigma^2(1-\rho^2)}\} = \frac{1}{(1-\rho)(9-7\rho)} + \frac{3}{4(1+\rho^2)} = V_0(d_0).$$

$Max.\{\frac{V(\hat{y}_{\mathbf{x}})}{\sigma^2(1-\rho^2)}\} = [2(1+\rho^2)(4+\rho^2) - 12\rho(1-\rho)(1+\rho^2) + 6(1-\rho)(9-7\rho+4\rho^2-3\rho^3) + 6\rho^2(1-\rho)(5-4\rho)]/R_1$, if $-1 < \rho < 0$, and $Max.\{\frac{V(\hat{y}_{\mathbf{x}})}{\sigma^2(1-\rho^2)}\} = [2(1+\rho^2)(4+\rho^2)+12\rho(1-\rho)(1+\rho^2) + 6(1-\rho)(9-7\rho+4\rho^2-3\rho^3) + 6\rho^2(1-\rho)(5-4\rho)]/R_1 = V_1(d_k)$; if $0 < \rho < 1, 1 \le k \le 4$.

Take $c_0 = c_1 = c_2 = 1$ unit, $C_T(d_0) = 34$, $C_T(d_k) = 31$, $1 \le k \le 4$. Computation of robust first-order rotatability measures and comparison of efficiency for different possible values of ρ are given in Table 5.1. From Table 5.1, we have the following: Weak rotatability region (WRR) of strength $\alpha = 5\%$ or 1%, $\beta = 5\%$ or 1% and $\gamma = 0\%$ is $R_{d_k(0.05,0.05,0.0)}(\rho) = (-0.9, 0.9)$ and $R_{d_k(0.01,0.01,0.0)}(\rho) = [(-0.9, -0.8) \cup (-0.4, 0.3)]$. WRR: $R^*_{d_k(.95)}(\rho) = (-0.5, 0.9)$; $R^*_{d_k(.99)}(\rho) = (-0.2, 0.6)$; Desirable region: $D_{d_k}(\rho) = (-0.2, 0.2)$; $1 \le k \le 4$.

5.4   WEAKLY ROBUST SECOND-ORDER ROTATABLE DESIGNS

The present section introduces a degree of robust second-order rotatability measure under the correlated regression model (3.1) (Chapter 3). Weakly robust second-order rotatable design is defined below. For the sake of simplicity let us consider only the autocorrelated error structure.

Definition 5.8 Weakly robust second-order rotatable design: When $\rho \neq 0$ in the autocorrelated error structure (2.40), a derived design which is *not* robust second-order rotatable but is very *near* to robust second-order rotatable one (in some sense) for a certain range of correlation parameter ρ, is called a *weakly robust second-order rotatable design* (WRSORD) under the autocorrelated structure.

Similarly, the same can be defined for other correlation structures. Below we asses the degree of weak robust second-order rotatability based on the moment matrix (Das, 1999).

5.4.1 Robust second-order rotatability measure based on moment matrix

The present section introduces a measure of robust second-order rotatability (Das, 1999) based on moment matrix following Draper and Pukelsheim (1990). The traditional representation of a second-order model is such that a row of the design matrix (X) consists of the terms

$$(1; x_1, x_2, ..., x_k, x_1^2, x_2^2, ..., x_k^2, x_1 x_2, ..., x_{k-1} x_k). \tag{5.13}$$

Here we are interested in a special type of representation of design matrix (X) following Draper and Pukelsheim (1990). The notation, which we shall use here, is the following. Let $\mathbf{x} = (\mathbf{x}_1, \mathbf{x}_2, ..., \mathbf{x}_k)'$ and $\mathbf{x}_i = (x_{1i}, x_{2i}, ..., x_{Ni})'$; $1 \leq i \leq k$. We shall denote the terms in the second-order model by a matrix $Z(\mathbf{x})$ of order $N \times (1 + k + k^2)$ which is given by

$$Z(\mathbf{x}) = (\mathbf{1} : \mathbf{x}' : \mathbf{x}' \otimes \mathbf{x}') \tag{5.14}$$

where the symbol $\otimes$ denotes the Kronecker product.

Thus, there are $(1 + k + k^2)$ terms in a row, say the uth row looks like

$$(1; x_{u1}, x_{u2}, ..., x_{uk}; x_{u1}^2, x_{u1} x_{u2}, ..., x_{u1} x_{uk}; x_{u2} x_{u1}, x_{u2}^2, ..., x_{u2} x_{uk};$$

$$...; x_{uk} x_{u1}, x_{uk} x_{u2}, ..., x_{uk}^2); 1 \leq u \leq N. \tag{5.15}$$

An obvious disadvantage of (5.15) is that all cross product terms occur twice, so the corresponding $Z'(\mathbf{x}) Z(\mathbf{x})$ matrix is singular. A suitable generalized inverse is obvious, however, and this notation is very easily extended to higher orders.

Let us consider any robust second-order rotatable design under the correlated regression model (3.1) (Chapter 3) with "normed" moment matrix (as in (5.2)) V of order $(1 + k + k^2) \times (1 + k + k^2)$ where

$$V = \frac{Z(\mathbf{x})' W^{-1} Z(\mathbf{x})}{\mathbf{1}' W^{-1} \mathbf{1}}. \tag{5.16}$$

Normed moment matrix V in (5.16) of order $(1 + k + k^2) \times (1 + k + k^2)$ can be written in the form

$$V = V_0 + \left(\frac{a_0}{v_{00}}\right)(2k)^{1/2} V_1 + \left(\frac{\frac{1}{e}}{v_{00}}\right)(k)^{1/2} V_2 + \left(\frac{d}{v_{00}}\right)\{k(k-1)\}^{1/2} V_3$$

$$+ \left\{\frac{(\frac{2}{c} + d)}{v_{00}}\right\}(k)^{1/2} V_4 + \left(\frac{\frac{1}{c}}{v_{00}}\right)\{2k(k-1)\}^{1/2} V_5 \tag{5.17}$$

where $v_{00} = \mathbf{1}' W^{-1} \mathbf{1}$, a_0, d, $\frac{1}{c}, \frac{1}{e}$ and $(\frac{2}{c} + d)$ are as in (3.3) (Chapter 3), and each V_i is of order $(1 + k + k^2) \times (1 + k + k^2)$, $0 \leq i \leq 5$.

In the above, V_0 consists of a one in $(1,1)$ position and zeros elsewhere; V_1 consists of $(2k)^{-1/2}$ in each of the $2k$ positions viz. $(1, j(k + 1) + 1)$ and $(j(k + 1) + 1, 1)$;

$1 \leq j \leq k$ and zeros elsewhere; V_2 consists of $(k)^{-1/2}$ in each of the k diagonal positions viz. (i, i), $2 \leq i \leq (k+1)$ and zeros elsewhere; V_3 consists of $\{k(k-1)\}^{-1/2}$ in each of the $k(k-1)$ positions corresponding to mixed *even* fourth-order moments i.e., $(\mathbf{x}_i \otimes_1 \mathbf{x}_i)' W^{-1} (\mathbf{x}_j \otimes_1 \mathbf{x}_j)$, $1 \leq i \neq j \leq k$ in V, and zeros elsewhere; V_4 consists of $(k)^{-1/2}$ in each of the k positions corresponding to pure fourth-order moments i.e., $(\mathbf{x}_i \otimes_1 \mathbf{x}_i)' W^{-1} (\mathbf{x}_i \otimes_1 \mathbf{x}_i)$ $1 \leq i \leq k$ in V, and zeros elsewhere, and finally V_5 consists of $\{2k(k-1)\}^{-1/2}$ in each of the $2k(k-1)$ positions corresponding to twisted mixed *even* fourth-order moments i.e., $(\mathbf{x}_i \otimes_1 \mathbf{x}_j)' W^{-1} (\mathbf{x}_i \otimes_1 \mathbf{x}_j)$, $1 \leq i \neq j \leq k$ in V, and zeros elsewhere. Here $\otimes_1$ denotes the Hadamard product is defined as in Section 3.2 (Chapter 3). The explicit expressions for V_i's, $1 \leq i \leq 5$ for $k = 3$ are given at the end of this chapter for illustration. Note that V_i's, $0 \leq i \leq 5$ are symmetric and orthogonal so that $V_i V_j = 0$, and also each V_i has norm $\| V_i \| = [tr(V_i V_i)]^{1/2} = 1$.

Let A_d be the moment matrix of a derived design $d \in D(d_0)$. Following Draper and Pukelsheim (1990), we regress A_d on V_0, V_1, V_2, V_3, V_4 and V_5 to yield the fitted equation

$$\bar{A}_d = \sum_{i=0}^{5} \alpha_i V_i \tag{5.18}$$

with regression coefficients $\alpha_0, \alpha_1, \alpha_2, \alpha_3, \alpha_4$ and α_5. These coefficients are determined by multiplying equation (5.18) in turn by $V_0, V_1, V_2, V_3, V_4, V_5$ and taking traces

$$\alpha_0 = tr(A_d V_0) = 1$$

$$\alpha_i = tr(A_d V_i); \; 1 \leq i \leq 5. \tag{5.19}$$

Hence we obtain the fitted regression equation as

$$\bar{A}_d = V_0 + \sum_{i=1}^{5} V_i tr(A_d V_i) \tag{5.20}$$

and $\bar{A}_d$ is called the rotatable component of A_d.

Two measures based on A_d and $\bar{A}_d$ are defined as follows:
(a) the measure of robust rotatability:

$$Q_d = \| \bar{A}_d - V_0 \|^2 \; / \; \| A_d - V_0 \|^2$$

$$= \{tr(\bar{A}_d - V_0)^2\} / \{tr(A_d - V_0)^2\} \tag{5.21}$$

($Q_d \leq 1$, with equality if and only if A_d is second-order robust rotatable), and (b) the distance between A_d and $\bar{A}_d$:

$$\delta = \| A_d - \bar{A}_d \| = \{tr(A_d - \bar{A}_d)^2\}^{1/2} \tag{5.22}$$

with smaller δ meaning more robust rotatable. So,

$$Q_d = \frac{\| \bar{A}_d - V_0 \|^2}{\delta^2 + \| \bar{A}_d - V_0 \|^2}. \tag{5.23}$$

Definition 5.9 Weakly robust second-order rotatable design: A derived design $d \in D(d_0)$ is said to be weakly robust second-order rotatable design (WRSORD) of strength ξ if

$$Q_d \geq \xi. \tag{5.24}$$

Note that Q_d and ξ satisfy the same conditions as in Definition 5.7. More specifically, Q_d involves the correlation parameter ρ and as such, $Q_d \geq \xi$ for *all* ρ is too strong a demand to be met with. For a given ξ (around 0.95), the range of values of ρ for which $Q_d \geq \xi$ is called the *weak rotatability region (WRR)* $(R^*_{d(\xi)}(\rho))$ of the design d.

5.4.2 Robust second-order rotatability measure based on polar transformation

The present section introduces a new measure (Das and Park, 2007) of robust second-order rotatability following Park *et al.* (1993). The second-order response surface model with correlated errors is given in (3.1). Following (3.1), we can write

$$y(\mathbf{x}) = \eta(\mathbf{x}) + \mathbf{e},$$

$$\text{or,} \quad y_u(\mathbf{x}) = \eta(\mathbf{x}_u) + e_u,$$

where

$$\eta(\mathbf{x}_u) = \beta_0 + \sum_{i=1}^{k} \beta_i x_{ui} + \sum \sum_{i \le j=1}^{k} \beta_{ij} x_{ui} x_{uj}, \tag{5.25}$$

which may be written in matrix notation as

$$\eta(\mathbf{x}) = \mathbf{x}'_s \boldsymbol{\beta}, \tag{5.26}$$

in which the $1 \times m$ vector $\mathbf{x}'_s = (1, x_1^2, x_2^2, ..., x_k^2, x_1, x_2, ..., x_k, x_1 x_2, x_1 x_3, ..., x_{k-1} x_k)$ and $\boldsymbol{\beta}$ is the $m \times 1$ column vector of unknown regression coefficients given in (3.1) and $m = \binom{k+2}{2}$.

For a known variance-covariance matrix W of errors, the best linear unbiased estimate of $\boldsymbol{\beta}$ assuming $(X'W^{-1}X)$ is positive definite, is

$$\hat{\boldsymbol{\beta}} = (X'W^{-1}X)^{-1}(X'W^{-1}\mathbf{Y}).$$

Therefore, the fitted response at $\mathbf{x}_s$ is

$$\hat{y}(\mathbf{x}) = \mathbf{x}'_s \hat{\boldsymbol{\beta}}.$$

When the fitted response $\hat{y}(\mathbf{x}) = \mathbf{x}'_s \hat{\boldsymbol{\beta}}$ is to be used to estimate $\eta(\mathbf{x})$, it is well known that

$$\text{Var}[\hat{y}(\mathbf{x})] = \mathbf{x}'_s (X'W^{-1}X)^{-1}\mathbf{x}_s = V(\mathbf{x}), \text{ say.} \tag{5.27}$$

Thus, $\text{Var}[\hat{y}(\mathbf{x})]$ depends on the particular values of the independent variables through the vectors $\mathbf{x}'_s$. It also depends on the design and the correlation parameter involved in W through the matrix $(X'W^{-1}X)^{-1}$.

In the k dimensional space ($k \ge 2$), $V(\mathbf{x})$ can be expressed in terms of spherical co-ordinates of $(r, \phi_1, \phi_2, ..., \phi_{k-2}, \theta)$ where

$$x_1 = r \cos \phi_1$$

$$x_2 = r \sin \phi_1 \cos \phi_2$$

$$\vdots$$

$$\vdots$$

$$x_{k-1} = r \sin \phi_1 \sin \phi_2 ... \sin \phi_{k-2} \cos \theta$$

$$x_k = r \sin \phi_1 \sin \phi_2 ... \sin \phi_{k-2} \sin \theta, \tag{5.28}$$

and $r \ge 0$, $0 \le \phi_1$, $\phi_2, ..., \phi_{k-2} \le \pi$, $0 \le \theta \le 2\pi$, (See Edwards, 1973).

The absolute value of the Jacobian of this transformation is

$$|J| = r^{k-1} \sin^{k-2} \phi_1 \sin^{k-3} \phi_2 ... \sin^2 \phi_{k-3} \sin \phi_{k-2}.$$

If we substitute (5.28) into (5.27), then (5.27) will be expressed as a function of r, ϕ_1, $\phi_2,...,\phi_{k-2}$, θ and the correlation parameter (ρ) involved in W, i.e.,

$$V(\mathbf{x}) = \omega(r, \phi_1, \phi_2, ..., \phi_{k-2}, \theta, \rho) = \omega, \text{ say.} \tag{5.29}$$

Let $\bar{\omega}(r) = \frac{1}{T_k} \int_0^{2\pi} \int_0^{\pi} \cdots \int_0^{\pi} \omega(r, \phi_1, \phi_2, ..., \phi_{k-2}, \theta, \rho) \, d\Omega$ (5.30)

where

$$d\Omega = \sin^{k-2}\phi_1 \sin^{k-3}\phi_2 ... \sin^2\phi_{k-3} \sin\phi_{k-2} \, d\phi_1 \, d\phi_2 ... d\phi_{k-2} \, d\theta,$$

and

$$T_k = \int_0^{2\pi} \int_0^{\pi} \cdots \int_0^{\pi} d\Omega = \frac{2\pi^{k/2}}{\Gamma(k/2)}.$$

$\bar{\omega}(r)$ means the averaged value of $V(\mathbf{x})$ overall the points on the hypersphere of radius r centered at the origin. To be robust rotatable, $\omega(r, \phi_1, \phi_2, ..., \phi_{k-2}, \theta, \rho) = \bar{\omega}(r, \rho) = \bar{\omega}$, say for all r, ϕ_i, θ, ρ.

For a given design d, the discrepancy from rotatability at r can be expressed as

$$h_d(r) = \int_0^{2\pi} \int_0^{\pi} \cdots \int_0^{\pi} (\omega_d - \bar{\omega})^2 \, d\Omega, \tag{5.31}$$

where ω_d is the value of $V(\mathbf{x}) = \omega$ as in (5.29) for the design d. If the region of interest is $0 \leq r \leq 1$, the proposed measure of robust rotatability for a design d, will be

$$P_k(d) = \frac{1}{1 + G_k(d)}, \tag{5.32}$$

where

$$G_k(d) = \frac{1}{S_k} \int_0^1 r^{k-1} h_d(r) dr, \tag{5.33}$$

and S_k is a positive constant depending only on k.

Let us take S_k to be

$$S_k = \int_0^1 r^{k-1} T_k dr = \frac{T_k}{k},$$

for convenience. By this way, $G_k(d)$ represents the average of $(\omega - \bar{\omega})^2$ over the region of integration.

Note that $P_k(d) \leq 1$, with equality if and only if the design d is robust rotatable, and it is smaller than one for a non-rotatable design. Also note that $P_k(d)$ is invariant with respect to the rotation of the co-ordinate axes, since $\bar{\omega}$, $h_d(r)$ and $G_k(d)$ are invariance with respect to the rotation of the co-ordinate axes.

Definition 5.10 Weakly robust second-order rotatable design: A design d or a derived design $d \in D(d_0)$ is said to be weakly robust second-order rotatable design (WR-SORD) of strength ν if

$$P_k(d) \geq \nu. \tag{5.34}$$

Note that $P_k(d)$ involves the correlation parameter ρ in W and as such, $P_k(d) \geq \nu$ for *all* ρ is too strong to be met with. $P_k(d)$ also satisfies the same conditions as $Q_k(d)$ (Definition 5.7). For a given ν (a very high value around 0.95), it is required to find the range of values of ρ for which $P_k(d) \geq \nu$. We will call this range as the *weak rotatability region* (WRR)

$(R_{d(\nu)}(\rho))$ of the design d. Naturally, the desirability of using d will rest on the wide nature of $WRR(R_{d(\nu)}(\rho))$ along with its strength ν.

Now, we introduce the following fact which is useful for evaluating our measure.

(1)
$$\int d\Omega = \frac{2\pi^{k/2}}{\Gamma(k/2)},$$

(2)
$$\int x_i^2 \, d\Omega = r^2 \frac{T_k}{k},$$

(3)
$$\int x_i^2 \, x_j^2 \, d\Omega = \frac{1}{3} \int x_i^4 \, d\Omega = r^4 \frac{T_k}{k(k+2)} \quad (i \neq j),$$

(4)
$$\int x_i^2 \, x_j^2 \, x_l^2 \, d\Omega = \frac{1}{3} \int x_i^4 \, x_j^2 d\Omega = \frac{1}{15} \int x_i^6 \, d\Omega = r^6 \frac{T_k}{k(k+2)(k+4)} \quad (i \neq j \neq l),$$

(5)
$$\int x_i^2 \, x_j^2 \, x_l^2 \, x_m^2 \, d\Omega = \frac{1}{3} \int x_i^4 \, x_j^2 \, x_l^2 d\Omega = \frac{1}{9} \int x_i^4 x_j^4 \, d\Omega = \frac{1}{15} \int x_i^6 x_j^2 \, d\Omega =$$

$$\frac{1}{105} \int x_i^8 \, d\Omega = r^8 \frac{T_k}{k(k+2)(k+4)(k+6)} \quad (i \neq j \neq l \neq m),$$

where i, j, l, m could be $1, 2, ..., k$ and $\int$ means $\int_0^{2\pi} \int_0^{\pi} ... \int_0^{\pi}$ over ϕ_1, ϕ_2, ..., ϕ_{k-2}, θ. The values of other integrals where at least one x_i has an odd exponent are *all* zeros.

The above ideas for weakly robust second-order rotatability are critically studied with example given in illustration (Section 5.4.4).

5.4.3 Comparison between robust second-order rotatable and weakly robust rotatable designs

As in Section 5.3.3, suppose $C_T(d)$ is the total cost of an experimental design d of k factors with N design points. Suppose c_0 is the overhead cost, c_1 is the cost of altering the level of a factor from one run to the next and c_2 is the cost of conducting the experiment and collecting data for *each* level combination of the k factors (i.e., in each run).

The cost function of a robust second-order rotatable design d_0 of Section 3.8.2 in Chapter 3, under autocorrelated structure, with N design points and k factors may be taken as

$$C_T(d_0) = c_0 + n_1 c_1 + N c_2 \tag{5.35}$$

where n_1 is the total number of *changes* on the levels of the factors of the design d_0 in the runs 1 through N.

Remark 5.2: The components c_0 and $N c_2$ of the cost function (5.35) of the robust second-order rotatable design remain unaltered for *all* competing designs. But the component $n_1 c_1$ can be reduced to some extent by a suitable sequencing of the design points in the successive runs of the robust rotatable design d_0.

Let d be a derived design where $d \in D(d_0)$. In order to compare d and d_0, we now introduce two criteria. Let $V_0(d_0)(\bar{V}(d_0))$ be the maximum variance (average variance) of the design d_0 under correlated regression model (3.1) having (for) the design points in the description of d_0. Recall that $C_T(d_0)$ is its total cost. Similarly, for a derived design $d \in D(d_0)$, having total cost $C_T(d)$, we define $V_1(d)$ and $(\bar{V}_1(d))$.

Criteria 1:

$$\text{If} \quad V_1(d)C_T(d) < V_0(d_0)C_T(d_0), \quad \text{for } \rho \text{ in } D_d^{(1)}(\rho), \tag{5.36}$$

the derived design d is better than d_0 in $D_d^{(1)}(\rho)$, and $D_d^{(1)}(\rho)$ is called the *desirable* region of Type I of the design d, where ρ represents the set of correlation coefficients for a *fixed* pattern of correlation structure W.

Criteria 2:

$$\text{If} \quad \bar{V}_1(d)C_T(d) < \bar{V}_0(d_0)C_T(d_0), \quad \text{for } \rho \text{ in } D_d^{(2)}(\rho), \tag{5.37}$$

the derived design d is better than d_0 in $D_d^{(2)}(\rho)$, and $D_d^{(2)}(\rho)$ is called the *desirable region* of Type II of the design d.

5.4.4   Illustrations

The present section considers an example of a nearly robust second-order rotatable design with autocorrelated errors. A general method of construction of robust second-order rotatable designs with autocorrelated errors is given in Section 3.8. A derived design d which is obtained by permutation of design points of d_0 (given below), where the derived design $d \in D(d_0)$ (class of all derived designs obtained from d_0), is displayed below.

Example 5.2: Following the method of Section 3.8.2 (Chapter 3), under autocorrelated structure (2.40), we take a RSORD with 2 factors ($k = 2$) and 17 ($N = 17$) design points. We start with a usual SORD having 8 non-central and 8 central design points involving 2 factors. The design (RSORD) (denoted by d_0) is displayed below (rows being factors and column being runs).

d_0	1	2	3	4	5	6	7	8	9	10	11	12	13	14	15	16	17
x_1	0	-1	0	1	0	-1	0	1	0	-1.414	0	1.414	0	0	0	0	0
x_2	0	-1	0	-1	0	1	0	1	0	0	0	0	0	-1.414	0	1.414	0

A derived design $d \in D(d_0)$ is displayed below (rows being factors and column being runs).

d	1	2	3	4	5	6	7	8	9	10	11	12	13	14	15	16	17
x_1	0	0	-1	0	1	0	-1	0	1	0	-1.414	0	1.414	0	0	0	0
x_2	0	0	-1	0	-1	0	1	0	1	0	0	0	0	0	-1.414	0	1.414

For the RSORD d_0, $N = 16$, $k = 2$, $N_1 = 17$, $\lambda_4 = 0.25$, $\lambda_2 = 0.4992$, and noting the variance function of this design in Section 3.8.2, we have

$$\text{Max.} \left\{ \frac{\text{Var}(\hat{y}_\mathbf{x})}{\sigma^2(1-\rho^2)} \right\} = \frac{1+\rho^2}{(1-\rho)T_1}$$

$$+ \frac{2[(1+\rho^2)(17-15\rho) - 32(0.249)(1-\rho)\{(1-\rho)^2 + (1+\rho^2)\}]}{16\lambda_2(1+\rho^2)T_1}$$

$$+ \frac{4[0.75(1+\rho^2)(17-15\rho) - 16(0.249)(1-\rho)^3]}{8(1+\rho^2)T_1} = V_0(d_0) \text{ say,}$$

where $T_1 = (1+\rho^2)(17-15\rho) - 32(0.249)(1-\rho)^3$ and $C_T(d_0) = c_0 + 24c_1 + 17c_2$.

TABLE 5.2: Robust second-order rotatability measure (Q_d) and efficiency for different values of ρ. Das, R.N. (1999). "Robust Second Order Rotatable Designs: Part-II (RSORD)," Cal. Statist. Assoc. Bull, 49, pp. 65-78.

ρ	Q_d	$V_0(d_0)C_T(d_0)$	$V_1(d)C_T(d)$	$V_0(d_0)C_T(d_0)$	$V_1(d)C_T(d)$
-.9	.9913	85.5617	32.5542	78.6021	17.7451
-.8	.9928	49.1451	30.8443	41.7008	17.4008
-.7	.9942	35.2615	29.0280	27.2517	16.6762
-.6	.9956	29.4370	27.5266	20.7914	15.8906
-.5	.9968	26.9636	26.5163	17.6503	15.2501
-.4	.9979	26.0508	25.9460	16.0826	14.8181
-.3	.9988	25.8812	25.6877	15.3214	14.5734
-.2	.9994	26.0154	25.6109	14.9808	14.4633
-.1	.9998	26.1936	25.6079	14.8507	14.4397
0.0	1.0000	26.2711	25.6455	14.8221	14.4778
0.1	.9998	26.2017	25.5967	14.8619	14.5544
0.2	.9991	26.0362	25.5922	15.0080	14.7574
0.3	.9977	25.9272	25.6496	15.3775	15.1957
0.4	.9954	26.1524	26.1006	16.1990	16.1036
0.5	.9921	27.2001	27.6709	17.9076	17.9272
0.6	.9878	30.0436	31.5536	21.4265	21.6789
0.7	.9828	37.0837	41.0533	29.1111	30.1089
0.8	.9772	56.3124	69.1260	48.9078	54.0370
0.9	.9713	133.3445	245.4334	126.3583	199.0028

For the derived design d,

$$\text{Max.}\left\{\frac{\text{Var}(\hat{y}_\mathbf{x})}{\sigma^2(1-\rho^2)}\right\} = V_1(d) \text{ say,}$$

$$Q_d = \frac{(480 - 960\rho + 1764\rho^2 - 840\rho^3 + 390\rho^4)}{(8\rho^2 - 8\rho^3 + 30\rho^4) + (480 - 960\rho + 1764\rho^2 - 840\rho^3 + 390\rho^4)},$$

$C_T(d) = c_0 + 23c_1 + 17c_2$; we take $c_0 = c_1 = c_2 = 1$ unit; $C_T(d_0) = 42$ units and $C_T(d) = 41$ units. Computation of robust second-order rotatability measure (Q_d) and the efficiency of the design d are given in Table 5.2. From Table 5.2, we have the following: WRR: $R^*_{d(0.99)}(\rho) = (-0.9, 0.6)$; $R^*_{d(0.95)}(\rho) = (-0.9, 0.9)$; Desirable region: $D_d^{(1)}(\rho) = (-0.9, 0.4)$; $D_d^{(2)}(\rho) = (-0.9, 0.4)$.

To compute the other robust second-order rotatability measure $(P_k(d))$ (Das and Park, 2007) with the $(Q_k(d))$ (Das, 1999), we have to compute the following for the design d. For the above design d, the moment matrix $(X'W^{-1}X)$ is given by

$$\begin{pmatrix} v_{00} & 0 & v_{0.1} & v_{0.2} & v_{0.22} & 0 \\ 0 & v_{1.1} & 0 & 0 & 0 & 0 \\ v_{1.0} & 0 & v_{2.2} & 0 & -2.827\rho^2 & 0 \\ v_{2.0} & 0 & 0 & 11.992(1+\rho^2) & 4(1+\rho^2) & 0 \\ v_{0.22} & 0 & -2.827\rho^2 & 4(1+\rho^2) & 7.996\rho^2 + 11.992 & 0 \\ 0 & 0 & 0 & 0 & 0 & 4(1+\rho^2) \end{pmatrix},$$

where $v_{00} = (15\rho^2 - 32\rho + 17)$, $v_{0.1} = 1.414\rho(1-\rho)$, $v_{0.2} = 7.998(1-\rho)^2$, $v_{1.1} = 7.998(1+\rho^2)$, $v_{2.2} = (5.999\rho^2 + 7.998)$ and $v_{0.22} = (5.999\rho^2 - 13.997\rho + 7.998)$.

Now it is very difficult to find $\text{Dis}(\hat{\beta}) = \{\sigma^2(1-\rho^2)\}^{-1}(X'W^{-1}X)^{-1}$ explicitly in terms of ρ. Without any loss of generality, we assume $\sigma^2 = 1$ for notational convenience. So, we compute numerically $(1 - \rho^2)^{-1}(X'W^{-1}X)^{-1}$ for different values of $\rho = -0.9, -0.8,...,0.0,$ $0.1,...,0.9$. It is seen (based on computation) that the inverse has the following form:

$$
\begin{pmatrix}
v^{00} & v^{0.1}=0 & v^{0.2}\simeq 0 & v^{0.11} & v^{0.22} & v^{0.12}=0 \\
v^{1.0}=0 & v^{1.1} & v^{1.2}=0 & v^{1.11}=0 & v^{1.22}=0 & v^{1.12}=0 \\
v^{2.0}\simeq 0 & v^{2.1}=0 & v^{2.2} & v^{2.11}\simeq 0 & v^{2.22}\simeq 0 & v^{2.12}=0 \\
v^{11.0} & v^{11.1}=0 & v^{11.2}\simeq 0 & v^{11.11} & v^{11.22} & v^{11.12}=0 \\
v^{22.0} & v^{22.1}=0 & v^{22.2}\simeq 0 & v^{22.11} & v^{22.22} & v^{22.12}=0 \\
v^{12.0}=0 & v^{12.1}=0 & v^{12.2}=0 & v^{12.11}=0 & v^{12.22}=0 & v^{12.12}
\end{pmatrix},
$$

where v^m's are the corresponding elements of the inverse matrix as in (3.2).

A comparison between $v^{1.1}$ and $v^{2.2}$, and also between $v^{11.11}$ and $v^{22.22}$, is given in Table 5.3. From Table 5.3, it is clear that $v^{1.1} \simeq v^{2.2}$ and $v^{11.11} \simeq v^{22.22}$ and also from the variance-covariance matrix of $\text{Dis}(\hat{\boldsymbol{\beta}})$ as the form is given above, it is seen that some covariance components are exactly 0 (zero) and some are approximately 0 (zero). We delete those components from the variance function. Therefore, from (5.29) $\text{Var}(\hat{y}_{\mathbf{x}})$ is given below.

$$
V(\mathbf{x}) = \text{Var}(\hat{y}_{\mathbf{x}}) = Var(\hat{\beta}_0 + \hat{\beta}_1 x_1 + \hat{\beta}_2 x_2 + \hat{\beta}_{11}x_1^2 + \hat{\beta}_{22}x_2^2 + \hat{\beta}_{12}x_1 x_2)
$$

$$
= v^{00} + v^{1.1}x_1^2 + v^{2.2}x_2^2 + v^{11.11}x_1^4 + v^{22.22}x_2^4 + v^{12.12}x_1^2 x_2^2 + 2v^{0.11}x_1^2 + 2v^{0.22}x_2^2 + 2v^{11.22}x_1^2 x_2^2
$$

$$
= v^{00} + 2(v^{1.1} + v^{0.11})r^2 + v^{11.11}r^4 + \{v^{12.12} - d(\rho)\}x_1^2 x_2^2 + 2g(\rho)x_2^2, \tag{5.38}
$$

where $d(\rho) = 2(v^{11.11} - v^{11.22})$ and $g(\rho) = v^{0.22} - v^{0.11}$.

From (5.30),

$$
\bar{\omega}(r) = \frac{1}{T_k} \int V(\mathbf{x})d\Omega
$$

$$
= v^{00} + 2(v^{1.1} + v^{0.11})r^2 + v^{11.11}r^4 + \{v^{12.12} - d(\rho)\}\frac{r^4}{8} + 2g(\rho)\frac{r^2}{2}. \tag{5.39}
$$

Also,

$$
[\omega_d - \bar{\omega}(r)]^2 = [\{v^{12.12} - d(\rho)\}(x_1^2 x_2^2 - \frac{r^4}{8}) + 2g(\rho)(x_2^2 - \frac{r^2}{2})]^2
$$

$$
= \{v^{12.12} - d(\rho)\}^2\ (x_1^4 x_2^4 + \frac{r^8}{64} - 2\frac{r^4}{8}x_1^2 x_2^2) + 4(g(\rho))^2\ (x_2^4 + \frac{r^4}{4} - r^2 x_2^2)
$$

$$
+ 4\ g(\rho)\ \{v^{12.12} - d(\rho)\}\ (x_1^2 x_2^4 - \frac{r^2}{2}x_1^2 x_2^2 - \frac{r^4}{8}x_2^2 + \frac{r^6}{16}).
$$

From (5.31),

$$
h_d(r) = [\{v^{12.12} - d(\rho)\}^2\frac{r^8}{128} + 4\ (g(\rho))^2\frac{r^4}{8} + 4\ g(\rho)\ \{v^{12.12} - d(\rho)\}\frac{r^6}{48}]T_k. \tag{5.40}
$$

Also from (5.33),

$$
G_k(d) = \frac{2}{T_k}[\ \frac{\{v^{12.12} - d(\rho)\}^2}{128 \times 9} + 4\frac{(g(\rho))^2}{8 \times 5} + 4\frac{g(\rho)\ \{v^{12.12} - d(\rho)\}}{48 \times 7}]T_k,
$$

$$
= [\ \frac{\{v^{12.12} - d(\rho)\}^2}{576} + \frac{(g(\rho))^2}{5} + \frac{g(\rho)\ \{v^{12.12} - d(\rho)\}}{42}].
$$

Therefore (from (5.32)),

$$
P_k(d) = \frac{1}{1 + [\ \frac{\{v^{12.12}-d(\rho)\}^2}{576} + \frac{(g(\rho))^2}{5} + \frac{g(\rho)\ \{v^{12.12}-d(\rho)\}}{42}]}, \tag{5.41}
$$

where $d(\rho)$ and $g(\rho)$ as in (5.38).

TABLE 5.3: Comparisons for $v^{1.1}$ and $v^{2.2}$, and for $v^{11.11}$ and $v^{22.22}$, and values of $P_k(d)$ and $Q_k(d)$ for different values of ρ. Das, R.N. Park, S. H. (2007). "A Measure of Robust Rotatability for Second Order Response Surface Designs," Journal of the Korean Statistical Society, 36(4), pp. 557-578.

ρ	$v^{2.2}-v^{1.1}$	$v^{22.22}-v^{11.11}$	$P_k(d)$	$Q_k(d)$
-0.9	0.002	0.003	0.9999999	.9913*
-0.8	0.004	0.002	0.9999987	.9928
-0.7	0.004	0.005	0.9999954	.9942
-0.6	0.005	0.009	0.9999916	.9956
-0.5	0.004	0.010	0.9999894	.9968
-0.4	0.004	0.010	0.9999897	.9979
-0.3	0.004	0.009	0.9999927	.9988
-0.2	0.001	0.005	0.9999958	.9994
-0.1	0.0003	0.004	0.9999988	.9998
0.0	0.000	0.000	1.0000000	1.0000
0.1	0.0003	0.003	0.9999985	.9998
0.2	0.001	0.005	0.9999940	.9991
0.3	0.002	0.006	0.9999868	.9977
0.4	0.006	0.007	0.9999780	.9954
0.5	0.004	0.007	0.9999691	.9921
0.6	0.005	0.006	0.9999622	.9878*
0.7	0.005	0.005	0.9999596	.9828
0.8	0.004	0.004	0.9999642	.9772
0.9	0.002	0.002	0.9999795	.9713

The values of $P_k(d)$ and $Q_k(d)$ for different values of ρ are given in Table 5.3. From Table 5.3, Weak Rotatability Region (WRR) based on $P_k(d)$ is $R_{d(0.99)}(\rho)=(-0.9,0.9)$, and based on $Q_k(d)$ are $R^*_{d(0.99)}(\rho)=(-0.9,0.6)$, $R^*_{d(0.95)}(\rho)=(-0.9,0.9)$. Therefore, $P_k(d)$ gives more wide range of variation of ρ than $Q_k(d)$.

5.4.5   Comparison between two robust second-order rotatability measures $Q_k(d)$ and $P_k(d)$

This section compares between two robust second-order rotatability measures $Q_k(d)$ and $P_k(d)$. The measure $P_k(d)$ is used only for second ($d=2$) order model but the measure $Q_k(d)$ can be used for any model ($d \geq 1$). However, both the measures do not provide information about variance contour shape. For the usefulness of our proposed measures, we want to mention the following facts which are given in Table 5.4.

TABLE 5.4: Comparison between two robust second-order rotatability measures $Q_k(d)$ and $P_k(d)$. Das, R.N. Park, S. H. (2007). "A Measure of Robust Rotatability for Second Order Response Surface Designs," Journal of the Korean Statistical Society, 36(4), pp. 557-578.

Criteria	$Q_k(d)$	$P_k(d)$
Applicability to asymmetric design	Yes	Yes
Invariance w.r.t. the design rotation	Yes	Yes
Information about variance contour shape	No	No
Range	0 to 1	0 to 1
Order of the model to which the measure applies	$d \geq 1$	d=2
Weak Rotatability Region of ρ	Smaller	Greater

5.5 CONCLUDING REMARKS

This chapter introduces the concept of weakly robust first- and second-order rotatability (for correlated observations). For first-order regression designs, we have derived two measures of degree of robust first-order rotatability. One measure is based on the dispersion matrix of regression coefficients and the other measure is based on the moment matrix, following Draper and Pukelsheim (1990). For second-order regression design, we have derived two measures of degree of robust second-order rotatability based on the moment matrix, following Draper and Pukelsheim (1990) and Park *et al.* (1993). These two measures of robust second-order rotatability are $Q_k(d)$ (Das, 1999) and $(P_k(d))$ (Das and Park, 2007). One can easily compute the variance contour $V(\mathbf{x})$ from the equations (5.27) and (5.29). The measure $P_k(d)$ is illustrated with an example (derived design d) which is very nearly robust second-order rotatable design under autocorrelated errors. This measure $(P_k(d))$ is also compared with the measure $Q_k(d)$. This measure $P_k(d)$ seems to give more wide range of ρ than the measure $Q_k(d)$ from the example considered. With the help of this measure one can examine the *robust rotatability* of a second-order design with respect to any variance-covariance structure of errors.

This chapter also develops the concept of cost function in conducting an experiment. With the help of the cost function and variance function, one can compare between a robust rotatable design and a weakly robust rotatable design. We have studied comparison rules separately for first- and second-order designs. All the concepts of weakly robust rotatability and comparison rules have been critically studied with examples for illustrations. Let us close the chapter with an example of the expressions of V_i's when $k = 3$ as an illustration.

THE VALUES OF V_1, V_2, V_3, V_4, V_5 **FOR** $k = 3$

The matrices $V_1, V_2, V_3, V_4,$ and V_5 for $k = 3$ in second-order case are of sizes 13×13, [in general, the dimensions are $(1 + k + k^2) \times (1 + k + k^2)$], with columns and rows designated by the elements in

$$
V_1 = \begin{pmatrix}
0 & 0 & 0 & 0 & g_1 & 0 & 0 & 0 & g_1 & 0 & 0 & 0 & g_1 \\
0 \\
0 \\
0 \\
g_1 \\
0 \\
0 \\
0 \\
g_1 \\
0 \\
0 \\
0 \\
g_1
\end{pmatrix}.
$$

In $V_1, g_1 = 6^{-1/2}$ and *all* unfilled positions are zeros.

$$V_2 = \begin{pmatrix} 0 & 0 & 0 & 0 & 0 & 0 & 0 & 0 & 0 & 0 & 0 & 0 & 0 \\ 0 & g_2 \\ 0 & & g_2 \\ 0 & & & g_2 \\ 0 \\ 0 \\ 0 \\ 0 \\ 0 \\ 0 \\ 0 \\ 0 \\ 0 \\ 0 \end{pmatrix}.$$

In $V_2, g_2 = 3^{-1/2}$ and *all* unfilled positions are zeros.

$$V_3 = \begin{pmatrix} 0 & 0 & 0 & 0 & 0 & 0 & 0 & 0 & 0 & 0 & 0 & 0 & 0 & 0 \\ 0 & & & & \cdot & & & & \cdot & & & & & \cdot \\ 0 & & & & \cdot & & & & \cdot & & & & & \cdot \\ 0 & & & & \cdot & & & & \cdot & & & & & \cdot \\ 0 & \cdot & \cdot & \cdot & \cdot & \cdot & \cdot & \cdot & g_3 & \cdot & \cdot & \cdot & \cdot & g_3 \\ 0 & & & & \cdot & & & & \cdot & & & & & \cdot \\ 0 & & & & \cdot & & & & \cdot & & & & & \cdot \\ 0 & & & & \cdot & & & & \cdot & & & & & \cdot \\ 0 & \cdot & \cdot & \cdot & g_3 & \cdot & \cdot & \cdot & \cdot & \cdot & \cdot & \cdot & \cdot & g_3 \\ 0 & & & & \cdot & & & & \cdot & & & & & \cdot \\ 0 & & & & \cdot & & & & \cdot & & & & & \cdot \\ 0 & & & & \cdot & & & & \cdot & & & & & \cdot \\ 0 & \cdot & \cdot & \cdot & g_3 & \cdot & \cdot & \cdot & g_3 & \cdot & \cdot & \cdot & \cdot & \cdot \end{pmatrix}$$

In $V_3, g_3 = 6^{-1/2}$ and *all* unfilled positions are zeros.

$$V_4 = \begin{pmatrix} 0 & 0 & 0 & 0 & 0 & 0 & 0 & 0 & 0 & 0 & 0 & 0 & 0 & 0 \\ 0 & & & & \cdot & & & & \cdot & & & & & \cdot \\ 0 & & & & \cdot & & & & \cdot & & & & & \cdot \\ 0 & & & & \cdot & & & & \cdot & & & & & \cdot \\ 0 & \cdot & \cdot & \cdot & g_4 & \cdot & \cdot & \cdot & \cdot & \cdot & \cdot & \cdot & \cdot & \cdot \\ 0 & & & & \cdot & & & & \cdot & & & & & \cdot \\ 0 & & & & \cdot & & & & \cdot & & & & & \cdot \\ 0 & & & & \cdot & & & & \cdot & & & & & \cdot \\ 0 & \cdot & \cdot & \cdot & \cdot & \cdot & \cdot & \cdot & g_4 & \cdot & \cdot & \cdot & \cdot & \cdot \\ 0 & & & & \cdot & & & & \cdot & & & & & \cdot \\ 0 & & & & \cdot & & & & \cdot & & & & & \cdot \\ 0 & & & & \cdot & & & & \cdot & & & & & \cdot \\ 0 & \cdot & \cdot & \cdot & \cdot & \cdot & \cdot & \cdot & \cdot & \cdot & \cdot & \cdot & \cdot & g_4 \end{pmatrix}$$

In $V_4, g_4 = 3^{-1/2}$ and *all* unfilled positions are zeros.

$$
V_5 = \begin{pmatrix}
0 & 0 & 0 & 0 & 0 & 0 & 0 & 0 & 0 & 0 & 0 & 0 & 0 \\
0 & & & & & \cdot & & \cdot & & \cdot & & \cdot & & \cdot \\
0 & & & & & \cdot & & \cdot & & \cdot & & \cdot & & \cdot \\
0 & & & & & \cdot & & \cdot & & \cdot & & \cdot & & \cdot \\
0 & & & & & \cdot & & \cdot & & \cdot & & \cdot & & \cdot \\
0 & \cdot & \cdot & \cdot & \cdot & g_5 & \cdot & g_5 & \cdot & \cdot & & \cdot & & \cdot \\
0 & \cdot & \cdot & \cdot & \cdot & \cdot & g_5 & \cdot & \cdot & \cdot & g_5 & \cdot & & \cdot \\
0 & \cdot & \cdot & \cdot & \cdot & g_5 & \cdot & g_5 & \cdot & \cdot & & \cdot & & \cdot \\
0 & & & & & \cdot & & \cdot & & \cdot & & \cdot & & \cdot \\
0 & \cdot & \cdot & \cdot & \cdot & \cdot & \cdot & \cdot & \cdot & g_5 & \cdot & g_5 & \cdot \\
0 & \cdot & \cdot & \cdot & \cdot & \cdot & g_5 & \cdot & \cdot & \cdot & g_5 & \cdot & & \cdot \\
0 & \cdot & \cdot & \cdot & \cdot & \cdot & \cdot & \cdot & \cdot & g_5 & \cdot & g_5 & \cdot \\
0 & & & & & \cdot & & \cdot & & \cdot & & \cdot & & \cdot
\end{pmatrix}
$$

In V_5, $g_5 = 12^{-1/2}$; and *all* unfilled positions are zeros.

Chapter 6

ROBUST SECOND-ORDER SLOPE ROTATABILITY

Slope-rotatability is also a natural and highly desirable property as rotatability in response surface methodology. The present chapter considers a class of multifactor designs for estimating the slope of second-order response surfaces with correlated errors. The concepts of *robust slope-rotatability* and *modified robust slope-rotatability along axial directions* are introduced. For robust slope-rotatability along axial directions, it requires that the variance of the estimated slope at a point to be only a function of the distance of the point from the design origin, and independent of correlation parameter or parameters involved in the errors variance-covariance matrix. General conditions of robust second-order *slope-rotatability* and *modified slope-rotatability along axial directions* are derived for a general correlated error structure. These conditions are simplified for intra-class correlation structure. Some optimal robust second-order slope-rotatable designs (over all directions or A-optimal, D-optimal and with equal maximum directional variance in Chapter 7) are examined with respect to modified slope-rotatability. It is observed that robust second-order slope-rotatable designs over all directions, or with equal maximum directional variance slope, or D-optimal slope are *not* generally modified robust second-order slope-rotatable designs. An example of modified robust second-order slope-rotatable design under tri-diagonal correlation structure is displayed.

6.1   INTRODUCTION AND OVERVIEW

The study of response surface designs has been exceptional in that it has primarily emphasized the estimation of absolute response although other related areas also of interest have attracted the notice of a few authors as well. Estimation of differences in response at different points in the factor space will often be of great importance. If differences at points close together are involved, estimation of the local slopes (the rate of change) of the response is of interest. This problem i.e., the estimation of slopes, occurs frequently in practical situations. For instance, there are the cases in which one wants to estimate rate of reaction in chemical experiments, rates of change in the yield of a crop to various fertilizers, rates of disintegration of radioactive material in an animal, and so forth. The problem considered in this chapter is, therefore, that of the choice of the experimental design so as to achieve useful information on the estimated slope of the response surface for correlated errors.

Much of the literature on response surface analysis has dealt with the variance and bias properties of estimated response. Some authors, however, have focused attention on the estimation of partial derivatives of the response function with respect to the independent variables. Atkinson (1970a), Murty and Studden (1972), Ott and Mendenhall (1972), Myers and Lahoda (1975), Hader and Park (1978), Mukherjee and Huda (1985), and others have considered problems associated with estimation of derivatives of response. In particular, Hader and Park (1978) suggested the concept of *slope-rotatability*, that is, where the variance of the estimated derivative of a predicted response is constant for all points equidistant from the design center, and studied slope rotatable central composite designs, assuming errors are uncorrelated and homoscedastic. Park (1987) studied *slope-rotatability over all directions* and examined the class of designs having this property for uncorrelated and homoscedastic errors.

Several authors have studied slope-rotatable designs assuming errors to be uncorrelated and homoscedastic. Some practical situations indicate that the errors are correlated, violating the usual assumptions of uncorrelatedness (Mayers *et al.*, 2002). Many authors such as Kiefer and Wynn (1984), Gennings *et al.* (1989), Bischoff (1996), Palta (2003), Das (2003a), Das and Park (2006) etc., mentioned some references for the correlated error situations. Analogous to rotatability, Das (1997) introduced robust second-order rotatable design (RSORD), and derived the conditions of rotatability for second-order regression designs for a general correlated error structure (Chapter 3).

Analogous to rotatability, Hader and Park (1978) and Park (1987) suggested the concept of *slope-rotatability* for uncorrelated and homoscedastic errors. Very recently Huda (2006) and Park (2006) reviewed the literature on experimental designs for the estimation of the differences between responses, and most of the relevant references are cited there. Analogous to robust rotatability as introduced by Das (1997), the concept of *robust slope-rotatability* has been introduced by Das (2003b), which requires that the variance of the estimated derivative of predicted response at a point to be constant, for all points equidistant from the design center, independent of correlation parameter involved in the error variance-covariance structure. Das, Pal and Park (2014) also studied modified robust second-order slope-rotatability.

For a general correlated error variance covariance structure (W), second-order *slope-rotatability* and *modified slope-rotatability* conditions along axial directions are derived. These conditions are further simplified under intra-class error structure. General conditions of second-order slope-rotatability and modified slope-rotatability with correlated errors are verified with the respective conditions for uncorrelated errors. Some robust second-order slope-rotatable designs over all directions (or A-optimal slope), D-optimal slope and with equal maximum directional variance are examined with respect to modified robust slope-rotatability along axial directions. The designs are examined under the following error variance-covariance structures: intra-class, inter-class, compound symmetry, tri-diagonal and autocorrelated error structure.

6.2 SECOND-ORDER SLOPE-ROTATABILITY WITH UNCORRELATED ERRORS

This section focuses on second-order slope-rotatability and modified slope-rotatability conditions for uncorrelated and homoscedastic errors.

6.2.1 Second-order slope-rotatability conditions for uncorrelated errors

Following Hader and Park (1978), a design d is said to be second-order slope-rotatable design (SOSRD) if the design points satisfy the following conditions

(**I**) $\sum_{u=1}^{N} \prod_{i=1}^{k} x_{ui}^{\alpha_i} = 0$; for α_i's non-negative integers, any α_i odd and $\sum_{i=1}^{4} \alpha_i \leq 4$,

(**II**) :(*i*) $\sum_{u=1}^{N} x_{ui}^2 = \text{constant} = N\lambda_2; 1 \leq i \leq k$,

   (*ii*) $\sum_{u=1}^{N} x_{ui}^4 = \text{constant} = cN\lambda_4; 1 \leq i \leq k$,

(**III**) $\sum_{u=1}^{N} x_{ui}^2 x_{uj}^2 = \text{constant} = N\lambda_4; 1 \leq i, j \leq k, \ i \neq j$,

(**IV**) $(c + k - 1)\lambda_4 > k\lambda_2^2$,

(**V**) $\lambda_4[k(5 - c) - (c - 3)^2] + \lambda_2^2[k(c - 5) + 4] = 0$, $\qquad$ (6.1)

where c, λ_2 and λ_4 are constants.

The conditions are as given in Section 6.3 with the simplification derived when $W = \sigma^2 I_N$.

6.2.2 Modified second-order slope-rotatability conditions for uncorrelated errors

For a second-order design, it is difficult to get the unknown levels satisfying slope-rotatability conditions ((**I**) through (**V**)) in (6.1). In such cases, generally, solution of unknown levels are obtained on imposing some restrictions to some equations involving the unknown parameters. To derive the second-order slope-rotatability conditions, the restriction $\text{Var}(\hat{\beta}_{ii}) = \frac{1}{4} \text{Var}(\hat{\beta}_{ij})$, i.e., $v^{ii.ii} = \frac{1}{4} v^{ij.ij}$ is used, resulting the equation (**V**) in (6.1). To find the value of the constant c, some other additional restriction is required. A design d which is constructed satisfying (6.1) and the other additional restriction for determining c is known as *modified slope-rotatable design*. Victorbabu (2005) studied modified slope-rotatability using the restriction $\lambda_2^2 = \lambda_4$, consequently (**V**) in (6.1) reduces to

$$c^2 - 6c + 5 = 0, \qquad (6.2)$$

yielding $c = 1$ or $c = 5$. Victorbabu and his co-researchers suggested different methods of construction of modified slope-rotatable designs using the above additional restriction (Victorbabu, 2005, 2006, 2009; Victorbabu and Vasundharadevi, 2005).

6.3   ROBUST SECOND-ORDER SLOPE ROTATABILITY CONDITIONS

The present section focuses on second-order slope-rotatability conditions along axial directions for a general correlated error structure W. As pointed by Hader and Park (1978), in second-order response surface designs with k independent variables, for estimating the first-order derivative of response with respect to each independent variable, the variance of the estimated derivative is a function of the coordinates of the point at which the derivative is evaluated and is also a function of the design parameters. By a proper choice of the design, it is possible to make this variance constant for all points equidistant from the design origin. This property is called *slope-rotatability* by analogy with the corresponding property of rotatability for the variance of the estimated response ($\hat{y}_{\mathbf{x}}$).

Definition 6.1 Robust second-order slope rotatable design: A design d on k factors under the model (3.1) which remains second-order slope-rotatable for all the variance-covariance matrices belonging to a well defined class $W_0 = \{W$ positive definite: $W_{N \times N}$ defined by a particular correlation structure neatly specified$\}$ is called a *robust second-order slope rotatable design* (RSOSRD), with reference to the variance covariance class W_0.

Second-order correlated response surface model (3.1) is described in Chapter 3. For ready reference it is reproduced below

$$y_u = \beta_0 + \sum_{i=1}^{k} \beta_i x_{ui} + \sum_{i \leq j=1}^{k} \sum \beta_{ij} x_{ui} x_{uj} + e_u; \ 1 \leq u \leq N.$$

The estimated response at $\mathbf{x}$ is given by

$$\hat{y}_{\mathbf{x}} = \hat{\beta}_0 + \sum_{i=1}^{k} \hat{\beta}_{ii} x_i^2 + \sum_{i=1}^{k} \hat{\beta}_i x_i + \sum_{i<j=1}^{k} \sum \hat{\beta}_{ij} x_i x_j. \tag{6.3}$$

For the second-order model as in (3.1), and the estimated response at $\mathbf{x}$ as in (6.3), we have

$$\frac{\partial \hat{y}_{\mathbf{x}}}{\partial x_i} = \hat{\beta}_i + 2\hat{\beta}_{ii} x_i + \sum_{j=1; \ j\neq i}^{k} \hat{\beta}_{ij} x_j. \tag{6.4}$$

According to (3.2), the variance of $\frac{\partial \hat{y}_{\mathbf{x}}}{\partial x_i}$ is given by

$$\text{Var}(\frac{\partial \hat{y}_{\mathbf{x}}}{\partial x_i}) = \text{Var}(\hat{\beta}_i) + 4x_i^2 \text{Var}(\hat{\beta}_{ii}) + 4x_i \text{Cov}(\hat{\beta}_i, \hat{\beta}_{ii}) + \sum_{j=1; \ j\neq i}^{k} x_j^2 \text{Var}(\hat{\beta}_{ij})$$

$$+ \sum_{j=1s=1; \ j\neq s\neq i}^{k} \sum x_j x_s \text{Cov}(\hat{\beta}_{ij}, \hat{\beta}_{is}) + 2 \sum_{j=1; \ j\neq i}^{k} x_j \text{Cov}(\hat{\beta}_i, \hat{\beta}_{ij}) + 4 \sum_{j=1; \ j\neq i}^{k} x_i x_j \text{Cov}(\hat{\beta}_{ii}, \hat{\beta}_{ij})$$

$$= v^{i.i} + 4x_i^2 v^{ii.ii} + 4x_i v^{i.ii} + \sum_{j=1; \ j\neq i}^{k} x_j^2 v^{ij.ij} + \sum_{j=1s=1; \ j\neq s\neq i}^{k} \sum x_j x_s v^{ij.is}$$

$$+ 2 \sum_{j=1; \ j\neq i}^{k} x_j v^{i.ij} + 4 \sum_{j=1; \ j\neq i}^{k} x_i x_j v^{ii.ij} = V(x_i) \text{ say.} \tag{6.5}$$

The variance of estimated first-order derivative with respect to each independent variable x_i as in (6.5) will be a function of $\sum_{i=1}^{k} x_i^2$ if and only if

(i) $v^{i.ii} = 0$; $1 \le i \le k$; $v^{i.ij} = 0$; $1 \le i, j \le k$, $i \ne j$,

(ii) $v^{ij.ij'} = 0$; $1 \le i \ne j \ne j' \le k$, $v^{ii.ij} = 0$; $1 \le i, j \le k$, $i \ne j$,

(iii) $v^{i.i} = $ constant; $1 \le i \le k$,

(iv) $v^{ii.ii} = $ constant; $1 \le i \le k$,

(v) $v^{ij.ij} = $ constant; $1 \le i < j \le k$,

and (vi) $v^{ii.ii} = \frac{1}{4}v^{ij.ij}$; $1 \le i < j \le k$. \hfill (6.6)

Equivalent conditions of (i) through (v) in (6.6) of *robust second-order slope-rotatability along axial directions* with correlated error model (3.1) in terms of the elements of moment matrix are given below in the following theorem (proof is immediate from (6.6)).

Theorem 6.1 (Das, 2003b) *The necessary and sufficient conditions of a robust second-order slope-rotatable design along axial directions for a general correlated error variance-covariance structure W in the model (3.1) are*

(I)*: (i) $v_{0.j} = v_{0.jl} = 0$; $1 \le j < l \le k$,

(ii) $v_{i.j} = 0$; $1 \le i, j \le k$, $i \ne j$,

(iii): (1) $v_{ii.j} = 0$; $1 \le i, j \le k$,

(2) $v_{i.jl} = 0$; $1 \le i, j < l \le k$,

(3) $v_{ii.jl} = 0$; $1 \le i, j < l \le k$, $(j,l) \ne (i,i)$,

(4) $v_{ij.lt} = 0$; $1 \le i, l < j, t \le k$, $(i,j) \ne (l,t)$,

(II)*: (i) $v_{0.jj} = $ constant $= a_0$, say; $1 \le j \le k$,

(ii) $v_{i.i} = $ constant $= \frac{1}{e}$, say; $1 \le i \le k$,

(iii) $v_{ii.ii} = $ constant $= \eta(\frac{2}{c} + d)$, say; $1 \le i \le k$,

(III)*: (i) $v_{ii.jj} = $ constant $= d$; $1 \le i, j \le k$, $i \ne j$,

(ii) $v_{ij.ij} = $ constant $= \frac{1}{c}$; $1 \le i < j \le k$,

(IV)* $[v_{00}\{\eta(\frac{2}{c} + d) + (k-1)d\} - ka_0^2] > 0$,

(V)* $\eta(\frac{2}{c} + d)[4v_{00} - cv_{00}\eta(\frac{2}{c} + d) - cv_{00}d(k-1) + cka_0^2 + v_{00}dc] + v_{00}d\{4(k-2) + (k-1)dc\} - a_0^2\{4(k-1) + kdc\} = 0$, \hfill (6.7)

for all values of correlation parameter ρ in W, where the v's are as in (3.2) and a_0, $\frac{1}{c}$, d, $\frac{1}{e}$ are constants as in (3.3) and η is a constant.

Equivalent condition of (vi) in (6.6) is $(\mathbf{V})^*$ in (6.7) which is derived below. For second-order robust slope-rotatability, dispersion matrix of $\hat{\boldsymbol{\beta}}$ reduces to the following form:

$$\mathrm{Dis}(\hat{\boldsymbol{\beta}}) = \begin{pmatrix} A_1 & O & O \\ & D_1 & O \\ Sym. & & D_2 \end{pmatrix}^{-1} = \begin{pmatrix} A_1^{-1} & O & O \\ & D_1^{-1} & O \\ Sym. & & D_2^{-1} \end{pmatrix}$$

where

$$A_{1_{(k+1)\times(k+1)}} = \begin{pmatrix} v_{00} & a_0 & a_0 & . & . & . & a_0 \\ a_0 & \eta(\frac{2}{c}+d) & d & . & . & . & d \\ a_0 & d & \eta(\frac{2}{c}+d) & . & . & . & d \\ . & . & . & . & . & . & . \\ a_0 & d & d & . & . & . & \eta(\frac{2}{c}+d) \end{pmatrix},$$

$$D_{1_{k\times k}} = \mathrm{Diag}(\frac{1}{e}, \frac{1}{e}, ..., \frac{1}{e}), \qquad D_{2_{\binom{k}{2}\times\binom{k}{2}}} = \mathrm{Diag}(\frac{1}{c}, \frac{1}{c}, ..., \frac{1}{c}),$$

$$A_{1_{(k+1)\times(k+1)}}^{-1} = \begin{pmatrix} b_{00} & b_1 & b_1 & . & . & . & b_1 \\ b_1 & q & g_1 & . & . & . & g_1 \\ b_1 & g_1 & q & . & . & . & g_1 \\ . & . & . & . & . & . & . \\ b_1 & g_1 & g_1 & . & . & . & q \end{pmatrix},$$

$$b_{00} = \frac{\eta(\frac{2}{c}+d)+(k-1)d}{T}, \quad b_1 = -a_0/T, \quad g_1 = \frac{(a_0^2 - dv_{00})}{T\{\eta(\frac{2}{c}+d)-d\}},$$

$$q = \frac{v_{00}\{\eta(\frac{2}{c}+d)+(k-2)d\}-(k-1)a_0^2}{T\{\eta(\frac{2}{c}+d)-d\}}, \quad T = v_{00}\{\eta(\frac{2}{c}+d)+(k-1)d\}-ka_0^2, \quad (6.8)$$

and v_{00}, a_0, $\frac{1}{c}$, $\frac{1}{e}$, d are as in (3.2), (3.3) and η is as in (6.7).

The variance-covariances of the estimated parameters of the model (3.1) under robust second-order slope-rotatability are

$$\mathrm{Var}(\hat{\beta}_0) = \frac{\eta(\frac{2}{c}+d)+(k-1)d}{T}; \quad \mathrm{Var}(\hat{\beta}_i) = e; \quad \mathrm{Var}(\hat{\beta}_{ij}) = c; \ 1 \le i < j \le k,$$

$$\mathrm{Var}(\hat{\beta}_{ii}) = \frac{v_{00}\{\eta(\frac{2}{c}+d)+(k-2)d\}-(k-1)a_0^2}{T\{\eta(\frac{2}{c}+d)-d\}}; \ 1 \le i \le k,$$

$$\mathrm{Cov}(\hat{\beta}_0, \ \hat{\beta}_{ii}) = -a_0/T; \quad \mathrm{Cov}(\hat{\beta}_{ii}, \ \hat{\beta}_{jj}) = \frac{(a_0^2 - dv_{00})}{T\{\eta(\frac{2}{c}+d)-d\}}; \ 1 \le i \ne j \le k, \quad (6.9)$$

where $T = v_{00}\{\eta(\frac{2}{c}+d)+(k-1)d\}-ka_0^2$ and other covariances are zeros.

The robust second-order slope-rotatable design is non-singular when the matrix

$$\begin{pmatrix} A_1 & O & O \\ O & D_1 & O \\ O & O & D_2 \end{pmatrix}$$

is non-singular, i.e., A_1 is non-singular, which means $\mid A_1 \mid > 0$. Note that D_1 and D_2 are diagonal matrices as in (6.8). Now

$$\mid A_1 \mid = \{\eta(\frac{2}{c}+d)-d\}^{k-1}[v_{00}\{\eta(\frac{2}{c}+d)+(k-1)d\}-ka_0^2].$$

Therefore, the non-singularity condition $((\mathbf{IV})^*$ in (6.7)) of a robust second-order slope rotatable design is

$$[v_{00}\{\eta(\frac{2}{c}+d)+(k-1)d\}-ka_0^2] > 0,$$

where v_{00}, a_0, $\frac{1}{c}$, d and η are as in (3.2) and (6.7).

For second-order robust slope-rotatability,

$$\text{Var}(\hat{\beta}_{ii}) = \frac{1}{4} \text{Var}(\hat{\beta}_{ij}) \text{ i.e., } v^{ii.ii} = \frac{1}{4} v^{ij.ij}$$

which is equivalently $((\mathbf{V})^*$ in (6.7)),

$$\eta(\frac{2}{c} + d)[4v_{00} - cv_{00}\eta(\frac{2}{c} + d) - cv_{00}d(k-1) + cka_0^2 + v_{00}dc]$$

$$+ v_{00}d\{4(k-2) + (k-1)dc\} - a_0^2\{4(k-1)+kdc\} = 0.$$

From (6.5), using robust second-order slope-rotatability conditions as in (6.7), (6.8) and (6.9), we have

$$\text{Var}(\frac{\partial \hat{y}_\mathbf{x}}{\partial x_i}) = e + 4x_i^2(\frac{c}{4}) + \sum_{j=1;\ j\neq i}^{k} x_j^2 c$$

$$= e + c\sum_{i=1}^{k} x_i^2 = e + cr^2, \qquad (6.10)$$

where $r^2 = \sum_{i=1}^{k} x_i^2$ and c, e are as in (6.7).

Remark 6.1 Conditions of robust second-order *rotatability* and *slope-rotatability* under correlated error structure (W) are *not* same. Robust second-order rotatability conditions (I), $(II) : (i)$, (ii) and (III) in (3.3) are exactly *same* as robust second-order slope-rotatability conditions $(I)^*$, $(II)^* : (i)$, (ii) and $(III)^*$ in (6.7), but the other conditions are different.

6.4 MODIFIED SECOND-ORDER SLOPE ROTATABLE DESIGN WITH CORRELATED ERRORS

In Chapter 3, the best linear unbiased estimator of $\boldsymbol{\beta}$ (of the model (3.1) $\mathbf{Y} = X\boldsymbol{\beta} + \mathbf{e}$) is given by $\hat{\boldsymbol{\beta}} = (X'W^{-1}X)^{-1} (X'W^{-1}\mathbf{Y})$ with $\text{Dis}(\hat{\boldsymbol{\beta}}) = (X'W^{-1}X)^{-1}$. Normed moment matrix (defined in Chapter 5) is given below

$$M = \frac{X'W^{-1}X}{\mathbf{1}'W^{-1}\mathbf{1}}$$

where X is the design matrix under the second-order correlated regression model (3.1). For $W = \sigma^2 I_N$, "normed" moment matrix reduces to the usual moment matrix. For convenience, the normed moment matrix is used in deriving modified slope-rotatability conditions. Let

$$M^{-1} = (\frac{X'W^{-1}X}{\mathbf{1}'W^{-1}\mathbf{1}})^{-1} = \begin{pmatrix} A & B & C \\ B' & P & Q \\ C' & Q' & R \end{pmatrix}_{m\times m}^{-1} \text{ say, } m = \binom{k+2}{2},$$

where A, B, C, P, Q and R as in (3.2) are slightly changed, i.e., each element of the matrices is divided by $\mathbf{1}'W^{-1}\mathbf{1}$, and the matrices are given below

$$A_{(k+1)\times(k+1)} = \begin{pmatrix} 1 & v_{0.11} & v_{0.22}\cdots & v_{0.kk} \\ v_{11.0} & & & \\ v_{22.0} & & ((v_{ii.jj});\ 1\leq i,j\leq k) & \\ .. & .. & .. & .. \\ v_{kk.0} & & & \end{pmatrix}$$

$$v_{0.jj} = \frac{\mathbf{1}'W^{-1}(\mathbf{x}_j \otimes_1 \mathbf{x}_j)}{\mathbf{1}'W^{-1}\mathbf{1}};\quad v_{ii.jj} = \frac{(\mathbf{x}_i \otimes_1 \mathbf{x}_i)'W^{-1}(\mathbf{x}_j \otimes_1 \mathbf{x}_j)}{\mathbf{1}'W^{-1}\mathbf{1}};\ 1\leq i,\ j\leq k,$$

$$B_{(k+1)\times k} = \begin{pmatrix} v_{0.1} & v_{0.2}\cdots & v_{0.k} \\ & ((v_{ii.j})_{k\times k};\ 1\leq i,\ j\leq k) & \end{pmatrix}$$

$$v_{0.j} = \frac{\mathbf{1}'W^{-1}\mathbf{x}_j}{\mathbf{1}'W^{-1}\mathbf{1}};\quad v_{ii.j} = \frac{(\mathbf{x}_i \otimes_1 \mathbf{x}_i)'W^{-1}\mathbf{x}_j}{\mathbf{1}'W^{-1}\mathbf{1}};\ 1\leq i,j\leq k,$$

$$C_{(k+1)\times\binom{k}{2}} = \begin{pmatrix} v_{0.12} & v_{0.13}\cdots & v_{0.(k-1)k} \\ & ((v_{ii.jl})_{k\times\binom{k}{2}};1\leq i,j<l\leq k) & \end{pmatrix}$$

$$v_{0.jl} = \frac{\mathbf{1}'W^{-1}(\mathbf{x}_j \otimes_1 \mathbf{x}_l)}{\mathbf{1}'W^{-1}\mathbf{1}};\quad v_{ii.jl} = \frac{(\mathbf{x}_i \otimes_1 \mathbf{x}_i)'W^{-1}(\mathbf{x}_j \otimes_1 \mathbf{x}_l)}{\mathbf{1}'W^{-1}\mathbf{1}};1\leq i,j<l\leq k,$$

$$P_{k\times k} = ((v_{i.j})),\quad v_{i.j} = \frac{\mathbf{x}_i'W^{-1}\mathbf{x}_j}{\mathbf{1}'W^{-1}\mathbf{1}};\ 1\leq i,j\leq k,$$

$$Q_{k\times\binom{k}{2}} = ((v_{i.jl})),\quad v_{i.jl} = \frac{\mathbf{x}_i'W^{-1}(\mathbf{x}_j \otimes_1 \mathbf{x}_l)}{\mathbf{1}'W^{-1}\mathbf{1}};\ 1\leq i,j<l\leq k,$$

$$R_{\binom{k}{2}\times\binom{k}{2}} = ((v_{ij.lt})),\quad v_{ij.lt} = \frac{(\mathbf{x}_i \otimes_1 \mathbf{x}_j)'W^{-1}(\mathbf{x}_l \otimes_1 \mathbf{x}_t)}{\mathbf{1}'W^{-1}\mathbf{1}};\ 1\leq i,l<j,t\leq k.$$

Even for the uncorrelated and homoscedastic errors, it is very difficult to construct a second-order slope-rotatable design. Thus, it is mentioned in Section 6.2, for a SOSRD determination of the unknown constant c is not easy unless some sort of restriction is imposed; and Victorbabu investigated the restriction $\lambda_2^2 = \lambda_4$ to determine c.

For correlated errors, Das (2003b) introduced second-order slope-rotatability conditions along axial directions, and the corresponding designs are termed as robust second-order slope-rotatable designs (RSOSRDs) along axial directions (Section 6.3). Das and Park (2006) studied RSOSRDs over all directions (Chapter 7). Das *et al.* (2010a, 2010b) studied D-optimal and with equal maximum directional variance RSOSRDs (Chapter 7). Constructions of RSOSRDs over all directions, D-optimal and with equal maximum directional variance under different correlated error structures are studied by Das and Park, (2006) and Das *et al.*, (2010a, 2010b) (Chapter 7). RSOSRDs along axial directions are *not* still constructed for any error correlation structure. Thus, some measures of degree of robust slope-rotatability have been introduced by Das and Park (2009) and Park *et al.*, (2009) (Chapter 8).

Analogous to modified second-order slope-rotatability (for uncorrelated and homoscedastic errors), Das *et al.* (2014) derived some such sort of additional restriction for slope-rotatable design with correlated errors, for the determination of the unknown constant 'η', and the corresponding design henceforth would be termed as the *modified robust second-order slope-rotatable design* (MRSOSRD). Conditions $(I)^*$ to $(IV)^*$ in (6.7) for RSOSRD remain same for MRSOSRD. In addition, denoting $3\eta = \alpha$, and using the extra restrictions $\frac{1}{3}(\frac{2}{c}+d) = d = a_0^2 = \xi_0$, the condition $(V)^*$ in (6.7) reduces to

$$[4\alpha - c\xi_0\alpha^2 - cd(k-1)\alpha + cka_0^2\alpha + cd\alpha] + [4(k-2) + (k-1)cd] - [4(k-1) + kcd] = 0$$

or,

$$c\xi_0\alpha^2 - 2(cd+2)\alpha + (4+cd) = 0, \qquad (6.11)$$

which is equivalently a quadratic equation of unknown constant α (or η). Note that for a second-order design, c, d (or ξ_0) are known, only α is unknown. Also, note that $\frac{1}{3}(\frac{2}{c}+d) = d$ gives $\frac{1}{c} = d$, i.e., $cd = 1 = c\xi_0$. Thus, (6.11) reduces to

$$\alpha^2 - 6\alpha + 5 = 0, \tag{6.12}$$

which is exactly the same as (6.2), whatever be the design parameters a_0, c, d and e, satisfying as in above.

Robust second-order slope-rotatability and modified slope-rotatability conditions are examined below under the intra-class correlated error structure (2.6).

6.5 ROBUST SECOND-ORDER SLOPE ROTATABLE AND MODIFIED SLOPE ROTATABLE DESIGNS UNDER INTRA-CLASS STRUCTURE

The simplest variance-covariance structure is intra-class structure given in (2.6). Let ρ be the correlation between any two errors, and each has the same variance σ^2. Then intra-class variance covariance structure of errors is given by

$$\text{Dis}(\mathbf{e}) = \sigma^2[(1-\rho)I_N + \rho E_{N \times N}] = W_{N \times N}(\rho) \text{ say}; \ \sigma > 0, -(N-1)^{-1} < \rho < 1.$$

From Section 6.3, first three conditions ((I)* to (III)*) of the necessary and sufficient conditions of robust second-order slope-rotatability and modified slope-rotatability under the intra-class structure (2.6), after some simplifications, turn out to be

(I)* $\sum_{u=1}^{N} \prod_{i=1}^{k} x_{ui}^{\alpha_i} = 0$; for α_i's non-negative integers, any α_i odd and $\sum_{i=1}^{4} \alpha_i \leq 4$,

(II)* $:(i)$ $\sum_{u=1}^{N} x_{ui}^2 = $ constant; $1 \leq i \leq k$,

(ii) $\sum_{u=1}^{N} x_{ui}^4 = $ constant; $1 \leq i \leq k$,

(III)* $:\sum_{u=1}^{N} x_{ui}^2 x_{uj}^2 = $ constant; $1 \leq i, j \leq k$, $i \neq j$. $\tag{6.13}$

Using $\sum_{u=1}^{N} x_{ui}^2 = N\lambda_2$ and $\sum_{u=1}^{N} x_{ui}^2 x_{uj}^2 = N\lambda_4$; $1 \leq i, j \leq k$, $i \neq j$, the second-order slope-rotatable and modified slope-rotatable design parameters under the intra-class structure (2.6) are the following:

$$a_0 = \lambda_2, \ d = \frac{\{1+(N-1)\rho\}N\lambda_4 - \rho N^2 \lambda_2^2}{N(1-\rho)}, \ \frac{1}{e} = \frac{\lambda_2\{1+(N-1)\rho\}}{(1-\rho)},$$

$$\frac{1}{c} = \frac{\lambda_4\{1+(N-1)\rho\}}{(1-\rho)}, \ \eta(\frac{2}{c}+d) = \eta\frac{\{1+(N-1)\rho\}3N\lambda_4 - \rho N^2 \lambda_2^2}{N(1-\rho)}. \tag{6.14}$$

Note that if $\rho = 0$, the conditions in (6.13) reduce to

(I) : $\sum_{u=1}^{N} x_{ui_1}^{\alpha_1} x_{ui_2}^{\alpha_2} x_{ui_3}^{\alpha_3} x_{ui_4}^{\alpha_4} = 0$; for α_i's non-negative integers, any α_i odd and $\sum_{i=1}^{4} \alpha_i \leq 4$.

(II) $:(i)$ $\sum_{u=1}^{N} x_{ui}^2 = $ constant $= N\lambda_2$; $1 \leq i \leq k$,

(ii) $\sum_{u=1}^{N} x_{ui}^4 = $ constant $= \alpha N\lambda_4$; $1 \leq i \leq k$,

(III) : $\sum_{u=1}^{N} x_{ui}^2 x_{uj}^2 = $ constant $= N\lambda_4$; $1 \leq i, j \leq k$, $i \neq j$, $\tag{6.15}$
where $\alpha = 3\eta$.

Note that (I), (II) and (III) as in (6.15) are the first three conditions of usual second-order slope-rotatability and modified slope-rotatability conditions (Section 6.2).

Using (6.14), the expression $[\{\eta(\frac{2}{c} + d) + (k-1)d\} - ka_0^2]$ simplifies to

$$[\{\alpha + (k-1)\}\lambda_4\{1 + (N-1)\rho\} - \{\eta + (k-1)\}\rho N\lambda_2^2 - k(1-\rho)\lambda_2^2].$$

Hence the non-singularity condition $(IV)^*$ in (6.7) under the intra-class structure (2.6) is

$$[\{\alpha + (k-1)\}\lambda_4\{1 + (N-1)\rho\} - \{\eta + (k-1)\}\rho N\lambda_2^2 - k(1-\rho)\lambda_2^2] > 0, \qquad (6.16)$$

where $\alpha = 3\eta$.

If $\rho = 0$, (6.16) reduces to

$$\frac{\lambda_4}{\lambda_2^2} > \frac{k}{\alpha + k - 1} \qquad (6.17)$$

which is the non-singularity condition (IV) in (6.1) of usual second-order slope-rotatable design.

Using (6.14), the condition $(V)^*$ in (6.7) simplifies to

$$[(1-\rho)\alpha^2 - N\rho(1-2\alpha)]\lambda_2^2 + [1 + (N-1)\rho]\lambda_4(5 - 6\alpha) = 0 \qquad (6.18)$$

For $\rho = 0$, (6.18) reduces to

$$\lambda_4[k(5-\alpha) - (\alpha-3)^2] + \lambda_2^2[k(\alpha-5) + 4] = 0 \qquad (6.19)$$

where $\alpha = 3\eta$. This is the same condition (V) of usual second-order slope-rotatability in (6.1).

It is also to be noted that for $\rho = 0$, we have

$$c = \frac{1}{\lambda_4}, \; e = \frac{1}{\lambda_2}, a_0 = \lambda_2, d = \lambda_4, (\frac{2}{c} + d) = 3\lambda_4,$$

so that the modified robust second-order slope-rotatability conditions simplify as

$$\xi_0 = \frac{1}{3}(\frac{2}{c} + d) = d = \lambda_4 = a_0^2 = \lambda_2^2, \text{equivalently,} \lambda_4 = \lambda_2^2,$$

which is the *same* condition as used by Victorbabu to obtain the modified second-order slope-rotatable designs. Equation (6.19) thus reduces to

$$\alpha^2 - 6\alpha + 5 = 0 \qquad (6.20)$$

yielding $\alpha = 1$ or $\alpha = 5$.

Condition (6.20) is the modified second-order slope-rotatability condition (6.2) for uncorrelated and homoscedastic errors. Note that the robust second-order slope-rotatability and modified slope-rotatability conditions under the intra-class structure (2.6) reduce to usual modified second-order slope-rotatable conditions for $\rho = 0$.

6.6   ILLUSTRATIONS

This section examines modified robust second-order slope rotatability conditions of some robust second-order slope-rotatable designs over all directions (Das and Park, 2006), *D*-optimal slope (Das *et al.*, 2010a), and with equal maximum directional variance slope (Das *et al.*, 2010b). Das and Park (2006) have shown that the robust second-order rotatable designs are also robust second-order slope-rotatable designs over all directions. Two factors robust second-order rotatable designs are always robust second-order D-optimal slope (Das *et al.* 2010a), and with equal maximum directional variance slope (Das *et al.* 2010b). Therefore, two factors robust second-order rotatable designs are robust slope-rotatable designs over all directions, D-optimal slope, and with equal maximum directional variance slope. This section examines modified robust second-order slope-rotatability conditions of two factors robust second-order rotatable designs under intra-class, inter-class, compound symmetry, autocorrelated and tri-diagonal correlation structures (Das and Park, 2006; Das *et al.*, 2010a, 2010b).

Example 6.1 A RSORD under intra-class structure: An usual SORD (two factors) is a RSORD (or RSOSRD over all directions, or D-optimal slope, or with equal maximum directional variance slope) under the intra-class structure $(\text{Dis}(\mathbf{e}) = \sigma^2[(1-\rho)I_N + \rho E_{N\times N}] = \sigma^2 A_0$ say, $\sigma > 0, -(N-1)^{-1} < \rho < 1)$, which is given below. An example of a SORD with 2 factors and 14 observations, which is a rotatable central composite design, is displayed below (rows being factors and columns being runs).

d_1 :	1	2	3	4	5	6	7	8	9	10	11	12	13	14
$\mathbf{x}_1$	-1	1	-1	1	$-2^{1/2}$	$2^{1/2}$	0	0	0	0	0	0	0	0
$\mathbf{x}_2$	-1	-1	1	1	0	0	$-2^{1/2}$	$2^{1/2}$	0	0	0	0	0	0

To examine the robust second-order slope-rotatability and modified slope-rotatability conditions, one needs to compute the normed moment matrix of the design d_1. Without any loss of generality, assume $\sigma^2 = 1$ for notational convenience. The normed moment matrix $(M(d_1))$ of the design d_1 is given below

$$\frac{1+13\rho}{14} \begin{pmatrix} \frac{14}{1+13\rho} & 0 & 0 & \frac{8}{1+13\rho} & \frac{8}{1+13\rho} & 0 \\ 0 & \frac{8}{1-\rho} & 0 & 0 & 0 & 0 \\ 0 & 0 & \frac{8}{1-\rho} & 0 & 0 & 0 \\ \frac{8}{1+13\rho} & 0 & 0 & \frac{4(3+23\rho)}{(1-\rho)(1+13\rho)} & \frac{4(1-3\rho)}{(1-\rho)(1+13\rho)} & 0 \\ \frac{8}{1+13\rho} & 0 & 0 & \frac{4(1-3\rho)}{(1-\rho)(1+13\rho)} & \frac{4(3+23\rho)}{(1-\rho)(1+13\rho)} & 0 \\ 0 & 0 & 0 & 0 & 0 & \frac{4}{1-\rho} \end{pmatrix}.$$

Here $a_0 = \frac{4}{7}$; $d = \frac{2(1-3\rho)}{7(1-\rho)}$; $\frac{1}{c} = \frac{2(1+13\rho)}{7(1-\rho)}$; $\frac{2}{c} + d = \frac{2(3+23\rho)}{7(1-\rho)}$. Therefore, $\frac{1}{3}(\frac{2}{c} + d) \neq d \neq a_0^2$. Thus, the design d_1 does not satisfy the robust second-order slope-rotatability and modified slope-rotatability conditions.

Example 6.2 A RSORD with 2 groups under inter-class structure: An example of a RSORD with $k = 2$ factors, $m = 2$ groups, each of $n = 9$ observations for inter-class structure $(\text{Dis}(\mathbf{e}) = \sigma^2[I_m \otimes A_0]$, where $A_0 = [(1-\rho)I_n + \rho E_{n\times n}])$, is given below. The RSORD (denoted by 'd_2') is displayed below (rows being factors and columns being runs).

d_2 :	1	2	3	4	5	6	7	8	9
$\mathbf{x}_1$	-1	1	-1	1	$-2^{1/2}$	$2^{1/2}$	0	0	0
$\mathbf{x}_2$	-1	-1	1	1	0	0	$-2^{1/2}$	$2^{1/2}$	0

d_2 :	10	11	12	13	14	15	16	17	18
$\mathbf{x}_1$	-1	-1	1	1	0	0	$-2^{1/2}$	$2^{1/2}$	0
$\mathbf{x}_2$	-1	1	-1	1	$-2^{1/2}$	$2^{1/2}$	0	0	0

Assuming $\sigma^2 = 1$, the normed moment matrix $(M(d_2))$ of the design d_2 is given below

$$\frac{1+8\rho}{18}\begin{pmatrix} \frac{18}{1+8\rho} & 0 & 0 & \frac{16}{1+8\rho} & \frac{16}{1+8\rho} & 0 \\ 0 & \frac{16}{1-\rho} & 0 & 0 & 0 & 0 \\ 0 & 0 & \frac{16}{1-\rho} & 0 & 0 & 0 \\ \frac{16}{1+8\rho} & 0 & 0 & \frac{8(3+8\rho)}{(1-\rho)(1+8\rho)} & \frac{8(1-8\rho)}{(1-\rho)(1+8\rho)} & 0 \\ \frac{16}{1+8\rho} & 0 & 0 & \frac{8(1-8\rho)}{(1-\rho)(1+8\rho)} & \frac{8(3+8\rho)}{(1-\rho)(1+8\rho)} & 0 \\ 0 & 0 & 0 & 0 & 0 & \frac{8}{1-\rho} \end{pmatrix}.$$

Here $a_0 = \frac{8}{9}$; $d = \frac{4(1-8\rho)}{9(1-\rho)}$; $\frac{1}{c} = \frac{4(1+8\rho)}{9(1-\rho)}$; $\frac{2}{c} + d = \frac{4(3+8\rho)}{9(1-\rho)}$. Therefore, $\frac{1}{3}(\frac{2}{c} + d) \neq d \neq a_0^2$. Thus, the design d_2 does not satisfy the robust second-order slope-rotatability and modified slope-rotatability conditions.

Note that this RSORD under the inter-class structure is also a RSORD under the compound symmetry structure $(\text{Dis}(\mathbf{e}) = \sigma^2[I_m \otimes (A_0 - B_0) + E_{m \times m} \otimes B_0]$, where $A_0 = [(1-\rho)I_n + \rho E_{n \times n}]$ and $B_0 = \rho_1 E_{n \times n})$.

Assuming $\sigma^2 = 1$, the normed moment matrix $(M(d_2))$ of the design d_2 under the compound symmetry structure is given below

$$\frac{1}{18b}\begin{pmatrix} 18b & 0 & 0 & 16b & 16b & 0 \\ 0 & 16b_1 & 0 & 0 & 0 & 0 \\ 0 & 0 & 16b_1 & 0 & 0 & 0 \\ 16b & 0 & 0 & 8b_2 & 8b_3 & 0 \\ 16b & 0 & 0 & 8b_3 & 8b_2 & 0 \\ 0 & 0 & 0 & 0 & 0 & 8b_1 \end{pmatrix}.$$

where $b = (\delta_2 + 8\gamma_2 + 9\delta_3)$, $b_1 = (\delta_2 - \gamma_2)$, $b_2 = (3\delta_2 + 13\gamma_2 - 16\delta_3)$, $b_3 = (\delta_2 + 15\gamma_2 - 16\delta_3)$, $\delta_2 = \gamma_2 + 1/(1-\rho)$, $\gamma_2 = [(m-1)n\rho_1^2 - (m-2)n\rho\rho_1 - \{1 + (n-1)\rho\}\rho]/R$, $\delta_3 = \{1 - \delta_2 - (n-1)\rho\gamma_2\}/(m-1)n\rho_1$, and $R = (1-\rho)[(m-2)n\rho_1\{1 + (n-1)\rho\} + \{1 + (n-1)\rho\}^2 - (m-1)n^2\rho_1^2]$.

Under the compound symmetry structure, for the design d_2, $a_0 = \frac{8}{9}$; $d = \frac{4(1-8\rho-7\rho_1)}{9(1-\rho)}$; $\frac{1}{c} = \frac{4(1+8\rho+9\rho_1)}{9(1-\rho)}$; $\frac{2}{c} + d = \frac{4(3+8\rho+11\rho_1)}{9(1-\rho)}$. Therefore, $\frac{1}{3}(\frac{2}{c} + d) \neq d \neq a_0^2$. Thus, the design d_2 under the compound symmetry structure, does not satisfy the robust second-order slope-rotatability and modified slope-rotatability conditions.

Example 6.3 A RSORD under autocorrelated structure: An example of a RSORD with 2 factors and 17 design points under autocorrelated structure $(\text{Dis}(\mathbf{e}) = \sigma^2\{\rho^{|i-j|}\}\ 1 \leq i, j \leq N)$, is given below. We start with a usual SORD having 8 non-central design points and 9 central points involving 2 factors. The RSORD (denoted by 'd_3') is displayed below (rows being factors and columns being runs).

d_3 :	1	2	3	4	5	6	7	8	9	10	11	12	13	14	15	16	17
$\mathbf{x}_1$	0	-1	0	1	0	-1	0	1	0	$-2^{1/2}$	0	$2^{1/2}$	0	0	0	0	0
$\mathbf{x}_2$	0	-1	0	-1	0	1	0	1	0	0	0	0	0	$-2^{1/2}$	0	$2^{1/2}$	0

Assuming $\sigma^2 = 1$, the normed moment matrix $(M(d_3))$ of the design d_3 under the autocorrelated structure is given below

$$\frac{1}{v_{00}}\begin{pmatrix} v_{00} & 0 & 0 & 8(1-\rho)^2 & 8(1-\rho)^2 & 0 \\ 0 & 8(1+\rho^2) & 0 & 0 & 0 & 0 \\ 0 & 0 & 8(1+\rho^2) & 0 & 0 & 0 \\ 8(1-\rho)^2 & 0 & 0 & 12(1+\rho^2) & 4(1+\rho^2) & 0 \\ 8(1-\rho)^2 & 0 & 0 & 4(1+\rho^2) & 12(1+\rho^2) & 0 \\ 0 & 0 & 0 & 0 & 0 & 4(1+\rho^2) \end{pmatrix},$$

where $v_{00} = (1 - \rho)(17 - 15\rho)$.

For this design d_3, $a_0 = \frac{8(1-\rho)}{17-15\rho}$; $d = \frac{4(1+\rho^2)}{(1-\rho)(17-15\rho)}$; $\frac{1}{c} = \frac{4(1+\rho^2)}{(1-\rho)(17-15\rho)}$; $\frac{2}{c} + d = \frac{12(1+\rho^2)}{(1-\rho)(17-15\rho)}$. Therefore, $\frac{1}{3}(\frac{2}{c} + d) = d \neq a_0^2$. Thus, the design d_3 under the autocorrelated structure, does not satisfy the robust second-order slope-rotatability and modified slope-rotatability conditions.

Example 6.4 A RSORD with 2 groups under tri-diagonal structure: An example of a RSORD with $k = 2$ factors, $m = 2$ groups, each of $n = 9$ observations under tri-diagonal structure

$$(\text{Dis}(\mathbf{e}) = \sigma^2 [\begin{pmatrix} I_n & I_n \\ I_n & I_n \end{pmatrix} \times \frac{1+\rho}{2} + \begin{pmatrix} I_n & -I_n \\ -I_n & I_n \end{pmatrix} \times \frac{1-\rho}{2}] = W_3, \text{say}),$$

is given below. The RSORD (denoted by 'd_4') is displayed below (rows being factors and columns being runs).

d_4 :	1	2	3	4	5	6	7	8	9
$\mathbf{x}_1$	-1	1	-1	1	$-2^{1/2}$	$2^{1/2}$	0	0	0
$\mathbf{x}_2$	-1	-1	1	1	0	0	$-2^{1/2}$	$2^{1/2}$	0

d_4 :	10	11	12	13	14	15	16	17	18
$\mathbf{x}_1$	0	0	0	0	0	0	0	0	0
$\mathbf{x}_2$	0	0	0	0	0	0	0	0	0

Assuming $\sigma^2 = 1$, the normed moment matrix $(M(d_4))$ of the design d_4 under tri-diagonal structure is given below

$$\frac{1}{18(1-\rho)} \begin{pmatrix} 18(1-\rho) & 0 & 0 & 8(1-\rho) & 8(1-\rho) & 0 \\ 0 & 8 & 0 & 0 & 0 & 0 \\ 0 & 0 & 8 & 0 & 0 & 0 \\ 8(1-\rho) & 0 & 0 & 12 & 4 & 0 \\ 8(1-\rho) & 0 & 0 & 4 & 12 & 0 \\ 0 & 0 & 0 & 0 & 0 & 4 \end{pmatrix}.$$

For the design d_4, $a_0 = \frac{4}{9}$; $d = \frac{2}{9(1-\rho)}$; $\frac{1}{c} = \frac{2}{9(1-\rho)}$; $\frac{2}{c} + d = \frac{2}{3(1-\rho)}$. Therefore, $\frac{1}{3}(\frac{2}{c} + d) = d \neq a_0^2$. Thus, the design d_4 under tri-diagonal structure, does not satisfy the robust second-order slope-rotatability and modified slope-rotatability conditions.

Example 6.5 A RSORD with 2 groups under tri-diagonal structure: An example of a RSORD with $k = 2$ factors, $m = 2$ groups, each of $n = 9$ observations under tri-diagonal structure as in Example 6.4, is given below. The RSORD (denoted by 'd_5') is displayed below (rows being factors and columns being runs).

d_5 :	1	2	3	4	5	6	7	8	9
$\mathbf{x}_1$	-1	1	-1	1	$-2^{1/2}$	$2^{1/2}$	0	0	0
$\mathbf{x}_2$	-1	-1	1	1	0	0	$-2^{1/2}$	$2^{1/2}$	0

d_5 :	10	11	12	13	14	15	16	17	18
$\mathbf{x}_1$	-1	1	-1	1	$-2^{1/2}$	$2^{1/2}$	0	0	0
$\mathbf{x}_2$	-1	-1	1	1	0	0	$-2^{1/2}$	$2^{1/2}$	0

Assuming $\sigma^2 = 1$, the normed moment matrix $(M(d_5))$ of the design d_5 under tri-diagonal structure is given below

$$\frac{(1+\rho)}{18} \begin{pmatrix} \frac{18}{(1+\rho)} & 0 & 0 & \frac{16}{(1+\rho)} & \frac{16}{(1+\rho)} & 0 \\ 0 & \frac{16}{(1+\rho)} & 0 & 0 & 0 & 0 \\ 0 & 0 & \frac{16}{(1+\rho)} & 0 & 0 & 0 \\ \frac{16}{(1+\rho)} & 0 & 0 & \frac{24}{(1+\rho)} & \frac{8}{(1+\rho)} & 0 \\ \frac{16}{(1+\rho)} & 0 & 0 & \frac{8}{(1+\rho)} & \frac{24}{(1+\rho)} & 0 \\ 0 & 0 & 0 & 0 & 0 & \frac{8}{(1+\rho)} \end{pmatrix}.$$

For the design d_5, $a_0 = \frac{8}{9}$; $d = \frac{4}{9}$; $\frac{1}{c} = \frac{4}{9}$; $\frac{2}{c} + d = \frac{12}{9}$. Therefore, $\frac{1}{3}(\frac{2}{c} + d) = d \neq a_0^2$. Thus, the design d_5 under tri-diagonal structure, does not satisfy the robust second-order slope-rotatability and modified slope-rotatability conditions.

Examples 6.1 to 6.5 are robust second-order slope-rotatable designs over all directions, D-optimal slope, and with equal maximum directional variance slope, but they are not robust second-order slope-rotatable and modified slope-rotatable designs (clear from the above illustration).

Example 6.6 A modified robust second-order slope rotatable design with 2 groups under tri-diagonal structure: An example of modified robust second-order slope rotatable design with $k = 2$ factors, $m = 2$ groups, each of $n = 13$ observations under tri-diagonal structure, is given below. The MRSOSRD (denoted by d_6) is displayed below (rows being factors and columns being runs).

d_6 :	1	2	3	4	5	6	7	8	9	10	11	12	13
$\mathbf{x}_1$	$-a$	a	$-a$	a	$-b$	b	0	0	$-f$	f	0	0	0
$\mathbf{x}_2$	$-a$	$-a$	a	a	0	0	$-b$	b	0	0	$-f$	f	0

d_6 :	14	15	16	17	18	19	20	21	22	23	24	25	26
$\mathbf{x}_1$	$-a$	a	$-a$	a	$-b$	b	0	0	$-f$	f	0	0	0
$\mathbf{x}_2$	$-a$	$-a$	a	a	0	0	$-b$	b	0	0	$-f$	f	0

Elements of the moment matrix of the design d_6 can be calculated from the simplified forms of $(\mathbf{I})^*$, $(\mathbf{II})^*$, $(\mathbf{III})^*$ (under tri-diagonal structure as given above), which are given below

$(\mathbf{I})^*$: (i) $v_{0.j} = 0 \Leftrightarrow \sum_{u=1}^{N} x_{uj} = 0; 1 \leq j \leq k$,

$v_{0.jl} = 0 \Leftrightarrow \sum_{u=1}^{N} x_{uj} x_{ul} = 0; 1 \leq j < l \leq k$,

(ii) $v_{i.j} = 0 \Leftrightarrow \sum_{u=1}^{N} x_{ui} x_{uj} - \rho\{\sum_{u=1}^{n} x_{(n+u)i} x_{uj} + \sum_{u=1}^{n} x_{ui} x_{(n+u)j}\} = 0; 1 \leq i, j \leq k, \ i \neq j$,

(iii): (1)$v_{ii.j} = 0 \Leftrightarrow \sum_{u=1}^{N} x_{ui}^2 x_{uj} - \rho\{\sum_{u=1}^{n} x_{(n+u)i}^2 x_{uj} + \sum_{u=1}^{n} x_{ui}^2 x_{(n+u)j}\} = 0; 1 \leq i \leq k, \ 1 \leq j \leq k$,

(2) $v_{ij.l} = 0 \Leftrightarrow \sum_{u=1}^{N} x_{ui} x_{uj} x_{ul} - \rho(\sum_{u=1}^{n} x_{(n+u)i} x_{(n+u)j} x_{ul} + \sum_{u=1}^{n} x_{ui} x_{uj} x_{(n+u)l}) = 0; 1 \leq i < j \leq k, \ 1 \leq l \leq k$,

(3) $v_{ii.jl} = 0 \Leftrightarrow \sum_{u=1}^{N} x_{ui}^2 x_{uj} x_{ul} - \rho(\sum_{u=1}^{n} x_{(n+u)i}^2 x_{uj} x_{ul} + \sum_{u=1}^{n} x_{ui}^2 x_{(n+u)j} x_{(n+u)l}) = 0; 1 \leq i \leq k, \ 1 \leq j < l \leq k$,

(4) $v_{ij.lt} = 0 \Leftrightarrow \sum_{u=1}^{N} x_{ui}x_{uj}x_{ul}x_{ut} - \rho(\sum_{u=1}^{n} x_{(n+u)i}x_{(n+u)j}x_{ul}x_{ut} + \sum_{u=1}^{n} x_{ui}x_{uj}$
$x_{(n+u)l}x_{(n+u)t}) = 0; 1 \leq i < j \leq k, \ 1 \leq l < t \leq k, \ (i,j) \neq (l,t),$

(II)*: (i) $v_{0.jj} = constant \Leftrightarrow \frac{(1-\rho)\{\sigma^2(1-\rho^2)\}^{-1}}{v_{00}} \sum_{u=1}^{N} x_{uj}^2 = a_0(> 0); 1 \leq j \leq k,$

(ii) $v_{i.i} = constant \Leftrightarrow \frac{\{\sigma^2(1-\rho^2)\}^{-1}}{v_{00}}[\sum_{u=1}^{N} x_{ui}^2 - 2\rho \sum_{u=1}^{n} x_{ui}x_{(n+u)i}] = \frac{1}{e}(> 0); 1 \leq i \leq k,$

(iii) $v_{ii.ii} = constant \Leftrightarrow \frac{\{\sigma^2(1-\rho^2)\}^{-1}}{v_{00}}[\sum_{u=1}^{N} x_{ui}^4 - 2\rho \sum_{u=1}^{n} x_{ui}^2 x_{(n+u)i}^2] = (\frac{2}{c} + d) (> 0); 1 \leq i \leq k,$

(III)*: (i) $v_{ii.jj} = constant \Leftrightarrow \frac{\{\sigma^2(1-\rho^2)\}^{-1}}{v_{00}}[\sum_{u=1}^{N} x_{ui}^2 x_{uj}^2 - \rho(\sum_{u=1}^{n} x_{(n+u)i}^2 x_{uj}^2 + \sum_{u=1}^{n} x_{ui}^2 x_{(n+u)j}^2)] = d(> 0); 1 \leq i, j \leq k, \ i \neq j,$

(ii) $v_{ij.ij} = constant \Leftrightarrow \frac{\{\sigma^2(1-\rho^2)\}^{-1}}{v_{00}}[\sum_{u=1}^{N} x_{ui}^2 x_{uj}^2 - 2\rho \sum_{u=1}^{n} x_{ui}x_{uj}x_{(n+u)i}x_{(n+u)j}] = \frac{1}{c}(> 0); 1 \leq i < j \leq k,$
where $v_{00} = \mathbf{1}'W_3^{-1}\mathbf{1}$.

Assuming $\sigma^2 = 1$, the normed moment matrix $(M(d_6))$ of the design d_6 under tridiagonal structure is given below

$$
\begin{pmatrix}
1 & 0 & 0 & \frac{2(2a^2+b^2+f^2)}{13} & \frac{2(2a^2+b^2+f^2)}{13} & 0 \\
0 & \frac{2(2a^2+b^2+f^2)}{13} & 0 & 0 & 0 & 0 \\
0 & 0 & \frac{2(2a^2+b^2+f^2)}{13} & 0 & 0 & 0 \\
\frac{2(2a^2+b^2+f^2)}{13} & 0 & 0 & \frac{2(2a^4+b^4+f^4)}{13} & \frac{4a^4}{13} & 0 \\
\frac{2(2a^2+b^2+f^2)}{13} & 0 & 0 & \frac{4a^4}{13} & \frac{2(2a^4+b^4+f^4)}{13} & 0 \\
0 & 0 & 0 & 0 & 0 & \frac{4a^4}{13}
\end{pmatrix}.
$$

For the design d_6, $a_0 = \frac{2(2a^2+b^2+f^2)}{13}$; $d = \frac{4a^4}{13}$; $\frac{1}{c} = \frac{4a^4}{13}$; $\frac{2}{c} + d = \frac{2(2a^4+b^4+f^4)}{13}$.
Examining the modified robust second-order slope-rotatability condition $d = a_0^2$, we get

$$b^2 + f^2 = ((13)^{\frac{1}{2}} - 2)a^2. \tag{6.21}$$

Also the modified robust second-order slope-rotatability condition $\frac{2/c+d}{3} = d$ yields

$$b^4 + f^4 = 4a^4. \tag{6.22}$$

Considering $a = 1$, and solving (6.21) and (6.22), we get $b = 1.4025$ and $f = 0.6012i$, where $i = (-1)^{\frac{1}{2}}$. In addition, eigenvalues of the moment matrix $(M(d_6))$ are 5.4736, 4.6920, 0.8338, 0.6097, 0.6097, 0.3077, indicating that $(M(d_6))$ is non-singular. Therefore, $\frac{1}{3}(\frac{2}{c} + d) = d = a_0^2 = 0.3077$. Thus, the design d_6 under tri-diagonal structure, satisfies the modified robust second-order slope-rotatability conditions. Note that some star points (containing $+f$ or $-f$) are imaginary. In practice, these imaginary points can not be located. Future researches will try to focus on the construction of real designs for the present situations.

6.7   CONCLUDING REMARKS

For quality improvement experiments, response surface methodology is generally used. If differences at points close together are involved, estimation of the local slopes (the rate of change) of the response is of interest. This problem, estimation of slopes, occurs frequently in many practical situations. Generally, observations are correlated in most of the practical situations (Myers *et al.*, 2002). Several well-known correlation structures are considered in Chapter 2. The present chapter considers robust second-order slope-rotatable and modified slope rotatable designs. Robust second-order slope-rotatability and modified slope-rotatability conditions are derived for a general correlated error structure. These conditions are simplified for intra-class error structure, and are compared with the usual (uncorrelated and homoscedastic errors) modified second-order slope-rotatability conditions. Some robust second-order slope-rotatable designs over all directions, D-optimal slope, and with equal maximum directional variance slope are examined with respect to modified robust second-order slope-rotatability. It is observed that robust second-order slope-rotatable designs over all directions, D-optimal slope, and with equal maximum directional variance slope are *not* modified robust second-order slope-rotatable designs. Second-order slope-rotatable designs along axial directions and modified second-order slope-rotatable designs with correlated errors are unknown in the literature. Construction of robust second-order slope-rotatable and modified slope-rotatable designs along axial directions under different error structures is currently under investigation. It is expected that in the next edition of this book, some constructions of robust second-order slope-rotatable and modified slope-rotatable designs along axial directions will be included.

Chapter 7

OPTIMAL ROBUST SECOND-ORDER SLOPE ROTATABLE DESIGNS

Second-order slope-rotatability and modified slope-rotatability conditions *along axial directions* are derived in Chapter 6 for a general variance-covariance structure of errors. The present chapter introduces some concepts of optimal robust second-order slope-rotatable designs (over all directions (or *A*-optimal slope), with equal maximum directional variance and *D*-optimal slope) for correlated errors. Second-order slope-rotatability conditions *over all directions, with equal maximum directional variance* and *D-optimal slope* are derived for a general correlated error structure. It is examined that robust second-order rotatable designs are also robust slope-rotatable over all directions, with equal maximum directional variances and *D*-optimal slope. Equal maximum directional variance slope-rotatable design requires that the maximum variance of the estimated slope over all possible directions to be only a function of the distance of the point from the design origin, and independent of correlation parameter or parameters involved in the variance-covariance matrix of errors. In the process, robust second-order symmetric balanced designs are introduced. A class of robust second-order slope-rotatable designs over all directions, with equal maximum directional variance and *D*-optimal slope-rotatable designs is investigated for some special correlation structures of errors, i.e., intra-class, inter-class, compound symmetry, tri-diagonal and autocorrelated structures.

7.1 INTRODUCTION AND SUMMARY

In studying the slope of a second-order response surface design, slope rotatability over all directions (SROAD) is a useful concept. SROAD designs ensure that knowledge of the slope is acquired symmetrically, whatever direction later becomes of more interest as the data are analyzed. The present chapter focuses on the estimation of differences, or slopes rather than the absolute value of the response variable in view of some practical applications of it in response surface methodology. It is then natural to consider the variance measure for the slope of the fitted surface at any given point. For equal information in all directions about the design origin, Park (1987) introduced the concept of second-order *slope rotatability over all directions* (SROAD), and derived the necessary and sufficient conditions for a design to have this property based on the precision matrix. Ying *et al.*, (1995a, 1995b) studied SROAD designs for two, three and more dimensions. Park and Kwon (1998) have introduced the concept of robust slope-rotatability with equal maximum directional variance for uncorrelated and homoscedastic errors. Huda *et al.* (2007) studied $A-$ and D–slope-rotatabily for uncorrelated and homoscedastic errors. The necessary and sufficient conditions of D–slope-rotatability have been derived by Huda *et al.* (2008). Recently, Huda and Benkherouf (2010) have shown that rotatability is a sufficient condition for $A-$ and D–slope-rotatabily for uncorrelated and homoscedastic errors.

Slope-rotatability and modified slope-rotatability *along axial directions* with correlated errors are described in Chapter 6. Das and Park (2006) have introduced robust slope-rotatable designs over all directions which is A-optimal with correlated errors. In the present chapter, the concept of slope-rotatability along axial directions has been extended to over all directions assuming correlated errors. For a general correlated error structure (given by the variance covariance matrix of errors, W), Das and Park (2006) have derived a necessary and sufficient condition of second-order *slope rotatability over all directions*. It is shown that robust second-order rotatable designs are also robust slope-rotatable over all directions (RSROAD).

This present chapter also develops the concept of *robust second-order slope-rotatable designs with equal maximum directional variance* for correlated errors. A design is said to be robust slope-rotatable with equal maximum directional variance if the maximum variance of the estimated slope over all possible directions is only a function of the distance of the point from the design origin, and independent of correlation parameter or parameters involved in the variance-covariance structure of errors. The concept of robust symmetric balanced designs is developed. A design is said to be *robust second-order symmetric balanced* if *all* the *odd moments* up to order *four* are zero, and the *moments are invariant* with respect to permutation of the factors, and independent of correlation parameter involved in the variance-covariance structure of errors (Das *et al.*, 2010a, 2010b). It is established that a robust second-order design with 2 factors is robust slope-rotatable with equal maximum directional variance if and only if it is robust second-order rotatable (Park *et al.*, 2009; Das *et al.*, 2010a). Robust second-order slope-rotatable designs with equal maximum directional variance for $k(> 2)$ factors are investigated within the class of robust second-order symmetric balanced designs. A necessary and sufficient condition for a design ($k > 2$ factors) to be robust slope-rotatable with equal maximum directional variance is derived for correlated errors. It is established that a robust second-order symmetric balanced design ($k > 2$ factors) is robust slope-rotatable with equal maximum directional variance if and only if it is robust second-order rotatable (Das *et al.*, 2010a).

Huda (2006) discussed various concepts of optimality for estimating the slopes of a response surface in the context of uncorrelated and homoscedastic errors. The present chapter also develops the concept of *D-optimal robust slope-rotatable* designs. A design is said to be D-optimal robust slope-rotatable if the determinant of the variance-covariance matrix of the estimated slopes at a point is only a function of distance of the point from the origin, independent of correlation parameter or parameters involved in the variance-covariance structure of errors. D-optimal robust second-order slope-rotatable designs have been investigated within the class of robust second-order symmetric balanced designs. A necessary and sufficient condition for a design to be D-optimal slope-rotatable with correlated errors has been derived by Das *et al.* (2010b). It is established that, a robust second-order symmetric balanced design is D-optimal robust slope-rotatable if and only if it is robust second-order rotatable. A class of robust second-order slope-rotatable designs over all directions, D-optimal slope and with equal maximum directional variance slope-rotatability are investigated for different correlated error structures.

7.2 ESTIMATION OF DERIVATIVES

The second-order response surface model with correlated errors is given in (3.1) (Chapter 3). Following (3.1), second-order correlated response surface model can be written as

$$y(x) = \eta(x) + e,$$

$$\text{or, } y_u(x) = \eta_u(x) + e_u,$$

$$\text{where } \eta_u(x) = \beta_0 + \sum_{i=1}^{k} \beta_i x_{ui} + \sum \sum_{i \le j=1}^{k} \beta_{ij} x_{ui} x_{uj}, \tag{7.1}$$

which may be written in matrix notation as

$$\eta(x) = \mathbf{x}'_s \boldsymbol{\beta}, \tag{7.2}$$

in which the $1 \times m$ vector $\mathbf{x}'_s = (1, x_1^2, x_2^2, ..., x_k^2, x_1, x_2, ..., x_k, x_1 x_2, x_1 x_3, ..., x_{k-1} x_k)$ and $\boldsymbol{\beta}$ is the $m \times 1$ column vector of unknown regression coefficients given in (3.1) and $m = \binom{k+2}{2}$.

For known variance-covariance matrix W of errors, the best linear unbiased estimate of $\boldsymbol{\beta}$ assuming $(X'W^{-1}X)$ is positive definite, is $\hat{\boldsymbol{\beta}} = (X'W^{-1}X)^{-1}(X'W^{-1}\mathbf{Y})$. Therefore, the fitted response at $\mathbf{x}_s$ is $\hat{y}(x) = \mathbf{x}'_s \hat{\boldsymbol{\beta}}$. When the fitted response $\hat{y}(x) = \mathbf{x}'_s \hat{\boldsymbol{\beta}}$ is to be used to estimate $\eta(x)$, it is well known that

$$\text{Var}[\hat{y}(x)] = \mathbf{x}'_s (X'W^{-1}X)^{-1} \mathbf{x}_s. \tag{7.3}$$

Thus, $\text{Var}[\hat{y}(x)]$ depends on the particular values of the independent variables through the vector $\mathbf{x}'_s$. It also depends on the design and the correlation parameter or parameters involved in W through the matrix $(X'W^{-1}X)^{-1}$.

For uncorrelated and homoscedastic observations, Box and Hunter (1957) suggested that, subject to a suitable scaling of the independent variables with respect to each other, it would be desirable to have equally reliable estimates of $\eta(x)$ for all points $\mathbf{x}' = (x_1, x_2, ..., x_k)$ equidistant from the origin, that is to have the variance of $\hat{y}(x)$ as a function of $r = (x_1^2 + x_2^2 + ... + x_k^2)^{1/2}$. This in turn requires $(X'W^{-1}X)$ to be invariant under rotation of axes, and therefore, designs having the property are called rotatable or robust rotatable (independent of correlation parameter or parameters involved in W) designs (Das, 1997).

Suppose now that the estimation of the first derivatives of $\eta(x)$ is of interest. For the second-order correlated response surface model

$$\frac{\partial \hat{y}(x)}{\partial x_i} = \hat{\beta}_i + 2\hat{\beta}_{ii} x_i + \sum_{j=1; \; j \ne i}^{k} \hat{\beta}_{ij} x_j, \tag{7.4}$$

the variance of this derivative is a function of the point $\mathbf{x}$ at which the derivative is estimated and also a function of the design and the correlation parameter or parameters in W through the relationship $\text{Var}(\hat{\beta}) = (X'W^{-1}X)^{-1}$. Das (2003b) proposed robust slope-rotatable designs (Chapter 6) analogous to the Hader and Park (1978) slope-rotatability criterion- which requires that the variance of $\frac{\partial \hat{y}(x)}{\partial x_i}$ to be constant, independent of correlation parameter or parameters involved in W. Estimates of the derivative along axial directions would then be equally reliable for all points $\mathbf{x}$ equidistant and for each of the partial derivatives along axial directions, from the design origin. Das (2003b) referred to this property under correlated error structure as *robust slope rotatability* and the rotatability property analogous to the Box-Hunter property as $\hat{y}$-robust rotatability (Das, 1997, 1999, 2003a, 2004).

7.3   ROBUST SECOND-ORDER SLOPE-ROTATABILITY OVER ALL DIRECTIONS

In practice, it is often of interest to estimate the slope of the response surface at a point $\mathbf{x}$, not only along the axial directions as done in Chapter 6, but also over any specified direction. The present section develops the concept of robust slope-rotatability over all directions (Das and Park, 2006; Park *et al.*, 2009). Let

$$\mathbf{g}(\mathbf{x}) = \left(\frac{\partial \hat{y}(\mathbf{x})}{\partial x_1}, \frac{\partial \hat{y}(\mathbf{x})}{\partial x_2}, ..., \frac{\partial \hat{y}(\mathbf{x})}{\partial x_k}\right)' = D\hat{\beta} \text{ say,} \qquad (7.5)$$

where D is the matrix arising from the differentiation of $\mathbf{x}'_s\hat{\beta}$ with respect to each of the k independent variables, where $\mathbf{x}'_s = (1, x_1^2, x_2^2,...,x_k^2, x_1, x_2,...,x_k, x_1x_2, x_1x_3,...,x_{k-1}x_k)$. For $k = 3$, "D" matrix is given below as an illustration

$$D = \begin{pmatrix} 0 & 2x_1 & 0 & 0 & 1 & 0 & 0 & x_2 & x_3 & 0 \\ 0 & 0 & 2x_2 & 0 & 0 & 1 & 0 & x_1 & 0 & x_3 \\ 0 & 0 & 0 & 2x_3 & 0 & 0 & 1 & 0 & x_1 & x_2 \end{pmatrix}.$$

The estimated derivative at any point $\mathbf{x}$ in the direction specified by the $k \times 1$ vector of direction cosines

$$\mathbf{p}' = (p_1, p_2, ..., p_k) \qquad (7.6)$$

is $\mathbf{p}'\mathbf{g}(\mathbf{x})$, where $\sum_{i=1}^{k} p_i^2 = 1$. The variance of this slope is

$$V_{\mathbf{p}}(\mathbf{x}) = \text{Var}[\mathbf{p}'\mathbf{g}(\mathbf{x})] = \mathbf{p}'D\text{Var}(\hat{\beta})D'\mathbf{p} = \mathbf{p}'D(X'W^{-1}X)^{-1}D'\mathbf{p}. \qquad (7.7)$$

If we are interested in all possible directions of $\mathbf{p}$, we want to consider the average of $V_{\mathbf{p}}(\mathbf{x})$ over all possible directions. The following lemma is needed to derive the necessary and sufficient conditions of a roust second-order slope-rotatable design over all directions.

Lemma 7.1 The average of $V_{\mathbf{p}}(\mathbf{x})$ over all possible directions (for known W) is

$$\bar{V}(\mathbf{x}) = \frac{1}{k} \, tr. \, [D(X'W^{-1}X)^{-1}D']. \qquad (7.8)$$

Proof. Suppose $M = D\text{Var}(\hat{\beta})D'$. The average of $V(\mathbf{x})$ over all directions is carried out in such a way that the distribution over all directions is uniform is

$$\bar{V}(\mathbf{x}) = avg_p(p'Mp) = avg_p[tr. \, (p'Mp)] = avg_p[tr. \, (Mpp')] = tr. \, [M \times avg_p(pp')].$$

Therefore, we have to calculate

$$avg_p(pp') = C \int pp' dA,$$

where dA is the area element of the hypersphere with unit radius, and C is the reciprocal of the surface area of the unit space in a k-dimensional space, namely,

$$U_s = \frac{2\pi^{k/2}}{\Gamma\frac{k}{2}}.$$

Park (1987) has shown that $C \int pp' dA$ is a diagonal matrix whose diagonal elements are all equal. Let $C \int pp' dA = \nu I_k$, then

$$\bar{V}(\mathbf{x}) = tr. (M\nu I_k) = tr. (\nu M) = \nu\, tr. (M) = \nu \sum_{i=1}^{k} \lambda_i,$$

where λ_i's are the eigenvalues of M. Note that ν is independent of M and is associated with $C \int pp' dA$. To find the constant ν, assume M is the identity matrix. The average of $p'Mp$ is 1 and $\lambda_i = 1$ for all i. Therefore, $\nu = \frac{1}{k}$ and $\bar{V}(\mathbf{x})$ becomes

$$\bar{V}(\mathbf{x}) = \frac{1}{k}\, tr. (M) = \frac{1}{k}\, tr. [D(X'W^{-1}X)^{-1}D']. \tag{7.9}$$

Note that $\bar{V}(\mathbf{x})$ is a function of $\mathbf{x}$, the point at which the derivative is being estimated, the correlation parameter or parameters involved in W, and also a function of the design through the design matrix X. By choice of design, independent of correlation parameter or parameters involved in W, it is possible to make this variance $\bar{V}(\mathbf{x})$ constant for all points equidistant from the design origin. This property will henceforth be called *robust slope rotatability over all directions*. In this sense, the slope rotatability by Das (2003b) considered in Chapter 6, is called *robust slope rotatability along axial directions*. The general conditions of second-order slope-rotatability over all directions under the correlated error structure (W) are given below.

Theorem 7.1 (Das and Park, 2006) *The necessary and sufficient conditions for a design to be slope-rotatable over all directions under the correlated error structure W (for all ρ in W and $1 \leq i, j \leq k$) are*

(i) $2\text{Cov}(\hat{\beta}_i, \hat{\beta}_{ii}) + \sum\limits_{j=1;\, j\neq i}^{k} \text{Cov}(\hat{\beta}_j, \hat{\beta}_{ij}) = 0$ *for all i,*

(ii) $2[\text{Cov}(\hat{\beta}_{ii}, \hat{\beta}_{ij}) + \text{Cov}(\hat{\beta}_{jj}, \hat{\beta}_{ij})] + \sum\limits_{t=1; t\neq i,j}^{k} \text{Cov}(\hat{\beta}_{it}, \hat{\beta}_{jt}) = 0$ *for all $(i,j), i \neq j$,*

(iii) $4\text{Var}(\hat{\beta}_{ii}) + \sum\limits_{j=1; j\neq i}^{k} \text{Var}(\hat{\beta}_{ij}) = constant = \delta,$ *say, for all i.* $\tag{7.10}$

Proof. From Lemma 7.1, the average $V_{\mathbf{p}}(\mathbf{x})$ over all directions is

$$\bar{V}(\mathbf{x}) = \frac{1}{k}\, tr. [D(X'W^{-1}X)^{-1}D'] = \frac{1}{k}\, tr. [(X'W^{-1}X)^{-1}D'D]$$

$$= \frac{1}{k}\sum_{i=1}^{k} v^{i.i} + \frac{2}{k}\sum_{i=1}^{k} x_i [2v^{ii.i} + \sum_{j=1;\, j\neq i}^{k} v^{j.ij}]$$

$$+ \frac{2}{k} \sum_{i=1}^{k}\sum_{j=1;\ i\neq j}^{k} x_i x_j [2v^{ii.ij} + \sum_{t=1;\ t\neq i,j}^{k} v^{it.jt}] + \frac{1}{k}\sum_{i=1}^{k} x_i^2 [4v^{ii.ii} + \sum_{j=1;\ j\neq i}^{k} v^{ij.ij}]. \quad (7.11)$$

It is immediately clear that the stated three conditions in (7.10) are sufficient to insure that $\bar{V}(\mathbf{x})$ in (7.11) will be a function of $r^2 = \sum_{i=1}^{k} x_i^2$ for slope-rotatability over all directions for *all* $\mathbf{x}$ & ρ, that is

$$\bar{V}(\mathbf{x}) = \frac{1}{k}\sum_{i=1}^{k} \mathrm{Var}(\hat{\beta}_i) + \frac{\delta}{k} r^2. \quad (7.12)$$

Again if any of the three conditions as in (7.10) are *not* satisfied, $\bar{V}(\mathbf{x})$ may be written as

$$= \frac{1}{k}\sum_{i=1}^{k} v^{i.i} + \frac{2}{k}\sum_{i=1}^{k} x_i d_i^* + \frac{2}{k} \sum_{i=1}^{k}\sum_{j=1;\ i\neq j}^{k} x_i x_j e_{ij}^* + \frac{1}{k}\sum_{i=1}^{k} x_i^2 f_i^*, \quad (7.13)$$

where d_i^*, e_{ij}^* and f_i^* are arbitrary constants. Then $\bar{V}(\mathbf{x})$ can not be written as a function of r^2 and also are different at two equidistant points from the design center. Therefore, the three conditions in (7.10) are necessary for a design to be slope-rotatable over all directions *for all* values of ρ in W.

Equivalent conditions of (i) through (iii) in (7.10) of second-order slope-rotatability over all directions in terms of the elements of the moment matrix are given in the following theorem (proof is immediate from (7.10)).

Theorem 7.2 (Das and Park, 2006) *The necessary and sufficient conditions for a design to be slope-rotatable over all directions under the correlated error structure W are*

(I) : (i) $v_{0.j} = v_{0.jl} = 0;\ 1 \leq j < l \leq k$,

(ii) $v_{i.j} = 0;\ 1 \leq i,\ j \leq k,\ i \neq j$,

(iii): (1) $v_{ii.j} = 0;\ 1 \leq i,\ j \leq k$,

(2) $v_{i.jl} = 0;\ 1 \leq i,\ j < l \leq k$,

(3) $v_{ii.jl} = 0;\ 1 \leq i,\ j < l \leq k,\ (j,l) \neq (i,i)$,

(4) $v_{ij.lt} = 0;\ 1 \leq i,\ l < j,\ t \leq k,\ (i,j) \neq (l,t)$,

(II): (i) $v_{0.jj} = $ constant; $1 \leq j \leq k$,

(ii) $v_{i.i} = $ constant; $1 \leq i \leq k$,

(iii) $v_{ii.ii} = $ constant; $1 \leq i \leq k$,

(III): (i) $v_{ii.jj} = $ constant; $1 \leq i,\ j \leq k,\ i \neq j$,

(ii) $v_{ij.ij} = $ constant; $1 \leq i < j \leq k$. \quad (7.14)

Definition 7.1 Robust second-order slope rotatable designs over all directions:
A design d on k factors under the model (3.1) which remains second-order slope-rotatable over all directions for all the variance covariance matrices belonging to a well defined class

$W_0 = \{W$ positive definite: $W_{N \times N}$ defined by a particular correlation structure possessing a definite pattern$\}$ is called a *Robust Second-Order Slope Rotatable Design Over all Directions* (RSOSRDOAD), with reference to the variance covariance class W_0.

Corollary 7.1 All $\hat{y}$-robust rotatable designs in the Das (1997) sense are robust slope-rotatable over all directions.

Proof. It has been shown (in Chapter 3, (3.3) (Das, 1997)) that in addition to the three moment conditions in (7.14) to be $\hat{y}$-robust rotatabale, one more moment condition, $v_{ii.ii} = 2v_{ij.ij} + v_{ii.jj}$; $1 \le i < j \le k$, should also be satisfied. Therefore, it is obvious that $\hat{y}$-robust rotatable designs are robust slope rotatable over all directions. However, robust slope-rotatable designs over all directions are *not* necessarily $\hat{y}$-robust rotatable.

Note that a class of robust slope-rotatable designs over all directions is examined by Das and Park (2006) under different correlated error structures.

7.4 ROBUST SECOND-ORDER SYMMETRIC BALANCED DESIGN

The discussion in the present section does not involve slope-rotatability as such, but is utilized in finding the D-optimal slope and robust slope-rotatable designs with equal maximum directional variance from within the class of robust symmetric, balanced designs ξ's, when the number of factors $k > 2$. These are considered in Sections 7.5 and 7.6.

Second-order rotatability conditions and analysis with correlated errors are given in Das (1997) (Chapter 3). From the second-order correlated model (3.1) (Chapter 3), it is derived that $\text{Dis}(\hat{\beta}) = (X'W^{-1}X)^{-1} = M^{-1}(\xi)$ say, where ξ is any second-order design with model matrix X and $M(\xi)$ is known as the moment matrix of the design ξ. Here we use the same notations (v's) as in Chapter 3 (Das, 1997). For ready reference they are reproduced as $v_{00} = \mathbf{1}'W^{-1}\mathbf{1}$; $v_{0.jl} = \mathbf{1}'W^{-1}(\mathbf{x}_j \otimes_1 \mathbf{x}_l)$; $v_{i.j} = \mathbf{x}_i'W^{-1}\mathbf{x}_j$; $v_{i.jl} = \mathbf{x}_i'W^{-1}(\mathbf{x}_j \otimes_1 \mathbf{x}_l)$; $v_{ij.lt} = (\mathbf{x}_i \otimes_1 \mathbf{x}_j)'W^{-1}(\mathbf{x}_l \otimes_1 \mathbf{x}_t)$; $1 \le i,l < j,t \le k$.

Note that the moment matrix ($M(\xi)$) of an arbitrary design ξ is symmetric, but it is *not* algebraically feasible to find the inverse of $M(\xi)$ for any arbitrary design ξ, and for large k. In analogous to symmetric design with uncorrelated and homoscedastic errors, a second-order design ξ is said to be *robust symmetric* under the correlated model (3.1) (Chapter 3) if all the *odd moments* up to order *four* are zero for all values of ρ in W, where ρ is the correlation parameter in W. Thus, for a second-order robust symmetric design ξ under the model (3.1), the following hold:

(i) $v_{0.j} = v_{0.jl} = v_{i.j}(i \ne j) = v_{ii.j} = v_{i.jl} = 0$; $1 \le i,\ j < l \le k$,

(ii) $v_{ii.jl} = v_{ij.lt} = 0$; $1 \le i < j \le k,\ 1 \le l < t \le k,\ (i,j) \ne (l,t)$ $\qquad (7.15)$

for all values of ρ in W, and v_{ij}'s are as in (3.2).

In analogous to balanced (permutation invariant) design with independent and homoscedastic errors, a second-order design ξ is said to be *robust balanced* under the correlated model (3.1) if the *moments are invariant* with respect to permutation of the factors $x_1,...,x_k$, and for all values of ρ in W. Thus, for a second-order robust balanced design ξ under the correlated model (3.1), the following hold for all values of ρ in W:

(i) $v_{0.jj} = a_0(> 0)$ say; (ii) $v_{i.i} = \frac{1}{e}(> 0)$ say; (iii) $v_{ii.ii} = b(> 0)$ say; $1 \le i,\ j \le k$,

(iv) $v_{ii.jj} = d(> 0)$ $(i \ne j)$; (v) $v_{ij.ij} = \frac{1}{c}(> 0)$; $1 \le i < j \le k$, $\qquad (7.16)$

where a_0, b, $\frac{1}{c}$, d, $\frac{1}{e}$ are constants. Therefore, the moment matrix $M(\xi)$ of a second-order *robust symmetric balanced* design ξ under the correlated model (3.1) is given by $M(\xi) =$ Diag.$\{M_{11}(\xi), D_1, D_2\}$, where

$$M_{11}(\xi) = \begin{pmatrix} v_{00} & a_0 E_{1k} \\ a_0 E_{k1} & (b-d)I_k + dE_{kk} \end{pmatrix}_{(k+1)\times(k+1)}, \quad D_1 = (\frac{1}{e})I_k, \quad D_2 = (\frac{1}{c})I_{k^*},$$

$$M_{11}^{-1}(\xi) = \begin{pmatrix} v_1 & qE_{1k} \\ qE_{k1} & (s-t)I_k + tE_{kk} \end{pmatrix}, \quad v_1 = \frac{b+(k-1)d}{v_{00}\{b+(k-1)d\} - ka_0^2}, \quad s = t + \frac{1}{b-d},$$

$$q = \frac{-a_0}{v_{00}\{b+(k-1)d\} - ka_0^2}, \quad t = \frac{(a_0^2 - dv_{00})}{(b-d)[v_{00}\{b+(k-1)d\} - ka_0^2]}, \quad k^* = \binom{k}{2}, \quad (7.17)$$

where I_k is an identity matrix of order k, E_{mn} is an $m \times n$ matrix with all elements 1, and $v_{00} = \mathbf{1}'W^{-1}\mathbf{1}$, a_0, b, $\frac{1}{c}$, $\frac{1}{e}$ and d are as in (7.16).

Under robust second-order rotatability (Chapter 3), the dispersion matrix of $\hat{\boldsymbol{\beta}}$ reduces to the following form:

$$\text{Dis}(\hat{\boldsymbol{\beta}}) = \begin{pmatrix} A_1 & 0 & 0 \\ 0 & D_1 & 0 \\ 0 & 0 & D_2 \end{pmatrix}^{-1} = \begin{pmatrix} A_1^{-1} & O & O \\ O & D_1^{-1} & O \\ O & O & D_2^{-1} \end{pmatrix}$$

where D_1, D_2 are as in (7.17), and

$$A_1 = \begin{pmatrix} v_{00} & a_0 E_{1k} \\ a_0 E_{k1} & \frac{2}{c}I_k + dE_{kk} \end{pmatrix}, \quad A_1^{-1} = \begin{pmatrix} v^{00} & a_1 E_{1k} \\ a_1 E_{k1} & \frac{c}{2}I_k + d_1 E_{kk} \end{pmatrix},$$

$$v^{00} = \frac{(\frac{2}{c} + kd)}{\{v_{00}(\frac{2}{c} + kd) - ka_0^2\}}, a_1 = \frac{-a_0}{\{v_{00}(\frac{2}{c} + kd) - ka_0^2\}}, \quad d_1 = \frac{c(a_0^2 - dv_{00})}{2\{v_{00}(\frac{2}{c} + kd) - ka_0^2\}}.$$

From (7.16) and the conditions of rotatability (3.3) (Chapter 3), it is clear that a second-order *robust symmetric balanced design* ξ will be robust second-order rotatable if

$$b = \frac{2}{c} + d \Rightarrow t = d_1, \ v_1 = v^{00}, \ q = a_1, \text{ and } s = (d_1 + \frac{c}{2}), \qquad (7.18)$$

where b, $\frac{1}{c}$, d, d_1, q, s, t, v^{00}, and a_1 are as in above, and in (7.16), (7.17) (Das *et al.*, 2010a).

Remark 7.1 A robust second-order rotatable design is always a robust second-order symmetric balanced design *but* the converse is not always true.

7.5 ROBUST SLOPE-ROTATABILITY WITH EQUAL MAXIMUM DIRECTIONAL VARIANCE

In the present section, the concept of *robust second-order slope-rotatability with equal maximum directional variance* is developed for correlated errors. Robust slope-rotatability over all directions is given in Section 7.3, and we use the same notations (D, $\mathbf{p}$, $V_{\mathbf{p}}(\mathbf{x})$, and $\bar{V}(\mathbf{x})$) as in Section 7.3, in the present section.

For a general correlated error structure W, the concept of *robust slope-rotatability with equal maximum directional variance* has been developed (Park *et al.*, 2009; Das *et al.*, 2010a) with correlated errors, following Park and Kwon (1998). The average slope variance $\bar{V}(\mathbf{x})$ $(= \frac{1}{k} \, tr. \, [D(X'W^{-1}X)^{-1}D'])$ is the average variance of $V_{\mathbf{p}}(\mathbf{x})$ $(= \mathbf{p}'D(X'W^{-1}X)^{-1}D'\mathbf{p})$, the variance of the estimated derivative at a point $\mathbf{x}$ in the direction specified by a $k \times 1$ vector of direction cosines $\mathbf{p} = (p_1, p_2, ..., p_k)'$ (Section 7.3). Note that D is a $k \times p$ matrix whose i-th row is $\frac{\partial \hat{y}_{\mathbf{x}}}{\partial x_i}(i = 1, 2, ..., k)$ with $p = \frac{(k+2)(k+1)}{2}$, and $V_{\mathbf{p}}(\mathbf{x})$ is a function of $\mathbf{p}$ and ρ in W, so it is important to investigate the *maximum* and *minimum* of $V_{\mathbf{p}}(\mathbf{x})$. Let $\lambda_{max}(\mathbf{x})$ and $\lambda_{min}(\mathbf{x})$ be the maximum and minimum variance of $V_{\mathbf{p}}(\mathbf{x})$ at a given point $\mathbf{x}$, where

$$\lambda_{max}(\mathbf{x}) = max_{\mathbf{p}:\mathbf{p}'\mathbf{p}=1}V_{\mathbf{p}}(\mathbf{x}) = max_{\mathbf{p}:\mathbf{p}'\mathbf{p}=1}\mathbf{p}'D(X'W^{-1}X)^{-1}D'\mathbf{p}$$

$$\text{and} \quad \lambda_{min}(\mathbf{x}) = min_{\mathbf{p}:\mathbf{p}'\mathbf{p}=1}V_{\mathbf{p}}(\mathbf{x}) = min_{\mathbf{p}:\mathbf{p}'\mathbf{p}=1}\mathbf{p}'D(X'W^{-1}X)^{-1}D'\mathbf{p}, \quad (7.19)$$

which are called the maximum and the minimum directional variance at a given point $\mathbf{x}$ in the region of interest, respectively.

Let $\lambda_1 \geq \lambda_2 \geq ... \geq \lambda_k$ be the eigenvalues of $D(X'W^{-1}X)^{-1}D'$ and $\mathbf{e_1}, \mathbf{e_2}, ..., \mathbf{e_k}$ be the corresponding eigenvectors. Then it is well known that $\lambda_{max}(\mathbf{x}) = \lambda_1$ and $\lambda_{min}(\mathbf{x}) = \lambda_k$, and the maximum and the minimum values are attained when $\mathbf{p} = \mathbf{e_1}$ and $\mathbf{p} = \mathbf{e_k}$, respectively. Also the eigenvectors corresponding to different eigenvalues are orthogonal to each other. If we have a choice in response designs, it will be very nice to have a design with small $\lambda_{max}(\mathbf{x})$ as well as uniform $\lambda_{max}(\mathbf{x})$ for all points $\mathbf{x}$ equidistant from the design origin, and for *all* values of ρ in W. If $\lambda_{max}(\mathbf{x})$ is *constant* on the circle $(k = 2)$, sphere $(k = 3)$, or hypersphere $(k \geq 4)$ centered at the design origin for *all* values of ρ in W, the estimates of the derivative would then be equally reliable for all points $\mathbf{x}$ equidistant from the center point in the view point of maximum directional variance. This property will henceforth be called *robust slope-rotatable designs with equal maximum directional variance* with respect to correlation structure W.

The matrix $D(X'W^{-1}X)^{-1}D'$, the variance-covariance matrix of the estimated slope vector at a point $\mathbf{x}$, depends on the point $\mathbf{x}$, on the design, and ρ in W through $(X'W^{-1}X)^{-1}$. Note that $D(X'W^{-1}X)^{-1}D'$ is a symmetric $k \times k$ matrix in k independent variables regardless of the assumed order of the model. Let $M = D(X'W^{-1}X)^{-1}D' = ((m_{ij}); 1 \leq i, j \leq k)$. The eigenvalues of the matrix M are the roots of the characteristic equation for the matrix M, which is

$$f(\lambda) = |M - \lambda I| = (-\lambda)^k + q_{k-1}(-\lambda)^{k-1} + ... + q_1(-\lambda) + q_0 = 0, \quad (7.20)$$

where $|A|$ denotes the determinant of the matrix A and q_i's are coefficients. If the order of the characteristic equation is greater than 4, the general solution of (7.20) does not exist. Therefore, it is not easy to find the necessary and sufficient conditions for robust slope-rotatability with equal maximum directional variance for any dimensions. In order to study for more than two factors we restrict ourselves within the *robust second-order symmetric balanced designs*. Below we develop *robust slope-rotatability with equal maximum directional variance* for $k = 2$ and $k > 2$ factors (Das *et al.*, 2010a). Robust slope-rotatable designs with equal maximum directional variance for $k = 2$ and $k > 2$ factors are searched respectively within the general second-order designs, and robust second-order symmetric balanced designs.

Case I: $k = 2$ **factors**

$$M = D(X'W^{-1}X)^{-1}D'$$

$$= \begin{pmatrix} m_{11} & m_{12} \\ m_{21} & m_{22} \end{pmatrix} = D \begin{pmatrix} v^{00} & v^{0.1} & v^{0.2} & v^{0.11} & v^{0.22} & v^{0.12} \\ v^{1.0} & v^{1.1} & v^{1.2} & v^{1.11} & v^{1.22} & v^{1.12} \\ v^{2.0} & v^{2.1} & v^{2.2} & v^{2.11} & v^{2.22} & v^{2.12} \\ v^{11.0} & v^{11.1} & v^{11.2} & v^{11.11} & v^{11.22} & v^{11.12} \\ v^{22.0} & v^{22.1} & v^{22.2} & v^{22.11} & v^{22.22} & v^{22.12} \\ v^{12.0} & v^{12.1} & v^{12.2} & v^{12.11} & v^{12.22} & v^{12.12} \end{pmatrix} D'$$

where $D = \begin{pmatrix} 0 & 2x_1 & 0 & 1 & 0 & x_2 \\ 0 & 0 & 2x_2 & 0 & 1 & x_1 \end{pmatrix}$, and $v^{i.i} = \text{Var}(\hat{\beta}_i)$, $v^{ii.ii} = \text{Var}(\hat{\beta}_{ii})$, $v^{ij.ij} = \text{Var}(\hat{\beta}_{ij})$, $v^{i.ii} = \text{Cov}(\hat{\beta}_i, \hat{\beta}_{ii})$, $v^{i.jj} = \text{Cov}(\hat{\beta}_i, \hat{\beta}_{jj})$, $v^{i.ij} = \text{Cov}(\hat{\beta}_i, \hat{\beta}_{ij})$, $v^{ii.ij} = \text{Cov}(\hat{\beta}_{ii}, \hat{\beta}_{ij})$, $v^{ij.il} = \text{Cov}(\hat{\beta}_{ij}, \hat{\beta}_{il})$.

By straightforward calculation, it can be shown that

$$m_{11} = v^{1.1} + 4v^{1.11}x_1 + 2v^{1.12}x_2 + 4v^{11.11}x_1^2 + v^{12.12}x_2^2 + 4v^{11.12}x_1x_2,$$

$$m_{12} = m_{21} = v^{1.2} + (v^{1.12} + 2v^{2.11})x_1 + (2v^{1.22} + v^{2.12})x_2$$
$$+ 2v^{11.12}x_1^2 + 2v^{22.12}x_2^2 + (4v^{11.22} + v^{12.12})x_1x_2,$$

$$m_{22} = v^{2.2} + 4v^{2.22}x_2 + 2v^{2.12}x_1 + 4v^{22.22}x_2^2 + v^{12.12}x_1^2 + 4v^{22.12}x_1x_2.$$

From (7.20), the characteristic equation for the matrix M is

$$\lambda^2 - (m_{11} + m_{22})\lambda + m_{11}m_{22} - m_{12}^2 = 0.$$

Therefore, the maximum directional variances are equal at all points equidistant from the origin if and only if the larger value of the two roots of this equation,

$$\frac{(m_{11} + m_{22}) + [(m_{11} - m_{22})^2 + 4m_{12}^2]^{1/2}}{2}, \tag{7.21}$$

is a function of $r^2 = x_1^2 + x_2^2$ for *all* values of ρ in W.

Theorem 7.3 (Das *et al.* 2010a) *The necessary and sufficient conditions for a design to be robust slope rotatable with equal maximum directional variance for the second-order response surface models under correlated error structure W for two independent variables are as follows (for all ρ in W):*

1. $v^{i.j} = v^{i.ii} = v^{i.ij} = v^{i.jj} = v^{ii.ij} = 0; (i \neq j; i, j = 1, 2)$,

2. $v^{1.1} = v^{2.2}$,

3. $v^{11.11} = v^{22.22}$,

4. $2(v^{11.11})^2 - v^{11.11}v^{12.12} - 2(v^{11.22})^2 - v^{11.22}v^{12.12} = 0.$ \tag{7.22}

Proof. If the four conditions in (7.22) hold then

$$m_{11} + m_{22} = 2v^{1.1} + (4v^{11.11} + v^{12.12})r^2 \text{ and } (m_{11} - m_{22})^2 + 4m_{12}^2 = (4v^{11.11} - v^{12.12})^2 r^4,$$

so that the expression in (7.21) is a function of r^2 for all values of ρ in W. That is $\lambda_{max}(\mathbf{x})$ is a function of r^2 independent of ρ in W.

In general, the expressions for $(m_{11} + m_{22})$ and $(m_{11} - m_{22})^2 + 4m_{12}^2$ are given below:

$$(m_{11} + m_{22}) = (v^{1.1} + v^{2.2}) + (4v^{1.11} + 2v^{2.12})x_1 + (2v^{1.12} + 4v^{2.22})x_2$$

$$+ (4v^{11.11} + v^{12.12})x_1^2 + (4v^{22.22} + v^{12.12})x_2^2 + (4v^{11.12} + 4v^{22.12})x_1x_2,$$

$$= c + \sum_i (d_{1i}x_i + e_ix_i^2) + fx_1x_2 \text{ say},$$

and $(m_{11} - m_{22})^2 + 4m_{12}^2 = [(v^{1.1} - v^{2.2}) + (4v^{1.11} - 2v^{2.12})x_1 + (2v^{1.12} - 4v^{2.22})x_2$

$+(4v^{11.11}-v^{12.12})x_1^2+(v^{12.12}-4v^{22.22})x_2^2+(4v^{11.12}-4v^{22.12})x_1x_2]^2+4[v^{1.2}+(v^{1.12}+2v^{2.11})x_1$

$\qquad +(2v^{1.22}+v^{2.12})x_2+2v^{11.12}x_1^2+2v^{22.12}x_2^2+(4v^{11.22}+v^{12.12})x_1x_2]^2,$

$$= c' + \sum_i (d'_{1i}x_i + e'_ix_i^2) + f'x_1x_2 + \sum_i g'_ix_i^3 + \sum_{i \neq j} h'_{ij}x_i^2x_j + \sum_i s'_ix_i^4 + t'x_1^2x_2^2, \text{ say,}$$

where $c, d_1, e, f, c', d'_1, e', f', g', h', s'$ and t' are arbitrary constants.

The followings are a necessary so that the above two equations are to be a function of r^2 for all ρ in W. First, all the coefficients with respect to odd order terms, such as $d_{1i}, d'_{1i}, f, f', g'$ and h'_{ij} must be zero. Second, the coefficients corresponding to x_1^2 and x_1^4 must be identical to those for x_2^2 and x_2^4, respectively. Third, the coefficient of $x_1^2x_2^2$ has to be twice that of x_1^4. If any one of the four conditions in Theorem 7.3 are omitted, then the above requirements are not always satisfied. Hence the four conditions in (7.22) are necessary.

Now we have the following theorem for obtaining robust slope rotatable designs with equal maximum directional variance for the second-order response surface models under correlated error structure W for two independent variables.

Theorem 7.4 (Das *et al.* 2010a) *A design is robust slope rotatable with equal maximum directional variance for the second-order response surface models under correlated error structure W for two independent variables if it is robust second-order rotatable.*

Proof. The first three conditions (in Theorem 7.3) are always satisfied by RSORDs for a general correlation structure W for all values of $\rho \in W$, which is directly followed from the conditions of rotatability, given in Chapter 3 (Das, 1997). Also, $v^{11.11} = v^{22.22} = d_1 + \frac{c}{2}$, $v^{12.12} = c$, $v^{11.22} = d_1$ (from the conditions of rotatability, given in Chapter 3). On substitution of these values in condition 4 (in Theorem 7.3), it is verified that $2(v^{11.11})^2 - v^{11.11}v^{12.12} - 2(v^{11.22})^2 - v^{11.22}v^{12.12} = 0$, for all $\rho \in W$. Thus, a RSORD is always a robust slope-rotatable design with equal maximum directional variance (for the second-order response surface models) under the correlated error structure W for two independent variables.

Case II: $k > 2$ factors

It is very difficult and practically impossible to investigate robust slope-rotatable designs with equal maximum directional variance for $k > 2$ factors within general second-order designs. For a general second-order design, it is practically impossible to find the expression of $D(X'W^{-1}X)^{-1}D'$. Thus, we investigate these designs within the robust second-order symmetric balanced designs. For a second-order design ξ the variance-covariance matrix of $\frac{d\hat{y}}{d\mathbf{x}}$ is given by

$$V(\xi, \mathbf{x}) = M = D(X'W^{-1}X)^{-1}D'. \tag{7.23}$$

Therefore, for a *robust symmetric balanced* second-order design ξ with $k = 2$ factors

$$V(\xi, \mathbf{x}) = \begin{pmatrix} (4s-c)x_1^2 + (e+cr^2) & (4t+c)x_1x_2 \\ (4t+c)x_1x_2 & (4s-c)x_2^2 + (e+cr^2) \end{pmatrix},$$

which is equivalently written as

$$V(\xi, \mathbf{x}) = (e+cr^2)I_2 + \{4s - (4t+2c)\}\text{Diag.}\{x_1^2, x_2^2\} + (4t+c)\begin{pmatrix} x_1^2 & x_1x_2 \\ x_1x_2 & x_2^2 \end{pmatrix}. \tag{7.24}$$

Thus for an arbitrary k factors $\mathbf{x}$, for a *robust symmetric balanced* second-order design ξ, (7.24) can be easily written as given below.

$$V(\xi, \mathbf{x}) = M = (e + cr^2)I_k + \{4s - (4t + 2c)\}\text{Diag.}\{x_1^2, x_2^2, ..., x_k^2\} + (4t + c)\mathbf{xx}', \quad (7.25)$$

where I_k is an identity matrix of order k, and c, e, s, t as in (7.16) and (7.17), and $r^2 = \mathbf{x}'\mathbf{x}$.

Theorem 7.5 (Das *et al.*, 2010a) *For a second-order model, any robust rotatable symmetric balanced design is also robust slope-rotatable design with equal maximum directional variance for $k > 2$ factors.*

Proof. For a robust symmetric balanced second-order design ξ, it is seen from (7.25) that

$$V(\xi, \mathbf{x}) = M = (e + cr^2)I_k + \{4s - (4t + 2c)\}\text{Diag.}\{x_1^2, x_2^2, ..., x_k^2\} + (4t + c)\mathbf{xx}'.$$

If the robust symmetric balanced second-order design ξ is robust rotatable, $t = d_1 \Rightarrow$ $s = t + \frac{c}{2} = (d_1 + \frac{c}{2})$ as given in (7.18), the second term on the right hand side of (7.25) vanishes, thus (7.25) reduces to

$$V(\xi, \mathbf{x}) = (e + cr^2)I_k + (4d_1 + c)\mathbf{xx}'.$$

For a given direction $\mathbf{p}$ such that $\mathbf{p}'\mathbf{p} = 1$

$$\mathbf{p}'M\mathbf{p} = (e + cr^2)I_k + (4t + c)(\mathbf{x}'\mathbf{p})^2. \quad (7.26)$$

As the coefficient of $(\mathbf{x}'\mathbf{p})^2$ in (7.26) is positive it immediately follows that

$$Max_{\mathbf{p}:\mathbf{p}'\mathbf{p}=1} \ \mathbf{p}'M\mathbf{p} = (e + cr^2)I_k + (4t + c)r^2$$

attained at $\mathbf{p} = \frac{\mathbf{x}}{r}$ (or $-\frac{\mathbf{x}}{r}$). Hence the Theorem is proved.

7.6   *D*-OPTIMAL ROBUST SECOND-ORDER SLOPE-ROTATABLE DESIGNS

In this section, a necessary and sufficient condition is developed for a second-order design to be *D*-optimal robust slope-rotatable. The estimated response at $\mathbf{x}$ is $\hat{y}_\mathbf{x}$, and the column vector of estimated slopes along the factor axes at a point $\mathbf{x}$ is given by $\frac{d\hat{y}}{d\mathbf{x}} = (\frac{\partial \hat{y}_\mathbf{x}}{\partial x_1}, \frac{\partial \hat{y}_\mathbf{x}}{\partial x_2}, ..., \frac{\partial \hat{y}_\mathbf{x}}{\partial x_k})' = D\hat{\beta}$, where D is a $k \times p$ matrix whose i-th row consists of the elements $\frac{\partial \hat{y}_\mathbf{x}}{\partial x_i}(i = 1, 2, ..., k)$ and $p = \frac{(k+2)(k+1)}{2}$. For an example, D for $k = 3$ is given in (7.5). Therefore, for a second-order design ξ the variance-covariance matrix of $\frac{d\hat{y}}{d\mathbf{x}}$ is given by $V(\xi, \mathbf{x}) = DM^{-1}(\xi)D'$, where $M(\xi) = (X'W^{-1}X)$ (in (7.23)).

Note that $V(\xi, \mathbf{x})$ depends on the design ξ and the point $\mathbf{x}$ through D, and also the correlation parameter involved in W. A design ξ is said to be *D*-optimal or *A*-optimal robust slope-rotatable according as $|V(\xi, \mathbf{x})|$ or $tr\ V(\xi, \mathbf{x})$ is a function of $r^2 = \mathbf{x}'\mathbf{x}$ only, independent of correlation parameter ρ involved in W. *A*-optimal and *D*-optimal robust slope-rotatable designs have been studied by Das and Park (2006), Das *et al.* (2010b), respectively. Expressions of $V(\xi, \mathbf{x})$ for a *robust symmetric balanced* second-order design ξ with $k = 2$ factors, and for an arbitrary k factors $\mathbf{x}$ are given in (7.24) and (7.25), respectively.

Thus, from (7.24) the determinant of $V(\xi, \mathbf{x})$ is given by

$$|V(\xi, \mathbf{x})| = (e + cr^2)^2 + (e + cr^2)(4s - c)r^2 + \{(4s - c)^2 - (4t + c)^2\}x_1^2 x_2^2, \qquad (7.27)$$

which will be only a function of $r^2 = (x_1^2 + x_2^2)$ if and only if $s = t + \frac{c}{2}$. If $t = d_1 \Rightarrow s = d_1 + \frac{c}{2}$, which is the condition of robust second-order rotatability for a *robust symmetric balanced* second-order design as given in (7.18). Therefore, we have the following theorem.

Theorem 7.6 (Das *et al.*, 2010b) *A robust symmetric balanced second-order design ξ is D-optimal robust slope-rotatable if and only if it is robust second-order rotatable.*

Proof. For $k = 2$ factors the proof is immediate from (7.27). Now we will prove the above theorem for any k factors. For a *robust symmetric balanced* second-order design ξ (with k factors $\mathbf{x}$), $V(\xi, \mathbf{x})$ is reproduced (from (7.25)) as

$$V(\xi, \mathbf{x}) = (e + cr^2)I_k + \{4s - (4t + 2c)\} \operatorname{Diag.}\{x_1^2, x_2^2, ..., x_k^2\} + (4t + c)\mathbf{x}\mathbf{x}'.$$

where $r^2 = \mathbf{x}'\mathbf{x}$.

If the symmetric balanced second-order design ξ is robust rotatable, $t = d_1 \Rightarrow s = t + \frac{c}{2} = (d_1 + \frac{c}{2})$ as given in (7.18), the second term in the right hand side of the above $V(\xi, \mathbf{x})$ vanishes, thus it reduces to

$$V(\xi, \mathbf{x}) = (e + cr^2)I_k + (4d_1 + c)\mathbf{x}\mathbf{x}'. \qquad (7.28)$$

Therefore, from (7.28) it is clear that eigenvalues (e-values) of $V(\xi, \mathbf{x})$ are $e + (4d_1 + 2c)r^2$ corresponding to e-vector $\mathbf{x}$, and $(e + cr^2)$ with multiplicity $k - 1$ corresponding to e-vectors orthogonal to $\mathbf{x}$. Note that e-values depend on $\mathbf{x}$ through r^2, and $|V(\xi, \mathbf{x})|$ is the product of e-values, therefore, $|V(\xi, \mathbf{x})|$ is a function of r^2 for all $\rho \in W$. Thus, robust rotatability is a sufficient condition for D-optimal robust second-order slope-rotatable design.

In general, $V(\xi, \mathbf{x})$ as in (7.25) can be expressed as

$$V(\xi, \mathbf{x}) = \alpha_1 I_k + (\alpha_2 - \alpha_3) \operatorname{Diag.}\{x_1^2, x_2^2, ..., x_k^2\} + \alpha_3 \mathbf{x}\mathbf{x}', \qquad (7.29)$$

where $\alpha_1 = (e + cr^2)$, $(\alpha_2 - \alpha_3) = 4s - (4t + 2c)$ and $\alpha_3 = (4t + c)$.

Therefore, the determinant of $V(\xi, \mathbf{x})$ as in (7.29) is given by

$$|V(\xi, \mathbf{x})| = \alpha_1^k + \sum_{m=1}^{k} [\alpha_1^{k-m}(\alpha_2 - \alpha_3)^{m-1}\{\alpha_2 + (m-1)\alpha_3\}\{\sum_{1 \le i_1 < i_2 < ... < i_m}^{k} (\prod_{j=i_1}^{i_m} x_j^2)\}]. \quad (7.30)$$

From (7.30) it is clear that $|V(\xi, \mathbf{x})|$ will be a function of $r^2 = \mathbf{x}'\mathbf{x}$ only, and it is necessary that *all* the terms for $m \ge 2$ in the right hand side of (7.30) will be zero for all values of $\mathbf{x}$ and ρ in W, which is *only* possible when $(\alpha_2 - \alpha_3) = 4s - (4t + 2c) = 0$. This means that $s = t + \frac{c}{2}$. If $t = d_1 \Rightarrow s = t + \frac{c}{2} = (d_1 + \frac{c}{2})$, which is the condition of robust second-order rotatability of a robust symmetric balanced design (7.18). Therefore, robust second-order rotatability is a necessary condition of D-optimal robust slope-rotatable design.

7.7 ROBUST SLOPE ROTATABLE DESIGNS OVER ALL DIRECTIONS, WITH EQUAL MAXIMUM DIRECTIONAL VARIANCE AND D-OPTIMAL SLOPE

In the present section, some robust second-order slope-rotatable designs over all directions, with equal maximum directional variance and D-optimal slope are investigated under intra-class, inter-class, compound symmetry, tri-diagonal and autocorrelated error variance-covariance structures (Chapter 2). It is shown in Corollary 7.1 that all $\hat{y}$-robust second-order rotatable designs are also robust slope-rotatable over all directions. Also note that the robust second-order rotatable designs (Chapter 3) are robust second-order slope-rotatable designs with equal maximum directional variance (Theorem 7.5) and D-optimal slope (Theorem 7.6) which are given below.

It is known (in Chapter 3) that a second-order rotatable design (SORD) in the usual sense preserves the property of rotatability under the intra-class error structure (2.6) and vice-versa. This holds irrespective of the value of the intra-class correlation coefficient ρ in $W_{N \times N}(\rho)$ in (2.6). Therefore, the usual SORDs are robust rotatable under the intra-class error structure. Thus, the usual SORDs are robust second-order slope-rotatable designs over all directions, with equal maximum directional variance and D-optimal slope under the intra-class structure. The following well known response surface designs (or SORDs) given in Park (1987), are robust slope-rotatable designs over all directions, with equal maximum directional variance and D-optimal slope, and they are also $\hat{y}$-robust rotatable under the intra-class error variance covariance structure

(i) rotatable central composite designs, where $\alpha = n_c^{1/4}$ and n_c is the number of points in the "cube" part,

(ii) equiradial designs in two variables, where there are n_1 ($n_1 \geq 5$) equally spaced points on a circle of radius r, augmented by n_2 ($n_2 \geq 1$) center points,

(iii) combinations of equiradial designs in two variables such as s ($s \geq 2$) sets of equiradial points on concentric circles, the wth containing n_w points ($n_w \geq 5$) and having radius r_w,

(iv) for three variables, 12 vertices of the icosahedron $(0, \pm a, \pm b)$, $(\pm b, 0, \pm a)$, $(\pm a, 0, \pm b)$, where $a/b = 1.618$, plus some center points $(0, 0, 0)$,

(v) for three variables, 20 vertices of the dodecahedron $(0, \pm c^{-1}, \pm c)$, $(\pm c, 0, \pm c^{-1})$, $(\pm c^{-1}, \pm c, 0)$, $(\pm 1 \pm 1, \pm 1)$, where $c = 1.618$, plus some center points $(0, 0, 0)$,

(vi) simplex-sum second-order rotatable designs proposed by Box and Behnken (1960).

Example 7.1 A robust second-order slope-rotatable design under intra-class structure: An usual SORD is a robust second-order slope-rotatable design over all directions, with equal maximum directional variance and D-optimal slope under the intra-class structure (2.6) which is given below. An example of a SORD with 2 factors and 14 observations, which is a rotatable central composite design, is displayed below (rows being factors and columns being runs).

d_1 :	1	2	3	4	5	6	7	8	9	10	11	12	13	14
x_1	-1	1	-1	1	-1.414	1.414	0	0	0	0	0	0	0	0
x_2	-1	-1	1	1	0	0	-1.414	1.414	0	0	0	0	0	0

Several methods of construction of RSORDs under inter-class structure are described in Chapter 3. For example, if the design points of a SORD are divided into some disjoint sets and if the points of each set satisfy the second-order rotatability conditions in the usual sense, then the overall design is a robust second-order rotatable under the inter-class covariance structure W_1 (2.11). The result of inter-class variance-covariance structure also holds good in case of compound symmetry structure $W_{N \times N}(\rho, \rho_1)$ (2.23). These above results are shown by Das (2003a, 2004). Therefore, the related RSORDs under the inter-class structure are also the robust second-order slope-rotatable designs over all directions, with equal maximum directional variance and D-optimal slope under the inter-class and compound symmetry structure.

Example 7.2 A robust second-order slope-rotatable design under inter-class structure: An example of a robust second-order slope-rotatable design over all directions, with equal maximum directional variance and D-optimal slope with 2 factors, 2 groups, each of 9 observations for inter-class structure, is given below. The robust second-order slope-rotatable design (denoted by 'd_2') is displayed below (rows being factors and columns being runs).

d_2 :	1	2	3	4	5	6	7	8	9
x_1	-1	1	-1	1	-1.414	1.414	0	0	0
x_2	-1	-1	1	1	0	0	-1.414	1.414	0

d_2 :	10	11	12	13	14	15	16	17	18
x_1	-1	-1	1	1	0	0	-1.414	1.414	0
x_2	-1	1	-1	1	-1.414	1.414	0	0	0

Note that this robust second-order slope-rotatable design over all directions, with equal maximum directional variance and D-optimal slope under the inter-class structure is also a robust second-order slope-rotatable design in the *same* sense, under the compound symmetry structure.

For autocorrelated error structure (2.40), a construction method of RSORDs is given in Chapter 3. From a usual SORD one can construct a RSORD under the autocorrelated structure. For example, let us start with a usual SORD having n non-central design points involving k-factors. The set of n design points can be extended to $(2n + 1)$ points by incorporating $(n + 1)$ central points in the following way. One central point is placed in between each pair of non-central design points in the sequence, utilizing thus $(n - 1)$ such central points. Of the other two central points, one is placed at the beginning and one at the end. The number of central points of the usual SORD with which we started may be different from the number of central points required in the design so constructed in the autocorrelated error structure situation. If the number of central points of the usual SORD with which we started is greater than $(n + 1)$, the remaining central points are placed in any manner. If the number is less we need to include the requisite number of *additional* central points. Then the resulting design is a robust second-order slope-rotatable design over all directions, with equal maximum directional variance and D-optimal slope under the autocorrelated structure W_4 in (2.40).

Example 7.3 A robust second-order slope-rotatable design under autocorrelated structure: An example of a robust second-order slope-rotatable design over all directions, with equal maximum directional variance and D-optimal slope with 2 factors and 17 design points under autocorrelated structure, is given below. We start with a usual SORD having 8 non-central design points and 9 central points involving 2 factors. The robust second-order slope-rotatable design (denoted by 'd_3') is displayed below (rows being factors and columns being runs).

d_3 :	1	2	3	4	5	6	7	8	9	10	11	12	13	14	15	16	17
x_1	0	-1	0	1	0	-1	0	1	0	-1.414	0	1.414	0	0	0	0	0
x_2	0	-1	0	-1	0	1	0	1	0	0	0	0	0	-1.414	0	1.414	0

RSORDs for tri-diagonal structure (2.30) are given in Chapter 3. One can construct a robust second-order slope-rotatable design over all directions, with equal maximum directional variance and D-optimal slope under tri-diagonal structure from a RSORD under the same correlation structure. One can start with a usual SORD, having n design points involving k-factors. The set of n design points can be extended to $2n$ design points by incorporating n central points together just below or above the n design points of the original design with which we started, then the resulting design is a robust second-order slope-rotatable design over all directions, with equal maximum directional variance and D-optimal slope under the tri-diagonal variance-covariance structure W_3 in (2.30).

Example 7.4 A robust second-order slope-rotatable design under tri-diagonal structure: An example of a robust second-order slope-rotatable design over all directions, with equal maximum directional variance and D-optimal slope with 2 factors, 2 groups, each of 9 observations under tri-diagonal structure, is given below. A robust second-order slope-rotatable design (denoted by 'd_4') is displayed below (rows being factors and columns being runs).

d_4 :	1	2	3	4	5	6	7	8	9
x_1	-1	1	-1	1	-1.414	1.414	0	0	0
x_2	-1	-1	1	1	0	0	-1.414	1.414	0

d_4 :	10	11	12	13	14	15	16	17	18
x_1	0	0	0	0	0	0	0	0	0
x_2	0	0	0	0	0	0	0	0	0

Another construction method of robust second-order slope-rotatable design over all directions, with equal maximum directional variance and D-optimal slope under tri-diagonal structure (2.30) is given below. Let us start with a usual SORD, having n design points involving k-factors. Let $(x_1, x_2,...,x_k)$ be a usual SORD of k factors with n design points, where x_i is i-th explanatory variable or factor and x_i is a vector of order $n \times 1$, $1 \le i \le k$. The set of n design points can be extended to $2n$ points by repeating i-th factor column just below itself, i.e., $(x_i', \ x_i')' = z_i$ say; $1 \le i \le k$. Now z_i is a vector of order $2n \times 1$ and $(z_1, z_2,...,z_k)$ form a robust second-order slope-rotatable design over all directions, with equal maximum directional variance and D-optimal slope under the tri-diagonal error variance-covariance structure (W_3) in (2.30).

Example 7.5 A robust second-order slope-rotatable design under tri-diagonal structure: An example of a robust second-order slope-rotatable design over all directions, with equal maximum directional variance and D-optimal slope with 2 factors, 2 groups, each of 9 observations under tri-diagonal structure, is given below. A robust second-order slope-rotatable design (denoted by 'd_5') is displayed below (rows being factors and columns being runs).

d_5 :	1	2	3	4	5	6	7	8	9
x_1	-1	1	-1	1	-1.414	1.414	0	0	0
x_2	-1	-1	1	1	0	0	-1.414	1.414	0

d_5 :	10	11	12	13	14	15	16	17	18
x_1	-1	1	-1	1	-1.414	1.414	0	0	0
x_2	-1	-1	1	1	0	0	-1.414	1.414	0

7.8   CONCLUDING REMARKS

In the present chapter, the design problems associated with the estimation of derivatives of second-order polynomial response functions are considered assuming correlated errors. The variance of the derivative $V_{\mathbf{p}}(\mathbf{x})$ in equation (7.7) at any point $\mathbf{x}$ is a function of direction and a function of design. If $V_{\mathbf{p}}(\mathbf{x})$ is averaged over all directions, the averaged variance $\bar{V}(\mathbf{x})$ is only a function of the point $\mathbf{x}$ and the design and also the correlation parameter or parameters involved in the dispersion matrix W_0 of errors. By choice of design, it is possible to make $\bar{V}(\mathbf{x})$ constant for all points equidistant from the design origin and also independent of correlation parameter or parameters involved in W_0. The class of designs with such property is referred to as robust slope-rotatable over all directions. The necessary and sufficient conditions for this class of designs are derived for a general correlated error structure W, and several designs in this class are illustrated. It turns out that the most commonly used robust response surface designs belong to this class.

The necessary and sufficient conditions for robust second-order slope-rotatable designs with equal maximum directional variance and D-optimal slope with correlated errors have been developed. These results have been derived for a general variance-covariance structure. Furthermore, some robust second-order slope-rotatable designs with equal maximum directional variance and D-optimal slope have been investigated for some special correlation structures.

It is clear from the construction method (Section 7.7) that the RSOSROADs, the D-optimal RSOSRDs, and the designs with equal maximum directional variance slope under the autocorrelated structure as constructed is *not* invariant under some permutations (i.e., the order of the runs of the experiment) of the design points with respect to robust rotatability. It is also true for the optimal RSOSRDs under all other correlation structures except the intra-class structure. The order of the runs of the experiment is very important in case of any optimal RSOSRDs. The practitioners need to maintain the order of the runs of the experiment. Accordingly, the experimental level combinations of the optimal RSOSRDs are to be arranged for conducting the experiment. It is not possible to conduct the experiment (for any optimal RSOSRD) at a time (one after another) for the similar level combinations (say central points) to reduce the cost of the experiment (Das, 1999). In general, the number of design points of an optimal RSOSRD is more than a usual SORD. Consequently, the cost of the experiment of an optimal RSOSRD is greater than a usual SORD. Any design, which is an optimal RSOSRD under a particular structure, may not be an optimal RSOSRD for other correlation structure. An optimal RSOSRD depends on the pattern of the correlation structure but independent of the values of the correlation parameters involved in it. Note that the optimal RSOSRDs are always SORDs (for $\rho = 0$) but the converse is not always true.

Response surface designs are mostly used in industry for improving the quality of a product. If differences at points close together are involved, estimation of the local slopes (the rate of change) of the response is of interest. This problem, estimation of slopes, occurs frequently in many practical situations. Several causes of correlation that may occur in the observations are given in Palta (2003). The optimal designs under the tri-diagonal and the autocorrelated structures are studied by different authors such as Kiefer and Wynn (1981, 1984), Bischoff (1996), Box and Draper (2007), etc. The practitioners only need to know the pattern of the correlation structure for using these optimal RSOSRDs.

Myers and Montgomery (1995) (also reproduced in Myers *et al.*, 2002, Example 6.3, p. 220) presented an experimental data in a semiconductor plant. There is a lamination process, and the chamber measurement is made four times on the same device procedure.

The chamber measurement is known to be non-normal with a heavy right-tailed distribution. In addition it is clear that the measurement taken on the same device may well be correlated. As a result we have "repeated measures," which represent the source of the correlation. Recently Myers *et al.* (2002) analyzed "The Worsted Yarn Data" (Myers *et al.*, 2002, Table 2.7, p. 36) using a usual second-order response surface design, and treating response ($y = T$) as the cycles to failure (T), but they noticed that variance is nonconstant and the analysis is inappropriate. Then using log transformation of the cycles to failure (i.e., $y = \ln T$), the data were analyzed, and it was found that the log model, overall, is an improvement on the original quadratic fit. But there is still some indication of inequality of variance. Myers *et al.* (2002, p.128) also noticed that in *industrial applications* experimental units are not independent at times *by design*. At times this leads to correlation among observations via a *repeated measures* scenario as in *split plot* design. Finally Das and Lee (2009) showed that simple log transformation is insufficient to reduce the variance constant, and the researchers analyzed the data using joint generalized model, and found that the log-normal distribution is more appropriate. Thus, it is reasonable to consider $y = \ln T$ as the response, and correlated errors to the model of a lifetime distribution.

Evidence from the experimental data (stated above) shows that the experimental errors may be correlated, and they may have different pattern of correlation structures. Response surface designs are widely used in different fields such as quality engineering, chemical process improvement, industrial and engineering chemistry, lifetime improvement, biological experiments etc. Thus, it is reasonable to consider that the experimental responses may be correlated in many cases. Response surface practitioners need to know the pattern of the correlation structure depending on the experimental conditions, responses (replicated, non-replicated), etc. before conducting the experiment. Accordingly, appropriate optimal RSOSRDs may be used in every field to get better information about the unknown process parameters.

Chapter 8

ROBUST SECOND-ORDER SLOPE-ROTATABILITY MEASURES

Robust second-order slope-rotatability and modified slope-rotatability along axial directions have been introduced in Chapter 6. These designs have not been still developed to a large extent in literature. Optimal robust second-order slope-rotatable designs (over all directions or A-optimal, with equal maximum directional variance and with D-optimal slope) have been developed in Chapter 7. The present chapter describes some measures of robust second-order slope-rotatability along axial directions, over all directions and with equal maximum directional variance for correlated errors. These measures have been illustrated with different examples. In the process, weakly robust second-order slope-rotatability has been introduced.

8.1 INTRODUCTION AND OVERVIEW

Second-order slope-rotatability conditions (with correlated errors) along axial directions have been derived by Das (2003b). This paper could not specify any robust second-order slope-rotatable designs along axial directions for any correlation structure of errors. To solve the problems, Das, Pal and Park, (2013) have developed modified robust second-order slope-rotatable designs along axial directions. It is observed that construction of robust second-order slope-rotatable and modified slope-rotatable designs are really very difficult. Das, Pal and Park (2013) have developed an example of modified robust second-order slope-rotatable design under tri-diagonal correlation structure, which is a *complex design* (Chapter 6). Chapter 7 describes some concepts of optimal robust second-order slope-rotatable designs (over all directions or A-optimal, with equal maximum directional variance and with D-optimal slope). It has been identified that RSORDs are the robust second-order slope-rotatable designs over all directions, with equal maximum directional variance and D-optimal slope. In practice, construction of RSORDs are also very difficult under many correlation structures (Das, 1997, 2004).

Park and Kim (1992) developed a measure of second-order slope-rotatability with uncorrelated and homoscedastic errors. Jang and Park (1993) described a measure and a graphical method for evaluating slope-rotatability (with uncorrelated and homoscedastic errors) in response surface designs. However, no general measure has been introduced yet that represents the degree of *robust slope-rotatability* for a given response surface design. Recently, Das and Park (2009) have developed a measure of *robust slope-rotatability* along axial directions of a second-order regression design under a fixed pattern of correlation structure of errors. Robustness of usual second-order slope-rotatable designs (SOSRDs) and RSORDs can be examined with this measure. Some measures for assessing the degree of robust second-order slope-rotatability along axial directions, over all directions and with equal maximum directional variance have been developed by Park, Jung and Das (2009). These measures have been developed by following Park and Kim (1992) and Jang and Park (1993). These measures have been illustrated with one example of a RSORD, one with a *nearly* RSORD, one with Box-Behnken design, and the others with usual central composite designs (CCDs), under autocorrelated error structure.

8.2   ROBUST SECOND-ORDER SLOPE-ROTATABILITY MEASURES

The present section introduces some measures to assess the degree of *robust slope-rotatability* along axial directions, over all directions and with equal maximum directional variances of a second-order design with a general correlated error structure W.

8.2.1   Robust second-order slope-rotatability measure along axial directions

The present section introduces a measure of robust second-order slope-rotatability along axial directions, proposed by Das and Park (2009). The second-order response surface model with correlated errors is given in (3.1) (Chapter 3). To develop a measure of robust slope-rotatability along axial directions, one needs to combine the conditions of robust slope-rotatability as in (6.7) (Chapter 6), and the variance function of the first-order derivative of the estimated response, i.e., $\mathrm{Var}(\frac{\partial \hat{y}_{\mathbf{x}}}{\partial x_i}) = V(x_i)$ as in (6.5) (Chapter 6) into a concrete form of expression. The poler transformation (as in Park and Kim, 1992) of $\mathbf{x}$ into the spherical coordinates $(r, \phi_1, \phi_2, \ldots, \phi_{k-2}, \theta)$ is quite useful for this situation.

In the k-dimensional space $(k \geq 2)$, $V(x_i)$ as in (6.5) (Chapter 6) can be expressed in terms of spherical coordinates of $(r, \phi_1, \phi_2, \ldots, \phi_{k-2}, \theta)$ where

$$x_1 = r \cos \phi_1$$

$$x_2 = r \sin \phi_1 \cos \phi_2$$

$$\vdots$$

$$x_{k-1} = r \sin \phi_1 \sin \phi_2 \ldots \sin \phi_{k-2} \cos \theta$$

$$x_k = r \sin \phi_1 \sin \phi_2 \ldots \sin \phi_{k-2} \sin \theta, \tag{8.1}$$

and $r \geq 0, \quad 0 \leq \phi_1, \ \phi_2, \ldots, \phi_{k-2} \leq \pi, \quad 0 \leq \theta \leq 2\pi$, (Edwards, 1973).

The absolute value of the Jacobian of this transformation is

$$|J| = r^{k-1} \sin^{k-2} \phi_1 \sin^{k-3} \phi_2 \ldots \sin^2 \phi_{k-3} \sin \phi_{k-2}.$$

On substituting $\mathbf{x}$ into the spherical co-ordinates $(r, \phi_1, \ldots, \phi_{k-2}, \theta)$ in (6.5), $V(x_i)$ will be expressed as a function of $r, \phi_1, \ldots, \phi_{k-2}, \theta$ and correlation parameter $(\rho$ in $W)$, i.e.,

$$V(x_i) = \omega_i(r, \phi_1, \phi_2, \ldots, \phi_{k-2}, \theta, \rho) = \omega_i, \text{ say.} \qquad (8.2)$$

Thus, the robust second-order slope-rotatability along axial directions as in (6.6) can be redefined in terms of spherical co-ordinates as follows:

(1) For each $i = 1, 2, \ldots, k$, $\omega_i(r, \phi_1, \ldots, \phi_{k-2}, \theta, \rho)$ is *constant* with respect to $\phi_1, \ldots, \phi_{k-2}, \theta, \rho$, and is only a function of the distance r from the design center to a design point and the correlation coefficient ρ.

(2) For any point $(r, \phi_1, \ldots, \phi_{k-2}, \theta)$, $\omega_1(r, \phi_1, \ldots \phi_{k-2}, \theta, \rho) = \omega_2(r, \phi_1, \ldots, \phi_{k-2}, \theta, \rho) = \ldots = \omega_k(r, \phi_1, \ldots, \phi_{k-2}, \theta, \rho)$.

Let us define the following terminology for developing the measure of robust slope-rotatability along axial directions:

$$(i) \; \bar{\omega}_i(r) = \frac{1}{T_k} \int_0^{2\pi} \int_0^{\pi} \ldots \int_0^{\pi} \omega_i(r, \phi_1, \ldots, \phi_{k-2}, \theta, \rho) d\Omega \; (1 \leq i \leq k), \qquad (8.3)$$

where $\bar{\omega}_i(r)$ is the average variance of $\frac{\partial \hat{y}_x}{\partial x_i}$ over all the points on the hypersphere of radius r centered at the design origin, $d\Omega = \sin^{k-2}\phi_1 \sin^{k-3}\phi_2 \ldots \sin^2\phi_{k-3} \sin\phi_{k-2} \, d\phi_1 \, d\phi_2 \ldots d\phi_{k-2} \, d\theta$ and $T_k = \int_0^{2\pi} \int_0^{\pi} \ldots \int_0^{\pi} d\Omega = \frac{2\pi^{k/2}}{\Gamma(k/2)}$,

$$(ii) \qquad \bar{\omega}(r, \phi_1, \ldots, \phi_{k-2}, \theta, \rho) = \frac{1}{k} \sum_{i=1}^{k} \omega_i(r, \phi_1, \ldots, \phi_{k-2}, \theta, \rho), \qquad (8.4)$$

where $\bar{\omega}(r, \phi_1, \ldots, \phi_{k-2}, \theta, \rho)$ is the average variance of the first derivatives of $\hat{y}_\mathbf{x}$, at the point $(r, \phi_1, \ldots, \phi_{k-2}, \theta)$, over the k axial directions, and

$$(iii) \; \bar{\bar{\omega}}(r) = \frac{1}{k} \sum_{i=1}^{k} \bar{\omega}_i(r) = \frac{1}{kT_k} \sum_{i=1}^{k} \int_0^{2\pi} \int_0^{\pi} \ldots \int_0^{\pi} \omega_i(r, \phi_1, \ldots, \phi_{k-2}, \theta, \rho) d\Omega,$$

$$= \frac{1}{T_k} \int_0^{2\pi} \int_0^{\pi} \ldots \int_0^{\pi} \bar{\omega}(r, \phi_1, \ldots, \phi_{k-2}, \theta, \rho) d\Omega, \qquad (8.5)$$

where $d\Omega$ and T_k as in (8.3), and $\bar{\bar{\omega}}(r)$ is the average variance of the first derivatives of $\hat{y}_\mathbf{x}$ over the k axial directions and over all the points on the hypersphere of radius r centered at the design origin.

For a given design d, the discrepancy from robust slope-rotatability at r can be expressed as

$$h_d(r) = \int_0^{2\pi} \int_0^{\pi} \ldots \int_0^{\pi} g_d(r, \phi_1, \phi_2, \ldots, \phi_{k-2}, \theta, \rho) \, d\Omega,$$

$$= \sum_{i=1}^{k} \int_0^{2\pi} \int_0^{\pi} \ldots \int_0^{\pi} \omega_i^2(r, \phi_1, \phi_2, \ldots, \phi_{k-2}, \theta, \rho) d\Omega - kT_k\bar{\bar{\omega}}^2(r), \qquad (8.6)$$

where $g_d(r, \phi_1, \ldots, \phi_{k-2}, \theta, \rho) = \sum_{i=1}^{k} [\omega_i(r, \phi_1, \phi_2, \ldots, \phi_{k-2}, \theta, \rho) - \bar{\bar{\omega}}(r)]^2$ is a measure of variation among the ω_i's over the k axial directions at a fixed point $(r, \phi_1, \ldots, \phi_{k-2}, \theta)$, and $h_d(r)$ is the sum of the g_d's over all the points on the hypersphere of radius r centered at the design origin. If the region of interest is $0 \leq r \leq 1$, the proposed measure $(M_k(d))$ of robust slope-rotatability for a design d, will be

$$M_k(d) = \frac{1}{1 + Q_k(d)}, \qquad (8.7)$$

where $Q_k(d) = \frac{1}{S_k} \int_0^1 r^{k-1} h_d(r) dr,$

and S_k is a positive constant depending only on k and $M_k(d)$ is the proposed *measure of robust slope-rotatability*. Note that $0 \leq M_k(d) \leq 1$, and it can be easily shown that $M_k(d)$ is one if and only if the design d is robust slope-rotatable for *all* values of ρ, and $M_k(d)$ approaches to zero as the design d deviates from robust slope-rotatability.

Now, we introduce the following fact which is useful for evaluating our measure in a more concrete form.

$$(1) \int d\Omega = \frac{2\pi^{k/2}}{\Gamma(k/2)} = T_k, \quad (2) \int x_i^2 \, d\Omega = r^2 \frac{T_k}{k} \;\; (1 \leq i \leq k),$$

$$(3) \int x_i^4 d\Omega = \frac{3r^4 T_k}{k(k+2)} \;\; (1 \leq i \leq k), \quad (4) \int x_i^2 x_j^2 d\Omega = \frac{r^4 T_k}{k(k+2)} \;\; (1 \leq i \neq j \leq k), \qquad (8.8)$$

where $\int$ means $\int_0^{2\pi} \int_0^\pi \ldots \int_0^\pi$ over $\phi_1, \ldots, \phi_{k-2}, \theta$. The values of other integrals where at least one x_i has an odd exponent are *all* zeros.

Using (8.8), $h_d(r)$ in (8.6) can be simplified as given below

$$h_d(r) = T_k A_k + \frac{2r^2 T_k}{k} B_k + \frac{3r^4 T_k}{k(k+2)} C_k, \qquad (9.9)$$

where $A_k = \sum_{i=1}^{k} (v^{i.i})^2 - \frac{1}{k} (\sum_{i=1}^{k} v^{i.i})^2; \quad B_k = \sum_{i=1}^{k} v^{i.i} (4v^{ii.ii} + \sum_{j=1; \, j \neq i}^{k} v^{ij.ij})$

$$- \frac{1}{k} (\sum_{i=1}^{k} v^{i.i})[\sum_{i=1}^{k} (4v^{ii.ii} + \sum_{j=1; \, j \neq i}^{k} v^{ij.ij})] + 2 \sum_{i=1}^{k} \{4(v^{i.ii})^2 + \sum_{j=1; \, j \neq i}^{k} (v^{i.ij})^2\};$$

$$C_k = \sum_{i=1}^{k} \{16(v^{ii.ii})^2 + \sum_{j=1; \, j \neq i}^{k} (v^{ij.ij})^2\} - \frac{k+2}{3k^2} [\sum_{i=1}^{k} (4v^{ii.ii} + \sum_{j=1; \, j \neq i}^{k} v^{ij.ij})]^2 + \frac{2}{3} \sum_{i=1}^{k} (4v^{ii.ii} \times$$

$$\sum_{j=1; \, j \neq i}^{k} v^{ij.ij} + \sum_{j<l; \, j,l \neq i}^{k} v^{ij.ij} v^{il.il}) + \frac{4}{3} \sum_{i=1}^{k} \{4 \sum_{j=1; \, j \neq i}^{k} (v^{ii.ij})^2 + \sum_{j<l; \, j,l \neq i}^{k} (v^{ij.il})^2\}.$$

Therefore, $\int_0^1 r^{k-1} h_d(r) dr = \frac{T_k}{k} [A_k + \frac{2B_k}{k+2} + \frac{3C_k}{(k+2)(k+4)}], \qquad (8.10)$

where A_k, B_k and C_k as in (9.9), and T_k as in (8.8). Note that this quantity in (8.10) itself is a measure of how much the given design differs from a robust slope-rotatable design and hence could be used as a measure of robust slope-rotatability. But in order to simplify the actual computation, (8.10) will be divided by an appropriate constant S_k which depends only on k. To determine a suitable constant S_k, let us consider a design (for example, RSORD) having the following special cases:

$$v^{i.ii} = v^{i.ij} = v^{ii.ij} = v^{ij.il} = 0 \;\; (i \neq j \neq l \neq i); \; v^{i.i} \text{ are equal for all } i, \; v^{ii.ii} \text{ are equal for}$$
all i, and $v^{ij.ij}$ are equal for all $(i,j; \; i \neq j)$, and for all ρ. $\qquad (8.11)$

For the above case, A_k and B_k become zero and C_k is found to be $\frac{2}{3}(k-1)(4v^{11.11} - v^{12.12})^2$, resulting (8.10) reduces to

$$\int_0^1 r^{k-1} h_d(r) dr = \frac{2(k-1)T_k}{k(k+2)(k+4)}(4v^{11.11} - v^{12.12})^2. \tag{8.12}$$

Let us take

$$S_k = \frac{2(k-1)T_k}{k(k+2)(k+4)}\sigma^4, \tag{8.13}$$

then $Q_k(d)$ in (8.7) becomes

$$Q_k(d) = \frac{1}{\sigma^4}(4v^{11.11} - v^{12.12})^2, \tag{8.14}$$

which is a very simple form.

For such S_k as in (8.13) and for $k \geq 2$, $Q_k(d)$ as in (8.7), in all general cases becomes

$$Q_k(d) = \frac{1}{2(k-1)\sigma^4}\{(k+2)(k+4)A_k + 2(k+4)B_k + 3C_k\}, \tag{8.15}$$

where A_k, B_k and C_k are as in (8.9). On simplification of (8.15), $Q_k(d)$ reduces to the following form

$$Q_k(d) = \frac{1}{2(k-1)\sigma^4}\{(k+2)(k+4)\sum_{i=1}^{k}[(v^{i.i} - \bar{v}) + \frac{a_i - \bar{a}}{k+2}]^2 + \frac{4}{k(k+2)}\sum_{i=1}^{k}(a_i - \bar{a})^2$$

$$+ 2\sum_{i=1}^{k}[(4v^{ii.ii} - \frac{a_i}{k})^2 + \sum_{j=1; \, j\neq i}^{k}(v^{ij.ij} - \frac{a_i}{k})^2] + 4(k+4)(4(v^{i.ii})^2 + \sum_{j=1; \, j\neq i}^{k}(v^{i.ij})^2)$$

$$+ 4\sum_{i=1}^{k}(4\sum_{j=1; \, j\neq i}^{k}(v^{ii.ij})^2 + \sum_{j<l; \, j,l\neq i}^{k}(v^{ij.il})^2)\}, \tag{8.16}$$

where $\bar{v} = \frac{1}{k}\sum_{i=1}^{k}v^{i.i}$, $a_i = 4v^{ii.ii} + \sum_{j=1; \, j\neq i}^{k}v^{ij.ij}$ $(1 \leq i \leq k)$, and $\bar{a} = \frac{1}{k}\sum_{i=1}^{k}a_i$.

It can be easily shown that $Q_k(d)$ in (8.16) becomes zero for *all* values of ρ, if and only if the conditions (6.6) hold. Therefore, $Q_k(d)$ in (8.16) can be used as a measure of robust slope-rotatability along axial directions. For $k = 2$, $Q_k(d)$ in (8.16) simplifies to

$$Q_2(d) = \frac{1}{2\sigma^4}\{12[(v^{1.1} - v^{2.2}) + (v^{11.11} - v^{22.22})]^2 + 4(v^{11.11} - v^{22.22})^2 + (4v^{11.11} - v^{12.12})^2$$

$$+ (4v^{22.22} - v^{12.12})^2 + 96((v^{1.11})^2 + (v^{2.22})^2) + 24((v^{1.12})^2 + (v^{2.12})^2) + 16((v^{11.12})^2 + (v^{22.12})^2)\}. \tag{8.17}$$

Definition 8.1 Weakly robust slope-rotatable design: When ρ (correlation parameter involved in W) $\neq 0$, a design which is *not* robust slope-rotatable but is very *near* to robust slope-rotatable one (in some sense) for a certain range of correlation parameter ρ in W, under a *fixed* pattern of correlation structure W, is called a weakly robust slope-rotatable design (WRSRD), under a well defined correlation structure class $W_0 = \{W$ positive definite: $W_{N\times N}$ defined by a particular correlation structure neatly specified$\}$.

Definition 8.2 Weakly robust second-order slope-rotatable design along axial directions: Any design d is said to be weakly robust second-order slope-rotatable (WR-SOSR) of *strength* ν if

$$M_k(d) \geq \nu. \tag{8.18}$$

Note that $M_k(d)$ involves the correlation parameter ρ in W and as such, $M_k(d) \geq \nu$ for *all* ρ is too strong to be met with. On the other hand, for a given ν, we can possibly find the range of values of ρ for which $M_k(d) \geq \nu$. We will call this range as the *weak slope-rotatability region (WSRR)* $(R_{d(\nu)}(\rho))$ of the design d. Naturally, the desirability of using d will rest on the wide nature of $WSRR(R_{d(\nu)}(\rho))$ along with its *strength* ν. Generally, we would require ν to be very high say, around 0.95.

8.2.2   Robust second-order slope-rotatability measure over all directions

This section describes a measure of robust second-order slope-rotatability over all directions, proposed by Park, Jung and Das (2009). With this measure one can measure the degree of robustness of a design which is *not* robust slope-rotatable design over all directions with respect to a particular correlation structure. This measure is proposed based on Theorem 7.1 and the notations used in Section 8.2.1. In Theorem 7.1, it has been established that a design is slope-rotatable over all directions under the correlated error structure W (for all ρ in W and $1 \leq i, j \leq k$) if and only if

(i) $2\text{Cov}(\hat{\beta}_i, \hat{\beta}_{ii}) + \displaystyle\sum_{j=1;\ j\neq i}^{k} \text{Cov}(\hat{\beta}_j, \hat{\beta}_{ij}) = 0$ for all i,

(ii) $2[\text{Cov}(\hat{\beta}_{ii}, \hat{\beta}_{ij}) + \text{Cov}(\hat{\beta}_{jj}, \hat{\beta}_{ij})] + \displaystyle\sum_{t=1; t\neq i,j}^{k} \text{Cov}(\hat{\beta}_{it}, \hat{\beta}_{jt}) = 0$ for all $(i,j), i \neq j$,

(iii) $4\text{Var}(\hat{\beta}_{ii}) + \displaystyle\sum_{j=1; j\neq i}^{k} \text{Var}(\hat{\beta}_{ij}) = constant = \delta$, say, for all i.

Equivalently, $2v^{i.ii} + \sum_{j=1, j\neq i}^{k} v^{j.ij} = 0$, (for all i); $2[v^{ii.ij} + v^{jj.ij}] + \sum_{t=1, t\neq i,j}^{k} v^{it.jt} = 0$, (for all $i \neq j$); and $\{(4v^{ii.ii} + \displaystyle\sum_{j=1;\ j\neq i}^{k} v^{ij.ij}) - (4v^{jj.jj} + \displaystyle\sum_{i=1;\ i\neq j}^{k} v^{ji.ji})\} = 0$ (for all $i \neq j$).

Therefore, the proposed measure $(MO_k(d))$ of robust second-order slope-rotatability over all directions for a design d is

$$MO_k(d) = \frac{1}{1 + R_{0k}(d)}, \tag{8.19}$$

where $R_{0k}(d) = \sum_{i=1}^{k} C_{1i}^2 + \sum_{i=1}^{k}\sum_{j=1; j\neq i}^{k} B_{1ij}^2 + \sum_{i=1}^{k}\sum_{j=1; j\neq i}^{k}(a_i - a_j)^2$, and $C_{1i} = 2v^{i.ii} + \sum_{j=1, j\neq i}^{k} v^{j.ij}$, $B_{1ij} = 2[v^{ii.ij} + v^{jj.ij}] + \sum_{t=1, t\neq i,j}^{k} v^{it.jt}$, and a_i's are in (8.16).

For a robust second-order slope-rotatable design (d) over all directions with respect to correlation structure W, $R_{0k}(d)$ is always zero. If the design d deviates from robust slope-rotatability over all directions, $R_{0k}(d)$ will be a positive value, depending on ρ in W. Note that $0 \leq MO_k(d) \leq 1$, and $MO_k(d)$ is one if and only if the design d is a robust slope-rotatable design over all directions for all values of ρ in W and $MO_k(d)$ approaches to zero as the design d deviates from robust slope-rotatability over all directions.

8.2.3 Robust second-order slope-rotatability measure with equal maximum directional variance

The present section describes a measure of robust second-order slope-rotatability with equal maximum directional variance for two factors (Park, Jung and Das, 2009). This particular measure is used to measure the degree of robustness of a design which is *not* robust second-order slope-rotatable design with equal maximum directional variance with respect to a particular correlation structure. This measure is proposed based on Theorem 7.3 (Chapter 7) and the notations used in Section 8.2.1. In Theorem 7.3, it has been established that a two factor design is robust second-order slope-rotatable with equal maximum directional variance under the correlated error structure W (for all ρ in W) if and only if

1. $v^{i.j} = v^{i.ii} = v^{i.ij} = v^{i.jj} = v^{ii.ij} = 0; (i \neq j; i, j = 1, 2)$,

2. $v^{1.1} = v^{2.2}$,

3. $v^{11.11} = v^{22.22}$,

4. $2(v^{11.11})^2 - v^{11.11}v^{12.12} - 2(v^{11.22})^2 - v^{11.22}v^{12.12} = 0$.

Thus, the proposed measure $(ME_2(d))$ of robust second-order slope-rotatability with equal maximum directional variance for a design d with two factors is

$$ME_2(d) = \frac{1}{1 + G_2(d)}, \tag{8.20}$$

where $G_2(d) = C_2^2 + C_3^2 + C_4^2 + C_5^2$, $C_2^2 = [(v^{1.2})^2 + (v^{1.11})^2 + (v^{2.22})^2 + (v^{1.12})^2 + (v^{2.12})^2 + (v^{1.22})^2 + (v^{2.11})^2 + (v^{11.12})^2 + (v^{22.12})^2]$, $C_3 = (v^{1.1} - v^{2.2})$, $C_4 = (v^{11.11} - v^{22.22})$ and $C_5 = [2(v^{11.11})^2 - v^{11.11}v^{12.12} - 2(v^{11.22})^2 - v^{11.22}v^{12.12}]$.

It is clear (from Theorem 7.3) that $G_2(d)$ is always zero for a robust second-order slope-rotatable design (d) with equal maximum directional variance. If the design d deviates from robust second-order slope-rotatability with equal maximum directional variance, $G_2(d)$ will be positive. Note that $0 \leq ME_2(d) \leq 1$, and $ME_2(d)$ is one if and only if the design d is a robust second-order slope-rotatable design with equal maximum directional variance for all values of ρ in W and $ME_2(d)$ approaches to zero as the design d deviates from robust slope-rotatability with equal maximum directional variance.

8.3   ILLUSTRATIONS OF ROBUST SLOPE-ROTATABILITY MEASURES

This section examines the robustness of some well known second-order designs with the help of the measures of robust slope-rotatability along axial directions, over all directions, and with equal maximum directional variance as described in Section 8.2. It is well known that RSORDs are always robust slope-rotatable over all directions and with equal maximum directional variance. Note that there is not a single robust slope-rotatable design along axial directions in the literature. Let us examine the degree of robust slope-rotatability (along axial directions, over all directions, and with equal maximum directional variance) with one example of RSORD, one nearly RSORD, one Box-Behnken design, and one central composite design (CCD), under autocorrelated error structure which is given in Chapter 2. The following examples are considered for examining the robustness with respect to the measures developed herein.

Example 8.1 Robust second-order rotatable design (RSORD): A RSORD d_0 with 2 factors and 17 design points under autocorrelated error structure (following the method as in Das, 1997) is given below. The design (RSORD, denoted by d_0) is displayed below (column being runs).

d_0 :	1	2	3	4	5	6	7	8	9	10	11	12	13	14	15	16	17
x_1	0	-1	0	1	0	-1	0	1	0	-1.414	0	1.414	0	0	0	0	0
x_2	0	-1	0	-1	0	1	0	1	0	0	0	0	0	-1.414	0	1.414	0

Example 8.2 Nearly RSORD: Let us consider a derived design d (not a RSORD) which is obtained by permutation of design points of d_0 (as given in Example 8.1), where the derived design $d \in D(d_0)$ (class of all derived designs obtained from d_0), is displayed below (rows being factors and column being runs).

d :	1	2	3	4	5	6	7	8	9	10	11	12	13	14	15	16	17
x_1	0	0	-1	0	1	0	-1	0	1	0	-1.414	0	1.414	0	0	0	0
x_2	0	0	-1	0	-1	0	1	0	1	0	0	0	0	0	-1.414	0	1.414

Example 8.3 Central composite design (CCD): A CCD $d(\alpha)$ (with varying α) of second-order usual (i.e., with uncorrelated and homoscedastic errors) design with 2 factors and 17 design points (same as d_0 and d) is displayed below (rows being factors and column being runs).

$d(\alpha)$:	1	2	3	4	5	6	7	8	9	10	11	12	13	14	15	16	17
x_1	-1	1	-1	1	$-\alpha$	α	0	0	0	0	0	0	0	0	0	0	0
x_2	-1	-1	1	1	0	0	$-\alpha$	α	0	0	0	0	0	0	0	0	0

Example 8.4 Box-Behnken Design: A Box-Behnken usual second-order design d_B with 2 factors and 17 design points (same as d_0, d and $d(\alpha)$) is displayed below (rows being factors and column being runs).

d_B :	1	2	3	4	5	6	7	8	9	10	11	12	13	14	15	16	17
x_1	-1	-1	1	1	-1	-1	1	1	0	0	0	0	0	0	0	0	0
x_2	-1	1	-1	1	0	0	0	0	-1	-1	1	1	0	0	0	0	0

Moment matrices ($MO(d)$s, for the design ds) of the above designs are given below, and their measures of robust slope-rotatability along axial directions, over all directions, and with equal maximum directional variance, and weak slope-rotatability region are respectively given in Table 8.1, Table 8.2, Table 8.3 and Table 8.4. Note that the measure of robust slope-rotatability along axial directions of the design d_0 and the others (d, $d(\alpha)$ and d_B) are obtained by using $Q_k(d)$ as in (8.14) and (8.17), respectively. Also the measures of robust second-order slope-rotatability over all directions and with equal maximum directional variance are obtained from $MO_k(d)$ (8.19) and $ME_2(d)$ (8.20), respectively.

Moment Matrices: To compute the measure of robust slope-rotatability as developed in Section 8.2, of a design d, we have to compute its moment matrix ($MO(d) = (X'W^{-1}X)$) and $\mathrm{Dis}(\hat{\beta}) = \{\sigma^2(1 - \rho^2)\}^{-1}(X'W^{-1}X)^{-1}$. But it is very difficult to compute $\mathrm{Dis}(\hat{\beta})$ explicitly in terms of ρ. Without loss of generality, we will assume that $\sigma^2 = 1$ for notational convenience. So, we have computed numerically $(1 - \rho^2)^{-1}(X'W^{-1}X)^{-1}$ for different values

TABLE 8.1: $M_2(d)$'s values for the designs d's with different values of ρ. Das, R.N. and Park, S. H. (2009). "A Measure of Robust Slope Rotatability for Second Order Response Surface Designs," Journal of Applied Statistics, Vol. 36, No. 7, pp. 755-767.

ρ	d_0	d	$d(1)$	$d(2\frac{1}{2})$	$d(3\frac{1}{2})$	$d(2)$	$d(3)$	d_B
-0.90	0.198	0.958	0.834	0.947	0.975	0.985	0.995	0.513
-0.80	0.336	0.881	0.682	0.868	0.930	0.955	0.982	0.769
-0.70	0.482	0.827	0.641	0.835	0.903	0.934	0.968	0.851
-0.60	0.611	0.811	0.656	0.847	0.905	0.930	0.957	0.880
-0.50	0.715	0.822	0.681	0.886	0.925	0.938	0.950	0.885
-0.40	0.794	0.847	0.682	0.930	0.951	0.952	0.946	0.876
-0.30	0.852	0.876	0.645	0.962	0.973	0.965	0.945	0.852
-0.20	0.894	0.904	0.583	0.975	0.989	0.975	0.945	0.812
-0.10	0.925	0.928	0.514	0.969	0.997	0.983	0.949	0.753
0.00	0.947	0.947	0.454	0.947	1.000	0.990	0.954	0.673
0.10	0.963	0.961	0.407	0.916	0.997	0.995	0.961	0.571
0.20	0.975	0.972	0.372	0.879	0.991	0.998	0.969	0.453
0.30	0.983	0.979	0.346	0.839	0.980	0.998	0.976	0.329
0.40	0.989	0.984	0.323	0.796	0.966	0.996	0.983	0.212
0.50	0.993	0.988	0.301	0.756	0.953	0.993	0.989	0.118
0.60	0.996	0.991	0.277	0.726	0.945	0.990	0.993	0.054
0.70	0.998	0.995	0.259	0.725	0.949	0.990	0.997	0.020
0.80	0.999	0.997	0.267	0.780	0.967	0.994	0.999	0.005
0.90	1.000	0.999	0.387	0.905	0.989	0.998	1.000	0.001

TABLE 8.2: $MO_2(d)$'s values for the designs d's with different values of ρ. Park, S. H., Jung S. H. and Das, R.N. (2009). "Slope Rotatability of Second Order Response Surface Regression Models with correlated errors," Journal of Quality Technology and Quality Management, Vol. 6, No. 4, pp. 471-493.

ρ	(d_0)	(d)	$(d(1))$	$(d(2^{1/2}))$	$(d(3^{1/2}))$	$(d(2))$	$(d(3))$
-0.9	1	1.000	0.987	0.998	0.999	0.999	0.998
-0.8	1	1.000	0.997	0.999	0.999	0.998	0.998
-0.7	1	1.000	0.999	0.999	0.998	0.998	0.998
-0.6	1	1.000	0.999	0.999	0.999	0.999	0.999
-0.5	1	1.000	0.999	0.999	0.999	0.999	0.999
-0.4	1	1.000	0.999	0.999	0.999	0.999	0.999
-0.3	1	1.000	1.000	1.000	1.000	1.000	1.000
-0.2	1	1.000	1.000	1.000	1.000	1.000	1.000
-0.1	1	1.000	1.000	1.000	1.000	1.000	1.000
0.0	1	1.000	1.000	1.000	1.000	1.000	1.000
0.1	1	1.000	1.000	1.000	1.000	1.000	1.000
0.2	1	1.000	1.000	1.000	1.000	1.000	1.000
0.3	1	1.000	0.999	1.000	1.000	1.000	1.000
0.4	1	1.000	0.998	0.999	1.000	1.000	1.000
0.5	1	1.000	0.996	0.999	0.999	0.999	1.000
0.6	1	1.000	0.990	0.998	0.999	0.999	1.000
0.7	1	1.000	0.974	0.996	0.998	0.999	1.000
0.8	1	0.999	0.922	0.993	0.998	0.999	1.000
0.9	1	0.999	0.729	0.988	0.997	0.999	1.000

TABLE 8.3: $ME_2(d)$'s values for the designs d's with different values of ρ. Park, S. H., Jung S. H. and Das, R.N. (2009). "Slope Rotatability of Second Order Response Surface Regression Models with correlated errors," Journal of Quality Technology and Quality Management, Vol. 6, No. 4, pp. 471-493.

ρ	(d_0)	(d)	$(d(1))$	$(d(2^{1/2}))$	$(d(3^{1/2}))$	$(d(2))$	$(d(3))$
-0.9	1	0.998	0.163	0.402	0.596	0.723	0.929
-0.8	1	0.999	0.363	0.629	0.777	0.856	0.965
-0.7	1	0.999	0.581	0.787	0.878	0.923	0.981
-0.6	1	0.999	0.761	0.888	0.938	0.961	0.990
-0.5	1	0.999	0.879	0.946	0.971	0.982	0.995
-0.4	1	1.000	0.944	0.976	0.987	0.992	0.998
-0.3	1	1.000	0.975	0.991	0.995	0.997	0.999
-0.2	1	1.000	0.988	0.997	0.998	0.999	1.000
-0.1	1	1.000	0.990	0.999	1.000	1.000	1.000
0.0	1	1.000	0.983	1.000	1.000	1.000	1.000
0.1	1	1.000	0.968	1.000	1.000	1.000	1.000
0.2	1	1.000	0.936	0.999	0.999	1.000	1.000
0.3	1	1.000	0.876	0.997	0.999	0.999	1.000
0.4	1	1.000	0.765	0.993	0.998	0.999	1.000
0.5	1	1.000	0.579	0.984	0.997	0.999	1.000
0.6	1	1.000	0.337	0.963	0.995	0.998	1.000
0.7	1	0.999	0.134	0.911	0.991	0.996	0.999
0.8	1	0.999	0.034	0.795	0.983	0.994	0.999
0.9	1	0.998	0.005	0.569	0.964	0.989	0.999

TABLE 8.4: Weak slope-rotatability region values ($R_{d(\nu)}(\rho)$'s) with strength ($\nu = 0.95$) based on $M_2(d)$, $MO_2(d)$, $ME_2(d)$ for the designs d_0, d, $d(\alpha)$'s with $\alpha = 1, 2^{1/2}, 3^{1/2}, 2, 3$. Park, S. H., Jung S. H. and Das, R.N. (2009). "Slope Rotatability of Second Order Response Surface Regression Models with correlated errors," Journal of Quality Technology and Quality Management, Vol. 6, No. 4, pp. 471-493.

Design	$R_{d(0.95)}(\rho)$ on $M_2(d)$	$R_{d(0.95)}(\rho)$ on $MO_2(d)$	$R_{d(0.95)}(\rho)$ on $ME_2(d)$
d_0	$(0.0, 0.9)$	$(-0.9, 0.9)$	$(-0.9, 0.9)$
d	$(0.0, 0.9)$	$(-0.9, 0.9)$	$(-0.9, 0.9)$
$d(1)$	0	$(-0.9, 0.7)$	$(-0.3, 0.1)$
$d(2^{1/2})$	$(-0.3, 0.0)$	$(-0.9, 0.9)$	$(-0.4, 0.6)$
$d(3^{1/2})$	$(-0.4, 0.9)$	$(-0.9, 0.9)$	$(-0.5, 0.9)$
$d(2)$	$(-0.4, 0.9)$	$(-0.9, 0.9)$	$(-0.6, 0.9)$
$d(3)$	$(-0.9, 0.9)$	$(-0.9, 0.9)$	$(-0.8, 0.9)$

of $\rho = -0.9, -0.8, \ldots, 0.0, 0.1, \ldots, 0.9$. Below are given the moment matrices of the designs as given in Examples 8.1 through 8.4. as they are needed to find the measure
$$MO(d_0) =$$

$$
\begin{pmatrix}
15\rho^2 - 32\rho + 17 & 0 & 0 & 7.998(1-\rho)^2 & 7.998(1-\rho)^2 & 0 \\
0 & g & 0 & 0 & 0 & 0 \\
0 & 0 & g & 0 & 0 & 0 \\
7.998(1-\rho)^2 & 0 & 0 & 12(1+\rho^2) & 4(1+\rho^2) & 0 \\
7.998(1-\rho)^2 & 0 & 0 & 4(1+\rho^2) & 12(1+\rho^2) & 0 \\
0 & 0 & 0 & 0 & 0 & 4(1+\rho^2)
\end{pmatrix},
$$

where $g = 7.998(1 + \rho^2)$,
$MO(d) =$

$$
\begin{pmatrix}
15\rho^2 - 32\rho + 17 & 0 & 1.414\rho(1-\rho) & 7.998(1-\rho)^2 & v_{0.22} & 0 \\
0 & g & 0 & 0 & 0 & 0 \\
1.414\rho(1-\rho) & 0 & (5.999\rho^2 + 7.998) & 0 & -2.827\rho^2 & 0 \\
7.998(1-\rho)^2 & 0 & 0 & 11.992(1+\rho^2) & 4(1+\rho^2) & 0 \\
v_{0.22} & 0 & -2.827\rho^2 & 4(1+\rho^2) & v_{22.22} & 0 \\
0 & 0 & 0 & 0 & 0 & 4(1+\rho^2)
\end{pmatrix},
$$

where $g = 7.998(1 + \rho^2)$, $v_{0.22} = (5.999\rho^2 - 13.997\rho + 7.998)$ and $v_{22.22} = 7.996\rho^2 + 11.992$,
$MO(d(\alpha)) =$

$$
\begin{pmatrix}
v_{00} & \rho^2 - \rho & \rho^2 - \rho & v_{0.11} & v_{0.22} & \rho - \rho^2 \\
\rho^2 - \rho & v_{1.1} & v_{1.2} & \rho^2 + \rho\alpha - \rho\alpha^2 & \rho^2 + \rho\alpha - \rho\alpha^3 & \rho^2 + \rho\alpha \\
\rho^2 - \rho & v_{2.1} & v_{2.2} & \rho^2 - \rho\alpha^2 + \rho\alpha^3 & \rho^2 & \rho^2 \\
v_{11.0} & \rho^2 + \rho\alpha - \rho\alpha^2 & g_1 & v_{11.11} & v_{11.22} & 2\rho - \rho^2 - \rho\alpha^2 \\
v_{22.0} & \rho^2 + \rho\alpha - \rho\alpha^3 & \rho^2 & v_{22.11} & v_{22.22} & 2\rho - \rho^2 \\
\rho - \rho^2 & \rho^2 + \rho\alpha & \rho^2 & 2\rho - \rho^2 - \rho\alpha^2 & 2\rho - \rho^2 & 4 + 2\rho + 3\rho^2
\end{pmatrix},
$$

where $g_1 = \rho^2 - \rho\alpha^2 + \rho\alpha^3$, $v_{00} = (15\rho^2 - 32\rho + 17)$, $v_{0.11} = 4 - 7\rho + 3\rho^2 + 2\alpha^2(1-\rho)^2 = v_{11.0} = v_{0.22} - v_{22.0}$, $v_{1.1} = 4 + 6\rho + 3\rho^2 + 2\rho\alpha + 2\alpha^2(1+\rho+\rho^2)$, $v_{1.2} = -2\rho - \rho^2 + \rho\alpha + \rho\alpha^2 = v_{2.1}$, $v_{2.2} = 4 - 2\rho + 3\rho^2 + 2\alpha^2(1+\rho+\rho^2)$, $v_{11.11} = 4 - 6\rho + 3\rho^2 - 2\rho\alpha^2 + 2\alpha^4(1-\rho+\rho^2)$, $v_{11.22} = 4 - 6\rho + 3\rho^2 - \rho\alpha^2 - \rho\alpha^4 = v_{22.11}$ and $v_{22.22} = 4 - 6\rho + 3\rho^2 + 2\alpha^4(1-\rho+\rho^2)$,

$MO(d_B) =$

$$
\begin{pmatrix}
15\rho^2 - 32\rho + 17 & \rho^2 - \rho & \rho^2 - \rho & 8 - 15\rho + 7\rho^2 & 8 - 15\rho + 7\rho^2 & \rho - \rho^2 \\
\rho^2 - \rho & g_2 & -\rho^2 & \rho^2 & \rho^2 & \rho^2 + \rho \\
\rho^2 - \rho & -\rho^2 & g_3 & \rho^2 & \rho^2 & \rho^2 \\
8 - 15\rho + 7\rho^2 & \rho^2 & \rho^2 & 8 - 14\rho + 7\rho^2 & 4 - 8\rho + 3\rho^2 & \rho - \rho^2 \\
8 - 15\rho + 7\rho^2 & \rho^2 & \rho^2 & 4 - 8\rho + 3\rho^2 & 8 - 12\rho + 7\rho^2 & 2\rho - \rho^2 \\
\rho - \rho^2 & \rho + \rho^2 & \rho^2 & \rho - \rho^2 & 2\rho - \rho^2 & 4 + 2\rho + 3\rho^2
\end{pmatrix},
$$

where $g_2 = 8 - 2\rho + 7\rho^2$ and $g_3 = 8 + 2\rho + 7\rho^2$.

8.4   CONCLUDING REMARKS

In the present chapter, some general measures ($M_k(d)$, $MO_k(d)$ and $ME_2(d)$) of robust second-order slope-rotatability along axial directions, over all directions and with equal maximum directional variance have been discussed. These measures ($M_k(d)$, $MO_k(d)$ and $ME_2(d)$) are illustrated with few examples, out of which one is an exactly RSORD, and one is a very nearly RSORD with autocorrelated errors, one Box-Behnken design, and the others are usual CCDs. Table 8.4 shows that the design $d(3)$ gives more wider range of variation of ρ than the other designs d's with respect to all the measures of robust slope-rotatability as defined herein (i.e., $M_k(d)$, $MO_k(d)$ and $ME_2(d)$). It is also seen that the measures of robust slope-rotatability of a CCD (along axial directions, over all directions and with equal maximum directional variance) increase with the increased value of α. Therefore, the CCD with $\alpha = 3$ is more robust slope-rotatable than the RSORD and non-RSORD as in Table 8.4. Also the RSORD (d_0) remains robust slope-rotatable design over all directions and also with equal maximum directional variance (from Table 8.2 and Table 8.3), under autocorrelated errors. Based on these measures one can examine the *robust slope-rotatability* of any second-order design with respect to any error variance-covariance structure. These measures can be used to compare the degree of robust slope-rotatability of the designs with the same number of independent variables. Note that the degree of robust second-order slope-rotatability with equal maximum directional variance and D-optimal slope (for $k > 2$) can be examined based on the measures of robust second-order rotatability as given in Chapter 5. The advantages of these measures are the following:

(i) They can be used for any kind of designs with any number of independent variables and design points.

(ii) These measure can be used without scaling. It is invariant with respect to the scaling constant.

(iii) They are very easy to compute. For computation, one needs to find only the W^{-1} in general form with correlation coefficient ρ. If general form of W^{-1} is not available, with the help of computer, one can easily compute the measure.

(iv) Degree of robust slope-rotatability can be compared of a design with respect to number of design points, i.e., by changing only the design points for the fixed number of factors.

(v) Degree of robust slope-rotatability can be compared of a design with respect to different $W = \text{Dis}(\mathbf{e})$, i.e., different error variance-covariance structures.

Chapter 9

REGRESSION ANALYSES WITH CORRELATED ERRORS AND APPLICATIONS

In the current book, rotatable and slope-rotatable designs with correlated errors are discussed in Chapters 2 to 8. First- and second-order correlated regression models have been discussed under different error structures (intra-class, inter-class, compound symmetry, tridiagonal and autocorrelated). Note that intra-class and inter-class structures are some particular cases of compound symmetry structure (Chapter 2). Using robust designs (controlled setting) or environmental data, one may estimate the regression model parameters.

155

Different sources of arising correlation in the observations have been discussed in Chapter 2. For example, on the organismal level, there may be measurements on several tumors, both hands, all teeth, and so on. It is usually expected in such cases that the measurements on a given individual are more similar than those on different individuals. One may consider the regression analysis for some sets of observations such that within each set, observations have an autocorrelation structure, and any two observations from any two sets have a constant correlation coefficient, and the variance of all the observations are same. In general, the pattern (or form) of the correlation structure is known for a given situation of data set but the parameter (parameters) involved in the correlation structure is (are) always unknown. The present chapter describes different method of regression analyses on planned and unplanned data under the compound symmetry and compound autocorrelated error structures. We have discussed some *robust* methods of estimating the best linear unbiased estimators of all the regression parameters *except* the intercept, which is often unimportant in practice. In this connection, we have also described some testing procedures for any set of linear hypotheses regarding the unknown regression coefficients. Confidence interval (for an estimable linear function) and confidence ellipsoid (for a set of estimable linear functions) of regression parameters have been developed. Index of fit has also been described for the fitted regression models. Two applications of correlated regression analysis in block designs have been described. Some examples (with simulated data) illustrate the theories described in the present chapter.

9.1   INTRODUCTION AND SUMMARY

Regression technique is one of the most widely used statistical tools as it provides a simple method for establishing a fundamental relationship between a dependent variable and one or more explanatory variables. It is viewed that regression analysis as a set of data analytic techniques that are used to help understand the interrelationship among variables in a certain environment. A detailed discussion of regression analysis is given in Chatterjee and Price (2000), Draper and Smith (1998), Palta (2003), Box and Draper (2007) etc.

Regression technique used in design of experiments is known as response surface methodology. Data may be available either from the environment or the data may be collected in a controlled setting so that the factors that are not of primary interest can be held constant. Data in the earlier case is known as environmental data and in the later case is known as experimental data.

In general, it is assumed that the errors in the regression model are independently, identically distributed (IID) with a constant variance. In that case, ordinary least squares (OLS) method is used for estimating the regression parameters. If the errors are not IID but the dispersion matrix of error is known *except* the common error variance, generalized least squares (GLS) method is used for estimating the regression parameters. In general, the pattern of the error dispersion matrix can be realized from the nature of the data under consideration, but the correlation parameter (parameters) involved in the correlation structure is (are) always unknown. Correlation in the error terms may be introduced due to several causes (Chatterjee and Price, 2000).

Data arising in many situations in health research are mostly correlated. Several pattern of correlation structures are observed in case of health data. For example, decay on the tooth in a person may make decay on another tooth more likely. This type of correlation structure is called intra-class structure (Mukherjee, 1981) but in some books, it is termed as compound symmetry structure (Palta, 2003; Lee *et al.*, 2006). Other situations often encountered in population studies of health are data arising from clusters of individuals (for example, families), and data collected on a set of individuals longitudinally in time. The first type of correlated data often arises in survey sampling, when sampling units may contain a cluster of individuals. When a household has been selected and enrolled based on, for example, random dialing, the effort involved to obtain health data on all its members may be relatively minimal. However, members of a family may be similar in health habits, diet and health status and therefore do not well represent the correlation pattern of the population at large. The second type arises in cohort or panel studies. In health studies, we may enroll all patients with a given diagnosis, at the beginning of the disease, and repeatedly remeasure them to document treatment effect on progression of disease.

Yet another type of correlation between observations arises, even if there are no clusters or repeated measurements on the same individuals, because observations that are in the same organ of an individual or close in time may be more similar than those in the other organ or distant from each other in time. If such situations arise over time, we term this correlated structure as autocorrelated, and if it arises in different organs of an individual in the same time, we term the correlated structure as compound symmetry structure (Mukherjee, 1981). But if it is observed in different organs of an individuals in different times, the correlated structure has been termed as *compound autocorrelated structure* (Das, 2010) in analogy with compound symmetry structure (Mukherjee, 1981). For compound autocorrelated structure, it is assumed that the observations within any organ have an autocorrelated structure, and any two observations from any two organs have a constant correlation coefficient (ρ_1), and all the observations have the same variance σ^2.

The study of correlated observations and their analysis are given in Mukherjee (1981), Palta (2003), Lee *et al.* (2006). Panda and Das (1994) initiated regression designs with correlated errors. Das (1997, 2003a, 2004) and Das and Park (2006, 2007) studied robust designs under some pattern error variance covariance matrices. If the correlation parameter (parameters) in the error dispersion matrix is (are) unknown, GLS cannot be applied. In this case maximum likelihood estimation (MLE) method can be used. An explicit solutions of the maximum likelihood (ML) equations for estimating the unknown parameters of any positive definite variance covariance matrix or its inverse which can be spectrally decomposed is given in Mukherjee (1981). Different iterative solution methods of ML equations are given in Rubin and Szatrowski (1982), Rogers and Young (1977), Szatrowski (1978), Palta (2003) and Lee *et al.* (2006).

For correlated regression models, so far many authors have studied iterative methods for estimation of regression coefficients and asymptotic method for hypotheses testing, under the compound symmetry, inter-class and intra-class error structures. In literature, little is known about any *exact* method of estimation and testing procedure under the compound symmetry (Kim *et al.*, 2009) and compound autocorrelated error structures (Das, 2010). Note that the intra-class, inter-class, compound symmetry and autocorrelated correlation structures are some particular cases of compound autocorrelated error structure (CAES). In the present chapter, an *exact* theory of regression analysis technique has been described under the compound symmetry and the compound autocorrelated error structures. The best linear unbiased estimates (BLUE), possibly with a little prediction in efficiency have been derived for all the regression parameters *except* the intercept. The method of estimation is free from between groups correlation coefficient (ρ_1). In the process, a *robust* method of testing procedure has been derived for any set of linear hypotheses regarding the unknown regression coefficients. *Confidence interval* of an estimable linear function and *confidence ellipsoid* of a set of estimable linear functions of regression parameters have been described. *Index of fit* has been developed for the fitted regression models under the compound symmetry (Kim *et al.*, 2009) and the compound autocorrelated error structures (Das, 2010).

Correlated regression analyses (as developed in the present chapter under the compound symmetry and the compound autocorrelated error structures) have been used in the analysis of a randomized block design (RBD). For a randomized block design, it is assumed that any two observations of a block are correlated with a constant correlation coefficient (ρ) and any two observations from any two blocks are correlated with another constant correlation (ρ_1). These two correlation coefficients are known as within block correlation coefficient (ρ) and between block correlation coefficient (ρ_1) and they may be same or different. This type of correlation structure is known as compound symmetry error structure (CSES). If $\rho_1 = 0$ (i.e., the correlation coefficient of any two observations from any two blocks is zero), this type of correlation structure is known as inter-class structure. Compound symmetry structure is the most common situation in an RBD and inter-class structure is a particular case of compound symmetry structure. Here a method of analysis has been derived for an RBD with correlated errors, having compound symmetry and compound autocorrelated error structures. The testing procedures which are described (in this chapter) are *robust*, i.e., free from correlation coefficient or coefficients. The confidence ellipsoid of a set of linear estimable functions (or contrasts) of treatments and the confidence interval of a linear estimable function (or contrast) of treatments have been developed for an RBD with the compound symmetry and compound autocorrelated error structures. In analogy with Scheffe's method, the multiple comparison technique has been derived for judging all possible treatment contrasts (under the compound symmetry and the compound autocorrelated error structures). In analogy with Tukey's method, all the treatment contrasts have been compared under the compound symmetry structure. All these developed results have been illustrated by simulated examples.

9.2   REGRESSION ANALYSES WITH COMPOUND SYMMETRY ERROR STRUCTURE

9.2.1   Correlated error regression models

First- and second-order response surface models are described respectively in Chapter 2 and Chapter 3. For ready reference, these are reproduced below. Suppose there are p factors $x_1, x_2, \ldots, x_p$ and their u-th run $(x_{u1}, x_{u2}, \ldots, x_{up})$, $1 \leq u \leq N$, yields a response y_u on the study variable y. Assuming that the response surface is of first-order or linear, we adopt the model

$$y_u = \beta_0 + \sum_{k=1}^{p} \beta_k x_{uk} + e_u; \ 1 \leq u \leq N,$$

$$\text{or,} \quad \mathbf{Y} = X\boldsymbol{\beta} + \mathbf{e}, \tag{9.1}$$

where $\mathbf{Y} = (y_1, y_2, \ldots, y_N)'$ is the vector of recorded observations on the study variable y, $\boldsymbol{\beta} = (\beta_0, \beta_1, \ldots, \beta_p)'$ is the vector of regression coefficients $X = (\mathbf{1} : (x_{uk}); \ 1 \leq k \leq p, \ 1 \leq u \leq N)$ is the model matrix. Further, $\mathbf{e}$ is the vector of errors which are assumed to be normally distributed with $E(\mathbf{e}) = \mathbf{0}$ and $\text{Dis}(\mathbf{e}) = \sigma^2 W$ with $\text{rank}(W) = N$. The matrix W may represent any error correlation structure. In general, the matrix W is unknown but for all the calculations as usual, W is assumed to be known. In Chapter 2, W and W^{-1} are computed for different correlated error structures.

If there is a curvature in the system, then a polynomial of higher degree, such as second-order model can be used, as given below:

$$y_u = \beta_0 + \sum_{i=1}^{p} \beta_i x_{ui} + \sum_{i \leq j=1}^{p} \beta_{ij} x_{ui} x_{uj} + e_u; \ 1 \leq u \leq N,$$

$$\text{or} \quad \mathbf{Y} = X_1 \boldsymbol{\beta}^* + \mathbf{e}, \tag{9.2}$$

where $\mathbf{Y}$ is the vector of recorded observations on the study variable y, $\boldsymbol{\beta}^* = (\beta_0, \beta_1, \ldots, \beta_p,$ $\beta_{11}, \ldots, \beta_{pp}, \beta_{12}, \ldots, \beta_{1p}, \beta_{23}, \ldots \beta_{2p}, \ldots, \beta_{(p-1)p})'$ is the vector of regression coefficients of order $\binom{p+2}{2} \times 1$, and $X_1 = (\mathbf{1} : Z^*)$ is the model matrix, where Z^* is given below by using the Hadamard product $(\otimes_1)$ as

$$Z^* = (\mathbf{x}_1, \ldots, \mathbf{x}_p, \ \mathbf{x}_1 \otimes_1 \mathbf{x}_1, \ldots, \mathbf{x}_p \otimes_1 \mathbf{x}_p, \ \mathbf{x}_1 \otimes_1 \mathbf{x}_2, \mathbf{x}_1 \otimes_1 \mathbf{x}_3, \ldots, \mathbf{x}_{p-1} \otimes_1 \mathbf{x}_p),$$

where $\mathbf{x}_i = (x_{1i}, \ x_{2i}, \ldots, x_{Ni})'$, and $\mathbf{x}_i \otimes_1 \mathbf{x}_j = (x_{1i}x_{1j}, \ x_{2i}x_{2j}, \ldots, x_{Ni}x_{Nj})'$.

A method of estimation of regression parameters, correlation coefficient or coefficients, error variance has been described below for the correlated regression models ((9.1) and (9.2)) under the compound symmetry error structure (CSES).

9.2.2   Regression parameter estimation with CSES

In the present section, estimation of regression parameters has been described under the compound symmetry error structure (2.23), which has been described in Chapter 2. For ready reference, it is reproduced below.

$$\text{Dis}(\mathbf{e}) = \sigma^2 \begin{pmatrix} A & B & \cdots & B \\ B & A & \cdots & B \\ \vdots & \vdots & \vdots & \vdots \\ B & B & \cdots & A \end{pmatrix}, \ A = \begin{pmatrix} 1 & \rho & \cdots & \rho \\ \rho & 1 & \cdots & \rho \\ \vdots & \vdots & \vdots & \vdots \\ \rho & \rho & \cdots & 1 \end{pmatrix}_{r \times r}, \ B = \begin{pmatrix} \rho_1 & \rho_1 & \cdots & \rho_1 \\ \rho_1 & \rho_1 & \cdots & \rho_1 \\ \vdots & \vdots & \vdots & \vdots \\ \rho_1 & \rho_1 & \cdots & \rho_1 \end{pmatrix}_{r \times r}.$$

It is simply represented by

$$\text{Dis}(\mathbf{e}) = \sigma^2 W = \sigma^2 [I_g \otimes (A - B) + E_g \otimes B], \tag{9.3}$$

where I_g is an identity matrix of order $g \times g$, E_g is a $g \times g$ matrix with all elements 1, $A = (1 - \rho)I_r + \rho E_r$ and $B = \rho_1 E_r$. Here $\otimes$ denotes Kronecker product. Note that if $\rho_1 = 0$, $\text{Dis}(\mathbf{e}) = \sigma^2[I_g \otimes A]$ which is known as inter-class correlation structure, and if $\rho_1 = 0$, $g = 1$, $\text{Dis}(\mathbf{e}) = \sigma^2 A$ which is known as intra-class correlation structure.

A robust method (free of correlation coefficients ρ and ρ_1) of estimation of all the regression coefficients *except* the intercept (Kim *et al.*, 2009) is described below for a first-order model as in (9.1). Also the estimation of error variance (σ^2), and the correlation coefficients ρ and ρ_1 has been described. Similar method can be used for a second-order model as in (9.2). As it is stated in (9.3) that there are g groups each with r observations, so total number of observations is $N = gr$. Let y_{ij} be the j-th observation of the i-th group, $1 \leq i \leq g$, $1 \leq j \leq r$. Assuming a first-order model we have

$$y_{ij} = \beta_0 + \beta_1 x_{ij1} + \ldots + \beta_p x_{ijp} + e_{ij}, \ 1 \leq i \leq g, \ 1 \leq j \leq r,$$

$$\text{or,} \quad \mathbf{Y} = X\boldsymbol{\beta} + \mathbf{e}, \tag{9.4}$$

where $\mathbf{Y} = (y_{11}, \ldots, y_{1r}, \ldots, y_{g1}, \ldots, y_{gr})'$, e_{ij} is the corresponding error of y_{ij}, $\boldsymbol{\beta}$ and x_{ijk}'s are as in (9.1). Note that $E(\mathbf{e}) = \mathbf{0}$ and $\text{Dis}(\mathbf{e}) = \sigma^2 W$, where W as in (9.3).

Let us define

$$z_{iw} = y_{iw} - y_{ir}; \ 1 \le i \le g, \ 1 \le w \le (r-1),$$

$$\text{or,} \quad z_{iw} = \sum_{k=1}^{p} \beta_k (x_{iwk} - x_{irk}) + (e_{iw} - e_{ir}),$$

$$\text{or,} \quad z_{iw} = \sum_{k=1}^{p} \beta_k s_{iwk} + \epsilon_{iw} \ \text{say,} \ \ 1 \le w \le (r-1), \ 1 \le i \le g,$$

$$\text{or,} \quad \mathbf{Z}_i = S_i \boldsymbol{\eta} + \boldsymbol{\epsilon}_i; \ 1 \le i \le g,$$

$$\text{or,} \quad \mathbf{Z} = S \boldsymbol{\eta} + \boldsymbol{\epsilon}, \tag{9.5}$$

where $\mathbf{Z}_i = (z_{i1}, \ldots, z_{i(r-1)})'$, $\boldsymbol{\eta} = (\beta_1, \ldots, \beta_p)'$, $S_i = ((s_{iwk}); k = 1, \ldots, p \,; w = 1, \ldots, r-1)$, $s_{iwk} = x_{iwk} - x_{irk}$, $\epsilon_{iw} = e_{iw} - e_{ir}$, $\boldsymbol{\epsilon}_i = (\epsilon_{i1}, \ldots, \epsilon_{i(r-1)})'$, $\mathbf{Z} = (\mathbf{Z}_1', \mathbf{Z}_2', \ldots, \mathbf{Z}_g')'$, $S = (S_1', S_2', \ldots, S_g')'$, and $\boldsymbol{\epsilon} = (\boldsymbol{\epsilon}_1', \boldsymbol{\epsilon}_2', \ldots, \boldsymbol{\epsilon}_g')'$. Assume S has full column rank, so that when the dispersion matrix of $\boldsymbol{\epsilon}$ is non-singular, all linear functions of $\beta_1, \ldots, \beta_p$ are estiamble. This is the usual assumption in a general linear regression model involving environmental data.

On simplification, we have $E(\boldsymbol{\epsilon}) = \mathbf{0}$, $\text{Dis}(\boldsymbol{\epsilon}) = \sigma_1^2 W_2$ say, where $\sigma_1^2 = 2\sigma^2(1-\rho)$ and $W_2 = I_g \otimes \left(\frac{1}{2}I_{r-1} + \frac{1}{2}E_{r-1}\right)$. The model (9.5) is a generalized linear least squares model (with known W_2). Therefore, we have the following results for the reduced model (9.5) (proofs are immediate from (9.5)).

Theorem 9.1 *The best linear unbiased estimator (BLUE) of $\boldsymbol{\eta}$ is*

$$\hat{\boldsymbol{\eta}} = (S' W_2^{-1} S)^{-1} S' W_2^{-1} \mathbf{Z}. \tag{9.6}$$

Theorem 9.2 *An unbiased estimator (UE) of $\sigma_1^2 = 2\sigma^2(1-\rho)$ is*

$$\hat{\sigma}_1^2 = \frac{\mathbf{Z}' W_2^{-1} \mathbf{Z} - \hat{\boldsymbol{\eta}}'(S' W_2^{-1} S)\hat{\boldsymbol{\eta}}}{g(r-1)-p}, \tag{9.7}$$

where $W_2^{-1} = I_g \otimes \left(2I_{r-1} - \frac{2}{r}E_{r-1}\right)$.

Note that $\hat{\boldsymbol{\eta}} \sim \text{MN}(\boldsymbol{\eta}, \sigma_1^2(S' W_2^{-1} S)^{-1})$, and $\hat{\boldsymbol{\eta}}$ does not depend on the other unknown parameters ρ, ρ_1 and σ^2.

The scheme for calculations of other unknown parameters is given hereunder. From equation (9.4), one can find the estimate of β_0 as

$$\hat{\beta}_0 = \bar{y} - \hat{\beta}_1 \bar{x}_1 - \hat{\beta}_2 \bar{x}_2 - \ldots - \hat{\beta}_k \bar{x}_k, \tag{9.8}$$

where $\hat{\beta}_1, \hat{\beta}_2, \ldots, \hat{\beta}_k$ are as in (9.6), $\bar{y}, \bar{x}_1, \ldots, \bar{x}_k$ (respective means) are known.

Theorem 9.3 *An estimate of ρ (from (9.7)) for known σ^2 is*

$$\hat{\rho} = 1 - \frac{1}{2\sigma^2} \frac{(\mathbf{Z}' W_2^{-1} \mathbf{Z}) - \hat{\boldsymbol{\eta}}'(S' W_2^{-1} S)\hat{\boldsymbol{\eta}}}{g(r-1)-p}. \tag{9.9}$$

To estimate ρ for *unknown* σ^2, an estimate of σ^2 is required, which is clear from (9.9).

Theorem 9.4 *An estimate of σ^2 (from the full model (9.4)) is*

$$\hat{\sigma}^2 = \frac{\hat{\mathbf{e}}_0' \hat{\mathbf{e}}_0}{gr - p - 1}, \tag{9.10}$$

where $\hat{\mathbf{e}}_0 = W^{-\frac{1}{2}}(\mathbf{Y} - X\hat{\boldsymbol{\beta}})$, $\hat{\boldsymbol{\beta}} = (\hat{\beta}_0, \hat{\boldsymbol{\eta}}')'$ and W is obtained from the scheme given below.

The scheme for calculations of W (i.e., ρ, ρ_1 and σ^2) is given below.

1. Assume some value of $\rho_1 \in (-1, 1)$.

2. Compute $\hat{\rho}$ using (9.9) (taking $\sigma^2 = 1$ in the first iteration, and for any other iteration, by plugging $\hat{\sigma}^2$ for σ^2 obtained in step 5 just in the previous iteration).

3. With the assumed value of ρ_1, say $\hat{\rho}_1$, and the estimate of ρ, say $\hat{\rho}$ in step 2, compute W (examining W is non-singular) and W^{-1} as in (9.3).

4. Calculate $\hat{\boldsymbol{\beta}} = (X'W^{-1}X)^{-1}(X'W^{-1}\mathbf{Y})$, assuming X has full column rank.

5. Compute $\hat{\sigma}^2$ using (9.10).

6. Calculate $S_0(\hat{\rho}_1, \hat{\boldsymbol{\beta}}) = (Y - X\hat{\boldsymbol{\beta}})'W^{-1}(Y - X\hat{\boldsymbol{\beta}})$, where W^{-1} as in step 3 and $\hat{\boldsymbol{\beta}}$ is as in step 4.

The same routine of calculations 1 through 6 is to be followed for different permissible values of ρ_1 in its range. We select that value of ρ_1 as the final estimate of ρ_1 for which $S_0(\hat{\rho}_1, \hat{\boldsymbol{\beta}})$ is minimum. For the final estimate of ρ_1, we get the final estimate of ρ in step 2. Thus, for the final estimates of ρ and ρ_1, one can compute W (an estimate of W) in step 3.

9.2.3   Hypotheses testing of regression parameters with CSES

In regression analysis, interpretation about the regression coefficients is very important. In the present section, a *robust exact test* procedure is described (following Kim *et al.*, 2009) for testing any set of linear hypotheses regarding regression coefficients. A set of m linear independent hypotheses regarding unknown regression coefficients are given below

$$H_0 : \begin{cases} l_{11}\beta_1 + \ldots + l_{1p}\beta_p = l_{10}(\text{known}) \\ l_{21}\beta_1 + \ldots + l_{2p}\beta_p = l_{20}(\text{known}) \\ \quad\quad\quad \vdots \\ l_{m1}\beta_1 + \ldots + l_{mp}\beta_p = l_{m0}(\text{known}). \end{cases}$$

Equivalently, $H_0 : R\boldsymbol{\eta} = \mathbf{l}_0(\text{known})$ against $H_A : R\boldsymbol{\eta} \neq \mathbf{l}_0$, where $rank(R) = m$,

$$R = \begin{pmatrix} l_{11} & \cdots & l_{1p} \\ l_{21} & \cdots & l_{2p} \\ \cdots & \cdots & \cdots \\ l_{m1} & \cdots & l_{mp} \end{pmatrix} \quad \text{and} \quad \mathbf{l}_0 = (l_{10}, \ldots, l_{m0})'.$$

Note that $\hat{\boldsymbol{\eta}} \sim \text{MN}(\boldsymbol{\eta}, \sigma_1^2(S'W_2^{-1}S)^{-1})$ as given in Section 9.2.2, thus

$$R\hat{\boldsymbol{\eta}} \sim \text{MN}(R\boldsymbol{\eta}, \sigma_1^2 R(S'W_2^{-1}S)^{-1}R').$$

Under H_0,

$$R\hat{\boldsymbol{\eta}} \sim \text{MN}(\mathbf{l}_0, \sigma_1^2 R(S'W_2^{-1}S)^{-1}R').$$

Therefore,

$$(R\hat{\boldsymbol{\eta}} - \mathbf{l}_0)'[\sigma_1^2 R(S'W_2^{-1}S)^{-1}R']^{-1}(R\hat{\boldsymbol{\eta}} - \mathbf{l}_0) \sim \chi_m^2,$$

where the degree of freedom m is given by the number of independent linear hypotheses in the $R\boldsymbol{\eta}$ vector. Also for the model (9.5), $\hat{\boldsymbol{\epsilon}}_0 = W_2^{-\frac{1}{2}}(Z - S\hat{\boldsymbol{\eta}})$, $\hat{\boldsymbol{\epsilon}}_0'\hat{\boldsymbol{\epsilon}}_0/\sigma_1^2 \sim \chi_{g(r-1)-p}^2$, and it is independent of $R\hat{\boldsymbol{\eta}}$. Thus we have the following result

Theorem 9.5 *If $R\eta = l_0$ is true, the basic result is*

$$F = \frac{(R\hat{\eta} - l_0)'[R(S'W_2^{-1}S)^{-1}R']^{-1}(R\hat{\eta} - l_0)/m}{\hat{\epsilon}_0'\hat{\epsilon}_0/\{g(r-1)-p\}} \sim F_{m,\ g(r-1)-p}. \tag{9.11}$$

The test procedure is: reject H_0 at $100\alpha\%$ level of significance if observed $F > F_{\alpha;\ m,\ g(r-1)-p}$, and accept otherwise.

9.2.4   Confidence ellipsoid of regression parameters with CSES

In the present section, the confidence ellipsoid for a set of independent linear functions of regression coefficients and the confidence interval for a linear function of regression coefficients are described following Kim *et al.*, (2009). Note that S has full rank (assumed), so every linear function of $\beta_1, \ldots, \beta_p$ are estimable for the linear model (9.5).

Suppose $\psi_1, \psi_2, \ldots, \psi_\nu$ are ν independent linear functions of $\beta_1, \beta_2, \ldots, \beta_p$. Let $\boldsymbol{\psi} = (\psi_1, \psi_2, \ldots, \psi_\nu)'$ be a vector of order $\nu \times 1$. Then

$$\boldsymbol{\psi} = C\boldsymbol{\eta},$$

where $\boldsymbol{\eta} = (\beta_1, \ldots, \beta_p)'$ and C(known) is a $\nu \times p$ matrix whose rows are linearly independent. Then $\hat{\boldsymbol{\psi}} = AZ$ is the generalized least squares (GLS) estimate of $\boldsymbol{\psi}$ (for the model (9.5)), and $A = ((a_{ij}))$ (known matrix depending on ψ_is). The variance-covariance matrix of $\hat{\boldsymbol{\psi}}$ is then

$$\text{Dis}(\hat{\boldsymbol{\psi}}) = A\text{Dis}(\mathbf{Z})A' = \sigma_1^2 AW_2A',$$

where $W_2 = I_g \otimes \left(\frac{1}{2}I_{r-1} + \frac{1}{2}E_{r-1}\right)$. Note that $\hat{\boldsymbol{\psi}} \sim \text{MN}(\boldsymbol{\psi}, \sigma_1^2 AW_2A')$ and is independent of

$$\frac{\hat{\epsilon}_0'\hat{\epsilon}_0}{\sigma_1^2} = \frac{(\mathbf{Z} - S\hat{\eta})'W_2^{-1}(\mathbf{Z} - S\hat{\eta})}{\sigma_1^2} \sim \chi_{g(r-1)-p}^2.$$

Again

$$(\hat{\boldsymbol{\psi}} - \boldsymbol{\psi})'\{\sigma_1^2(AW_2A')\}^{-1}(\hat{\boldsymbol{\psi}} - \boldsymbol{\psi}) \sim \chi_\nu^2,$$

and independent of $\hat{\epsilon}_0'\hat{\epsilon}_0/\sigma_1^2$ (for the model (9.5)). Therefore, we have the following result.

Theorem 9.6 *The distribution of the test statistic is*

$$F = \frac{(\hat{\boldsymbol{\psi}} - \boldsymbol{\psi})'(AW_2A')^{-1}(\hat{\boldsymbol{\psi}} - \boldsymbol{\psi})/\nu}{(\hat{\epsilon}_0'\hat{\epsilon}_0)/(g(r-1)-p)} \sim F_{\nu,\ g(r-1)-p}. \tag{9.12}$$

Therefore,

$$(\hat{\boldsymbol{\psi}} - \boldsymbol{\psi})'(AW_2A')^{-1}(\hat{\boldsymbol{\psi}} - \boldsymbol{\psi}) \leq \nu s^2 F_{\alpha;\ \nu,\ g(r-1)-p}, \tag{9.13}$$

where $s^2 = \frac{\hat{\epsilon}_0'\hat{\epsilon}_0}{g(r-1)-p}$ which is an UE of σ_1^2 (9.7).

Inequality (9.13) determines an *ellipsoid* in the ν-dimensional $\boldsymbol{\psi}$- space with center $\hat{\boldsymbol{\psi}} = (\hat{\psi}_1, \hat{\psi}_2, \ldots, \hat{\psi}_\nu)'$, and the probability that this *random ellipsoid* covers the true parameter $\boldsymbol{\psi}$ is $(1-\alpha)$, no matter whatever be the values of unknown parameters.

We may obtain a confidence interval for a single linear function $\psi = \mathbf{c}'\boldsymbol{\eta}(\mathbf{c} \neq \mathbf{0})$ by specializing the above calculation to $\nu = 1$. The resulting confidence interval is given by

$$(\mathbf{a}'W_2\mathbf{a})^{-1}(\hat{\psi} - \psi)^2 \leq s^2 F_{\alpha;1,\ g(r-1)-p}, \tag{9.14}$$

where $\hat{\psi} = \mathbf{a}'\mathbf{Z}$ is the GLS estimate of ψ. Note that $\text{Var}(\hat{\psi}) = \mathbf{a}'\text{Dis}(\mathbf{Z})\mathbf{a} = \sigma_1^2\mathbf{a}'W_2\mathbf{a}$, and its unbiased estimate $\hat{\sigma}_{\hat{\psi}}^2 = s^2(\mathbf{a}'W_2\mathbf{a})$. We may write (9.14) as

$$\hat{\psi} - t_{\alpha/2;\ g(r-1)-p}\hat{\sigma}_{\hat{\psi}} \leq \psi \leq \hat{\psi} + t_{\alpha/2;\ g(r-1)-p}\hat{\sigma}_{\hat{\psi}}, \tag{9.15}$$

the probability that this random interval covers the unknown ψ is $(1-\alpha)$. The interval (9.15) could also be derived from the fact that $(\hat{\psi} - \psi)/\hat{\sigma}_{\hat{\psi}} \sim t_{g(r-1)-p}$.

9.2.5 Index of fit with CSES

Analogous to uncorrelated regression analysis, two criteria of judging the best fit of correlated error linear regression models are described below (Kim *et al.*, 2009; Das, 2010). For multiple regression with independent errors, the index of fit is measured by the multiple correlation coefficient (R^2) and adjusted multiple correlation coefficient (R_{adj}^2) of the fitted regression model. In analogy with uncorrelated case, we define the multiple correlation coefficients $R^2(\mathbf{Z})$, $R^2(\mathbf{Y})$ and adjusted multiple correlation coefficient $R_{adj}^2(\mathbf{Z})$, for the fitted model (9.5) and (9.4) as follow:

$$R^2(\mathbf{Z}) = 1 - \frac{\hat{\epsilon}_0'\hat{\epsilon}_0}{TSS_{\mathbf{Z}_0}} \quad \text{and} \quad R^2(\mathbf{Y}) = \text{Corr.}^2(\mathbf{Y}, \hat{\mathbf{Y}}), \tag{9.16}$$

$$R_{adj}^2(\mathbf{Z}) = 1 - \frac{g(r-1)-1}{g(r-1)-p}(1 - R^2(\mathbf{Z})), \tag{9.17}$$

where $\hat{\epsilon}_0 = W_2^{-\frac{1}{2}}(\mathbf{Z} - S\hat{\eta})$, $TSS_{\mathbf{Z}_0} = (Z - \bar{Z})'W_2^{-1}(Z - \bar{Z})$, $\bar{Z} = \sum_{i=1}^{g}\sum_{j=1}^{r-1}z_{ij}/g(r-1)$ (as in the model (9.5)). The fit is considered as very good, if R^2 and R_{adj}^2 as in (9.16), (9.17) are both quite close to unity.

9.2.6 Illustration of regression analysis with CSES

Example 9.1 The example (with simulated data) considered is the performance or quality of metalized glass plates which involves various factors. Main factors (i.e., exploratory variables) taken may be designated as $x_1 = $ Argon sputter pressure, $x_2 = $ Target current and $x_3 = $ Back ground pressure. Let y be the performance or quality or thickness of metalized glass plates in some convenient unit. The study conducted is to estimate the unknown parameters involved in the mean response function of y, error variance and correlation coefficients. Three factors as above are considered for this experiment.

Appropriate change of origin and scale is used for each exploratory variable so that it lies between -1 and $+1$ (the range within which the experimentation is conducted). We assume

$$y_{ij} = \beta_0 + \beta_1 x_{ij1} + \beta_2 x_{ij2} + \beta_3 x_{ij3} + e_{ij}, \ i = 1,\ldots,10; j = 1,\ldots,4. \tag{9.18}$$

The model matrix X (taken from Chapter 2 for CSES) used is

$$X = \begin{pmatrix} D \\ \vdots \\ D \end{pmatrix}, \quad \text{where } D = \begin{pmatrix} 1 & 1 & 1 & 1 \\ 1 & -1 & 1 & -1 \\ 1 & 1 & -1 & -1 \\ 1 & -1 & -1 & 1 \end{pmatrix}.$$

TABLE 9.1: Responses under the simulation setting of $(\sigma^2 = 2, \rho = 0.8, \rho_1 = 0.6)$ with CSES. Kim, J., Das, R.N., Sengupta, A., and Paul, J. (2009). "Regression Analysis for Correlated Data under Compound Symmetry Structure," Journal of Statistical Theory and Applications, Vol. 8, No. 3, pp. 269-282.

group	within group observations			
	1	2	3	4
1	52.307	43.693	57.153	48.990
2	51.652	45.250	57.478	48.823
3	51.990	42.612	55.388	49.065
4	52.672	45.418	58.365	51.081
5	51.546	44.989	55.536	49.242
6	50.303	44.643	56.644	47.035
7	52.905	44.368	55.563	49.991
8	51.250	43.407	56.986	49.249
9	54.474	46.910	58.384	52.334
10	51.010	43.950	56.331	48.823

In the absence of real data we generate observations (100 replications) according to formula (9.18) with $\beta_0 = 50.0$, $\beta_1 = 3.8$, $\beta_2 = -2.5$, $\beta_3 = 0.05$, $\sigma^2 = 2.0$, $\rho = 0.8$ and $\rho_1 = 0.6$ using the above model matrix 'X', and $\mathbf{e} \sim MN(\mathbf{0}, \sigma^2 W)$ where W is given in (9.3). The observations for a single replication so obtained are given in Table 9.1.

Computed estimates of unknown regression parameters, correlation coefficients and variance (based on the responses as in Table 9.1) are $\hat{\boldsymbol{\beta}} = (\hat{\beta}_0, \hat{\beta}_1, \hat{\beta}_2, \hat{\beta}_3)' = (50.695, 3.702, -2.428, 0.042)'$, $\hat{\rho}(\sigma^2) = 0.73$, $\hat{\rho}(\hat{\sigma}^2) = 0.71$, $\hat{\rho}_1 = 0.57$, and $\hat{\sigma}^2 = 1.520$, where $\hat{\rho}(\sigma^2)$ and $\hat{\rho}(\hat{\sigma}^2)$ are the estimate of ρ for known and unknown σ^2, respectively.

For the present simulation study, we consider the following eight combination of parameters, (σ^2, ρ, ρ_1) where $\sigma^2 = 1, 2$; $\rho = 0.4, 0.8$; $\rho_1 = 0.1, 0.6$, and with $\beta_0 = 50.0$, $\beta_1 = 3.8$, $\beta_2 = -2.5$, $\beta_3 = 0.05$.

For each simulation setting, each calculation is repeated 100 times. We compute the sample bias and sample variance for each estimate, where the sample bias and sample variance for the estimate θ are defined by

$$\text{Bias}(\hat{\theta}) = |\bar{\hat{\theta}} - \theta|, \quad \bar{\hat{\theta}} = \frac{\sum \hat{\theta}}{100} \quad \text{and} \quad \text{Var}(\hat{\theta}) = \frac{\sum (\hat{\theta} - \theta)^2}{100}.$$

Table 9.2 summarizes the simulation results.

The test results from 100 replications (with $\beta_0 = 50.0$, $\beta_1 = 3.8$, $\beta_2 = -2.5$, $\beta_3 = 0.05$, $\sigma^2 = 2.0$, $\rho = 0.8$ and $\rho_1 = 0.6$) for some hypotheses are given in Table 9.3.

Table 9.4 presents the values of two index of fit measures viz. $R^2(\mathbf{Z})$, $\text{Corr.}^2(\mathbf{Y}, \hat{\mathbf{Y}})$ and $R^2_{adj}(\mathbf{Z})$ as defined in Section 9.2.5 for the following four models:

$$M_1 : y = \beta_0 + \beta_1 x_1 + \beta_2 x_2 + \beta_3 x_3 + e,$$
$$M_2 : y = \beta_0 + \beta_1 x_1 + \beta_2 x_2 + e,$$
$$M_3 : y = \beta_0 + \beta_1 x_1 + \beta_3 x_3 + e,$$
$$M_4 : y = \beta_0 + \beta_1 x_1 + e.$$

In the above, we have considered the regression analysis with correlated error under the compound symmetry error structure. Similar techniques can be used for inter-class and intra-class correlated error structures, as these two correlation structures are some particular cases of compound symmetry structure (Section 9.2.2). In the process, the BLUE of all the

TABLE 9.2: Simulation results: $\hat{\beta}_0$, $\hat{\beta}_1$, $\hat{\beta}_2$, $\hat{\beta}_3$, $\hat{\rho}(\sigma^2)$, $\hat{\rho}(\hat{\sigma}^2)$, $\hat{\rho}_1$ and $\hat{\sigma}^2$ with CSES. Kim, J., Das, R.N., Sengupta, A., and Paul, J. (2009). "Regression Analysis for Correlated Data under Compound Symmetry Structure," Journal of Statistical Theory and Applications, Vol. 8, No. 3, pp. 269-282.

	$\hat{\beta}_0$	$\hat{\beta}_1$	$\hat{\beta}_2$	$\hat{\beta}_3$	$\hat{\rho}(\sigma^2)$	$\hat{\rho}(\hat{\sigma}^2)$	$\hat{\rho}_1$	$\hat{\sigma}^2$
			$\rho = 0.4, \rho_1 = 0.1, \sigma^2 = 1$					
Mean	50.026	3.809	−2.492	0.053	0.364	0.305	0.046	0.909
Bias($\hat{\theta}$)	0.026	0.009	0.008	0.003	0.036	0.095	0.054	0.091
Var($\hat{\theta}$)	0.128	0.015	0.013	0.015	0.007	0.028	0.001	0.058
			$\rho = 0.4, \rho_1 = 0.1, \sigma^2 = 2$					
Mean	49.966	3.786	−2.491	0.044	0.381	0.300	0.038	1.816
Bias ($\hat{\theta}$)	0.034	0.014	0.009	0.006	0.019	0.100	0.062	0.184
Var($\hat{\theta}$)	0.406	0.036	0.030	0.032	0.029	0.030	0.001	0.220
			$\rho = 0.4, \rho_1 = 0.6, \sigma^2 = 1$					
Mean	49.989	3.808	−2.495	0.062	0.327	0.318	0.487	0.486
Bias($\hat{\theta}$)	0.011	0.008	0.005	0.012	0.073	0.092	0.113	0.514
Var($\hat{\theta}$)	0.733	0.018	0.015	0.012	0.006	0.006	0.083	0.013
			$\rho = 0.4, \rho_1 = 0.6, \sigma^2 = 2$					
Mean	50.006	3.822	−2.523	0.040	0.390	0.339	0.502	1.015
Bias($\hat{\theta}$)	0.006	0.022	0.023	0.010	0.010	0.061	0.098	0.985
Var($\hat{\theta}$)	1.073	0.027	0.025	0.026	0.029	0.004	0.079	0.069
			$\rho = 0.8, \rho_1 = 0.1, \sigma^2 = 1$					
Mean	49.971	3.796	−2.500	0.043	0.897	0.748	0.050	0.944
Bias($\hat{\theta}$)	0.029	0.004	0.000	0.007	0.097	0.052	0.050	0.056
Var($\hat{\theta}$)	0.208	0.004	0.007	0.007	0.001	0.015	0.004	0.124
			$\rho = 0.8, \rho_1 = 0.1, \sigma^2 = 2$					
Mean	50.003	3.790	−2.496	0.034	0.801	0.748	0.053	1.802
Bias($\hat{\theta}$)	0.003	0.010	0.004	0.016	0.001	0.052	0.047	0.198
Var($\hat{\theta}$)	0.347	0.011	0.010	0.008	0.003	0.012	0.003	0.470
			$\rho = 0.8, \rho_1 = 0.6, \sigma^2 = 1$					
Mean	49.987	3.800	−2.504	0.050	0.902	0.723	0.449	0.411
Bias ($\hat{\theta}$)	0.013	0.000	0.004	0.000	0.102	0.077	0.151	0.589
Var ($\hat{\theta}$)	0.627	0.005	0.006	0.005	0.001	0.024	0.001	0.018
			$\rho = 0.8, \rho_1 = 0.6, \sigma^2 = 2$					
Mean	50.035	3.786	−2.502	0.048	0.795	0.632	0.435	1.362
Bias($\hat{\theta}$)	0.035	0.014	0.002	0.002	0.005	0.168	0.165	0.638
Var($\hat{\theta}$)	1.232	0.009	0.009	0.008	0.003	0.027	0.002	0.061

TABLE 9.3: Test results from 100 replications with $\alpha = 0.05$ under CSES. Kim, J., Das, R.N., Sengupta, A., and Paul, J. (2009). "Regression Analysis for Correlated Data under Compound Symmetry Structure," Journal of Statistical Theory and Applications, Vol. 8, No. 3, pp. 269-282.

Null Hypothesis	degree of freedom	Accepted cases	Rejected cases
$H_{01} : \beta_1 = \beta_2 = \beta_3 = 0$	(3, 27)	0	100
$H_{02} : \beta_1 = 0$	(1, 27)	0	100
$H_{03} : \beta_2 = 0$	(1, 27)	0	100
$H_{04} : \beta_3 = 0$	(1, 27)	97	3

TABLE 9.4: Index of fit measures with CSES. Kim, J., Das, R.N., Sengupta, A., and Paul, J. (2009). "Regression Analysis for Correlated Data under Compound Symmetry Structure," Journal of Statistical Theory and Applications, Vol. 8, No. 3, pp. 269-282.

Model	$R^2(\mathbf{Z})$	Corr.$^2(\mathbf{Y}, \hat{\mathbf{Y}})$	$R^2_{adj}(\mathbf{Z})$
M_1	0.9610	0.9348	0.9583
M_2	0.9608	0.9347	0.9594
M_3	0.3676	0.6536	0.3450
M_4	0.3572	0.6535	0.3343

regression parameters have been derived except the intercept for the reduced model (9.5). Analytically, we have derived the estimate of *all* the regression coefficients β's, σ^2, and ρ for known and unknown σ^2, except ρ_1. The estimation of ρ_1 has been done by trail and error method. From the present simulation study (in Table 9.2), it is clear that each estimated value is very close to its imputed value. Table 9.3 reveals the true situation of the regression coefficients, and Table 9.4 presents the original situations of the regression models. It is clear that the models M_1 and M_2 are equivalent which is reflected in Table 9.4. From Table 9.4 it is seen that the value of $R^2(\mathbf{Z})$ (or $R^2_{adj}(\mathbf{Z})$) is more (close to unity) than Corr.$^2(\mathbf{Y}, \hat{\mathbf{Y}})$ in the true situation as because that $R^2(\mathbf{Z})$ (or $R^2_{adj}(\mathbf{Z})$) is based on the model (9.5) where all the estimates of regression coefficients are BLUE, but it is not so for the model (9.4). Also for the wrong model, $R^2(\mathbf{Z})$ (or $R^2_{adj}(\mathbf{Z})$) is less than Corr.$^2(\mathbf{Y}, \hat{\mathbf{Y}})$, thus $R^2(\mathbf{Z})$ and $R^2_{adj}(\mathbf{Z})$ appear to be more sensitive index of fit measures. But as a measure of fit to the given data $R^2(\mathbf{Z})$ and $R^2_{adj}(\mathbf{Z})$ are more appropriate.

An application of the regression analysis (with the compound symmetry error structure (CSES)) in a randomized block design (RBD) is described below.

9.2.7   Randomized block design with CSES

In design of experiments, it is generally assumed that errors are *independent* and homoscedastic. However, in agricultural experiments, soil fertility is *not* distributed at random and nearby plots happen to be correlated. Here it is assumed that the errors of an RBD are correlated with a compound symmetry structure (9.3). In the present section, an analysis of an RBD with a CSES is described. Confidence ellipsoid of a set of linear estimable functions (or contrasts) and confidence interval of a linear estimable function (or contrast) of treatments are described. Analogous to Scheffe's and Tukey's methods, the multiple comparison technique is described for judging all possible treatment contrasts. All these developed results are illustrated with a simulated example.

9.2.7.1   Background of an RBD with CSES

It is well known that *experimental error* is the most important element in a design of experiment. Classical assumptions on experimental errors are that they are uncorrelated and homoscedastic. However, it is not always true. In case of agricultural experiments, soil fertility is not distributed at random and nearby plots happen to be correlated. The design problems with correlated errors resulting from the neighbouring plots which are arranged in rectangular arrays have been addressed by several researchers. See Berenblut and Webb (1974), Kiefer and Wynn (1981, 1983, 1984), Martin (1982, 1986, 1996), Gill and Shukla (1985), Kunert (1985, 1988), Martin and Eccleston (1993), Morgan (1990), Morgan and Uddin (1991, 1998), Uddin and Morgan (1997a, 1997b), Uddin (1997, 2000), Satpati *et al.* (2007) etc. and the related references are cited there. Most of the authors studied block designs assuming autocorrelated error structure. Kiefer and Wynn (1981, 1983, 1984) studied optimal block designs under autocorrelated error structure. Das and Park (2008c) have studied the analysis of an RBD with a compound symmetry error structure, which is the most common phenomenon of an RBD.

In the present section, it is assumed that the errors of an RBD are correlated with a compound symmetry structure. That is, any two observations of each block are correlated with a constant correlation coefficient (ρ) and any two observations from any two blocks are correlated with another constant correlation (ρ_1). These two correlation coefficients are known as within (ρ) and between (ρ_1) block correlation coefficient and they may be same or different. This type of correlation structure is known as compound symmetry structure (9.3). If $\rho_1 = 0$ (i.e., the correlation coefficient of any two observations from any two blocks is zero), this type of correlation structure is known as inter-class structure. Compound symmetry structure is the most common situation in an RBD and inter-class structure is a particular case of compound symmetry structure. Here a method of analysis of an RBD with a compound symmetry error structure is described. The testing procedures which are presented here are *robust*, i.e., free from both the correlation coefficients. The confidence ellipsoid of a set of linear estimable functions (or contrasts) and confidence interval of a linear estimable function (or contrast) of treatments are described. Analogous to Scheffe's and Tukey's methods, the multiple comparison technique for judging all possible treatment contrasts are described. A simulated example illustrates all the results of an RBD under the compound symmetry error structure.

9.2.7.2 Randomized block design model with CSES

Let us consider an RBD with v treatments, r blocks and v plots per block. Each treatment occurs once and only once in each block. Hence the fixed effect model is

$$y_{ij} = \mu + \beta_i + \tau_j + e_{ij}; \ 1 \leq i \leq r, \ 1 \leq j \leq v, \tag{9.19}$$

where y_{ij} is the observation for the jth treatment from the ith block and e_{ij} is the corresponding error, μ is the general effect, β_i is the effect due to ith block, τ_j is the effect due to jth treatment, $1 \leq i \leq r, \ 1 \leq j \leq v$. Here μ, β_i, τ_j's are unknown constants with $\sum_{i=1}^{r} \beta_i = \sum_{j=1}^{v} \tau_j = 0$. The e_{ij}'s are random error components which are not independent but correlated. That is $E(e_{ij}) = 0$; for all i, j and

$$\text{Cov}(e_{ij}, \ e_{i'j'}) = \begin{cases} \sigma^2; \text{if } 1 \leq i = i' \leq r; \ 1 \leq j = j' \leq v, \\ \rho\sigma^2; \text{if } 1 \leq i = i' \leq r; \ 1 \leq j \neq j' \leq v, \\ \rho_1\sigma^2; \text{if } 1 \leq i \neq i' \leq r; \ 1 \leq j, \ j' \leq v. \end{cases}$$

Therefore, in matrix and vector notation (9.19) can be written as

$$\mathbf{Y} = X\boldsymbol{\xi} + \mathbf{e}, \tag{9.20}$$

where $\mathbf{Y} = (y_{11}, \ldots, y_{1v}, y_{21}, \ldots, y_{2v}, \ldots, y_{r1}, \ldots, y_{rv})'$ is the vector of recorded observations on the study variable y from all r blocks; $\boldsymbol{\xi} = (\mu, \beta_1, \ldots, \beta_r, \tau_1, \ldots, \tau_v)' = (\mu, \boldsymbol{\beta}', \boldsymbol{\tau}')'$, is the vector of unknown parameters, $\boldsymbol{\beta} = (\beta_1, \ldots, \beta_r)'$, $\boldsymbol{\tau} = (\tau_1, \ldots, \tau_v)'$; X is the design matrix of order $N \times (1 + r + v)$ and $N = rv$. Note that rank of the design matrix X is rank$(X) = (r + v - 1)$, which is not full rank. Further, $\mathbf{e}$ is the vector of errors which are assumed to be normally distributed with $E(\mathbf{e}) = \mathbf{0}$ and $\text{Dis}(\mathbf{e}) = \sigma^2[I_r \otimes (A - B) + E_{r \times r} \otimes B] = \sigma^2 W$ (as in (9.3)), where $A = (1 - \rho)I_v + \rho E_{v \times v}$ and $B = \rho_1 E_{v \times v}$, that is $\mathbf{e} \sim \text{MN}(\mathbf{0}, \ \sigma^2 W)$. This structure W is known as compound symmetry structure (9.3). If $\rho_1 = 0$, $\text{Dis}(\mathbf{e}) = \sigma^2(I_r \otimes A)$, which is known as inter-class structure.

In general, the matrix W is unknown (as ρ and ρ_1 are unknown) but for all the calculations as usual, W is assumed to be known. In practice, however, W includes the unknown parameters ρ and ρ_1, and in the calculations which follow, the expressions for W and W^{-1} are replaced by those obtained after replacing the unknown parameters with suitable estimates or some assumed values based on experience.

9.2.7.3 Analysis of an RBD with CSES

In an RBD, the main hypothesis is to test the equality of v treatment effects. Here the null hypothesis $(H_0 : \tau_1 = \tau_2 = \ldots = \tau_v)$ is to be tested against the alternative hypothesis H_A: at least one of τ_i is different from others. The equivalent hypothesis of H_0 is $H_{0E} : \tau_{01} = \tau_{02} = \ldots = \tau_{0(v-1)} = 0$, where $\tau_{0j} = \tau_j - \tau_v$, $1 \le j \le (v-1)$. If H_0 (or H_{0E}) is accepted we stop analysis. If H_0 is rejected we test for equality of any two treatment means, that is $H_{0ij} : \tau_i = \tau_j$, against $H_{Aij} : \tau_i \ne \tau_j$. The equivalent hypothesis of H_{0ij} is $H_{0ijE} : \tau_{0i} = \tau_{0j}$ (if $j \ne v$) or $H_{0ijE} : \tau_{0i} = 0$ (if $j = v$).

Note that $\{\tau_{0j} = \tau_j - \tau_v : j = 1, 2, \ldots, (v-1)\}$ form a basis of the contrast space of the treatment effects $\tau_1, \tau_2, \ldots, \tau_v$. Therefore, a set of k independent treatment contrasts can be expressed as $C\tau$ with $C = \{(c_{ij} : i = 1, 2, \ldots, k; j = 1, 2, \ldots, v)\}$ is a matrix of order $k \times v$ (rank $(C) = k$) where $\sum_{j=1}^{v} c_{ij} = 0$; $(i = 1, 2, \ldots, k)$. As $\sum_{j=1}^{v} c_{ij} = 0$, one can express $c_{iv} = -\sum_{j=1}^{v-1} c_{ij}$ $(i = 1, 2, \ldots, k)$, resulting, $C\tau = C_1\tau_0$, where $\tau_0 = (\tau_{01}, \tau_{02}, \ldots, \tau_{0(v-1)})'$ and $C_1 = \{(c_{ij} : i = 1, 2, \ldots, k; j = 1, 2, \ldots, (v-1))\}$ with rank$(C) = $rank$(C_1) = k$.

Original model is stated as in (9.19) or (9.20). Let us consider the transformed model as

$$z_{ij} = y_{ij} - y_{iv}; \ 1 \le j \le (v-1), \ 1 \le i \le r,$$

$$\text{or,} \ \ z_{ij} = \tau_j - \tau_v + (e_{ij} - e_{iv}); \ 1 \le j \le (v-1), \ 1 \le i \le r,$$

$$\text{or,} \ \ z_{ij} = \tau_j - \tau_v + \epsilon_{ij} \text{ say}; \ 1 \le j \le (v-1), \ 1 \le i \le r,$$

$$\text{or,} \quad \mathbf{Z} = X_1\boldsymbol{\tau} + \boldsymbol{\epsilon}, \tag{9.21}$$

where $\mathbf{Z} = (z_{11}, z_{12}, \ldots, z_{1(v-1)}, z_{21}, z_{22}, \ldots, z_{2(v-1)}, \ldots, z_{r1}, z_{r2}, \ldots, z_{r(v-1)})'$, $\boldsymbol{\epsilon} = (\epsilon_{11}, \epsilon_{12}, \ldots, \epsilon_{1(v-1)}, \epsilon_{21}, \epsilon_{22}, \ldots, \epsilon_{2(v-1)}, \ldots, \epsilon_{r1}, \epsilon_{r2}, \ldots, \epsilon_{r(v-1)})'$, $\epsilon_{ij} = (e_{ij} - e_{iv})'$, $\boldsymbol{\tau}$ as in (9.20), $N_1 = r \times (v-1)$ and the design matrix X_1 of order $N_1 \times v$ with $rank(X_1) = v-1$ (which is not full rank) is given below

$$X_1 = \begin{pmatrix} D \\ D \\ . \\ . \\ D \end{pmatrix}, \text{ where } D = \begin{pmatrix} 1 & 0 & 0 & \ldots & 0 & -1 \\ 0 & 1 & 0 & \ldots & 0 & -1 \\ .. & .. & .. & \ldots & .. & .. \\ .. & .. & .. & \ldots & .. & .. \\ 0 & 0 & 0 & \ldots & 1 & -1 \end{pmatrix}_{v-1 \times v}.$$

As it is taken $\tau_{0j} = \tau_j - \tau_v$; $j = 1, 2, \ldots, (v-1)$, the transformed model of (9.21) is

$$z_{ij} = \tau_{0j} + \epsilon_{ij}; \ 1 \le j \le (v-1), \ 1 \le i \le r,$$

$$\text{or,} \quad \mathbf{Z} = X_2\boldsymbol{\tau}_0 + \boldsymbol{\epsilon}, \tag{9.22}$$

where $\boldsymbol{\tau}_0 = (\tau_{01}, \tau_{02}, \ldots, \tau_{0(v-1)})'$,

$$X_2 = \begin{pmatrix} D_1 \\ D_1 \\ . \\ . \\ D_1 \end{pmatrix}, \quad D_1 = \begin{pmatrix} 1 & 0 & 0 & \ldots & 0 \\ 0 & 1 & 0 & \ldots & 0 \\ .. & .. & .. & \ldots & .. \\ .. & .. & .. & \ldots & .. \\ 0 & 0 & 0 & \ldots & 1 \end{pmatrix}_{v-1 \times v-1},$$

and D_1 is obtained from D by deleting its last column, which is the sum of the first $(v-1)$ columns of D.

Note that $E(\epsilon_{ij}) = 0$, $\text{Var}(\epsilon_{ij}) = \text{Var}(e_{ij} - e_{iv}) = \text{Var}(e_{ij}) + \text{Var}(e_{iv}) - 2\text{Cov}(e_{ij}, e_{iv}) = 2\sigma^2(1-\rho)$, $1 \le i \le r$, $1 \le j \le (v-1)$; $\text{Cov}(\epsilon_{ij}, \epsilon_{i'j'}) = \text{Cov}(e_{ij} - e_{iv}, e_{i'j'} - e_{i'v}) = \text{Cov}(e_{ij}, e_{i'j'}) - \text{Cov}(e_{ij}, e_{i'v}) - \text{Cov}(e_{iv}, e_{i'j'}) + \text{Cov}(e_{iv}, e_{i'v}) = \rho\sigma^2 - \rho\sigma^2 - \rho\sigma^2 + \sigma^2 = \sigma^2(1-\rho)$, if $1 \le i = i' \le r$, $1 \le j \ne j' \le (v-1)$, and $\text{Cov}(\epsilon_{ij}, \epsilon_{i'j'}) = \rho_1\sigma^2 - \rho_1\sigma^2 - \rho_1\sigma^2 + \rho_1\sigma^2 = 0$, if $1 \le i \ne i' \le r$, $1 \le j, j' \le (v-1)$.

$$\text{Therefore, } \text{Corr}(\epsilon_{ij}, \epsilon_{i'j'}) = \begin{cases} \frac{1}{2}; & \text{if } 1 \le i = i' \le r; \ 1 \le j \ne j' \le (v-1), \\ 0; & \text{if } 1 \le i \ne i' \le r; \ 1 \le j, j' \le (v-1). \end{cases}$$

Therefore, $E(\epsilon) = 0$, $\text{Dis}(\epsilon) = 2\sigma^2(1-\rho)[I_r \otimes W_1] = \sigma_1^2 W_2$ say, where $W_1 = \frac{1}{2}I_{v-1} + \frac{1}{2}E_{v-1 \times v-1}$, $\sigma_1^2 = 2\sigma^2(1-\rho)$ and $W_2 = I_r \otimes W_1$. The following results are immediate for the model (9.22).

Theorem 9.7 *The best linear unbiased estimate (BLUE) of τ_0 (from (9.22)) is*

$$\hat{\tau}_0 = (X_2' W_2^{-1} X_2)^{-1} \ (X_2' W_2^{-1} \mathbf{Z}). \tag{9.23}$$

Theorem 9.8 *An unbiased estimate (UE) of σ_1^2 is*

$$\hat{\sigma}_1^2 = \frac{(\mathbf{Z}' W_2^{-1} \mathbf{Z}) - \hat{\tau}_0'(X_2' W_2^{-1} X_2)\hat{\tau}_0}{(N_1 - v + 1)}, \tag{9.24}$$

where $N_1 = r \times (v-1)$.

Note that $\hat{\tau}_0 \sim \text{MN}(\tau_0, \ \sigma_1^2(X_2' W_2^{-1} X_2)^{-1})$. It is aimed to test the hypothesis H_{0E} : $\tau_{01} = \tau_{02} = \ldots = \tau_{0v-1} = 0$ against the alternative hypothesis H_{AE} : at least one of τ_{0i} is different from 0 (zero). Equivalently, $H_{0E} : R\tau_0 = \mathbf{0}$, where R is an identity matrix of order $(v-1)$, rank $(R) = (v-1)$, and is given by

$$R = \begin{pmatrix} 1 & 0 & 0 & \ldots & 0 \\ 0 & 1 & 0 & \ldots & 0 \\ \ldots & \ldots & \ldots & \ldots & \ldots \\ 0 & 0 & 0 & \ldots & 1 \end{pmatrix}.$$

Therefore, $E(R\hat{\tau}_0) = R\tau_0$, $\text{Dis}(R\hat{\tau}_0) = \sigma_1^2 R(X_2' W_2^{-1} X_2)^{-1} R'$. Since $\hat{\tau}_0$ is distributed as Multivariate Normal distribution, then $R\hat{\tau}_0 \sim \text{MN}(R\tau_0, \ \sigma_1^2 R(X_2' W_2^{-1} X_2)^{-1} R')$. Under $H_{0E} : R\tau_0 = \mathbf{0}$, then $R\hat{\tau}_0 \sim \text{MN}(\mathbf{0}, \ \sigma_1^2 R(X_2' W_2^{-1} X_2)^{-1} R')$.

Therefore, under H_{0E}, $(R\hat{\tau}_0)'[\sigma_1^2 R(X_2' W_2^{-1} X_2)^{-1} R']^{-1}(R\hat{\tau}_0) \sim \chi_{v-1}^2$, where the *d.f.* $(v-1)$ is given by the number of elements in the $R\tau_0$ vector. Also from the model (9.22), $\hat{\epsilon}_0 = W_2^{-\frac{1}{2}}(Z - X_2\hat{\tau}_0)$, $\frac{\hat{\epsilon}_0'\hat{\epsilon}_0}{\sigma_1^2} \sim \chi_{N_1 - v + 1}^2$ and independent of $\hat{\tau}_0$, and hence independent of $R\hat{\tau}_0$. Now the following result is immediate.

Theorem 9.9 *If the basic result $R\tau_0 = \mathbf{0}$ is true, the test statistic is*

$$F = \frac{(R\hat{\tau}_0)'[R(X_2' W_2^{-1} X_2)^{-1} R']^{-1}(R\hat{\tau}_0)/(v-1)}{(\hat{\epsilon}_0'\hat{\epsilon}_0)/(N_1 - v + 1)} \sim F_{v-1, N_1 - v + 1}. \tag{9.25}$$

The test procedure is: reject H_{0E} at $100\alpha\%$ level of significance if observed $F > F_{\alpha; \ v-1, \ N_1 - v + 1}$, and accept otherwise. If the null hypothesis H_{0E} is accepted, we stop analysis. But the rejection of H_{0E} leads to test the equality of any two treatment means.

Suppose H_{0E} is rejected, we want to test for $H_{0ij} : \tau_i - \tau_j = 0$ against $H_{Aij} : \tau_i \neq \tau_j$. Equivalent hypothesis of H_{0ij} is $H_{0ijE} : R_1 \boldsymbol{\tau}_0 = 0$, where $R_1 = (0, \ldots, 0, 1, 0, \ldots, 0, -1, 0, \ldots, 0)$ (for $j \neq v$) is a row vector of order $1 \times v - 1$ with the ith position is unity (1) and the jth position is $-$unity (-1) and all other elements are zeros. For $j = v$, R_1 is a row vector of order $1 \times v - 1$ with the ith position is unity (1) and all other elements are zeros. Then the test statistic for H_{0ijE} is given below in Theorem 9.10.

Theorem 9.10 *If the basic result $R_1 \boldsymbol{\tau}_0 = 0$ is true, the test statistic is*

$$F_1 = \frac{(R_1 \hat{\boldsymbol{\tau}}_0)'[R_1(X_2' W_2^{-1} X_2)^{-1} R_1']^{-1}(R_1 \hat{\boldsymbol{\tau}}_0)/1}{(\hat{\boldsymbol{\epsilon}}_0' \hat{\boldsymbol{\epsilon}}_0)/(N_1 - v + 1)} \sim F_{1, N_1 - v + 1}. \tag{9.26}$$

The test procedure is: reject H_{0ijE} at $100\alpha\%$ level of significance if observed $F_1 > F_{\alpha; 1, N_1 - v + 1}$, and accept otherwise.

Note that the test statistics in (9.25) and (9.26) are independent of ρ and ρ_1. Hence the above two tests are robust.

9.2.7.4 Confidence ellipsoid of treatment contrasts with CSES

In Section 9.2.7.3, it is noted that the dimension of the contrast space of v treatment effects is $(v-1)$. Therefore, any treatment effect contrast is estimable. In the present section, the confidence ellipsoid of a set of treatment effect contrasts and the confidence interval of a treatment effect contrast are described.

Let $\psi_1, \psi_2, \ldots, \psi_q$ denote q linearly independent estimable functions of v treatment effects viz., $\tau_1, \tau_2, \ldots, \tau_v$ as in the model (9.19). They are q independent treatment contrasts also. Let $\boldsymbol{\psi} = (\psi_1, \psi_2, \ldots, \psi_q)'$ be a vector of order $q \times 1$. Then $\boldsymbol{\psi} = C\boldsymbol{\tau}_0$, where the rows of C (known) are linearly independent. Then $\hat{\boldsymbol{\psi}} = G\mathbf{Z}$, is the generalized least squares (GLS) estimate of $\boldsymbol{\psi}$, from the linear model (9.22), $\mathbf{Z}$ as in (9.21), $\hat{\boldsymbol{\psi}} = (\hat{\psi}_1, \hat{\psi}_2, \ldots, \hat{\psi}_q)'$ and $G = ((g_{ij}))$ (known matrix depending on ψ_i's) of order $q \times N_1$. The variance covariance of the estimates $\{\hat{\psi}_i\}$ is then $\mathrm{Dis}(\hat{\boldsymbol{\psi}}) = G\mathrm{Dis}(\mathbf{Z})G' = \sigma_1^2(GW_2 G')$, where W_2 as in (9.22).

Note that $\hat{\boldsymbol{\psi}} \sim \mathrm{MN}(\boldsymbol{\psi}, \sigma_1^2(GW_2 G'))$ and $\hat{\boldsymbol{\psi}}$ is statistically independent of $\frac{\hat{\boldsymbol{\epsilon}}_0' \hat{\boldsymbol{\epsilon}}_0}{\sigma_1^2} = \frac{(\mathbf{Z} - X_2 \hat{\boldsymbol{\tau}}_0)' W_2^{-1} (\mathbf{Z} - X_2 \hat{\boldsymbol{\tau}}_0)}{\sigma_1^2} \sim \chi_{N_1 - v + 1}^2$ (for the model (9.22)). Again, $(\hat{\boldsymbol{\psi}} - \boldsymbol{\psi})'\{\sigma_1^2(GW_2 G')\}^{-1}(\hat{\boldsymbol{\psi}} - \boldsymbol{\psi}) \sim \chi_q^2$ and independent of $\frac{\hat{\boldsymbol{\epsilon}}_0' \hat{\boldsymbol{\epsilon}}_0}{\sigma_1^2} \sim \chi_{N_1 - v + 1}^2$. Therefore, $F = \frac{(\hat{\boldsymbol{\psi}} - \boldsymbol{\psi})'(GW_2 G')^{-1}(\hat{\boldsymbol{\psi}} - \boldsymbol{\psi})/q}{(\hat{\boldsymbol{\epsilon}}_0' \hat{\boldsymbol{\epsilon}}_0)/N_1 - v + 1} \sim F_{q, N_1 - v + 1}$.

Equivalently, $F = \dfrac{(\hat{\boldsymbol{\psi}} - \boldsymbol{\psi})'(GW_2 G')^{-1}(\hat{\boldsymbol{\psi}} - \boldsymbol{\psi})}{qs^2} \sim F_{q, N_1 - v + 1}$, \hfill (9.27)

where $s^2 = \frac{(\hat{\boldsymbol{\epsilon}}_0' \hat{\boldsymbol{\epsilon}}_0)}{N_1 - v + 1}$ (for the model (9.22)) is an UE of σ_1^2 as in (9.24).

Theorem 9.11 *The desired confidence set falls out of (9.27), under the model (9.19), the probability is $(1 - \alpha)$ that the F-variable in (9.27) is $\leq F_{\alpha; q, N_1 - v + 1}$, or that*

$$(\hat{\boldsymbol{\psi}} - \boldsymbol{\psi})'(GW_2 G')^{-1}(\hat{\boldsymbol{\psi}} - \boldsymbol{\psi}) \leq qs^2 F_{\alpha; q, N_1 - v + 1}. \tag{9.28}$$

Inequality (9.28) determines an *ellipsoid* in the q-dimensional $\boldsymbol{\psi}$-space with center at $(\hat{\psi}_1, \hat{\psi}_2, \ldots, \hat{\psi}_q)$ and the probability that this *random ellipsoid* covers the true parameter point $(\psi_1, \psi_2, \ldots, \psi_q)$ is $(1 - \alpha)$ no matter whatever be the value of the unknown parameters $\mu, \beta_1, \beta_2, \ldots, \beta_r, \tau_1, \tau_2, \ldots, \tau_v, \rho, \rho_1$ and σ^2, and the equation (9.28) is known as *confidence ellipsoid*.

One can obtain a *confidence interval* for a single treatment effect contrast $\psi = \mathbf{c}'\tau_0$, $(\mathbf{c} \neq \mathbf{0})$ by specializing the above calculation $q = 1$. The resulting one-dimensional ellipsoid is the interval given by

$$b^{-1}(\hat{\psi} - \psi)^2 \leq s^2 F_{\alpha;\ 1,\ N_1 - v + 1}, \tag{9.29}$$

where $\hat{\psi} = \mathbf{a}'\mathbf{Z}$, is the GLS estimate of ψ from (9.22) and $\mathrm{Var}(\hat{\psi}) = \mathbf{a}'\mathrm{Dis}(\mathbf{Z})\mathbf{a} = \mathbf{a}'\sigma_1^2 W_2\mathbf{a}$ $= \sigma_1^2(\mathbf{a}'W_2\mathbf{a})$. Then $\mathrm{Var}(\hat{\psi})$ is an UE $\hat{\sigma}_{\hat{\psi}}^2 = s^2 b$ say, where s^2 as in (9.27) and $b = (\mathbf{a}'W_2\mathbf{a})$. Equivalently, (9.29) can be expressed as

$$\hat{\psi} - t_{\alpha/2;\ N_1 - v + 1}\ \hat{\sigma}_{\hat{\psi}} \leq \psi \leq \hat{\psi} + t_{\alpha/2;\ N_1 - v + 1}\ \hat{\sigma}_{\hat{\psi}}, \tag{9.30}$$

the probability that this random interval covers the unknown ψ is $(1 - \alpha)$. The interval (9.30) could also be derived from the fact that $(\hat{\psi} - \psi)/\hat{\sigma}_{\hat{\psi}}$ is $t_{N_1 - v + 1}$.

9.2.7.5 Multiple comparison of treatment contrasts with CSES

If the null hypothesis H_0 (or H_{0E}) (as stated in Section 9.2.7.3) is rejected for testing the equality of v treatment means in the original model (9.19) or transformed model (9.21) or (9.22) (model (9.21) or (9.22) is a one-way layout with known correlated error model), we are interested for judging all the treatment effect contrasts. In the present section, two methods of multiple comparison technique are described for judging *all* the treatment effect contrasts analogous to Scheffe's and Tukey's method.

Let L be a set of q linearly independent estimable functions (or contrasts) $\{\psi_i, i = 1, 2, \ldots, q$ where $\psi_i = \sum_{j=1}^{v} \lambda_{ij}\tau_j\}$, where L is called a q-dimensional space of linearly estimable functions (or contrasts) of v treatments. Then every ψ in L is of the form $\psi = \sum_{i=1}^{q} h_i\psi_i$, where $h_1, h_2, \ldots, h_q$ are known constant coefficients. That is L is the set of all linear combinations of $\psi_1, \psi_2, \ldots, \psi_q$. Consider a q-dimensional space L of linearly estimable functions (or contrasts) of v treatments generated by a set of q linearly independent estimable functions (or contrasts) $\{\psi_1, \psi_2, \ldots, \psi_q\}$. For $\psi \in L$, let $\hat{\psi} = \mathbf{a}'\mathbf{Z}$, be it's GLS estimate from the model (9.22). $\mathrm{Var}(\hat{\psi}) = \sigma_{\hat{\psi}}^2$ and the estimate of $\mathrm{Var}(\hat{\psi}) = \sigma_{\hat{\psi}}^2$ are given in (9.29). Then the proposed method of Multiple Comparison is based on the following theorems.

Theorem 9.12 *Under the model (9.19), the probability is $(1 - \alpha)$ that simultaneously for all $\psi \in L$*

$$\hat{\psi} - D\ \hat{\sigma}_{\hat{\psi}} \leq \psi \leq \hat{\psi} + D\ \hat{\sigma}_{\hat{\psi}}, \tag{9.31}$$

where the constant D is $D = (q\ F_{\alpha;\ q,\ N_1 - v + 1})^{1/2}$.

Proof. The theorem can be easily proved following in the same line as *S*-Method of Multiple Comparison Technique (see Scheffe, 1959, pp. 68-70), considering the ellipsoid as in (9.28).

For judging *all* the contrasts of v treatments, under the model (9.19), the proposed method is thus immediate from Theorem 9.12 to the following.

Theorem 9.13 *Under the model (9.19), the probability is $(1 - \alpha)$ that the values of all contrasts of v treatments $\tau_1, \tau_2, \ldots, \tau_v$ simultaneously satisfy the inequalities*

$$\hat{\psi} - D\ \hat{\sigma}_{\hat{\psi}} \leq \psi \leq \hat{\psi} + D\ \hat{\sigma}_{\hat{\psi}}, \tag{9.32}$$

where the constant D is $D = \{(v - 1)F_{\alpha;\ v-1,\ N_1 - v + 1}\}^{1/2}$.

The proof follows directly from Theorem 9.12 as a corollary, noting that dimension of the contrasts space is $(v - 1)$.

TABLE 9.5: Yields y_{ij} for different treatments in blocks with CSES. Kim, J., Das, R.N., Sengupta, A., and Paul, J. (2009). "Regression Analysis for Correlated Data under Compound Symmetry Structure," Journal of Statistical Theory and Applications, Vol. 8, No. 3, pp. 269-282.

Treatment:	T_1	T_2	T_3	T_4	T_5
Block : B_1	54.93316	52.61071	48.53662	50.89518	50.404113
B_2	53.50128	51.91155	46.79449	49.67082	48.91196
B_3	52.22097	50.54522	44.72411	47.76691	46.94799
B_4	49.89687	47.12505	43.02662	45.22032	44.60692

In Section 9.2.7.3, it has been derived that $\hat{\tau}_0 \sim \mathrm{MN}(\tau_0,\ \sigma_1^2(X_2'W_2^{-1}X_2)^{-1})$, where $(X_2'W_2^{-1}X_2)^{-1}$ is a known variance covariance matrix. On simplification it is seen that $\mathrm{Var}(\hat{\tau}_{0i}) = \text{constant} = a_1\sigma_1^2$ say, and $\mathrm{Cov}(\hat{\tau}_{0i},\ \hat{\tau}_{0j}) = \text{constant} = b_1\sigma_1^2$ say; $1 \leq i \neq j \leq (v-1)$. Again $\{\hat{\tau}_{0j}\}$'s are distributed independent of $s^2 = \frac{(\hat{\epsilon}_0'\hat{\epsilon}_0)}{N_1-v+1}$. Hence we have the following theorem analogous to Tukey's method (see Scheffe, 1959, p. 75, Theorem 3).

Theorem 9.14 *Under the model (9.19), the probability is $(1-\alpha)$ that all the contrasts of $\tau_1, \tau_2, \ldots, \tau_v$ (i.e., $\psi_i = \sum_{i=1}^{v} c_i\tau_i$ with $\sum_{i=1}^{v} c_i = 0$) simultaneously satisfy the following*

$$\hat{\psi} - T\,s\left(\frac{1}{2}\sum_{i=1}^{v}|c_i|\right) \leq \psi \leq \hat{\psi} + T\,s\left(\frac{1}{2}\sum_{i=1}^{v}|c_i|\right), \tag{9.33}$$

where $\hat{\psi} = \sum_{i=1}^{v} c_i\hat{\tau}_i$, $T = (a_1^2 - b_1)^{1/2}\ q_{\alpha;\ v-1,\ N_1-v+1}$ and $q_{\alpha;\ v-1,\ N_1-v+1}$ is the upper α point of the Studentized range $q_{\alpha;\ v-1,\ N_1-v+1}$.

9.2.7.6   Illustration of an RBD with CSES

Let us consider an RBD with 5 treatments, 4 blocks and each block contains 5 plots. The observations are generated by simulation from the following model

$$y_{ij} = \mu + \beta_i + \tau_j + e_{ij};\ 1 \leq j \leq 5,\ 1 \leq i \leq 4. \tag{9.34}$$

In the absence of real experimental data, we generate observations (100 replications) according to formula (9.34) with $\mu = 50.0$, $\beta_1 = 2.5$, $\beta_2 = 1.6$, $\beta_3 = -1.2$, $\beta_4 = -2.9$, $\tau_1 = 3.5$, $\tau_2 = 1.8$, $\tau_3 = -3.3$, $\tau_4 = -0.5$, $\tau_5 = -1.5$, $\rho = 0.80$, $\rho_1 = 0.60$ and $\sigma^2 = 4.0$. The generated observations for a single replication of the the column vector $\mathbf{Y}$ is given below in block wise (non-randomly) in Table 9.5.

With the observation vector $\mathbf{Y}$ and following the method of analysis as explained in Section 9.2.7.3, the estimate of $\tau_0 = (5.0, 3.3, -1.8, 1.0)'$ from (9.23) is given by $\hat{\tau}_0 = (4.92107, 2.83113, -1.94646, 0.671305)'$.

The test results and average estimates of τ_0 from 100 replications (with $\mu = 50.0$, $\beta_1 = 2.5$, $\beta_2 = 1.6$, $\beta_3 = -1.2$, $\beta_4 = -2.9$, $\tau_1 = 3.5$, $\tau_2 = 1.8$, $\tau_3 = -3.3$, $\tau_4 = -0.5$, $\tau_5 = -1.5$, $\rho = 0.80$, $\rho_1 = 0.60$ and $\sigma^2 = 4.0$) are given in Table 9.6 and Table 9.7, respectively.

Thus, the estimated values of elementary treatment contrasts are very close to their respective imputed values. The estimate of σ_1^2 is not so good. However, with 16 observations only we do not hope to estimate σ_1^2 very accurately.

Here we have considered the analysis of an RBD with correlated error. Correlation in the observations may be introduced due to several causes (Chapter 2). There may be different systematic pattern of correlation structure of errors (Chapter 2). In the above,

TABLE 9.6: Test results from 100 replications with $\alpha = 0.05$ of an RBD with CSES. Kim, J., Das, R.N., Sengupta, A., and Paul, J. (2009). "Regression Analysis for Correlated Data under Compound Symmetry Structure," Journal of Statistical Theory and Applications, Vol. 8, No. 3, pp. 269-282.

Null hypothesis	Accepted cases	Rejected cases
$H_{0E} : \tau_{01} = \tau_{02} = \tau_{03} = \tau_{04} = 0$	0	100
$H_{012} : \tau_1 - \tau_2 = 0$	0	100
$H_{013} : \tau_1 - \tau_3 = 0$	0	100
$H_{014} : \tau_1 - \tau_4 = 0$	0	100
$H_{015} : \tau_1 - \tau_5 = 0$	0	100
$H_{023} : \tau_2 - \tau_3 = 0$	0	100
$H_{024} : \tau_2 - \tau_4 = 0$	0	100
$H_{025} : \tau_2 - \tau_5 = 0$	0	100
$H_{034} : \tau_3 - \tau_4 = 0$	0	100
$H_{035} : \tau_3 - \tau_5 = 0$	0	100
$H_{045} : \tau_4 - \tau_5 = 0$	0	100

TABLE 9.7: Imputed values and average estimated values from 100 replications of an RBD with CSES. Kim, J., Das, R.N., Sengupta, A., and Paul, J. (2009). "Regression Analysis for Correlated Data under Compound Symmetry Structure," Journal of Statistical Theory and Applications, Vol. 8, No. 3, pp. 269-282.

Imputed values	Average Estimated values
$\mu = 50.00$	$\hat{\mu} = -$
$\beta_1 = 2.5$	$\hat{\beta}_1 = -$
$\beta_2 = 1.6$	$\hat{\beta}_2 = --$
$\beta_3 = -1.2$	$\hat{\beta}_3 = --$
$\beta_4 = -2.9$	$\hat{\beta}_4 = -$
$\tau_{01} = 5.0$	$\hat{\tau}_{01} = 4.9311280$
$\tau_{02} = 3.3$	$\hat{\tau}_{02} = 3.131137$
$\tau_{03} = -1.8$	$\hat{\tau}_{03} = -1.9547478$
$\tau_{04} = 1.0$	$\hat{\tau}_{04} = 0.9271487$
$\rho = 0.80$	$\hat{\rho} = -$
$\rho_1 = 0.60$	$\hat{\rho}_1 = -$
$\sigma^2 = 4.0$	$\hat{\sigma}^2 = -$
$\sigma_1^2 = 1.6$	$\hat{\sigma}_1^2 = 1.1984372$

we have considered the compound symmetry error structure, which is the most common phenomenon of errors in an RBD. For such an RBD, we have developed a method of analysis, and the method thus developed is free from correlation coefficients ρ and ρ_1. Therefore, the present analysis is recommended as a *robust* analysis. Again, the method of estimation of τ_0 is free from both the correlation coefficients ρ and ρ_1, and $\hat{\tau}_0$ is the BLUE of τ_0. Also it is seen that the estimates are very close to the imputed values (based on simulation study). The confidence ellipsoid of a set of estimable functions (or contrasts) of treatments and also the confidence interval of an estimable function (or a contrast) of treatments are derived here free from correlation parameters (ρ and ρ_1) and any other fixed effect parameters. In the process, a robust method of multiple comparison technique is described of comparing all treatment contrasts analogous to Scheffe's and Tukey's methods. The beauty of this method is that the same regression technique is used for analysis, testing of hypotheses, estimation of unknown parameters, deriving confidence ellipsoids and confidence intervals, and also for multiple comparison of treatments.

9.3   REGRESSION ANALYSES WITH COMPOUND AUTO-CORRELATED ERROR STRUCTURE

In the present section we turn our attention to a covariance structure of errors which is an extension of compound symmetry structure (9.3). Under the same situation as stated in compound symmetry structure (Chapter 2) here we assume a new correlation structure which is such that observations within each set have an autocorrelated structure $[\sigma^2\{\rho^{|j-j'|}\}$ $_{1 \leq j, j' \leq r}]$, and any two observations from any two groups are correlated with a constant correlation coefficient (ρ_1). These two correlation coefficients ρ and ρ_1 are known respectively as within and between group correlation coefficients. This type of correlation structure has been termed as *compound autocorrelated error structure* (CAES) (Das, 2010) in analogous to compound symmetry structure (Vataw, 1948). Writing ρ_2 for $\rho^{|j-j'|}$ for all $j \neq j'$, the compound autocorrelated structure will be reduced to compound symmetry structure. If $\rho_1 = 0$ (i.e., the correlation coefficient of any two observations from any two groups is zero), compound symmetry structure is known as inter-class structure. Compound autocorrelated structure is the most common situation for correlated observations, and compound symmetry, autocorrelated, inter-class, intra-class structures are the particular cases of compound autocorrelated structure. Compound autocorrelated error variance covariance structure is given below

$$
\text{Dis}(\mathbf{e}) = \sigma^2 \begin{pmatrix} A & B & \cdots & B \\ B & A & \cdots & B \\ \vdots & \vdots & \vdots & \vdots \\ B & B & \cdots & A \end{pmatrix}, \ A = \{\rho^{|j-j'|}\}_{1 \leq j, j' \leq r}, \ B = \begin{pmatrix} \rho_1 & \rho_1 & \cdots & \rho_1 \\ \rho_1 & \rho_1 & \cdots & \rho_1 \\ \vdots & \vdots & \vdots & \vdots \\ \rho_1 & \rho_1 & \cdots & \rho_1 \end{pmatrix}_{r \times r}.
$$

It is simply represented by

$$
\text{Dis}(\mathbf{e}) = \sigma^2 W = \sigma^2[I_g \otimes (A - B) + E_g \otimes B], \tag{9.35}
$$

where I_g is an identity matrix of order $g \times g$, E_g is a $g \times g$ matrix with all elements 1, A is as in above and $B = \rho_1 E_r$. Here $\otimes$ denotes Kronecker product.

9.3.1   Estimation of regression parameters with CAES

In the present section, a robust method of estimation of all the regression coefficients *except* the intercept is described for the correlated regression model (9.1) (Section 9.2.1) under the compound autocorrelated error structure (9.35) (Das, 2010). Estimation of variance (σ^2) and the correlation coefficients ρ and ρ_1 are described here (Das, 2010). The estimation method is considered for the first-order model as in (9.1), and the similar method can be used for the second-order model as in (9.2). As it is stated in (9.4) that there are g groups each with r observations, so total number of observations is $N = gr$. Let y_{ij} be the j-th observation of the i-th group, $1 \leq j \leq r$, $1 \leq i \leq g$. Assuming first-order model (as in (9.1)) we have

$$
y_{ij} = \beta_0 + \beta_1 x_{ij1} + \ldots + \beta_p x_{ijp} + e_{ij}, \ 1 \leq i \leq g, \ 1 \leq j \leq r,
$$

$$\text{or,} \quad \mathbf{Y} = X\boldsymbol{\beta} + \mathbf{e}, \tag{9.36}$$

where $\mathbf{Y} = (y_{11}, \ldots, y_{1r}, \ldots, y_{g1}, \ldots, y_{gr})'$, e_{ij} is the corresponding error of y_{ij}, $\boldsymbol{\beta}$ and x_{ijk}'s are as in (9.1). Note that $E(\mathbf{e}) = \mathbf{0}$ and $\text{Dis}(\mathbf{e}) = \sigma^2 W$, where W as in (9.35).

Let us define

$$t_{iw} = y_{iw} - \rho y_{i(w-1)}; \ 2 \leq w \leq r, \ 1 \leq i \leq g,$$

$$\text{or,} \quad t_{iw} = \beta_0(1-\rho) + \sum_{k=1}^{p} \beta_k(x_{iwk} - \rho x_{i(w-1)k}) + (e_{iw} - \rho e_{i(w-1)}),$$

$$\text{or,} \quad t_{iw} = \alpha_0 + \sum_{k=1}^{p} \beta_k v_{iwk} + \phi_{iw} \text{ say,} \ \ 2 \leq w \leq r, \ 1 \leq i \leq g,$$

$$\text{or,} \quad \mathbf{t}_i = V_i \boldsymbol{\tau} + \boldsymbol{\phi}_i; \ 1 \leq i \leq g,$$

$$\text{or,} \quad \mathbf{t} = V\boldsymbol{\tau} + \boldsymbol{\phi}, \tag{9.37}$$

where $\mathbf{t}_i = (t_{i2}, \ldots, t_{ir})'$, $\boldsymbol{\tau} = (\alpha_0, \beta_1, \ldots, \beta_p)'$, $\alpha_0 = \beta_0(1-\rho)$, $V_i = ((v_{iwk}); k = 1, \ldots, p; w = 2, \ldots, r)$, $v_{iwk} = x_{iwk} - \rho x_{i(w-1)k}$, $\phi_{iw} = e_{iw} - \rho e_{i(w-1)}$, $\boldsymbol{\phi}_i = (\phi_{i2}, \ldots, \phi_{ir})'$, $\mathbf{t} = (\mathbf{t}_1', \mathbf{t}_2', \ldots, \mathbf{t}_g')'$, $V = (V_1', V_2', \ldots, V_g')'$, and $\boldsymbol{\phi} = (\boldsymbol{\phi}_1', \boldsymbol{\phi}_2', \ldots, \boldsymbol{\phi}_g')'$.

On simplification, we have $E(\boldsymbol{\phi}) = \mathbf{0}$, $\text{Var}(\phi_{iw}) = \sigma^2(1-\rho^2)$; $2 \leq w \leq r$, $1 \leq i \leq g$,

$$\text{and} \quad \text{Cov}(\phi_{iw}, \phi_{i'w'}) = \begin{cases} 0; \text{ if } 2 \leq w \neq w' \leq r, \ 1 \leq i = i' \leq g, \\ \rho_1(1-\rho)^2\sigma^2 = \rho_3\sigma^2 \text{ say, if } 2 \leq w, \ w' \leq r, \ 1 \leq i \neq i' \leq g, \end{cases}$$

where $\rho_3 = \rho_1(1-\rho)^2$.

Let us also transform

$$z_{iw} = t_{iw} - t_{ir}; \ 2 \leq w \leq (r-1), \ 1 \leq i \leq g,$$

$$\text{or,} \quad z_{iw} = \sum_{k=1}^{p} \beta_k(v_{iwk} - v_{irk}) + (\phi_{iw} - \phi_{ir}),$$

$$\text{or,} \quad z_{iw} = \sum_{k=1}^{p} \beta_k s_{iwk} + \epsilon_{iw} \text{ say,} \ \ 2 \leq w \leq (r-1), \ 1 \leq i \leq g,$$

$$\text{or,} \quad \mathbf{Z}_i = S_i \boldsymbol{\eta} + \boldsymbol{\epsilon}_i; \ 1 \leq i \leq g,$$

$$\text{or,} \quad \mathbf{Z} = S\boldsymbol{\eta} + \boldsymbol{\epsilon}, \tag{9.38}$$

where $\mathbf{Z}_i = (z_{i2}, \ldots, z_{i(r-1)})'$, $\boldsymbol{\eta} = (\beta_1, \ldots, \beta_p)'$, $s_{iwk} = v_{iwk} - v_{irk}$, $\epsilon_{iw} = \phi_{iw} - \phi_{ir}$, $S_i = ((s_{iwk}); k = 1, \ldots, p; w = 2, \ldots, (r-1))$, $\boldsymbol{\epsilon}_i = (\epsilon_{i2}, \ldots, \epsilon_{i(r-1)})'$, $\mathbf{Z} = (\mathbf{Z}_1', \mathbf{Z}_2', \ldots, \mathbf{Z}_g')'$, $S = (S_1', S_2', \ldots, S_g')'$, and $\boldsymbol{\epsilon} = (\boldsymbol{\epsilon}_1', \boldsymbol{\epsilon}_2', \ldots, \boldsymbol{\epsilon}_g')'$. Assume S has full column rank. So, all linear parametric functions of the model (9.38) are estimable when the dispersion matrix $\text{Dis}(\boldsymbol{\epsilon})$ is non-singular.

Note that $E(\epsilon_{iw}) = 0$; $\text{Var}(\epsilon_{iw}) = 2\sigma^2(1-\rho^2) = \sigma_1^2$ say; $2 \leq w \leq r-1$, $1 \leq i \leq g$, where $\sigma_1^2 = 2\sigma^2(1-\rho^2)$,

$$\text{Cov}(\epsilon_{iw}, \epsilon_{i'w'}) = \begin{cases} \frac{\sigma_1^2}{2}; \text{ if } 2 \leq w \neq w' \leq (r-1), \ 1 \leq i = i' \leq g, \\ 0; \text{ if } 2 \leq w, \ w' \leq (r-1), \ 1 \leq i \neq i' \leq r, \end{cases}$$

$$\text{and,} \quad \text{Corr}(\epsilon_{iw}, \epsilon_{i'w'}) = \begin{cases} \frac{1}{2}; \text{ if } 2 \leq w \neq w' \leq (r-1), \ 1 \leq i = i' \leq g; \\ 0; \text{ if } 2 \leq w, \ w' \leq (r-1), \ 1 \leq i \neq i' \leq g. \end{cases}$$

Therefore, $E(\epsilon) = \mathbf{0}$, $\mathrm{Dis}(\epsilon) = \sigma_1^2[I_g \otimes W_1] = \sigma_1^2 W_2$ say, where $W_1 = \frac{1}{2}I_{(r-2)} + \frac{1}{2}E_{(r-2)\times(r-2)}$, and $W_2 = I_g \otimes W_1$. Now we have the following results for the model (9.38).

Theorem 9.15 *The best linear unbiased estimator of $\boldsymbol{\eta}$ of the model (9.38) for known ρ is*

$$\hat{\boldsymbol{\eta}} = (S'W_2^{-1}S)^{-1}S'W_2^{-1}\mathbf{Z}, \tag{9.39}$$

where S, W_2 and Z are in (9.38).

Theorem 9.16 *An unbiased estimator of $\sigma_1^2 = 2\sigma^2(1 - \rho^2)$ in (9.38) is*

$$\hat{\sigma}_1^2 = \frac{\mathbf{Z}'W_2^{-1}\mathbf{Z} - \hat{\boldsymbol{\eta}}'(S'W_2^{-1}S)\hat{\boldsymbol{\eta}}}{g(r-2) - p}, \tag{9.40}$$

where S, W_2 and Z are in (9.38) and $\hat{\boldsymbol{\eta}}$ is in (9.39).

The scheme for calculations of $\boldsymbol{\eta}$, ρ and σ^2 is given below.

1. Assume some value of $\rho \in (-1,\ 1)$.

2. For the assumed value of ρ, say $\hat{\rho}$, compute $\mathbf{t}$, $\mathbf{Z}$, V and S as in (9.37) and (9.38).

3. Calculate $\hat{\boldsymbol{\eta}}$ using (9.39).

4. Calculate $S(\hat{\rho},\ \hat{\boldsymbol{\eta}}) = (Z - S\hat{\boldsymbol{\eta}})'W_2^{-1}(Z - S\hat{\boldsymbol{\eta}})$, where $\hat{\boldsymbol{\eta}}$ as in step 3, and S, Z are in step 2, and W_2 in (9.38).

The same routine of calculations 1 through 4 (as in above) is to be followed for different permissible values of ρ in its range. We select that value of ρ as the final estimate of ρ for which $S(\hat{\rho},\ \hat{\boldsymbol{\eta}})$ is minimum. Using the final estimate of ρ, an unbiased estimate of $\boldsymbol{\eta}$ and σ_1^2 can be obtained from (9.38) and (9.40), respectively. Note that $\mathbf{Z}$ and S in (9.38) are computed based on the final estimate of ρ. From (9.40) we have the following result.

Theorem 9.17 *An estimate of σ^2 (as $\sigma_1^2 = 2\sigma^2(1 - \rho^2)$) is given by*

$$\hat{\sigma}^2 = \frac{(\mathbf{Z}'W_2^{-1}\mathbf{Z}) - \hat{\boldsymbol{\eta}}'(\mathbf{Z}'W_2^{-1}\mathbf{Z})\hat{\boldsymbol{\eta}}}{2(1 - \hat{\rho}^2)\{g(r-2) - p\}}. \tag{9.41}$$

We assume the final estimated value of ρ as its true value. For this true value of ρ, we compute $\mathbf{t}$ (in (9.37)), $\mathbf{Z}$, S (in (9.38)), $\hat{\boldsymbol{\eta}}$ (in (9.39)) and $\hat{\sigma}^2$ (in (9.41)). For all the subsequent derivations, we assume that ρ, $\mathbf{Z}$ and S are all known which are the computed values based on the final estimate of ρ. Thus, it is clear that $\hat{\boldsymbol{\eta}} \sim \mathrm{MN}(\boldsymbol{\eta},\ \sigma_1^2(S'W_2^{-1}S)^{-1})$, where $\boldsymbol{\eta}$, S and W_2 are in (9.38).

From equation (9.36), we can find the estimate of β_0 as

$$\hat{\beta}_0 = \bar{y} - \hat{\beta}_1\bar{x}_1 - \hat{\beta}_2\bar{x}_2 - \ldots - \hat{\beta}_k\bar{x}_k, \tag{9.42}$$

and also from equation (9.37), we can find the estimate of β_0 that is $\hat{\beta}_0 = \frac{\hat{\alpha}_0}{1-\hat{\rho}}$ where

$$\hat{\alpha}_0 = \bar{t} - \hat{\beta}_1\bar{v}_1 - \hat{\beta}_2\bar{v}_2 - \ldots - \hat{\beta}_k\bar{v}_k. \tag{9.43}$$

The scheme for calculations of ρ_1 is given below.

1. Assume some value of $\rho_1 \in (-1,\ 1)$.

2. With the assumed value of ρ_1, say $\hat{\rho}_1$, and the estimate of ρ, say $\hat{\rho}$ (as obtained in earlier scheme) compute W and W^{-1} as in (9.35).

3. Calculate $\hat{\hat{\beta}} = (X'W^{-1}X)^{-1}(X'W^{-1}\mathbf{Y})$, assume X has full column rank.

4. Calculate $S_0(\hat{\rho}_1, \hat{\hat{\beta}}) = (Y - X\hat{\hat{\beta}})'W^{-1}(Y - X\hat{\hat{\beta}})$, where $\hat{\hat{\beta}}$ as in step 3, W^{-1} as in step 2, Y and X are in (9.36).

The same routine of calculations 1 through 4 is to be followed for different permissible values of ρ_1 in its range. We select that value of ρ_1 as the final estimate of ρ_1 for which $S_0(\hat{\rho}_1, \hat{\hat{\beta}})$ is minimum.

9.3.2 Hypothesis testing of regression parameters with CAES

In the present section, an *exact test* procedure (as in Section 9.2.3) independent of ρ_1 but depends on the value of ρ is described here (following Das, 2010) for testing any set of linear hypotheses regarding unknown regression coefficients. A set of m linear independent hypotheses regarding unknown regression coefficients are given below

$$H_0 : \begin{cases} l_{11}\beta_1 + \ldots + l_{1p}\beta_p = l_{10}(\text{known}) \\ l_{21}\beta_1 + \ldots + l_{2p}\beta_p = l_{20}(\text{known}) \\ \vdots \\ l_{m1}\beta_1 + \ldots + l_{mp}\beta_p = l_{m0}(\text{known}) \end{cases}$$

or, $H_0 : R\boldsymbol{\eta} = \mathbf{l}_0(\text{known})$ against $H_A : R\boldsymbol{\eta} \neq \mathbf{l}_0$. Here $\text{rank}(R) = m$, where

$$R = \begin{pmatrix} l_{11} & \ldots & l_{1p} \\ l_{21} & \ldots & l_{2p} \\ \ldots & \ldots & \ldots \\ l_{m1} & \ldots & l_{mp} \end{pmatrix} \quad \text{and} \quad \mathbf{l}_0 = (l_{10}, \ldots, l_{m0})'.$$

Note that $\hat{\boldsymbol{\eta}} \sim \text{MN}(\boldsymbol{\eta}, \sigma_1^2(S'W_2^{-1}S)^{-1})$ (for the model (9.38)) as given in Section 9.3.1, thus

$$R\hat{\boldsymbol{\eta}} \sim \text{MN}(R\boldsymbol{\eta}, \sigma_1^2 R(S'W_2^{-1}S)^{-1}R').$$

Under H_0,

$$R\hat{\boldsymbol{\eta}} \sim \text{MN}(\mathbf{l}_0, \sigma_1^2 R(S'W_2^{-1}S)^{-1}R').$$

Therefore,

$$(R\hat{\boldsymbol{\eta}} - \mathbf{l}_0)'[\sigma_1^2 R(S'W_2^{-1}S)^{-1}R']^{-1}(R\hat{\boldsymbol{\eta}} - \mathbf{l}_0) \sim \chi_m^2,$$

where the degree of freedom m is given by the number of independent linear hypotheses in the $R\boldsymbol{\eta}$ vector. Also for the model (9.38), with known Z and S for the final estimated value of ρ, and $\hat{\boldsymbol{\epsilon}}_0 = W_2^{-\frac{1}{2}}(Z - S\hat{\boldsymbol{\eta}})$, $\hat{\boldsymbol{\epsilon}}_0'\hat{\boldsymbol{\epsilon}}_0/\sigma_1^2 \sim \chi_{g(r-2)-p}^2$, and it is independent of $R\hat{\boldsymbol{\eta}}$. Now we have the following result.

Theorem 9.18 *If the basic result $R\boldsymbol{\eta} = \mathbf{l}_0$ is true, then the test statistic is*

$$F = \frac{(R\hat{\boldsymbol{\eta}} - \mathbf{l}_0)'[R(S'W_2^{-1}S)^{-1}R']^{-1}(R\hat{\boldsymbol{\eta}} - \mathbf{l}_0)/m}{\hat{\boldsymbol{\epsilon}}_0'\hat{\boldsymbol{\epsilon}}_0/\{g(r-2)-p\}} \sim F_{m,\ g(r-2)-p}. \tag{9.44}$$

The test procedure is: reject H_0 at $100\alpha\%$ level of significance if observed $F > F_{\alpha;\ m,\ g(r-2)-p}$, and accept otherwise.

9.3.3 Confidence ellipsoid of regression parameters with CAES

In the present section, the confidence ellipsoid for a set of estimable functions of regression coefficients and the confidence interval for an estimable function of regression coefficients are derived under the compound autocorrelated error structure.

Suppose $\psi_1, \psi_2, \ldots, \psi_\nu$ are ν independent linear (also estimable) functions of $\beta_1, \beta_2, \ldots, \beta_p$. Let $\boldsymbol{\psi} = (\psi_1, \psi_2, \ldots, \psi_\nu)'$ be a vector of order $\nu \times 1$. Then

$$\boldsymbol{\psi} = C\boldsymbol{\eta},$$

where $\boldsymbol{\eta} = (\beta_1, \ldots, \beta_p)'$ and C(known) is a $\nu \times p$ matrix whose rows are linearly independent. Then $\hat{\boldsymbol{\psi}} = A\mathbf{Z}$ is the generalized least squares (GLS) estimate of $\boldsymbol{\psi}$ from (9.38), and $A = ((a_{ij}))$ (known matrix depending on ψ_is). The variance-covariance matrix of $\hat{\boldsymbol{\psi}}$ is then

$$\mathrm{Dis}(\hat{\boldsymbol{\psi}}) = A\mathrm{Dis}(\mathbf{Z})A' = \sigma_1^2 A W_2 A',$$

where $W_2 = I_g \otimes \left(\frac{1}{2} I_{r-2} + \frac{1}{2} E_{r-2}\right)$. Note that $\hat{\boldsymbol{\psi}} \sim \mathrm{MN}(\boldsymbol{\psi}, \sigma_1^2 A W_2 A')$ and is independent of

$$\frac{\hat{\boldsymbol{\epsilon}}_0' \hat{\boldsymbol{\epsilon}}_0}{\sigma_1^2} = \frac{(\mathbf{Z} - S\hat{\boldsymbol{\eta}})' W_2^{-1}(\mathbf{Z} - S\hat{\boldsymbol{\eta}})}{\sigma_1^2} \sim \chi_{g(r-2)-p}^2.$$

Again

$$(\hat{\boldsymbol{\psi}} - \boldsymbol{\psi})' \{\sigma_1^2 (A W_2 A')\}^{-1} (\hat{\boldsymbol{\psi}} - \boldsymbol{\psi}) \sim \chi_\nu^2,$$

and independent of $\hat{\boldsymbol{\epsilon}}_0' \hat{\boldsymbol{\epsilon}}_0 / \sigma_1^2$ (for the model (9.38)). The following result is now immediate for the model (9.38).

Theorem 9.19 *The distribution of the test statistic is*

$$F = \frac{(\hat{\boldsymbol{\psi}} - \boldsymbol{\psi})'(A W_2 A')^{-1}(\hat{\boldsymbol{\psi}} - \boldsymbol{\psi})/\nu}{(\hat{\boldsymbol{\epsilon}}_0' \hat{\boldsymbol{\epsilon}}_0)/(g(r-2)-p)} \sim F_{\nu, \, g(r-2)-p}. \tag{9.45}$$

Therefore,

$$(\hat{\boldsymbol{\psi}} - \boldsymbol{\psi})'(A W_2 A')^{-1}(\hat{\boldsymbol{\psi}} - \boldsymbol{\psi}) \leq \nu s^2 F_{\alpha; \, \nu, \, g(r-2)-p}, \tag{9.46}$$

where $s^2 = \frac{\hat{\boldsymbol{\epsilon}}_0' \hat{\boldsymbol{\epsilon}}_0}{g(r-2)-p}$ which is an UE of σ_1^2 (9.40).

Inequality (9.46) determines an *ellipsoid* in the ν-dimensional $\boldsymbol{\psi}$- space with center $\hat{\boldsymbol{\psi}} = (\hat{\psi}_1, \hat{\psi}_2, \ldots, \hat{\psi}_\nu)'$, and the probability that this *random ellipsoid* covers the true parameter $\boldsymbol{\psi}$ is $(1 - \alpha)$, no matter whatever be the values of unknown parameters.

A confidence interval for a single estimable function $\psi = \mathbf{c}'\boldsymbol{\eta}(\mathbf{c} \neq \mathbf{0})$ can be obtained by specializing the above calculation to $\nu = 1$. The resulting confidence interval is given by

$$(\mathbf{a}' W_2 \mathbf{a})^{-1}(\hat{\psi} - \psi)^2 \leq s^2 F_{\alpha; \, 1, \, g(r-2)-p}, \tag{9.47}$$

where $\hat{\psi} = \mathbf{a}'\mathbf{Z}$ is the GLS estimate of ψ for the model (9.38). Note that $\mathrm{Var}(\hat{\psi}) = \mathbf{a}'\mathrm{Dis}(\mathbf{Z})\mathbf{a} = \sigma_1^2 \mathbf{a}' W_2 \mathbf{a}$, and its unbiased estimate $\hat{\sigma}_{\hat{\psi}}^2 = s^2(\mathbf{a}' W_2 \mathbf{a})$. Equivalently, (9.47) can be expressed as

$$\hat{\psi} - t_{\alpha/2; \, g(r-2)-p}\hat{\sigma}_{\hat{\psi}} \leq \psi \leq \hat{\psi} + t_{\alpha/2; \, g(r-2)-p}\hat{\sigma}_{\hat{\psi}}, \tag{9.48}$$

the probability that this random interval covers the unknown ψ is $(1 - \alpha)$. The interval (9.48) could also be derived from the fact that $(\hat{\psi} - \psi)/\hat{\sigma}_{\hat{\psi}} \sim t_{g(r-2)-p}$.

9.3.4   Index of fit with CAES

In the present section, two criteria of judging the best fit of the regression model are described in case of compound autocorrelated error structure (9.35) in analogous to uncorrelated errors. For uncorrelated and homoscedastic errors under multiple regression, the index of fit is measured by the multiple correlation coefficient (R^2) and adjusted multiple correlation coefficient (R^2_{adj}) of the fitted regression model. Analogous to uncorrelated case, we define the multiple correlation coefficients $R^2(\mathbf{Z})$, $R^2(\mathbf{Y})$ and adjusted multiple correlation coefficients $R^2_{adj}(\mathbf{Z})$ for the fitted model (9.38) and (9.36) as follow:

$$R^2(\mathbf{Z}) = 1 - \frac{\hat{\boldsymbol{\epsilon}}_0' \hat{\boldsymbol{\epsilon}}_0}{TSS_{\mathbf{Z}_0}} \quad \text{and} \quad R^2(\mathbf{Y}) = \text{Corr.}^2(\mathbf{Y}, \hat{\mathbf{Y}}), \tag{9.49}$$

$$R^2_{adj}(\mathbf{Z}) = 1 - \frac{g(r-2)-1}{g(r-2)-p}(1 - R^2(\mathbf{Z})), \tag{9.50}$$

where $\hat{\boldsymbol{\epsilon}}_0 = W_2^{-\frac{1}{2}}(\mathbf{Z} - S\hat{\boldsymbol{\eta}})$, $TSS_{\mathbf{Z}_0} = (Z - \bar{Z})'W_2^{-1}(Z - \bar{Z})$, $\bar{Z} = \sum_{i=1}^{g}\sum_{j=2}^{r-1} z_{ij}/g(r-2)$ (for the model (9.38)). Generally, R^2 and R^2_{adj} as in (9.49) and (9.50) are both close to unity for good fitting (Section 9.3.5).

9.3.5   Illustration of regression analysis with CAES

Example 9.2 The example (with simulated data) considered is the pull strength (y) of a wire bond as an important characteristic. The data set gives information on pull strength (y), die height (x_1), wire length (x_2) and loop height (x_3). The pull strength (y) of a wire bond is recorded in some convenient unit. The study conducted is to estimate the unknown parameters involved in the mean response function of y, error variance and correlation coefficients. Three factors $(x_1, x_2$ and $x_3)$ as above are considered for this experiment. Appropriate change of origin and scale is used for each exploratory variable so that it lies between -1 and $+1$ (the range within which the experimentation is conducted). We assume

$$y_{ij} = \beta_0 + \beta_1 x_{ij1} + \beta_2 x_{ij2} + \beta_3 x_{ij3} + e_{ij}, \ j = 1, \dots, 15, \ i = 1, \dots, 4. \tag{9.51}$$

The model matrix (X) (selected from Chapter 2) used is

$$X = \begin{pmatrix} D \\ D \\ D \\ D \end{pmatrix}, \text{ where } D = \begin{pmatrix} 1 & 0 & 0 & 0 \\ 1 & 1 & 1 & 1 \\ 1 & -1 & 1 & -1 \\ 1 & 1 & -1 & -1 \\ 1 & -1 & -1 & 1 \\ 1 & 0 & 0 & 0 \\ 1 & 1 & 1 & 1 \\ 1 & -1 & -1 & -1 \\ 1 & -1 & 1 & -1 \\ 1 & 1 & -1 & 1 \\ 1 & 1 & -1 & -1 \\ 1 & -1 & 1 & 1 \\ 1 & -1 & -1 & 1 \\ 1 & 1 & 1 & -1 \\ 1 & 0 & 0 & 0 \end{pmatrix}.$$

In the absence of real data, observations are generated (100 replications) according to formula (9.51) with $\beta_0 = 50.0$, $\beta_1 = 3.8$, $\beta_2 = -2.5$, $\beta_3 = 0.05$, $\sigma^2 = 2.0$, $\rho = 0.8$ and ρ_1

TABLE 9.8: Responses under the simulation setting of $(\sigma^2 = 2, \rho = 0.8, \rho_1 = 0.6)$ with CAES

Observations	Groups			
	1	2	3	4
1	48.615	46.935	50.230	50.845
2	51.555	48.064	52.413	49.758
3	43.544	41.166	43.194	42.386
4	56.081	53.387	54.486	54.975
5	48.754	45.724	48.384	48.163
6	50.043	47.086	49.677	49.325
7	51.035	48.789	50.729	50.331
8	48.504	46.280	48.979	48.708
9	43.942	41.644	43.864	42.504
10	56.740	55.723	55.552	54.815
11	57.151	56.479	54.775	54.068
12	44.157	44.728	42.523	42.183
13	48.916	49.736	47.476	47.228
14	51.892	51.277	50.506	51.120
15	50.577	50.101	48.931	50.036

$= 0.6$ using the model matrix 'X', and $e \sim \text{MN}(\mathbf{0}, \sigma^2 W)$ where W is given in (9.35). The observations for a single replication so obtained are given in Table 9.8.

Using the responses as in Table 9.8, computed estimates of unknown parameters are $\hat{\boldsymbol{\beta}} = (\hat{\beta}_0, \hat{\beta}_1, \hat{\beta}_2, \hat{\beta}_3)' = (49.519, 3.716, -2.480, 0.042)'$, $\hat{\rho} = 0.811$, $\text{Min}.S(\hat{\rho}, \hat{\boldsymbol{\eta}}) = 27.775$, $\hat{\rho}_1 = 0.561$, and $\hat{\sigma}^2 = 1.437$.

For the simulation study, we consider the following eight combination of parameters, (σ^2, ρ, ρ_1) where $\sigma^2 = 1, 2$; $\rho = 0.4, 0.8$; $\rho_1 = 0.1, 0.6$, and with $\beta_0 = 50.0$, $\beta_1 = 3.8$, $\beta_2 = -2.5$, $\beta_3 = 0.05$.

For each simulation setting, each calculation is repeated 100 times. We compute the sample bias and sample variance for each estimate, where the sample bias and sample variance for the estimate θ are defined by

$$\text{Bias}(\hat{\theta}) = |\bar{\hat{\theta}} - \theta|, \quad \bar{\hat{\theta}} = \frac{\sum \hat{\theta}}{100} \quad \text{and} \quad \text{Var}(\hat{\theta}) = \frac{\sum (\hat{\theta} - \theta)^2}{100}.$$

Summarized simulation results are given in Table 9.9.

The test results from 100 replications (with $\beta_0 = 50.0$, $\beta_1 = 3.8$, $\beta_2 = -2.5$, $\beta_3 = 0.05$, $\sigma^2 = 2.0$, $\rho = 0.8$ and $\rho_1 = 0.6$) regarding the unknown regression coefficients for some hypotheses is given in Table 9.10.

Table 9.11 displays the values of two index of fit measures viz., $R^2(\mathbf{Z})$, $R^2(\mathbf{Y})$ and $R^2_{adj}(\mathbf{Z})$ as defined in Section 9.3.4 for the following four models:

$$M_1 : y = \beta_0 + \beta_1 x_1 + \beta_2 x_2 + \beta_3 x_3 + e,$$
$$M_2 : y = \beta_0 + \beta_1 x_1 + \beta_2 x_2 + e,$$
$$M_3 : y = \beta_0 + \beta_1 x_1 + \beta_3 x_3 + e,$$
$$M_4 : y = \beta_0 + \beta_1 x_1 + e.$$

Regression analysis is described above under the compound autocorrelated error structure. The best linear unbiased estimates of all the regression parameters have been derived except the intercept. The estimate of all the regression coefficients β's, σ^2, ρ and ρ_1 have been derived. It is observed that each estimated value is very close to its imputed value

TABLE 9.9: Simulation results: $\hat{\beta}_0$, $\hat{\beta}_1$, $\hat{\beta}_2$, $\hat{\beta}_3$, $\hat{\rho}(\sigma^2)$, $\hat{\rho}(\hat{\sigma}^2)$, $\hat{\rho}_1$ and $\hat{\sigma}^2$ with CAES

	$\hat{\beta}_0$	$\hat{\beta}_1$	$\hat{\beta}_2$	$\hat{\beta}_3$	$\hat{\rho}(\sigma^2)$	$\hat{\rho}(\hat{\sigma}^2)$	$\hat{\rho}_1$	$\hat{\sigma}^2$
	$\rho = 0.4, \rho_1 = 0.1, \sigma^2 = 1$							
Mean	50.026	3.809	−2.492	0.053	0.694	0.305	0.046	0.909
Bias($\hat{\theta}$)	0.026	0.009	0.008	0.003	0.294	0.095	0.054	0.091
Var($\hat{\theta}$)	0.128	0.015	0.013	0.015	0.007	0.028	0.001	0.058
	$\rho = 0.4, \rho_1 = 0.1, \sigma^2 = 2$							
Mean	49.966	3.786	−2.491	0.044	0.381	0.300	0.038	1.816
Bias ($\hat{\theta}$)	0.034	0.014	0.009	0.006	0.019	0.100	0.062	0.184
Var($\hat{\theta}$)	0.406	0.036	0.030	0.032	0.029	0.030	0.001	0.220
	$\rho = 0.4, \rho_1 = 0.6, \sigma^2 = 1$							
Mean	49.989	3.808	−2.495	0.062	0.710	−0.190	0.487	0.486
Bias($\hat{\theta}$)	0.011	0.008	0.005	0.012	0.310	0.590	0.113	0.514
Var($\hat{\theta}$)	0.733	0.018	0.015	0.012	0.006	0.006	0.083	0.013
	$\rho = 0.4, \rho_1 = 0.6, \sigma^2 = 2$							
Mean	50.006	3.822	−2.523	0.040	0.390	−0.197	0.502	1.015
Bias($\hat{\theta}$)	0.006	0.022	0.023	0.010	0.010	0.597	0.098	0.985
Var($\hat{\theta}$)	1.073	0.027	0.025	0.026	0.029	0.004	0.079	0.069
	$\rho = 0.8, \rho_1 = 0.1, \sigma^2 = 1$							
Mean	49.971	3.796	−2.500	0.043	0.897	0.748	0.050	0.944
Bias($\hat{\theta}$)	0.029	0.004	0.000	0.007	0.097	0.052	0.050	0.056
Var($\hat{\theta}$)	0.208	0.004	0.007	0.007	0.001	0.015	0.004	0.124
	$\rho = 0.8, \rho_1 = 0.1, \sigma^2 = 2$							
Mean	50.003	3.790	−2.496	0.034	0.801	0.748	0.053	1.802
Bias($\hat{\theta}$)	0.003	0.010	0.004	0.016	0.001	0.052	0.047	0.198
Var($\hat{\theta}$)	0.347	0.011	0.010	0.008	0.003	0.012	0.003	0.470
	$\rho = 0.8, \rho_1 = 0.6, \sigma^2 = 1$							
Mean	49.987	3.800	−2.504	0.050	0.902	0.493	0.449	0.411
Bias ($\hat{\theta}$)	0.013	0.000	0.004	0.000	0.102	0.307	0.151	0.589
Var ($\hat{\theta}$)	0.627	0.005	0.006	0.005	0.001	0.024	0.001	0.018
	$\rho = 0.8, \rho_1 = 0.6, \sigma^2 = 2$							
Mean	50.035	3.786	−2.502	0.048	0.795	0.432	0.435	0.762
Bias($\hat{\theta}$)	0.035	0.014	0.002	0.002	0.005	0.368	0.165	1.238
Var($\hat{\theta}$)	1.232	0.009	0.009	0.008	0.003	0.027	0.002	0.061

TABLE 9.10: Test result from 100 replications with $\alpha = 0.05$ under CAES. Kim, J., Das, R.N., Sengupta, A., and Paul, J. (2009). "Regression Analysis for Correlated Data under Compound Symmetry Structure," Journal of Statistical Theory and Applications, Vol. 8, No. 3, pp. 269-282.

Null Hypothesis	degree of freedom	Accepted cases	Rejected cases
$H_{01} : \beta_1 = \beta_2 = \beta_3 = 0$	(3, 127)	0	100
$H_{02} : \beta_1 = 0$	(1, 127)	0	100
$H_{03} : \beta_2 = 0$	(1, 127)	0	100
$H_{04} : \beta_3 = 0$	(1, 127)	96	4

TABLE 9.11: Index of fit measures with CAES. Kim, J., Das, R.N., Sengupta, A., and Paul, J. (2009). "Regression Analysis for Correlated Data under Compound Symmetry Structure," Journal of Statistical Theory and Applications, Vol. 8, No. 3, pp. 269-282.

Model	$R^2(\mathbf{Z})$	$R^2(\mathbf{Y})$	$R^2_{adj}(\mathbf{Z})$
M_1	0.9866	0.9212	0.9861
M_2	0.9591	0.8132	0.9582
M_3	0.3676	0.6536	0.3450
M_4	0.3572	0.6535	0.3343

(simulation study, Table 9.9). The true values of regression coefficients are justified from Table 9.10. The original model is examined from Table 9.11, which reveals that the models M_1 and M_2 are equivalent. The value of $R^2(\mathbf{Z})$ (or $R^2_{adj}(\mathbf{Z})$) is more (less) than $R^2(\mathbf{Y})$ for the correct (incorrect) model (Table 9.11), as $R^2(\mathbf{Z})$ is based on BLUEs. In practice, $R^2(\mathbf{Z})$ and $R^2_{adj}(\mathbf{Z})$ are the appropriate measures of index of fit. An application of regression analysis with CAES in an RBD is given below.

9.3.6 Randomized block design with CAES

In the present section, a randomized block design with *compound autocorrelated error structure* is considered. A method of analysis is derived for an RBD with a CAES. Confidence ellipsoid (for a set of independent treatment contrasts) and confidence interval (for a treatment contrast) are described. Analogous to Scheffe's method, the multiple comparison technique of judging all possible treatment contrasts is derived. An example (with simulated data) illustrates this approach.

9.3.6.1 A reinforced RBD with CAES

In Section 9.2.7.1, an RBD with the compound symmetry error structure has been described. The design problem with correlated errors and its related references is given in Section 9.2.7.1. Most of the authors have studied block design models assuming autocorrelated error structure. Kiefer and Wynn (1981, 1984) studied optimal block designs for autocorrelated error. Compound autocorrelated structure is one of the most common phenomenon of errors in an RBD. This type of situation may be observed, in general, when the fertility gradients are in two directions (i.e., along the blocks and along the plots within each block), and the fertility gradient is *decreasing* along the plots of each block. Under this situation, one may assume a first-order autocorrelation structure between the plots within each block (assuming decreasing fertility gradient along the plots), and a constant correlation coefficient of any two plots from any two blocks (assuming fertility gradient remains constant between the block distance). Little is known in literature about an RBD with a CAES (Das and Park, 2014). Here an analysis of an RBD under the above assumptions is described.

Here it is assumed that the errors of an RBD experiment are correlated according to a compound autocorrelated structure. That is, the errors within a block, for example ith block, have autocorrelated structure, $[\sigma^2 \{\rho^{|ij-ij'|}\}_{1 \le j, j' \le v}]$, and any two observations from any two blocks are correlated with a constant correlation coefficient (ρ_1). These two correlations are known as within and between block correlation coefficients. This type of correlation structure is termed as *compound autocorrelated structure* (Das, 2010), analogous to compound symmetry structure (Mukherjee, 1981). Note that the autocorrelation structure $[\sigma^2\{\rho^{|ij-ij'|}\}_{1\leq j, j'\leq v}]$ is *same* for all i, i.e., for all blocks. Thus, it is independent of i, but depends on the position of experimental unit (plot), where the jth treatment is placed (as the fertility gradient changes from plot to plot). So we can write this autocorrelation structure as $[\sigma^2\{\rho^{|j-j'|}\}_{1\leq j, j'\leq v}]$, where j and j' are two plots, ordered as jth and j'th, receiving the jth and j'th treatment respectively, in the same block. Considering ρ_2 for $\rho^{|j-j'|}$ for all $j\neq j'$, the compound autocorrelated structure will be reduced to compound symmetry structure. If $\rho_1 = 0$ (i.e., the correlation coefficient of any two observations from any two blocks is zero), compound symmetry structure is known as inter-class structure. Compound autocorrelated structure is the most common situation in an RBD, and compound symmetry structure, inter-class structure are the particular cases of compound autocorrelated structure.

In general, in an RBD each treatment occurs once and only once in each block. Some times some treatments may have greater importance than the others, in which case more information will be needed about them, and this will be achieved by the use of more units for these treatments in each block (Mead, 1994; Goon *et al.*, 2002, etc.). Such block designs are of interest in various fields of research, particularly in agriculture, medicine and serology (Rees, 1967) where often one (or more) treatment is used as a control or standard (Cutler, 1993). In the classic literature such designs are usually called *reinforced* (Das, 1958), *augmented* (Federer, 1956, 1961) or *supplemented* (Pearce, 1960). A detailed discussion of a reinforced incomplete block design is given in Das (1958).

To study an RBD with a CAES (in the present subsequent sections) only one treatment is repeated twice in each block. The usual randomization of an RBD is performed here, and the repeated treatment is placed in any plot of a block (according to randomization rule). For example, here the repeated treatment is assigned to *two neighbouring plots* in each block of the RBD. Such a reinforced RBD is considered for deriving the analysis and multiple comparison technique given in Section 9.3.6.3. The reason why we have considered a reinforced RBD instead of a usual RBD is that the design matrix X_{10} (singular) in (9.54) can be easily transformed into a non-singular matrix in order to derive the analysis and multiple comparison in an easy way for CAES.

Randomization is one of the three basic principles of experimental design (Yates, 1965). The procedure of randomization of placing the treatments to the experimental plots for any block is to select one plot at random for the first treatment, another plot at random for the second treatment, and so on. This random process of placing the treatments to the experimental plots is to be done afresh for each block (Finney, 1960, p. 23). For an RBD with a CAES, order of the position of plots is very important as the observations (or yields) from the experimental plots within a block satisfy autocorrelation structure as given above. Here we follow the same randomization procedure as an RBD for all the treatments, including the repeated treatment also. It is mentioned that only one treatment is to be repeated twice in each block. For the repeated treatment we select one plot at random as it is done in the process of randomization. For one more replication of the repeated treatment in the same block, one can place the repeated treatment in any plot according to the randomization rule. In the present analysis, we assume that the repeated treatment is placed in the next adjacent plot of the randomly selected plot for this treatment, such that all the two replications of the repeated treatment must occur in the two neighbouring plots (Das and Park, 2013). That is, we perform randomization of *all* the treatments with the assumption that *only* the two replications of the repeated treatment must occur in the two neighbouring plots in each block. For example, we have five treatments (T_1, T_2, T_3, T_4, T_5) and the $3rd$ treatment (T_3) is repeated twice in each block. Thus, we have six plots per block. Suppose a random permutation of six plots of a block is (3, 1, 4, 2, 5, 6). According to this assumed randomization, the layout of the treatments in that block is ($T_2, T_4, T_1, T_3, T_3, T_5$), i.e., T_2 is in first plot, T_4 is in second plot, and so on. A few more examples of such specific randomization are given in Section 9.3.6.6. This specific randomization is considered here as an example, just for presenting the analysis, but it is not essential for developing the theories in Section 9.3.6.3.

An analysis of an RBD with the compound autocorrelated error structure is described. The statistical inference procedures described herein are only free of correlation coefficient ρ_1 but depend on the correlation coefficient ρ. The confidence ellipsoid (for a set of independent treatment contrasts) and confidence interval (for a treatment contrast) are illustrated with a CAES. Analogous to Scheffe's method, the multiple comparison technique is described for judging all possible treatment contrasts.

9.3.6.2 An RBD model with CAES

Let us consider a reinforced RBD with v treatments, r blocks and $(v+1)$ plots per block. Each treatment occurs once and only once in each block, except the vth. Only the vth (it may be any) treatment occurs *twice* (in the two neighbouring plots) in each block. If we repeat any treatment more than once, but equal number of times in each block of an RBD, it does not distort its *orthogonality* property. This is an *orthogonal* randomized block design as it satisfies the necessary and sufficient condition, i.e., $n_{ij} = \frac{k_i \times r_j}{n}$ for *all* i, j, where n_{ij} is the number of replication of jth treatment in the ith block, k_i is the ith block size, r_j is the number of replication of jth treatment in the whole design and n is the total number of experimental units of the block design. Hence the fixed-effect model is

$$y_{ij} = \mu + \beta_i + \tau_j + e_{ij};\ 1 \le j \le (v-1),\ 1 \le i \le r,$$

$$y_{ivu} = \mu + \beta_i + \tau_v + e_{ivu};\ 1 \le u \le 2,\ 1 \le i \le r, \tag{9.52}$$

where y_{ij} is the observation for the jth treatment from the ith block and e_{ij} is the corresponding error, μ is the general effect, β_i is the effect due to ith block, τ_j is the effect due to jth treatment, and y_{ivu} is the yield due to the vth treatment in the ith block, receiving in uth plot and e_{ivu} is the corresponding error; $1 \le u \le 2$, $1 \le i \le r$. Here μ, β_i, τ_j's are unknown constants with $\sum_{i=1}^{r} \beta_i = \sum_{j=1}^{v} \tau_j = 0$.

The e_{ij}'s and e_{ivu}'s are random error components which are not independent but correlated. That is $E(e_{ij}) = E(e_{ivu}) = 0$; $\text{Var}(e_{ij}) = \text{Var}(e_{ivu}) = \sigma^2$; for all i, j and u and

$$\text{Cov}(e_{ij}, e_{i'j'}) = \begin{cases} \rho^{|j-j'|}\sigma^2; & \text{if } 1 \le j \ne j' \le (v-1),\ 1 \le i = i' \le r; \\ \rho_1\sigma^2; & \text{if } 1 \le j,\ j' \le (v-1),\ 1 \le i \ne i' \le r; \end{cases}$$

$$\text{Cov}(e_{ij}, e_{i'vu}) = \begin{cases} \rho^{|j-u|}\sigma^2; & \text{if } 1 \le j \le (v-1),\ 1 \le u \le 2,\ 1 \le i = i' \le r; \\ \rho_1\sigma^2; & \text{if } 1 \le j \le (v-1),\ 1 \le u \le 2,\ 1 \le i \ne i' \le r; \end{cases}$$

$$\text{Cov}(e_{ivu}, e_{i'vu'}) = \begin{cases} \rho^{|u-u'|}\sigma^2; & \text{if } 1 \le u \ne u' \le 2,\ 1 \le i = i' \le r; \\ \rho_1\sigma^2; & \text{if } 1 \le u,\ u' \le 2,\ 1 \le i \ne i' \le r. \end{cases}$$

Therefore, in matrix and vector notation (9.52) can be written as

$$\mathbf{Y} = X\boldsymbol{\xi} + \mathbf{e}, \tag{9.53}$$

where $\mathbf{Y} = (y_{11}, \ldots, y_{1(v-1)}, y_{1v1}, y_{1v2}, y_{21}, \ldots, y_{2(v-1)}, y_{2v1}, y_{2v2}, \ldots, y_{r1}, \ldots, y_{r(v-1)}, y_{rv1}, y_{rv2})'$ is the vector of recorded observations of the study variable y from all r blocks; $\boldsymbol{\xi} = (\mu, \beta_1, \ldots, \beta_r, \tau_1, \ldots, \tau_v)' = (\mu, \boldsymbol{\beta}', \boldsymbol{\tau}')'$, is the vector of unknown parameters, where $\boldsymbol{\beta} = (\beta_1, \ldots, \beta_r)'$ and $\boldsymbol{\tau} = (\tau_1, \ldots, \tau_v)'$; X is the model (design) matrix of order $N \times (1 + r + v)$, and $N = r(v+1)$. Note that rank of the design (or model) matrix X is rank$(X) =$

$(r+v-1)$ (not full rank). Further, $\mathbf{e}$ is the vector of errors which are assumed to be normally distributed with $E(\mathbf{e}) = \mathbf{0}$ and $\text{Dis}(\mathbf{e}) = \sigma^2[I_r \otimes (A - B) + E_{r \times r} \otimes B] = \sigma^2 W$ say, where $A = [\{\rho^{|j-j'|}\}_{1 \le j, \, j' \le (v+1)}]$ and $B = \rho_1 E_{(v+1) \times (v+1)}$, that is, $\mathbf{e} \sim \text{MN}(\mathbf{0}, \, \sigma^2 W)$. This structure W is termed as compound autocorrelated error structure (Das, 2010). Considering ρ_2 for $\rho^{|j-j'|}$ for all $j \ne j'$, $\text{Dis}(\mathbf{e}) = \sigma^2[I_r \otimes (A_0 - B) + E_{r \times r} \otimes B]$, where $A_0 = (1 - \rho_2)I_{v+1} + \rho_2 E_{(v+1) \times (v+1)}$, is known as compound symmetry structure. Das and Park (2008c) have studied RBD under the compound symmetry error structure. If $\rho_1 = 0$, compound symmetry structure reduces to $\text{Dis}(\mathbf{e}) = \sigma^2(I_r \otimes A_0)$, known as inter-class structure. In general, the matrix W is unknown (as ρ and ρ_1 are unknown) but for all the calculations, W is assumed to be known.

To derive the theories in Section 9.3.6.3, we assume some specific randomization in all the blocks of the RBD. Without any loss of generality, suppose the treatments are arranged to the experimental plots according to the *increasing order of their suffices*. This specific *non-random* layout assumption leads to use very simple notations for developing the theories in Section 9.3.6.3. In literature, such assumption leads to the analysis, known as *null analysis of variance* (Nelder, 1965a, p. 148). In this analysis, the treatments play only a nominal role (Nelder would call them null 'treatments', Nelder, 1965a). When going to the final analysis applicable to a real experiment, one would replace the null treatments by the experimental treatments, assigned to the experimental units (plots) according to the usual randomization of an RBD. For independent experimental errors, the validity of the null analysis applied to a real experimental data depends on two basic assumptions: (a) unit-treatment additivity, (b) a valid form of randomization (Nelder, 1965b, p. 168). In the context of the present study, the assumption (a) will mean that the dispersion parameters σ^2, ρ and ρ_1 do not depend on the treatments used in the experiment. The assumption (b) is here related to the matrix transforming the vector of observational data $\mathbf{Y}$ to the vector $\mathbf{T}$ (Section 9.3.6.3). In the null analysis it has the form $I_r \otimes P$ say, where

$$P = \begin{pmatrix} -\rho & 1 & 0 & \dots & 0 & \rho & -1 \\ 0 & -\rho & 1 & \dots & 0 & \rho & -1 \\ . & . & . & \dots & . & . & . \\ 0 & 0 & 0 & \dots & 1 & \rho & -1 \end{pmatrix}.$$

(Note that merging the last three columns of P one obtains the matrix D in Section 9.3.6.3). It has been shown that if $\text{Dis}(\mathbf{Y}) = \sigma^2 W$, then $\text{Dis}(\mathbf{T}) = \sigma^2(I_r \otimes P)W(I_r \otimes P') = \sigma_1^2 W_2$, where $\sigma_1^2 = 2\sigma^2(1 - \rho^2)$ and $W_2 = I_r \otimes W_1$, with $W_1 = 2^{-1}(I_{v-1} + E_{v-1 \times v-1})$ (Section 9.3.6.3). A valid randomization of the assignment of treatments to plots will here mean that the relevant permutations of the rows and columns of P will not change the dispersion matrix $\text{Dis}(\mathbf{T})$. More precisely, that $\text{Var}(\mathbf{c}'\boldsymbol{\tau})$ for any linear estimable function $\mathbf{c}'\boldsymbol{\tau}$ will not be changed after the usual randomization applied. Following this non-random layout and notations, all the derivations are given in Section 9.3.6.3.

9.3.6.3 A reinforced RBD analysis with CAES

Similarly as in Section 9.2.7.3, in an RBD, the main hypothesis is to test the equality of v treatment effects. Thus, the null hypothesis is $H_0 : \tau_1 = \tau_2 = \dots = \tau_v$, and it will be tested against the alternative hypothesis H_A : at least one of τ_i is different from others. The equivalent hypothesis of H_0 is $H_{0E} : \tau_{01} = \tau_{02} = \dots = \tau_{0(v-1)} = 0$, where $\tau_{0j} = \tau_j - \tau_v$, $1 \le j \le (v - 1)$. If H_0 (or H_{0E}) is accepted we stop analysis. If H_0 is rejected we test for equality of any two treatment means, that is $H_{0ij} : \tau_i = \tau_j$, against $H_{Aij} : \tau_i \ne \tau_j$. The equivalent hypothesis of H_{0ij} is $H_{0ijE} : \tau_{0i} = \tau_{0j}$ (if $j \ne v$) or $H_{0ijE} : \tau_{0i} = 0$ (if $j = v$). Original model is stated as in (9.52) or (9.53). It is assumed that the treatments are arranged to the experimental plots according to the increasing order of their suffices. Let us consider the following transformed model

$$z_{ij} = y_{ij} - \rho y_{i(j-1)}; \ 2 \le j \le (v-1); \quad z_{iv1} = y_{iv1} - \rho y_{i(v-1)};$$

$$\text{and } z_{iv2} = y_{iv2} - \rho y_{iv1}; \ 1 \le i \le r,$$

(Kmenta, 1971; Chatterjee and Price, 2000)

$$\text{or, } z_{ij} = \mu(1-\rho) + \beta_i(1-\rho) + (\tau_j - \rho\tau_{j-1}) + (e_{ij} - \rho e_{i(j-1)}); \ 2 \le j \le (v-1),$$

$$z_{iv1} = \mu(1-\rho) + \beta_i(1-\rho) + (\tau_v - \rho\tau_{v-1}) + (e_{iv1} - \rho e_{i(v-1)}),$$

$$\text{and } z_{iv2} = \mu(1-\rho) + \beta_i(1-\rho) + \tau_v(1-\rho) + (e_{iv2} - \rho e_{iv1}); \ 1 \le i \le r.$$

Let $\mathbf{Z} = (z_{12}, z_{13}, \ldots, z_{1(v-1)}, z_{1v1}, z_{1v2}, z_{22}, z_{23}, \ldots, z_{2(v-1)}, z_{2v1}, z_{2v2}, \ldots, z_{r2},$
$z_{r3}, \ldots, z_{r(v-1)}, z_{rv1}, z_{rv2})'$; $\phi_{ij} = (e_{ij} - \rho e_{i(j-1)})$, $\phi_{iv1} = (e_{iv1} - \rho e_{i(v-1)})$, $\phi_{iv2} = (e_{iv2} - \rho e_{iv1})$; $2 \le j \le (v-1)$, $1 \le i \le r$. Note that $E(\phi_{ij}) = E(\phi_{ivu}) = 0$; and
$\text{Var}(\phi_{ij}) = \text{Var}(\phi_{ivu}) = \sigma^2(1-\rho^2)$; $2 \le j \le (v-1)$, $1 \le u \le 2$, $1 \le i \le r$.
Also, $\text{Cov}(\phi_{ij}, \phi_{i'j'}) = \text{Cov}(\phi_{ij}, \phi_{i'vu}) = \text{Cov}(\phi_{ivu}, \phi_{i'vu'}) = 0$; $2 \le j \ne j' \le (v-1)$,
$1 \le u \ne u' \le 2$, $1 \le i = i' \le r$; and $\text{Cov}(\phi_{ij}, \phi_{i'j'}) = \text{Cov}(\phi_{ij}, \phi_{i'vu}) = \text{Cov}(\phi_{ivu}, \phi_{i'vu'})$
$= \rho_1(1-\rho)^2\sigma^2 = \rho_3\sigma^2$; say, $2 \le j$, $j' \le (v-1)$, $1 \le u$, $u' \le 2$, $1 \le i \ne i' \le r$; where
$\rho_3 = \rho_1(1-\rho)^2$.

Let us also transform

$$t_{ij} = z_{ij} - z_{iv2}; \ 2 \le j \le (v-1); \quad \text{and} \quad t_{iv1} = z_{iv1} - z_{iv2}; \ 1 \le i \le r,$$

$$\text{or, } t_{ij} = -\rho\tau_{(j-1)} + \tau_j - (1-\rho)\tau_v + \epsilon_{ij}; \ 2 \le j \le (v-1),$$

$$\text{and} \quad t_{iv1} = -\rho\tau_{(v-1)} + \rho\tau_v + \epsilon_{iv1}; \ 1 \le i \le r,$$

$$\text{or,} \quad \mathbf{T} = X_{10}\boldsymbol{\tau} + \boldsymbol{\epsilon}, \tag{9.54}$$

where $\mathbf{T} = (t_{12}, t_{13}, \ldots, t_{1(v-1)}, t_{1v1}, t_{22}, t_{23}, \ldots, t_{2(v-1)}, t_{2v1}, \ldots, t_{r2}, t_{r3}, \ldots, t_{r(v-1)}, t_{rv1})'$,
$\boldsymbol{\epsilon} = (\epsilon_{12}, \epsilon_{13}, \ldots, \epsilon_{1(v-1)}, \epsilon_{1v1}, \epsilon_{22}, \epsilon_{23}, \ldots, \epsilon_{2(v-1)}, \epsilon_{2v1}, \ldots, \epsilon_{r2}, \epsilon_{r3}, \ldots, \epsilon_{r(v-1)}, \epsilon_{rv1})'$, $\epsilon_{ij} = (\phi_{ij} - \phi_{iv2})$, $\epsilon_{iv1} = (\phi_{iv1} - \phi_{iv2})$; $2 \le j \le (v-1)$, $1 \le i \le r$, and $\boldsymbol{\tau}$ as in (9.53),
$N_1 = r \times (v-1)$ and the model matrix X_{10} is of order $N_1 \times v$ with rank $(X_{10}) = v - 1$ (not
full rank), is given below:

$$X_{10} = \begin{pmatrix} D \\ D \\ . \\ . \\ D \end{pmatrix}, \quad \text{where} \quad D = \begin{pmatrix} -\rho & 1 & 0 & \ldots & 0 & 0 & -(1-\rho) \\ 0 & -\rho & 1 & \ldots & 0 & 0 & -(1-\rho) \\ . & . & . & \ldots & . & . & . \\ 0 & 0 & 0 & \ldots & -\rho & 1 & -(1-\rho) \\ 0 & 0 & 0 & \ldots & 0 & -\rho & \rho \end{pmatrix}_{v-1 \times v}.$$

It is assumed that $\tau_{0k} = \tau_k - \tau_v$, $1 \le k \le v - 1$. Therefore, the transformed model (9.54) is

$$\text{or, } t_{ij} = -\rho\tau_{0(j-1)} + \tau_{0j} + \epsilon_{ij}; \ 2 \le j \le (v-1),$$

$$\text{and} \quad t_{iv1} = -\rho\tau_{0(v-1)} + \epsilon_{iv1}; \ 1 \le i \le r,$$

$$\text{or,} \quad \mathbf{T} = X_2\boldsymbol{\tau}_0 + \boldsymbol{\epsilon}, \tag{9.55}$$

where $\boldsymbol{\tau}_0 = (\tau_{01}, \tau_{02}, \ldots, \tau_{0(v-1)})'$,

$$X_2 = \begin{pmatrix} D_1 \\ D_1 \\ . \\ . \\ D_1 \end{pmatrix} \quad \text{and} \quad D_1 = \begin{pmatrix} -\rho & 1 & 0 & \ldots & 0 & 0 \\ 0 & -\rho & 1 & \ldots & 0 & 0 \\ . & . & . & \ldots & . & . \\ 0 & 0 & 0 & \ldots & -\rho & 1 \\ 0 & 0 & 0 & \ldots & 0 & -\rho \end{pmatrix}_{v-1 \times v-1},$$

and D_1 is obtained from D by deleting its last column. Note that the last column of D is the sum of the first $(v-1)$ columns. Design matrix X_2 (of order $N_1 \times v - 1$) for some specific random layouts is given in Section 9.3.6.6.

Note that $E(\epsilon_{ij}) = E(\epsilon_{iv1}) = 0$; $\text{Var}(\epsilon_{ij}) = \text{Var}(\epsilon_{iv1}) = 2\sigma^2(1-\rho^2) = \sigma_1^2$ say; $2 \leq j \leq v - 1$, $1 \leq i \leq r$, where $\sigma_1^2 = 2\sigma^2(1-\rho^2)$, and

$$\text{Cov}(\epsilon_{ij}, \epsilon_{i'j'}) = \text{Cov}(\epsilon_{ij}, \epsilon_{i'v1}) = \begin{cases} \frac{\sigma_1^2}{2}; & \text{if } 2 \leq j \neq j' \leq v - 1, \ 1 \leq i = i' \leq r, \\ 0; & \text{if } 2 \leq j, \ j' \leq v - 1, \ 1 \leq i \neq i' \leq r. \end{cases}$$

Therefore,

$$\text{Corr}(\epsilon_{ij}, \ \epsilon_{i'j'}) = \text{Corr}(\epsilon_{ij}, \epsilon_{i'v1}) = \begin{cases} \frac{1}{2}; & \text{if } 2 \leq j \neq j' \leq v - 1, \ 1 \leq i = i' \leq r; \\ 0; & \text{if } 2 \leq j, \ j' \leq v - 1, \ 1 \leq i \neq i' \leq r. \end{cases}$$

Therefore, $E(\boldsymbol{\epsilon}) = \mathbf{0}$, $\text{Dis}(\boldsymbol{\epsilon}) = 2\sigma^2(1-\rho^2)[I_r \otimes W_1] = \sigma_1^2 W_2$ say, where $W_1 = 2^{-1}(I_{v-1} + E_{v-1 \times v-1})$, $\sigma_1^2 = 2\sigma^2(1-\rho^2)$ and $W_2 = I_r \otimes W_1$. Now we have the following theorems, and the proofs are directly followed from the model (9.55) (Rao, 1971, 1973).

Theorem 9.20 *The best linear unbiased estimate of $\boldsymbol{\tau}_0$ (for the model (9.55)) for known ρ is*

$$\hat{\boldsymbol{\tau}}_0 = (X_2'W_2^{-1}X_2)^{-1} \ (X_2'W_2^{-1}\mathbf{T}). \tag{9.56}$$

Theorem 9.21 *An unbiased estimate of σ_1^2 (for the model (9.55)) is*

$$\hat{\sigma}_1^2 = \frac{(\mathbf{T}'W_2^{-1}\mathbf{T}) - \hat{\boldsymbol{\tau}}_0'(X_2'W_2^{-1}X_2)\hat{\boldsymbol{\tau}}_0}{(N_1 - v + 1)}, \tag{9.57}$$

where $N_1 = r \times (v-1)$.

The scheme for calculations of $\boldsymbol{\tau}$, ρ and σ^2 is given below.

1. Assume some value of $\rho \in (-1, 1)$.

2. For the assumed value of ρ, say $\hat{\rho}$, compute $\mathbf{Z}$, $\mathbf{T}$ and X_2 as in (9.54) and (9.55).

3. Calculate $\hat{\boldsymbol{\tau}}_0$ (for the model (9.55)) using (9.56).

4. Calculate $S(\hat{\rho}, \ \hat{\boldsymbol{\tau}}_0) = (\mathbf{T} - X_2\hat{\boldsymbol{\tau}}_0)'W_2^{-1}(\mathbf{T} - X_2\hat{\boldsymbol{\tau}}_0) = R_0^2$, say, (for the model (9.55)).

The same routine of calculations 1 through 4 is to be followed for different permissible values of ρ in its range. We select that value of ρ as the final estimate of ρ for which $S(\hat{\rho}, \ \hat{\boldsymbol{\tau}}_0)$ is minimum. Using the final estimate of ρ, an unbiased estimate of $\boldsymbol{\tau}_0$ and σ_1^2 can be obtained from (9.56) and (9.57), respectively. Note that $\mathbf{T}$ and X_2 in (9.54) are computed based on the final estimate of ρ. From (9.57) we have the following theorem.

Theorem 9.22 *An estimate of σ^2 ($\sigma_1^2 = 2\sigma^2(1-\rho^2)$) (for the model (9.52) (or 9.55) is*

$$\hat{\sigma}^2 = \frac{(\mathbf{T}'W_2^{-1}\mathbf{T}) - \hat{\boldsymbol{\tau}}_0'(X_2'W_2^{-1}X_2)\hat{\boldsymbol{\tau}}_0}{2(1-\hat{\rho}^2)(N_1 - v + 1)}. \tag{9.58}$$

A similar method of estimation procedure and the properties of the regression estimates and autocorrelation coefficient ρ are given in Kmenta (1971) and Chatterjee and Price (2000). This iterative estimators will have the same asymptotic properties as the maximum likelihood estimators; some evidence concerning their small sample properties is presented on page 293 (Kmenta, 1971). Note that the estimated value of ρ is numerically very close to the maximum likelihood estimate as well as the true value of ρ, for many computed examples (Kmenta, 1971). We assume the final estimated value of ρ as its true value. For this true value of ρ, we calculate $\mathbf{Z}$, $\mathbf{T}$ and X_2 (as in (9.54) and (9.55)). For all the subsequent derivations, we assume that ρ, $\mathbf{Z}$, $\mathbf{T}$ and X_2 are all known, and these are the calculated values based on the final estimate of ρ. This assumption will only affect on the observational vector $\mathbf{T}$ and the model matrix X_2. Consequently, it will only affect on the mean vector $(X_2\tau_0)$ of the distribution $\mathbf{T}$, but the dispersion parameters are always independent. As the estimated value of ρ is generally, very close to its true value, we may assume that the observational vector $\mathbf{T}$ and the model matrix X_2 will *not* be affected very much by $\hat{\rho}$.

From (9.57) it is clear that $\hat{\tau}_0 \sim MN(\tau_0, \sigma_1^2(X_2'W_2^{-1}X_2)^{-1})$. We have to test the hypothesis $H_0 : \tau_1 = \tau_2 = \ldots = \tau_v$, against the alternative hypothesis H_A : at least one of τ_i is different from others. Equivalently, H_0 can be written as $H_{0E} : \tau_{01} = \tau_{02} = \ldots = \tau_{0(v-1)} = 0$. Equivalently, $H_{0E} : R\tau_0 = \mathbf{0}$, where R is an identity matrix of order $(v-1)$ with rank $(R) = (v-1)$.

Therefore, $E(R\hat{\tau}_0) = R\tau_0$, $\text{Dis}(R\hat{\tau}_0) = \sigma_1^2 R(X_2'W_2^{-1}X_2)^{-1}R'$. As $\hat{\tau}_0$ is distributed as Multivariate Normal distribution, then $R\hat{\tau}_0 \sim MN(R\tau_0, \sigma_1^2 R(X_2'W_2^{-1}X_2)^{-1}R')$. Under $H_{0E} : R\tau_0 = \mathbf{0}$, then $R\hat{\tau}_0 \sim MN(\mathbf{0}, \sigma_1^2 R(X_2'W_2^{-1}X_2)^{-1}R')$. Therefore, under H_{0E}, $(R\hat{\tau}_0)'[\sigma_1^2 R(X_2'W_2^{-1}X_2)^{-1}R']^{-1}(R\hat{\tau}_0) \sim \chi_{v-1}^2$, where the degrees of freedom (d.f.) $(v-1)$ is the number of elements in the $R\tau_0$ vector. Also for the model (9.55), $\frac{(N_1-v+1)s^2}{\sigma_1^2} = \frac{R_0^2}{\sigma_1^2} \sim \chi_{N_1-v+1}^2$ (where $R_0^2 = (\mathbf{T} - X_2\hat{\tau}_0)'W_2^{-1}(\mathbf{T} - X_2\hat{\tau}_0)$) and is independent of $\hat{\tau}_0$, and hence independent of $R\hat{\tau}_0$, where $s^2 = \frac{R_0^2}{N_1-v+1}$ is an UE of σ_1^2, as in (9.57). Thus, from the above we have the following theorem (proof is directly followed).

Theorem 9.23 *The test statistic*

$$F = \frac{(R\hat{\tau}_0)'[R(X_2'W_2^{-1}X_2)^{-1}R']^{-1}(R\hat{\tau}_0)/(v-1)}{s^2} \quad (9.59)$$

follows F distribution with d.f. $v-1$, $N_1 - v + 1$ if the basic result $R\tau_0 = \mathbf{0}$ is true.

The test procedure is: reject H_{0E} at $100\alpha\%$ level of significance if observed $F > F_{\alpha; \, v-1, \, N_1-v+1}$, and accept otherwise. If the null hypothesis H_{0E} is accepted, we stop analysis. But the rejection of H_{0E} leads to test the equality of any two treatment means.

Suppose H_{0E} is rejected, we want to test for $H_{0ij} : \tau_i - \tau_j = 0$ against $H_{Aij} : \tau_i \neq \tau_j$. Equivalent hypothesis of H_{0ij} is $H_{0ijE} : R_1\tau_0 = 0$, where $R_1 = (0,\ldots,0,1,0,\ldots,0,-1,0,\ldots,0)$ (for $j \neq v$) is a row vector of order $1 \times v - 1$ with the ith position is unity (1) and the jth position is −unity (-1) and all other elements are zeros. For $j = v$, R_1 is a row vector of order $1 \times v - 1$ with the ith position is unity (1) and all other elements are zeros. Then the test statistic for H_{0ijE} is given below in Theorem 9.24. Therefore, from the above we have the following theorem (proof is directly followed).

Theorem 9.24 *The test statistic*

$$F_1 = \frac{(R_1\hat{\tau}_0)'[R_1(X_2'W_2^{-1}X_2)^{-1}R_1']^{-1}(R_1\hat{\tau}_0)/1}{s^2} \quad (9.60)$$

follows F distribution with d.f. 1, $N_1 - v + 1$ if $H_{0ijE} : R_1\tau_0 = 0$ is true.

The test procedure is: reject H_{0ijE} at $100\alpha\%$ level of significance if observed $F_1 > F_{\alpha; \, 1, \, N_1-v+1}$, and accept otherwise. Note that the test statistics in (9.59) and (9.60) are independent of ρ_1 but depend on ρ. Therefore, the above two tests are robust of ρ_1 only.

Note that exact F statistic may be derived for testing the same hypotheses ($H_0 : \tau_1 = \ldots = \tau_v = 0$) as in Theorem 9.23 (with different R), following Kenward and Rozer (1997), based on the restricted MLE (RMLE) of σ^2, ρ and ρ_1 of $\Sigma = \sigma^2 W$, for original observational vector $\mathbf{Y}$, assuming $\mathbf{Y} \sim MN(X\xi, \sigma^2 W)$, where W is as in Section 9.3.6.2 (or in (9.53)), and W contains two unknown parameters ρ and ρ_1. Note that W may be a *singular* matrix for many possible combinations of ρ and ρ_1. Consequently, RMLEs of ρ and ρ_1 may not be derived for many cases. Thus, it will be difficult to derive exact F statistic for this case following Kenward and Rozer (1997). Theorem 9.23 describes F statistic based on $\mathbf{T} \sim \mathrm{MN}(X_2 \boldsymbol{\tau}_0, \sigma_1^2 W_2)$, where W_2 is a known matrix with elements 1 (all diagonal elements) and 1/2 & 0 (off-diagonal elements) (9.55). Also note that the design matrix X_2 (unknown) (9.55) and the observational vector $\mathbf{T}$ (unknown) (9.54) contain only one unknown parameter ρ. Therefore, this problem is completely different from Kenward and Rozer (1997, p. 984) assumptions. Moreover, Kenward and Rozer (1997) have assumed that Σ is unknown (with $r \geq 2$ unknown parameters) but $\mathbf{Y}$ and X are known, which are completely different from the present context.

9.3.6.4 Confidence ellipsoid of treatment contrasts with CAES

Roy and Bose (1953) first presented a paper on simultaneous confidence intervals following Roy's (1953) 'Union-Intersection Principle'. Simultaneous confidence intervals for a set of general parametric functions are given in Roy and Bose (1953, p. 514). As an application to univariate simultaneous estimation problems, Roy and Bose (1953) have suggested simultaneous confidence intervals for a set of linearly estimable parametric functions which is essentially Scheffe's (1953) result. Scheffe (1953) and Tukey (1952) have considered some special cases of simultaneous confidence intervals for a finite and infinite set of linearly parametric functions which are some particular cases of Roy and Bose's (1953) results. Some detailed results of simultaneous confidence intervals of a set of general and special parametric functions are given in Roy and Bose (1953).

In this section we describe the confidence ellipsoid (simultaneous confidence intervals) of a set of linearly estimable treatment parametric functions, and the confidence interval of a linearly estimable treatment parametric function (Das and Park, 2013), which are also some special cases of Roy and Bose's (1953) results.

Let $\psi_1, \psi_2, \ldots, \psi_q$ denote q independent estimable linear functions of v treatment effects viz., $\tau_1, \tau_2, \ldots, \tau_v$ as in the model (9.52). They are also q independent treatment contrasts. Let $\boldsymbol{\psi} = (\psi_1, \psi_2, \ldots, \psi_q)'$ be a vector of order $q \times 1$. Then $\boldsymbol{\psi} = C\boldsymbol{\tau}_0$, where the rows of C (known) are linearly independent. Then $\hat{\boldsymbol{\psi}} = G\mathbf{T}$ is the generalized least squares (GLS) estimate (or BLUE) of $\boldsymbol{\psi}$, from the linear model (9.55), $\mathbf{T}$ as in (9.54) based on the final estimated value of ρ, $\hat{\boldsymbol{\psi}} = (\hat{\psi}_1, \hat{\psi}_2, \ldots, \hat{\psi}_q)'$ and $G = ((g_{ij})) = C(X_2' W_2^{-1} X_2)^{-1} X_2' W_2^{-1}$ (as $\hat{\boldsymbol{\psi}} = C\hat{\boldsymbol{\tau}}_0$ and $\hat{\boldsymbol{\tau}}_0$ is as in (9.56)) (known matrix depending on ψ_i's) of order $q \times N_1$. The variance covariance of the estimates $\{\hat{\psi}_i\}$'s is then $\mathrm{Dis}(\hat{\boldsymbol{\psi}}) = G\mathrm{Dis}(\mathbf{T})G' = \sigma_1^2(GW_2G')$, where W_2 as in (9.55).

Note that $\hat{\boldsymbol{\psi}} \sim \mathrm{MN}(\boldsymbol{\psi}, \sigma_1^2(GW_2G'))$, and $\hat{\boldsymbol{\psi}}$ is statistically independent of $\frac{R_0^2}{\sigma_1^2} = \frac{(\mathbf{T}-X_2\hat{\boldsymbol{\tau}}_0)'W_2^{-1}(\mathbf{T}-X_2\hat{\boldsymbol{\tau}}_0)}{\sigma_1^2} \sim \chi_{N_1-v+1}^2$ (for the model (9.55)). Again, $(\hat{\boldsymbol{\psi}} - \boldsymbol{\psi})'\{\sigma_1^2(GW_2G')\}^{-1}(\hat{\boldsymbol{\psi}} - \boldsymbol{\psi}) \sim \chi_q^2$ and is independent of $\frac{R_0^2}{\sigma_1^2} \sim \chi_{N_1-v+1}^2$. Then

$$F = \frac{(\hat{\boldsymbol{\psi}} - \boldsymbol{\psi})'(GW_2G')^{-1}(\hat{\boldsymbol{\psi}} - \boldsymbol{\psi})}{qs^2} \sim F_{q,\ N_1-v+1}, \qquad (9.61)$$

where $s^2 = \frac{R_0^2}{N_1-v+1}$. Consequently, we have the following theorem (proof is directly followed).

Theorem 9.25 *The desired confidence set falls out of (9.61), under the model (9.52), the probability is $(1 - \alpha)$ that the F-variable in (9.61) is $\leq F_{\alpha;\ q,\ N_1-v+1}$, or that*

$$(\hat{\boldsymbol{\psi}} - \boldsymbol{\psi})'(GW_2G')^{-1}(\hat{\boldsymbol{\psi}} - \boldsymbol{\psi}) \leq qs^2 F_{\alpha;\ q,\ N_1-v+1}. \qquad (9.62)$$

Inequality (9.62) determines an *ellipsoid* in the q-dimensional $\boldsymbol{\psi}$-space with center at

$(\hat{\psi}_1, \hat{\psi}_2, \ldots, \hat{\psi}_q)$ and the probability that this *random ellipsoid* covers the true treatment parameter point $(\psi_1, \psi_2, \ldots, \psi_q)$ is $(1 - \alpha)$ no matter whatever be the values of the unknown parameters $\mu, \beta_1, \beta_2, \ldots, \beta_r, \tau_1, \tau_2, \ldots, \tau_v, \rho_1$ and σ^2, and the equation (9.62) is known as *confidence ellipsoid*. We may obtain a *confidence interval* for a single estimable function $\psi = \mathbf{c}'\tau_0$, $(\mathbf{c} \neq \mathbf{0})$ by specializing the above calculation $q = 1$. The resulting one-dimensional ellipsoid is the interval given by

$$b^{-1}(\hat{\psi} - \psi)^2 \leq s^2 F_{\alpha;\ 1,\ N_1 - v + 1}, \tag{9.63}$$

where $\hat{\psi} = \mathbf{a}'\mathbf{T}$, is the GLS estimate (or BLUE, i.e., $\hat{\psi} = \mathbf{c}'\hat{\tau}_0 = \mathbf{c}'(X_2'W_2^{-1}X_2)^{-1} (X_2'W_2^{-1}\mathbf{T})$ $= \mathbf{a}'\mathbf{T}$, say) of ψ from (9.55) and $\mathrm{Var}(\hat{\psi}) = \mathbf{a}'\mathrm{Dis}(\mathbf{T})\mathbf{a} = \sigma_1^2(\mathbf{a}'W_2\mathbf{a})$. Then $\mathrm{Var}(\hat{\psi})$ has an UE $\hat{\sigma}_{\hat{\psi}}^2 = s^2 b$ say, where s^2 as in above of (9.59), and $b = (\mathbf{a}'W_2\mathbf{a})$. This gives directly the following.

Theorem 9.26 *Under the model (9.52), the probability that the following random interval*

$$\hat{\psi} - t_{\alpha/2;\ N_1 - v + 1}\ \hat{\sigma}_{\hat{\psi}} \leq \psi \leq \hat{\psi} + t_{\alpha/2;\ N_1 - v + 1}\ \hat{\sigma}_{\hat{\psi}}, \tag{9.64}$$

covers the unknown ψ is $(1 - \alpha)$.

Note that the interval (9.64) could also be derived from the fact that $(\hat{\psi} - \psi)/\hat{\sigma}_{\hat{\psi}}$ is $t_{N_1 - v + 1}$.

9.3.6.5   Multiple comparison of treatment contrasts with CAES

If the null hypothesis H_0 (or H_{0E}) (as stated in Section 9.3.6.3) is rejected for testing the equality of v treatment means in the original model (9.52) or (9.53), or transformed model (9.54) or (9.55) (model (9.54) or (9.55) is a one-way layout with known correlated error model), we are interested for judging all the treatment contrasts.

Let L be a set of q independent linearly estimable parametric functions $\{\psi_i, i = 1, 2, \ldots, q$ where $\psi_i = \sum_{j=1}^{v} \lambda_{ij} \tau_j\}$, where L is called a q-dimensional space of linearly estimable parametric functions (or contrasts) of v treatments. Then every ψ in L is of the form $\psi = \sum_{i=1}^{q} h_i \psi_i$, where $h_1, h_2, \ldots, h_q$ are known constant coefficients. That is, L is the set of all linear combinations of $\psi_1, \psi_2, \ldots, \psi_q$. Consider a q-dimensional space L of linearly estimable parametric functions (or contrasts) of v treatments generated by a set of q independent linearly estimable parametric functions $\{\psi_1, \psi_2, \ldots, \psi_q\}$. For $\psi \in L$, let $\hat{\psi} = \mathbf{a}'\mathbf{T}$ be its GLS estimate (or BLUE) from the model (9.55). $\mathrm{Var}(\hat{\psi}) = \sigma_{\hat{\psi}}^2$ and its estimate are given in (9.63). Then the proposed method of Multiple Comparison is based on the following theorems.

Theorem 9.27 *Under the model (9.52), the probability is $(1 - \alpha)$ that simultaneously for all $\psi \in L$*

$$\hat{\psi} - D_2\ \hat{\sigma}_{\hat{\psi}} \leq \psi \leq \hat{\psi} + D_2\ \hat{\sigma}_{\hat{\psi}}, \tag{9.65}$$

where the constant D_2 is $D_2 = (q\ F_{\alpha;\ q,\ N_1 - v + 1})^{1/2}$.

Proof. The theorem can be easily proved following Roy and Bose's (1953, p. 517) results (or in the same line as *S*-Method of Multiple Comparison Technique (see Scheffe, 1959, pp. 68-70), considering the ellipsoid as in (9.62)).

For judging *all* the contrasts of v treatment parameters, under the model (9.52), the proposed method is thus immediate from Theorem 9.27 to the following.

Theorem 9.28 *Under the model (9.52), the probability is $(1 - \alpha)$ that the values of all the contrasts of v treatments $\tau_1, \tau_2, \ldots, \tau_v$ simultaneously satisfy the inequalities*

$$\hat{\psi} - D_3\ \hat{\sigma}_{\hat{\psi}} \leq \psi \leq \hat{\psi} + D_3\ \hat{\sigma}_{\hat{\psi}}, \tag{9.66}$$

TABLE 9.12: Specific random layouts of a randomized block design with CAES. Kim, J., Das, R.N., Sengupta, A., and Paul, J. (2009). "Regression Analysis for Correlated Data under Compound Symmetry Structure," Journal of Statistical Theory and Applications, Vol. 8, No. 3, pp. 269-282.

Block	Treatments					
B_1	T_5	T_5	T_2	T_1	T_3	T_4
B_2	T_2	T_1	T_4	T_5	T_5	T_3
B_3	T_3	T_1	T_5	T_5	T_2	T_4
B_4	T_3	T_4	T_2	T_1	T_5	T_5

where the constant D_3 is $D_3 = \{(v-1)F_{\alpha;\ (v-1),\ N_1-v+1}\}^{1/2}$.

Proof is directly followed from Theorem 9.27 as a corollary, noting that dimension of the contrasts space is $(v-1)$.

In Section 9.3.6.3, it has been derived that $\hat{\tau}_0 \sim \text{MN}(\tau_0,\ \sigma_1^2(X_2'W_2^{-1}X_2)^{-1})$, where $(X_2'W_2^{-1}X_2)^{-1}$ is a known variance covariance matrix (for known ρ). Again, marginal distribution of any set of $\hat{\tau}_0$'s are also multivariate normal distribution whose dispersion matrix can be obtained from the original dispersion matrix with a suitable partitioning. In the compound autocorrelated error structure, design matrix X_2 depends on the layout of the treatments, as the position of placing the treatment is very important in this context. Therefore, $\text{Var}(\hat{\tau}_{0i})$'s are different and also $\text{Cov}(\hat{\tau}_{0i}, \hat{\tau}_{0j})$'s are different due to the design matrix X_2. Thus, it does not satisfy Tukey's method of multiple comparison technique.

9.3.6.6   Illustration of a reinforced RBD with CAES

In the present section, analysis of a reinforced RBD under the compound autocorrelated error structure (9.53) is illustrated with simulated observations. Let us consider an RBD with 5 treatments, 4 blocks and each block contains 6 plots. Only the 5th treatment occurs twice in each block, and every other treatment occurs once and only once in each block. The observations are generated by simulation. Observations of the column vector **Y** is generated in block wise from the specific random layouts (as in Section 9.3.6.1) of a randomized block design as given in Table 9.12.

Model is

$$y_{ijk} = \mu + \beta_i + \tau_j + e_{ijk}; \quad 1 \le j \le 4,\ 1 \le i \le 4,$$

$$y_{i5u} = \mu + \beta_i + \tau_5 + e_{i5u};\ 1 \le u \le 2,\ 1 \le i \le 4, \tag{9.67}$$

where y_{ijk} is the yield due to the jth treatment from kth plot of ith block, and also y_{i5u} is the yield due to the 5th treatment from uth plot of ith block. Note that $k = 1$ for all the treatments, except the 5th. Model matrix (X_2) for the specific random layouts (in Table 9.12) is given below

$$X_2 = \begin{pmatrix} D_1^* \\ D_2^* \\ D_3^* \\ D_4^* \end{pmatrix}, \quad \text{where } D_1^* = \begin{pmatrix} 0 & 1 & 0 & 0 \\ 1 & -\rho & 0 & 0 \\ -\rho & 0 & 1 & 0 \\ 0 & 0 & -\rho & 1 \end{pmatrix}, \quad D_2^* = \begin{pmatrix} 1 & -\rho & 0 & 0 \\ -\rho & 0 & 0 & 1 \\ 0 & 0 & 0 & -\rho \\ 0 & 0 & 1 & 0 \end{pmatrix},$$

$$D_3^* = \begin{pmatrix} 1 & 0 & -\rho & 0 \\ -\rho & 0 & 0 & 0 \\ 0 & 1 & 0 & 0 \\ 0 & -\rho & 0 & 1 \end{pmatrix}, \quad D_4^* = \begin{pmatrix} 0 & 0 & -\rho & 1 \\ 0 & 1 & 0 & -\rho \\ 1 & -\rho & 0 & 0 \\ -\rho & 0 & 0 & 0 \end{pmatrix}.$$

Note that the design matrix (X_2) is derived as in Section 9.3.6.3, but for computing

TABLE 9.13: Imputed values and average estimated values from 100 replications of an RBD with CAES for Set I. Kim, J., Das, R.N., Sengupta, A., and Paul, J. (2009). "Regression Analysis for Correlated Data under Compound Symmetry Structure," Journal of Statistical Theory and Applications, Vol. 8, No. 3, pp. 269-282.

Imputed values	Estimated values
$\tau_{01} = 7.15$	$\hat{\bar{\tau}}_{01} = 7.1350247$
$\tau_{02} = 0.10$	$\hat{\bar{\tau}}_{02} = 0.0926543$
$\tau_{03} = 7.25$	$\hat{\bar{\tau}}_{03} = 7.1895637$
$\tau_{04} = 3.75$	$\hat{\bar{\tau}}_{04} = 3.6632156$
$\rho = 0.99$	$\hat{\bar{\rho}} = 0.9887000$
$\sigma^2 = 1.69$	$\hat{\bar{\sigma}}^2 = 1.8352747$

t_{ijk}'s, we always subtract one of the z_{ijk} which is coming from the repeated treatment. For example, we transform $z_{ijk} = y_{ijk} - \rho y_{ij_1(k-1)}$; $2 \le k \le 6$ to compute z_{ijk}'s for the ith block, $i = 1,2,3,4$. To compute t_{ijk}'s, we transform $t_{1jk} = z_{1jk} - z_{152}$; $3 \le k \le 6$ for Block 1, for Block 2, we transform $t_{2jk} = z_{2jk} - z_{255}$, $2 \le k \le 6$; $k \ne 5$, and so on.

Set I: In the absence of real experimental data, we generate observations (100 replications) according to formula (9.67) and following the specific random layouts as in Table 9.12, with $\mu = 30$, $\beta_1 = 2.5$, $\beta_2 = -3.0$, $\beta_3 = -2.5$, $\beta_4 = 3.0$, $\tau_1 = 3.5$, $\tau_2 = -3.55$, $\tau_3 = 3.6$, $\tau_4 = 0.1$, $\tau_5 = -3.65$, $\rho = 0.99$, $\rho_1 = 0.10$ and $\sigma^2 = 1.69$. Average estimates and test results from 100 replications are given in Table 9.13 and Table 9.14, respectively.

Set II: In the absence of real experimental data, we generate observations (100 replications) according to formula (9.67) and following the specific random layouts as in Table 9.12, with $\mu = 20$, $\beta_1 = 3.2$, $\beta_2 = -2.0$, $\beta_3 = -3.2$, $\beta_4 = 2.0$, $\tau_1 = 2.5$, $\tau_2 = -2.6$, $\tau_3 = 2.55$, $\tau_4 = 0.05$, $\tau_5 = -2.5$, $\rho = 0.90$, $\rho_1 = 0.20$ and $\sigma^2 = 2.25$. Average estimates and test results from 100 replications are given in Table 9.15 and Table 9.16, respectively.

Thus, the estimated (leading parameters) unknown elementary treatment contrasts, the correlation coefficient ρ and σ^2 are very close to the true values with the help of which the data have been simulated. Also the test results reflect the true relationships among the treatment parameters. This concludes that simulation results clearly support the theories that are developed in the above.

TABLE 9.14: Test results from 100 replications with $\alpha = 0.05$ of an RBD with CAES for Set I. Kim, J., Das, R.N., Sengupta, A., and Paul, J. (2009). "Regression Analysis for Correlated Data under Compound Symmetry Structure," Journal of Statistical Theory and Applications, Vol. 8, No. 3, pp. 269-282.

Null hypothesis	Accepted cases	Rejection cases
$H_{0E} : \tau_{01} = ... = \tau_{04} = 0$	0	100
$H_{012} : \tau_1 - \tau_2 = 0$	0	100
$H_{013} : \tau_1 - \tau_3 = 0$	95	05
$H_{014} : \tau_1 - \tau_4 = 0$	0	100
$H_{015} : \tau_1 - \tau_5 = 0$	0	100
$H_{023} : \tau_2 - \tau_3 = 0$	0	100
$H_{024} : \tau_2 - \tau_4 = 0$	0	100
$H_{025} : \tau_2 - \tau_5 = 0$	93	07
$H_{034} : \tau_3 - \tau_4 = 0$	0	100
$H_{035} : \tau_3 - \tau_5 = 0$	0	100
$H_{045} : \tau_4 - \tau_5 = 0$	0	100

TABLE 9.15: Imputed values and average estimated values from 100 replications of an RBD with CAES for Set II. Kim, J., Das, R.N., Sengupta, A., and Paul, J. (2009). "Regression Analysis for Correlated Data under Compound Symmetry Structure," Journal of Statistical Theory and Applications, Vol. 8, No. 3, pp. 269-282.

Imputed values	Estimated values
$\tau_{01} = 5.0$	$\hat{\bar{\tau}}_{01} = 4.9118325$
$\tau_{02} = -0.10$	$\hat{\bar{\tau}}_{02} = -0.0916249$
$\tau_{03} = 5.05$	$\hat{\bar{\tau}}_{03} = 5.0089017$
$\tau_{04} = 2.55$	$\hat{\bar{\tau}}_{04} = 2.5208364$
$\rho = 0.90$	$\hat{\bar{\rho}} = 0.8975000$
$\sigma^2 = 2.25$	$\hat{\bar{\sigma}}^2 = 2.226257$

TABLE 9.16: Test results from 100 replications with $\alpha = 0.05$ of an RBD with CAES for Set II. Kim, J., Das, R.N., Sengupta, A., and Paul, J. (2009). "Regression Analysis for Correlated Data under Compound Symmetry Structure," Journal of Statistical Theory and Applications, Vol. 8, No. 3, pp. 269-282.

Null hypothesis	Accepted cases	Rejected cases
$H_{0E} : \tau_{01} = ... = \tau_{04} = 0$	0	100
$H_{012} : \tau_1 - \tau_2 = 0$	0	100
$H_{013} : \tau_1 - \tau_3 = 0$	95	05
$H_{014} : \tau_1 - \tau_4 = 0$	0	100
$H_{015} : \tau_1 - \tau_5 = 0$	0	100
$H_{023} : \tau_2 - \tau_3 = 0$	0	100
$H_{024} : \tau_2 - \tau_4 = 0$	0	100
$H_{025} : \tau_2 - \tau_5 = 0$	96	04
$H_{034} : \tau_3 - \tau_4 = 0$	0	100
$H_{035} : \tau_3 - \tau_5 = 0$	0	100
$H_{045} : \tau_4 - \tau_5 = 0$	0	100

9.4 CONCLUDING REMARKS

In the present chapter, we have considered the regression analysis with correlated errors under the compound symmetry and compound autocorrelated structures. Similar technique can be used for inter-class and intra-class error structures, as these two correlation structures are some particular cases of compound symmetry and compound autocorrelated structures. The best linear estimates of all the regression coefficients have been derived *except* the intercept. Simulation studies show that the estimated values are very close to the imputed values in both the cases. Hypothesis testing and index of fit procedures reflect the original situations of the models (from simulation study).

Here we have considered the compound symmetry and compound autocorrelated error structures, which are the most common phenomenon of errors in an RBD. However, we have repeated only one treatment twice in each block, and every other treatment occurs once and only once in each block of an RBD (under the compound autocorrelated error structure). For such a reinforced RBD, we have developed a method of analysis, and the method thus developed is free from correlation coefficient ρ_1, and depends only on ρ. Thus, this analysis is *robust* of ρ_1 only. Also the method of estimation of τ_0 is free from correlation coefficient of ρ_1, and $\hat{\boldsymbol{\tau}}_0$ is the BLUE of $\boldsymbol{\tau}_0$. All the results are derived (Section 9.3.6.3) based on the estimates of ρ, $\boldsymbol{\tau}_0$ and σ^2 only.

For an RBD under the compound symmetry and a reinforced RBD under the compound autocorrelated error structure, it is seen that the leading estimates are very close to the imputed values with the help of which the data have been simulated (Sections 9.2.7.6 and 9.3.6.6). For an RBD under the CSES, hypothesis testing, parameters estimation, confidence ellipsoid, confidence interval, multiple comparison based on Scheffe's and Tukey's methods have been derived free from both the correlation coefficients ρ and ρ_1. All the results related to a reinforced RBD under the CAES depend on ρ but free from ρ_1. Throughout the whole chapter, regression technique is used based on linear models. The same regression technique is used for analysis, hypotheses testing, unknown parameters estimation, deriving confidence ellipsoids and confidence intervals, and also for multiple comparison technique.

Note that for a reinforced RBD under the compound autocorrelated error structure, the estimates $\hat{\boldsymbol{\tau}}_0$, $\hat{\sigma}_1^2$ and $\hat{\rho}$ are used in everywhere for deriving all the results, whereas the other estimates ($\hat{\mu}$, $\hat{\boldsymbol{\beta}}$ and $\hat{\rho}_1$) are redundant. Estimation of $\boldsymbol{\tau}_0$, σ_1^2 and ρ have been done according to Kmenta (1971), and the estimates thus obtained have many desirable properties (Kmenta, 1971; Chatterjee and Price, 2000). Estimation of μ, $\boldsymbol{\beta}$ and ρ_1 (which are unimportant in the text) can be done following Kmenta (1971). They can also be estimated using the restricted maximum likelihood (REML) method (Pawitan, 2001; Lee *et al.*, 2006) which is not discussed here. Note that the compound autocorrelated structure is *not* a non-singular matrix for *all* possible values of ρ and ρ_1. Even though this structure is singular for many values of ρ and ρ_1, but the present approach easily performs the analysis of an RBD under the compound autocorrelated error structure. The present approach estimates the leading parameters ρ, $\boldsymbol{\tau}_0$ and σ^2 in any situation, but may miss the estimate of ρ_1 in some odd situations, which is unimportant in the text. It is observed that for $\rho > 0$, compound autocorrelated structure is non-singular for many negative and all positive values of ρ_1. In practice, autocorrelation coefficient ρ is positive in most of the cases (Kmenta, 1971; Chatterjee and Price, 2000). Thus, the present approach is an appropriate method for the analysis of an RBD under the compound autocorrelated error structure.

Chapter 10

POSITIVE DATA ANALYSES VIA LOG-NORMAL AND GAMMA MODELS

Generally, experimental responses are positive which may be analysed either by log-normal or gamma models. Some continuous positive process variables may have non-normal error distributions, and the class of generalized linear models includes distributions useful for the analysis of such process characteristic data. In regression models with multiplicative error, estimation is often based on either the log-normal or the gamma model. It is well known that the gamma model with constant coefficient of variation and the log-normal model with constant variance give almost *same* analysis. However, in the analysis of data from quality improvement experiments neither the coefficient of variation nor the variance needs to be constant, so that the two models do not necessarily give similar results. A choice needs to be made between the gamma and the log-normal models. The present chapter focuses the discrepancies of the regression estimates between the two models based on real examples. It is identified that even though the variance or the coefficient of variation remains constant, but regression estimates may be *different* between the two models. For non-constant variance, regression estimates may be different between the two models. It is also identified that for the *same* positive data set, the variance is constant under the log-normal model but non-constant under the gamma model. For this data set, the regression estimates are completely *different* between the two models. It is shown that for non-constant variance, even though the measures of fitting criteria and estimates are almost same in both the models, but the fittings may *not* always be identical. Moreover, this chapter points that some *insignificant* effects may also be sometimes very important in fitting. In the process, the causes of discrepancies have been explained between the two models. Some examples illustrate these points. A few applications of both the log-normal and the gamma models are illustrated with real examples in the different fields of science.

10.1   INTRODUCTION AND OVERVIEW

In regression models for positive observations analysis can often be based on either the log-normal or the gamma model. There is a well known correspondence between multiplicative regression models and additive models of their logarithms (Firth, 1988). Sources in literature indicate that for positive observations, an analysis assuming a log-normal distribution and an analysis assuming a gamma distribution usually give the same results (Atkinson, 1982b; Firth, 1988; McCullagh and Nelder, 1989, p. 289 and p. 293; Myers *et al.*, 2002). Recently, log-normal and gamma models are of interest in fitting data arising from quality improvement experiments (Myers *et al.*, 2002). Some continuous positive process characteristics in practice have non-normal error distributions, and the class of generalized linear models includes distributions useful for the analysis of such data. The simplest examples are perhaps the exponential and gamma distributions, which are often useful for modeling positive data (McCullagh and Nelder, 1989). The classical linear models assume that the variance of the response (Y) is constant over the entire range of parameter values.

Log-normal (LN) and gamma (GA) distributions are widely applied in lifetime distributions. Special attention has been given to discriminate some specific lifetime distribution functions. The problem of testing whether some given observations follow one of the two probability distributions is quite old in literature (Atkinson, 1969, 1970b; Chen, 1980; Cox, 1961, 1962). Firth (1988) and Wiens (1999) discussed the problems of discriminating between the log-normal and the gamma distributions. Atkinson (1982b) and Firth (1988) pointed that if variance is non-constant, regression estimates may be different between the two models. Recently, Das and Lee (2009) have shown that for non-constant variance, regression estimates may be quite different between the two models. Das (2012) has shown the discrepancy in fitting between the log-normal and gamma model fits, for non-constant variance, even though the measures of fitting criteria and regression estimates are almost *identical*, based on a real example. Das and Park (2012) have shown the discrepancies of the regression estimates between the log-normal and gamma models based on real examples for constant variance.

These two models may provide similar data fit for moderate sample sizes (Firth, 1988). Therefore, practitioners in general, select any one of these two models for analysing skewed positive data. But it is still desirable to choose the correct or nearly correct order model, because the inference based on particular model will often involve tail probabilities where the effect of the model assumption will be more crucial. Therefore, even if large sample sizes are not available, it is very important to make the best possible decision based on the given observations. So far many authors have discussed about the discrepancy of regression coefficient estimates between the log-normal and the gamma model fits (Atkinson, 1982b; Firth, 1988; Wiens, 1999; Dick, 2004; Brynjarsdottir and Stefansson, 2004; Kundu *et al.*, 2005; Das and Lee, 2009; Das and Park, 2012, Das, 2012). Recently, log-normal and gamma model fittings are widely used in quality engineering (Das and Lee, 2008, 2009, 2010; Das 2011a, 2011d, 2013a), Medical science (Das, 2011b, 2011c, 2013b; Das and Mukhopadhya, 2009; Das and Sarkar, 2012), Demography (Das and Dihidar, 2013; Das, Dihidar and Verdugo, 2011), Hydrogeology (Das and Kim, 2012), Fisheries (Dick, 2004; Brynjarsdottir and Stefansson, 2004) etc.

The present chapter describes the discrepancies of the regression estimates between the log-normal and the gamma models based on real examples for constant and non-constant variances (following Das, 2012; Das and Lee, 2009; Das and Park, 2012). For both these models, two real examples have been provided with constant variance having discrepancy in fittings. For one real example (described in Section 10.2), variance is constant under the log-normal distribution assumption but non-constant under the gamma distribution. It is shown that there is discrepancy in fitting between the log-normal and the gamma model fits, for non-constant variance, even though the measures of fitting criteria and regression estimates are almost *identical*, based on a real example. In the process, it is shown that some *insignificant* effects may be sometimes very important for fitting the model. Some applications of both the model fittings are illustrated in quality engineering, demography, physical science and medical science.

10.2   DISCREPANCY IN REGRESSION ESTIMATES BETWEEN LOG-NORMAL AND GAMMA MODELS FOR CONSTANT VARIANCE

In the present section, log-normal and gamma models are described for constant and non-constant variances. This section focuses the discrepancies of the regression estimates (for constant variance) between the two models based on real examples (following Das and Park, 2012). Here it is described that even though the variance or the coefficient of variation remains constant, but regression estimates may be different between the two models. It is described that for the *same* positive data set, the variance is constant under the log-normal model but non-constant under the gamma model. For this data set, the regression estimates are completely *different* between the two models. The causes of discrepancies are explained between the two models.

10.2.1   Log-normal and gamma models for constant variance

Log-normal (with constant variance) and gamma (with constant coefficient of variation) models are studied by different authors (Atkinson, 1982b; Firth, 1988; McCullagh and Nelder, 1989; Myers *et al.*, 2002; Wiens, 1999; Park, 1999). For non-constant variance, these two models are studied by Das and Lee (2009, 2010) and Das (2012). Generally, regression analysis of positive observations can often be based on either the log-normal or the gamma model (Firth, 1988). There is a well known correspondence between multiplicative regression models and additive models of their logarithms. In classical linear models, it is assumed that the variance of the response (Y) is constant over the entire range of parameter values. For constant coefficient of variation (i.e., the standard deviation increases proportionally with the mean), we have

$$\text{Var}(Y) = \sigma^2 \mu_Y^2,$$

where σ is the coefficient of variation of Y and $\mu_Y = E(Y)$. In generalized linear models (GLMs; McCullagh and Nelder, 1989) the gamma model satisfies the above mean and variance relationship. For small σ, the variance-stabilizing transformation, $Z = \log(Y)$, has approximate moments

$$E(Z) = \log \mu_Y - \sigma^2/2 \quad \text{and} \quad \text{Var}(Z) \simeq \sigma^2.$$

If the systematic part of the model is multiplicative on the original scale, and hence additive on the log scale, then

$$Y_i = \mu_{Y_i} \epsilon_i \quad (i = 1, 2, \ldots, n) \tag{10.1}$$

with

$$\eta_i = \log \mu_{Y_i} = \mathbf{x}_i^t \boldsymbol{\beta} = \beta_0 + x_{i1}\beta_1 + \ldots + x_{ip}\beta_p$$

and $\{\epsilon_i\}$'s are independent identically distributed (IID) with $E(\epsilon_i) = 1$. In GLMs μ_{Y_i} is the scale parameter and $\text{Var}(\epsilon_i) = \sigma^2$ is the shape parameter. Then

$$Z_i = \log Y_i = \mu_{Z_i} + \delta_i \quad (i = 1, 2, \ldots, n) \tag{10.2}$$

with

$$\mu_{Z_i} = [\beta_0 + E\{\log(\epsilon_i)\}] + x_{i1}\beta_1 + \ldots + x_{ip}\beta_p$$

and $\{\delta_i = \log \epsilon_i - E\{\log(\epsilon_i)\}\}$'s are IID with $E(\delta_i) = 0$.

Conversely, if Y_i follows a log-normal distribution, i.e. $Z_i \backsim N(\mu_{Z_i}, \sigma^2)$ then

$$\mu_{Y_i} = E(\exp Z_i) = \exp(\mu_{Z_i} + \sigma^2/2) \neq \exp(\mu_{Z_i}).$$

Thus, with the exception for the intercept term, the remaining parameters $\beta_1, \beta_2, \ldots, \beta_p$ can be estimated either from the constant coefficient of variation model (10.1) or linear model for the transformation of the original data to log scale (10.2). The intercept parameters in the models (10.1) and (10.2) are not the same, but they will often be unimportant in practice (see discussion in Myers et al., 2002; p. 169).

Firth (1988) presented a comparison of the efficiencies of the maximum-likelihood (ML) estimators from the gamma model (the constant coefficient of variation model) when the errors are in fact log-normal with those of the log-normal model when the errors have a gamma distribution. The researcher also concluded that the ML estimators from the gamma model perform slightly better under reciprocal misspecification. For small σ^2 it is likely to be difficult to discriminate between normal-theory linear models for $\log Y$ and gamma-theory multiplicative models for Y.

10.2.2   Log-normal and gamma models for non-constant variance

When the standard deviation does not increase proportionally with the mean (i.e., σ^2 is not a constant), i.e., ϵ_i is not identically distributed with a common $E(\log \epsilon_i)$, parameter estimates from one model may have no interpretation on the other model. For analysis of data from quality-improvement experiments the aim is to minimize the variance using covariates while controlling the mean to the target. Thus, in the analysis of data from quality-improvement experiments, σ^2 is often not a constant. For these situations, Lee and Nelder (1998, 2003) proposed to use joint GLMs (JGLMs) to allow for structured dispersions. A detailed discussion on JGLMs is given in Lee, Nelder and Pawitan (2006) (see also Qu, Tan and Rybicky, 2000; Lesperance and Park, 2003; and Das and Lee, 2008, 2009, 2010).

Consider a JGLM for the multiplicative model (10.1)

$$E(Y_i) = \mu_{Y_i} \quad \text{and} \quad \text{Var}(Y_i) = \sigma_{Y_i}^2 \mu_{Y_i}^2,$$

where

$$\eta_i = \log(\mu_{Y_i}) = \mathbf{x}_i^t \boldsymbol{\beta}_Y \quad \text{and} \quad \xi_i = \log(\sigma_{Y_i}^2) = \mathbf{g}_i^t \boldsymbol{\gamma}_Y, \tag{10.3}$$

where $\mathbf{x}_i$ and $\mathbf{g}_i$ are respectively the row vectors of the model matrix used in the mean and the dispersion models. The ML estimators for $\boldsymbol{\beta}_Y$ are obtained by maximizing the log-likelihood

$$\ell_g(\theta) = \sum_i \log f(y_i, \theta) \tag{10.4}$$

and restricted ML (REML) estimators for $\boldsymbol{\gamma}_Y$ are estimated by maximizing the (log) adjusted profile likelihood (Cox and Reid, 1987; Lee and Nelder, 1998; Pawitan, 2001)

$$p_{\beta_Y}(\ell_g(\theta)) = \{\ell_g(\theta) - \{\log \det(E(-\partial \ell_g^2(\theta)/\partial \beta_Y^2)/2\pi)\}/2\}|_{\beta_Y = \hat{\beta}_Y}. \tag{10.5}$$

The whole estimation process is done iteratively by using two interconnected iterative weighted least squares (Lee et al., 2006).

Consider a JGLM for the log normal model (10.2)

$$E(Z_i) = \mu_{Z_i} \quad \text{and} \quad \text{Var}(Z_i) = \sigma_{Z_i}^2,$$

where

$$\mu_{Z_i} = \mathbf{x}_i^t \boldsymbol{\beta}_Z \quad \text{and} \quad \xi_i = \log(\sigma_{Z_i}^2) = \mathbf{g}_i^t \boldsymbol{\gamma}_Z. \tag{10.6}$$

The ML estimators for $\boldsymbol{\beta}_Z$ under the log-normal model are obtained by maximizing the log-likelihood

$$\ell_l(\theta) = \sum_i \{\log f(z_i, \theta) - z_i\} \qquad (10.7)$$

and REML estimators for $\boldsymbol{\gamma}_Z$ are estimated by maximizing the adjusted profile likelihood

$$p_{\beta_Z}(\ell_l(\theta)) = \{\ell_l(\theta) - \{\log \det(E(-\partial\ell_l^2(\theta)/\partial\beta_Z^2)/2\pi)\}/2\}|_{\beta_Z = \hat{\beta}_\gamma}, \qquad (10.8)$$

where $-z_i = \log|dz_i/dy_i| = -\log y_i$ is logarithm of Jacobian of the transformation.

To select a better model, generally, following three information criteria are considered:

  Akaike information criterion (AIC) $= -L_n(\theta) + 2p_0$,

  Bayesian information criterion (BIC) $= -L_n(\theta) + p_0\ln(n)$,

  Hannan-Quinn information criterion (HQIC) $= -L_n(\theta) + 2p_0\ln(\ln(n))$,

where n is the sample size, $L_n(\theta) = \ell_g(\theta)$ (10.4) and p_0 (the number of parameters in the fitted model) $= p_g$ say, or $L_n(\theta) = \ell_l(\theta)$ (10.7) and $p_0 = p_l$ say, according as the fitted model is gamma or log-normal. A model is selected with a smaller information criterion value (Akaike, 1977; Hannan and Quinn, 1979; Shwarz, 1978). For recent works on the performance evaluation of these criteria, see (Egrioglu and Gunay, 2010; Hacker and Hatemi-J, 2008; Wang and Bessler, 2009), for example.

If one needs to compare models with the same number of parameters $(p_g = p_l)$, only the maximized likelihood values ($\ell_g(\theta)$ and $\ell_l(\theta)$) to be compared. To compare models with different scales of response variables (y for the gamma model and $\log y$ for the log-normal model) we need the Jacobian term in (10.7).

Below we describe the discrepancy of regression estimates between the log-normal (with constant variance) and the gamma model (with constant coefficient of variance) fittings (Das and Park, 2012). In Section 10.3 and Section 10.4, we describe respectively the discrepancy of regression estimates (Das and Lee, 2009) and the discrepancy in fittings (Das, 2012) between the log-normal and the gamma models for non-constant variance, based on real examples.

10.2.3 Motivating example

Firth (1988) concludes that the estimates of regression parameters of the gamma model (10.1) and the log-normal model (10.2) give almost same results. In this section we consider an example (Das and Park, 2012), where the estimates of the two models are similar, but the model selection criteria are *not* same. However, in Section 10.2.4, we present two examples (Das and Park, 2012), where the two models give quite *different* results, so that the model selection is very important between the log-normal model (with constant variance) and the gamma model (with constant coefficient of variation).

Example 10.1 Roman catapult data: Luner (1994) presented an experimental setup of data for the Roman catapult. The aim of the catapult experiment is to deliver a projectile to a designated target (80 inches) with a high degree of accuracy and precision. A second-order central composite design with three factors is used for the catapult experiment. Factors are arm length (x_1), stop angle (x_2), and pivot height (x_3). The model matrix, factors and the responses are given in Luner (1994, pp. 695-698). For ready reference, experimental layout and resulting responses (outcomes) are reproduced in Appendix (Table Ap.1) (taken from Luner, 1994, p. 695-698).

TABLE 10.1: Results for mean models (10.1) and (10.2) of catapult data from log-normal and gamma fits. Das, R.N. and Park, J.S. (2012). "Discrepancy in regression estimates between Log-normal and Gamma: Some case studies," Journal of Applied Statistics, Vol. 39, No.1, pp. 97-111

| | Covar. | log-normal model | | | | gamma model | | | |
		estimate	s.e.	t	P-value	estimate	s.e.	t	P-value
Mean	Const.	4.29	0.03	167.57	<0.01	4.31	0.02	173.91	<0.01
model	x_1	0.19	0.03	6.02	<0.01	0.19	0.03	6.27	<0.01
	x_2	0.02	0.03	0.62	0.54	0.02	0.03	0.85	0.40
	x_3	0.25	0.03	8.03	<0.01	0.25	0.03	8.36	<0.01
	$x_2 x_3$	−0.08	0.04	−2.01	0.05	−0.08	0.04	−1.91	0.06
Const. Var model	Const.	−3.24	0.19	−16.96	<0.01	−3.31	0.19	−17.35	<0.01
AIC		486.2				482.6			

We analyse the catapult data using the constant coefficient of variation model (i.e., gamma model (10.1)), and linear model for the transformation of the original data to log scale (i.e., log-normal model (10.2)). The selected models have the smallest Akaike information criterion (AIC) value in each class, as AIC selects a model which minimizes the predicted additive errors and the squared error loss (Hastie et al., 2001, p. 203-204). We retain some insignificant effects in the model in order to respect the marginality rule, namely that when an interaction term is significant all related lower-order interactions and main effects should be included in the model (Nelder, 1994). We fit both of the models ((10.1) and (10.2)), and the results are reported in Table 10.1.

In Figures 10.1(a) and 10.1(b), we plot respectively the absolute residuals with respect to the fitted values for the log-normal and gamma models in Table 10.1. If the transformation and gamma model with constant coefficient of variation are satisfactory, they should have a

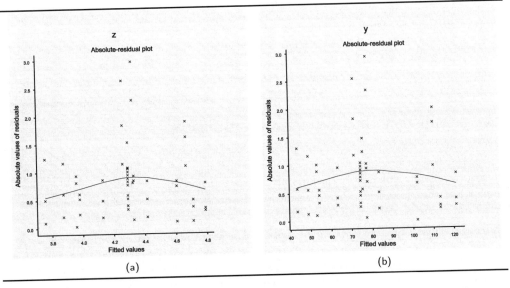

FIGURE 10.1: The absolute residuals plots with respect to fitted values for (a) LN and (b) GA fits (Table 10.1) of catapult data. Das, R.N. and Park, J.S. (2012). "Discrepancy in regression estimates between Log-normal and Gamma: Some case studies," Journal of Applied Statistics, Vol. 39, No.1, pp. 97-111

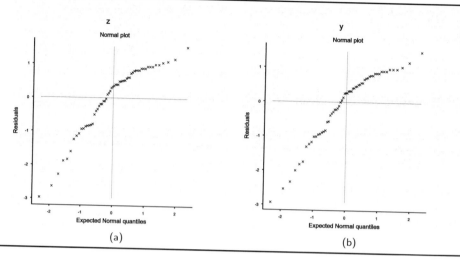

FIGURE 10.2: The normal probability plots for mean model (Table 10.1) of catapult data for (a) LN and (b) GA fits. Das, R.N. and Park, J.S. (2012). "Discrepancy in regression estimates between Log-normal and Gamma: Some case studies," Journal of Applied Statistics, Vol. 39, No.1, pp. 97-111

flat running means. Here both the plots show flat running means, an indication that variance is constant, so that the simple log transformation or gamma model with constant coefficient of variation is sufficient to remove heteroscedasticity. In addition, both Figures 10.1(a) and 10.1(b) show that the original data set contains three heterogeneous groups of observations in magnitude. But the fitted residuals seem to be randomly distributed within each group. In Figures 10.2(a) and 10.2(b), we plot respectively the normal probability plots (vertical axis presents the ordered response values and horizontal axis presents the log-normal (for fitted log-normal model) or the gamma (for fitted gamma model) order statistics i.e., medians or means) for the mean model of log-normal and gamma fits in Table 10.1. Figures 10.2(a) and 10.2(b) do not show any systematic departures, indicating no lack of fit of the final selected models (in Table 10.1). Even though the estimates and their standard errors (in Table 10.1) are identical in both the models, satisfying Firth's (1988) conjecture, the AIC (in Table 10.1) shows that gamma model gives a better fit, which is also clear from the plots, as there is no gap in Figure 10.2(b) but there is a gap in Figure 10.2(a).

10.2.4   Examples of different regression estimates

Here we present two examples where the estimates of regression parameters between the gamma (with constant coefficient of variation) and the log-normal (with constant variance) models are completely different with different interpretations (Das and Park, 2012). These two analyses show that the model selection is very important between these two models with constant variance.

Example 10.2 Lifetime data of a cutting tool: Watkins *et al.* (1994) discussed an experiment to reduce the tooling costs of cutting threads on pipes to make electrical conduit. The cutting tool is called a chaser and the angle at which it approaches the pipe is called the rake angle. The experimental factors are chaser type (A), coolant type (B), speed (C), conduit manufacturer (D), and rake angle (E). Two levels for each factor were investigated using a 2^{5-1} design with $E = ABCD$. The response is the time until poor thread quality is observed or the product has to be scrapped. The threading machine had two spindles (left and right denoted by L and R); each spindle can handle a pipe. The design matrix

TABLE 10.2: Results for mean models (10.1) and (10.2) of threading machine lifetime data from log-normal and gamma fits. Das, R.N. and Park, J.S. (2012). "Discrepancy in regression estimates between Log-normal and Gamma: Some case studies," Journal of Applied Statistics, Vol. 39, No.1, pp. 97-111

	Covar.	log-normal model				gamma model			
		estimate	s.e.	t	P-value	estimate	s.e.	t	P-value
Mean	Const.	4.10	0.11	37.46	<0.01	4.37	0.11	39.73	<0.01
model	A	−0.84	0.11	−7.66	<0.01	−0.84	0.11	−7.55	<0.01
	B	0.48	0.11	4.32	<0.01	0.47	0.11	4.24	<0.01
	C	0.44	0.11	4.01	<0.01	0.42	0.11	3.83	<0.01
	D	−0.18	0.11	−1.66	0.10	−0.14	0.11	−1.28	0.21
	E	0.04	0.11	0.38	0.71	0.07	0.11	0.59	0.55
	CE	−0.36	0.11	−3.30	<0.01	−0.31	0.11	−2.76	0.01
	DE	0.22	0.11	1.96	0.05	0.18	0.11	1.65	0.10
	AB	−0.19	0.11	−1.71	0.09	−0.17	0.11	−1.58	0.12
Const. Var model	Const.	−0.44	0.20	−2.19	0.03	−0.54	0.20	−2.64	0.01
AIC		599.0				604.4			

and lifetime data are given in Watkins *et al.* (1994). For ready reference, it is reproduced in Appendix (Table Ap.2) (taken from Watkins *et al.*, 1994). Note that no observation was censored, setting number 4 has three runs, and setting number 15 has two runs.

We analyse the threading machine lifetime data using the gamma and the log-normal models. The selected models have the smallest AIC value in each class. We fit both of the models ((10.1) and (10.2)), and the results are displayed in Table 10.2. In Figures 10.3(a) and 10.3(b), we plot respectively the absolute residuals with respect to the fitted values for the log-normal and gamma models in Table 10.2. Here both the plots show flat running means, an indication that the variance is constant under each of the fitted model. In Figures 10.4(a) and 10.4(b), we plot respectively the normal probability plots for the mean model

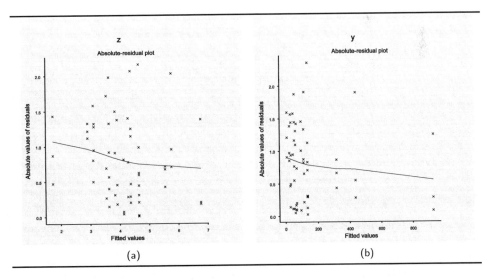

(a)   (b)

FIGURE 10.3: The absolute residuals plot w.r.to fitted values for (a) LN and (b) GA fits (Table 10.2) of threading machine lifetime data. Das, R.N. and Park, J.S. (2012). "Discrepancy in regression estimates between Log-normal and Gamma: Some case studies," Journal of Applied Statistics, Vol. 39, No.1, pp. 97-111

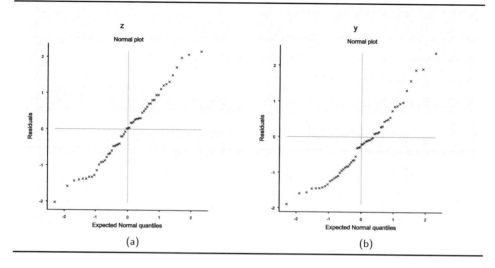

FIGURE 10.4: The normal probability plots for mean model (Table 10.2) of threading machine lifetime data for (a) LN and (b) GA fits. Das, R.N. and Park, J.S. (2012). "Discrepancy in regression estimates between Log-normal and Gamma: Some case studies," Journal of Applied Statistics, Vol. 39, No.1, pp. 97-111

of log-normal and gamma fits in Table 10.2. These two figures do not show any lack of fit of the final selected models in Table 10.2.

From Table 10.2, it is clear that the estimates are *not* same in both the models. Numerically only the estimates A and B are almost same, whereas the estimates C, D, E, CE, DE, and AB are not same in both the models. Note that the estimate DE is statistically *significant* in the log-normal model, but it is *insignificant* in the gamma model. Also note that AB is included in both the models for better fit, whereas the estimate AB is marginally *significant* in the log-normal model, but it is not so in the gamma model. Even though the estimates are *not* same but their standard errors (Table 10.2) are identical in both the models. It is well-known that if the AIC difference between the two models is larger than two, it is considered as significant and less than one is insignificant (Sakamoto *et al.*, 1986, p. 84). Under the AIC (Table 10.2) rule, the log-normal model gives a better fit, which is also clear from the plots, as there are two gaps in Figure 10.4(b), but there is no gap in Figure 10.4(a). Thus, for this example, regression estimates are quite *different* between the two models.

Example 10.3 Hardness of water data: Mumtazuddin *et al.* (2009) have studied physico-chemical parameters of bored tube well water samples at different sites of Muzaffarpur Town, India. The parameters such as pH (x_1), electrical conductivity (EC or x_2), temperature (TEP or x_3), total dissolved solids (TDS or x_4), dissolved oxygen (DO or x_5), total alkalinity (TAK or x_6), total hardness (THD or $y1$), calcium (Ca or x_8), magnesium (Mg or x_9), cloride (CLD or x_{10}) and chemical oxygen demand (COD or x_{11}) have been studied. The water samples were selected in the consecutive months of October, November and December 2008 form six bored tube wells in the selected sites of different regions of Muzaffarpur Town. A detailed method of collection and the data is given in Mumtazuddin *et al.* (2009). For ready reference this data set is reproduced in Appendix (Table Ap.3) (taken from Mumtazuddin *et al.*, 2009).

We analyse the hardness of water data by using the gamma (10.1) and the log-normal (10.2) models and the results are displayed in Table 10.3. It is observed that the variance

TABLE 10.3: Results for mean models ((10.1) and (10.2)) of hardness of water data from log-normal and gamma fits. Das, R.N. and Park, J.S. (2012). "Discrepancy in regression estimates between Log-normal and Gamma: Some case studies," Journal of Applied Statistics, Vol. 39, No.1, pp. 97-111

		log-normal model				gamma model			
	Covar.	estimate	s.e.	t	P-value	estimate	s.e.	t	P-value
Mean	Const.	4.7302	0.14	33.92	<0.01	12.968	3.98	3.25	<0.01
model	x_2	−0.0002	0.00*	−1.94	0.08	0.01	0.00*	3.84	<0.01
	x_3	−0.0050	0.00*	−1.28	0.22	−0.33	0.11	−2.99	0.01
	x_5	−0.0079	0.01	−1.08	0.30	−0.79	0.21	−3.78	<0.01
	x_8	0.0064	0.00*	5.19	<0.01	0.93	0.03	26.79	<0.01
	x_9	0.0132	0.00*	22.14	<0.01	4.11	0.02	199.76	<0.01
Const.	Const.	−6.9050	0.41	−16.91	<0.01	−11.39	0.41	−27.89	<0.01
Var model									
AIC				121.800				41.122	

TABLE 10.4: Results for mean and variance models (10.3) and (10.6) of hardness of water data from log-normal and gamma fits. Das, R.N. and Park, J.S. (2012). "Discrepancy in regression estimates between Log-normal and Gamma: Some case studies," Journal of Applied Statistics, Vol. 39, No.1, pp. 97-111

		log-normal model				gamma model			
	Covar.	estimate	s.e.	t	P-value	estimate	s.e.	t	P-value
Mean	Const.	4.7302	0.14	33.88	<0.01	11.58	2.02	5.74	<0.01
model	x_2	−0.0002	0.00*	−1.94	0.08	0.01	0.00*	6.67	<0.01
	x_3	−0.0051	0.00*	−1.28	0.22	−0.30	0.06	−5.39	<0.01
	x_5	−0.0079	0.01	−1.08	0.30	−0.76	0.11	−7.16	<0.01
	x_8	0.0064	0.00*	5.20	<0.01	0.94	0.02	53.70	<0.01
	x_9	0.0132	0.00*	22.13	<0.01	4.10	0.01	372.17	<0.01
Disper.	Const.	−6.4190	13.92	−0.46	0.65	−46.21	14.87	−3.11	0.01
Var model	x_1	−0.0650	1.86	−0.03	0.97	4.54	1.96	2.31	0.04
AIC				121.800				25.802	

of hardness of water data is non-constant under the gamma model and it is constant under the log-normal model. Thus, the hardness of water data is further analysed using the joint gamma GLMs (10.3) and the joint log-normal GLMs (10.6). Results for JGLMs are displayed in Table 10.4. In Tables 10.3 and 10.4, we use '0.00*' in the standard error column to represent the standard error whose first two decimal places are zeros. From Table 10.4, it is observed that there is only one *significant* factor in the dispersion model of gamma fit but none in log-normal fit. Therefore, the variance of hardness of water is constant under the log-normal model assumptions but it is non-constant under the gamma model (Table 10.4). The selected models have the smallest AIC value in each class and AIC selects the gamma model as it is smaller than the log-normal model.

In Figures 10.5(a) and 10.5(b), we plot respectively the absolute residuals with respect to the fitted values and the normal probability plot for the mean model of gamma fit (models (10.3); Table 10.4). Figure 10.5(a) shows a flat running means, indicating that variance is constant under the structure dispersion joint gamma GLMs (10.3). Figure 10.5(b) does not show any lack of fit. Figures 10.6(a) and 10.6(b) display respectively the absolute residuals plot with respect to the fitted values and the normal probability plot for the mean model of log-normal fit (model (10.2); Table 10.3). Figures 10.6(a) and 10.6(b) do not show any lack of fit, indicating that constant variance log-normal model (10.2) (Table 10.3) is appropriate

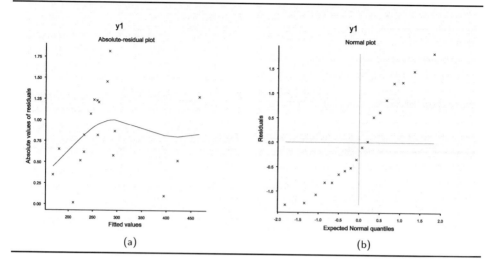

FIGURE 10.5: (a) The absolute residuals plot w.r.to fitted values and (b) normal probability plot for mean model (Table 10.4) of water data for GA fit. Das, R.N. and Park, J.S. (2012). "Discrepancy in regression estimates between Log-normal and Gamma: Some case studies," Journal of Applied Statistics, Vol. 39, No.1, pp. 97-111

for this data set. But the AIC difference between these two models is very high, so the gamma model is more better than the log-normal model. This is also clear from Figure 10.5(b) and Figure 10.6(b), as there are two gaps in Figure 10.6(b) but there is no gap in Figure 10.5(b).

From Table 10.3 and Table 10.4, it is found that the regression estimates are completely different in numerically and statistically between the two models. Many parameters (x_1, x_2,

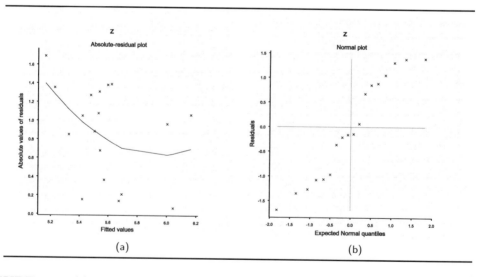

FIGURE 10.6: (a) The absolute residuals plot w.r.to fitted values and (b) normal probability plot for mean model (Table 10.3) of water data for LN fit. Das, R.N. and Park, J.S. (2012). "Discrepancy in regression estimates between Log-normal and Gamma: Some case studies," Journal of Applied Statistics, Vol. 39, No.1, pp. 97-111

x_3 and x_5) are *insignificant* in the log-normal model but *significant* in the gamma model. Thus, the interpretation of parameters between the two models are completely different. This shows that wrong assumption of model selection treats significant factors as insignificant which is a serious error in any data analysis. This leads to wrong decision in every case. For example, in quality improvement experiments, it leads to select unimportant process parameters, resulting poor quality of product. In medical treatment process, it selects wrong treatment process, increasing hazards in human life, etc.

10.2.5   Discussion about discrepancy

Residuals plot: Generally, heteroscedasticity is examined by residuals plot. An increasing nature of the residuals plot indicates heteroscedasticity in the given data set. If the residuals plot shows an outward-opening funnel, indicating that variance is non-constant (Myers *et al.*, 2002, p. 37, Figure 2.5). For a non-constant variance, the residuals plot shows an increasing or decreasing spread with the running means (Das and Lee, 2009, p. 82, Figure 1; p. 84, **Figure 3**). A flat diagram of the residuals plot with the running means **presents** a constant variance in the given data set (Das and Lee, 2009, p. 82, Figure 2; p. 84, Figure 4; Lee *et al.*, 2006). In each of the present cases, the residuals plot is almost flat with the running means, indicating that variance is constant under the respective fitting.

Heteroscedasticity test: Heteroscedasticity tests are known under normality assumptions. For non-normality, heteroscedasticity is generally examined by residuals analysis. Three examples on replicated measures are considered in Section 10.2. In the case of replicated data, the general model is assumed as

$$Y_{ij} = \beta_0 + \beta_1 x_{i1} + \beta_2 x_{i2} + \ldots + \beta_p x_{ip} + u_{ij}; j = 1, 2, \ldots, n_i; \ i = 1, 2, \ldots, m,$$

where Y_{ij} $(j = 1, 2, \ldots, n_i)$ denotes the resultant set of n_i yields for the ith level combination at $\mathbf{x}_i = (x_{i1}, x_{i2}, \ldots, x_{ip})$ of p explanatory variables , and u_{ij}'s are random errors. Denoting the vector of disturbances for the ith level combination at $\mathbf{x}_i$ by $\mathbf{u}_i$, we make the conventional assumptions $E(\mathbf{u}_i) = \mathbf{0}$ and $E(\mathbf{u}_i\mathbf{u}_i') = \sigma_i^2 I_{n_i}$; $i = 1, 2, \ldots, m$, and $\mathbf{u}_i \sim N(\mathbf{0}, \sigma_i^2 I_{n_i})$.

To test the null hypothesis $H_0: \ \sigma_1^2 = \sigma_2^2 = \ldots = \sigma_m^2 = \sigma^2$, say, against the alternative hypothesis H : at least one of σ_i^2 is different from σ^2, generally, the test statistic (Chatterjee and Price, 2000) used is

$$Q_1 = \nu \ln s^2 - \sum_{i=1}^{m} \nu_i \ln s_i^2,$$

where $s_i^2 = \frac{1}{\nu_i} \sum_{j=1}^{n_i} (Y_{ij} - \bar{Y}_i)^2; i = 1, 2, ..., m; \nu_i = n_i - 1, \bar{Y}_i = \sum_{j=1}^{n_i} Y_{ij}/n_i; s^2 = (\sum_{i=1}^{m} \nu_i s_i^2)/\nu; \nu = \sum_{i=1}^{m} (\nu_i - 1)$.

Under the null hypothesis Q_1 will be approximately distributed as χ^2 with d.f. $(m-1)$. Approximation will be improved by dividing Q_1 by the scaling constant C, where $C = 1 + \frac{1}{3(m-1)}(\sum_{i=1}^{m} \frac{1}{\nu_i} - \frac{1}{\nu})$ to give $Q = \frac{Q_1}{C}$. If $Q > \chi^2_{\alpha; (m-1)}$, say, then the null hypothesis H_0 will be rejected at $100\alpha\%$ level of significance, and accept otherwise. Computed values of Q for these three examples are given in Table 10.5. Q values show that variances are non-constant for all these three examples, under normality assumption. Originally, data sets are positive which are non-normal. Thus, the Q values may not reflect the real situations. For the above three examples, we examine the heteroscedasticity under the log-normal and gamma distributions.

Transformation of response variable: Appropriate modeling is the principal part of data analysis. Generally, estimated models are selected using all possible model checking tools. Box (1976, p. 792) pointed that "all models are wrong". Yet, each model provides some information regarding the data set. Log-transformation is often recommended for heteroscedastic data to stabilize the variance. It is selected based on the estimated value of λ in power transformation ($y^{(\lambda)} = \frac{y^\lambda - 1}{\lambda y_0^{\lambda-1}}$ for $\lambda \neq 0$ and $y^{(\lambda)} = y_0 \ln y$ for $\lambda = 0$, where $y_0 = \ln^{-1}[\frac{1}{n}\sum \ln y]$) (Box and Cox, 1964). If the value of $\hat\lambda$ is very close to zero for a given data set, log-transformation is appropriate (Box and Cox, 1964). For Example 10.1 (catapult data), $\hat\lambda$ is 0.65 (approx.), indicating that log-transformation is not appropriate to stabilize the variance. The present analysis shows that variance of catapult data is constant under both the distributions. For Example 10.2 (lifetime data of cutting tool machine), $\hat\lambda$ is 0.08, indicating that log-transformation is appropriate to stabilize the variance. The present analysis shows that variance of lifetime data of cutting tool machine is constant under both the distributions. For Example 10.3 (hardness of water data), $\hat\lambda$ is -0.2, indicating that log-transformation is appropriate to stabilize the variance. The present analysis shows that variance of hardness of water data is *non-constant* under the gamma distribution but *constant* under the log-normal distribution. Here it is observed that the log-transformation may stabilize the variances of Example 10.1 and Example 10.2, but not Example 10.3. Note that the log-transformation may not always stabilize the variance (Myers *et al.*, 2002, p. 36).

Correct selection probability test between two distributions: Kundu *et al.*, (2005) provided the probability of correct selection (PCS) of log-normal versus gamma distributions based on extensive computer simulations. Let $\text{LN}(\sigma, \theta)$ be a log-normal distribution with shape parameter σ and scale parameter θ, and $\text{GA}(\alpha, \delta)$ be a gamma distribution with shape parameter α and scale parameter δ. Let $Y_1, Y_2, \ldots, Y_n$ be n independent identically distributed (iid) random variables from $\text{LN}(\sigma, \theta)$ (with probability density function (pdf) $f_{LN}(y: \sigma, \theta)$) or $\text{GA}(\alpha, \delta)$ (with pdf $f_{GA}(y: \alpha, \delta)$), where

$$f_{LN}(y : \sigma, \theta) = \frac{1}{(2\pi)^{1/2} y \sigma} \exp\left[-\frac{(\ln(\frac{y}{\theta}))^2}{2\sigma^2}\right]; y,\ \theta,\ \sigma > 0,$$

and

$$f_{GA}(y : \alpha, \delta) = \frac{1}{\delta \Gamma \alpha}\left(\frac{y}{\delta}\right)^{\alpha-1} \exp\left[-\frac{y}{\delta}\right]; y, \alpha, \delta > 0.$$

The likelihood functions assuming that the data are coming from $\text{LN}(\sigma, \theta)$ or $\text{GA}(\alpha, \delta)$ are

$$L_{LN}(\sigma, \theta) = \prod_{i=i}^{n} f_{LN}(y : \sigma, \theta) \ \text{ and } \ L_{GA}(\alpha, \delta) = \prod_{i=i}^{n} f_{GA}(y : \alpha, \delta),$$

respectively. The ratio of the maximized likelihoods (RML) is defined as

$$L = \frac{L_{LN}(\hat\sigma, \hat\theta)}{L_{GA}(\hat\alpha, \hat\delta)},$$

where $(\hat\sigma, \hat\theta)$ and $(\hat\alpha, \hat\delta)$ are maximum likelihood estimates of (σ, θ) and (α, δ), respectively, based on the sample observations $Y_1, Y_2, \ldots, Y_n$. The natural logarithm of RML can be written as

$$T = n[\ln(\frac{\Gamma\hat\alpha}{\hat\sigma}) - \hat\alpha \ln(\frac{\bar Y_0}{\hat\delta}) + (\frac{\bar Y}{\hat\delta}) - \frac{1}{2\hat\sigma^2 n}\sum_{i=1}^{n}(\ln(\frac{Y_i}{\theta}))^2 - \frac{1}{2}\ln(2\pi)],$$

where $\bar Y$ and $\bar Y_0$ are arithmetic and geometric means of $Y_1, Y_2, \ldots, Y_n$, respectively. Note that for log-normal distribution $\hat\theta = \bar Y_0$ and $\hat\sigma^2 = \frac{1}{n}\sum_{i=1}^{n}(\ln(\frac{Y_i}{\theta}))^2$, respectively. Also $\hat\alpha$ and $\hat\delta$ satisfy the following relation $\hat\alpha = \frac{\bar Y}{\hat\delta}$. The discrimination rule between log-normal and gamma, suggested by Kundu *et al.*, (2005) is to select the log-normal distribution if $T > 0$, otherwise to select the gamma distribution. Computed T values and the related parameters for three examples (Sections 10.2.3 and 10.2.4) are given in Table 10.5. Therefore, computed

TABLE 10.5: Shape and scale parameters, T values and K-S distances of three examples. Das, R.N. and Park, J.S. (2012). "Discrepancy in regression estimates between Log-normal and Gamma: Some case studies," Journal of Applied Statistics, Vol. 39, No.1, pp. 97-111

Distribution	Parameter	Example 10.1	Example 10.2	Example 10.3
Log-normal	Scale (θ)	73.130855	62.515966	266.124764
	Shape (σ)	0.324169	1.320332	0.2585428
Gamma	Scale (δ)	7.477233	287.679500	22.584160
	Shape (α)	10.286697	0.474940	12.20125
	T	-0.698676	7.203072	0.605699
	$\sup_y |F(y) - G(y)|$	0.022676	0.144589	0.033153
	Q	31.88925	75.18184	5.74163
	n	60	57	18

T value suggests gamma distribution for Example 10.1 and log-normal for Examples 10.2 and 10.3. However, we have found that the T value correctly specifies (based on AIC and graphical analysis) the first two examples but misleads Example 10.3.

Kolmogorov-Smirnov (K-S) distance test: Kolmogorov-Smirnov (K-S) distance (Kundu *et al.*, 2005) measures the closeness or the distance between two distribution functions. Kundu *et al.* (2005) have computed K-S distance between the log-normal and the gamma for different value of sample sizes (n) and shape parameter σ. The K-S distance between two distribution functions, say $F(y)$ (gamma) and $G(y)$ (log-normal) is defined as $\sup_y |F(y) - G(y)|$, and its values for three examples are given in Table 10.5. On the basis of K-S distance (Table 10.5), the differences between the two distributions are *insignificant* for Examples 10.1 and 10.3, and *significant* for Example 10.2. Thus, K-S distance supports our analysis for Examples 10.1 and 10.2, but completely misleads for Example 10.3. Therefore, T value and K-S distance both mislead Example 10.3 due to its small sample size (18).

Regression with non-constant variance: It is well known that the estimates of regression parameters between the log-normal and the gamma models may *not* have a common interpretation for non-constant variance (Das and Lee, 2009) (Section 10.3). Atkinson (1982b) suggests that the two analyses (log-normal and gamma) should provide similar results if σ^2 is as large as 0.6. However, Das and Lee (2009) have shown that these two models give quite different results for the estimate of σ^2, i.e., $\hat{\sigma}^2 = 0.20$ under the log-normal model and $\hat{\sigma}^2 = 0.18$ under the gamma model. Das and Lee (2009) have also shown that under these situations variance is non-constant and thus, the researchers perform structured analysis (Das and Lee, 2009, Example 3). Moreover, they have shown that even $\hat{\sigma}^2 = 0.037$ (Das and Lee, 2009, Example 1) and $\hat{\sigma}^2 = 0.012$ (Das and Lee, 2009, Example 2), variances are non-constant. Evidence (Das and Lee, 2009) shows that variance may be non-constant if σ^2 is smaller than 0.6 under the log-normal and gamma distributions. Recently, Wiens (1999) has also shown that the two models (log-normal and gamma) give different results where the estimate of σ^2 (from the two combined data group) is 0.45. Several authors (Dick, 2004; Brynjarsdottir and Stefansson, 2004) have shown that the two models (log-normal and gamma) give different results when the variance may be non-constant. Das and Lee (2009) have explained the discrepancy of the estimates between the two models for non-constant variance.

In Section 10.2.3, for Example 10.1, the estimate of σ^2 is $\hat{\sigma}^2 = 0.039$ under the log-normal model or $\hat{\sigma}^2 = 0.036$ under the gamma model. Figures 10.1(a) and 10.1(b) show that variance is constant for both the models. Also for Example 10.2, the estimate of σ^2 is $\hat{\sigma}^2 = 0.644$ under the log-normal model or $\hat{\sigma}^2 = 0.583$ under the gamma model. Figures 10.3(a) and 10.3(b) show that variance is constant for both the models. Note that $\hat{\sigma}^2$ ($= 0.644$) for the log-normal model does not meet the Atkinson's (1982b) specification (0.6), even though the variance is constant, which is clear from Figure 10.3(a). For Example 10.3, value of $\hat{\sigma}^2$ is very small under both the log-normal ($\hat{\sigma}^2 = e^{-6.9050}$) and gamma ($\hat{\sigma}^2 = e^{-46.21+4.54x_1}$, where $x_1 \in (7.12, 8.23)$) models. For this example, variance is constant under the log-normal distribution but non-constant under the gamma distribution.

When the AIC difference between the two models is greater than two, the model difference is considered as *significant*, and less than one is *insignificant* (Sakamoto *et al.*, 1986, p. 84). Based on the AIC rule, model differences are *significant* for the above three examples. For the first example, estimates, AIC value and graphical analysis show that the gamma model fit is slightly better than the log-normal. In case of the second example, the estimates are different in both numerically and statistically. The AIC difference between the two models is significant. Also graphical analysis (Figure 10.4) shows that there are two gaps in the gamma fit and no gap in the log-normal fit. Thus, for Example 10.2, estimates, AIC value and graphical analysis show that the log-normal fit is more better than the gamma fit, even though the variance is constant. That is, the log-normal fit is *significantly* different from the gamma fit, and it is more better than the gamma fit. For Example 10.3, analysis shows that the gamma fit is absolutely better (in respect of AIC, graphical analysis and number of active factors) than the log-normal fit, whereas the log-normal fit completely misleads the model.

The present section focuses the discrepancies of the regression estimates between the log-normal (constant variance) and the gamma (constant coefficient of variation) models based on real data analyses. First two examples show that variances are constant under both the models whereas, Example 10.3 shows that variance is constant under the log-normal distribution but non-constant under the gamma distribution. For the first two examples, it is observed that the two models with constant coefficient of variation (gamma) or constant variance (log-normal) still provide similar standard error of estimates, even though the estimates and AIC are quite different. But for Example 10.3, standard error of estimates, estimated values, AIC, and interpretation of parameters are *all* different. Therefore, parameter estimates may be quite different to give quite different conclusions between the two models. In such circumstances, the information criteria and model checking plots are useful to select an appropriate model.

10.3   DISCREPANCY OF REGRESSION PARAMETERS BETWEEN LOG-NORMAL AND GAMMA MODELS FOR NON-CONSTANT VARIANCE

Section 10.2 describes the discrepancy of regression estimates between the log-normal (with constant variance) and the gamma (with constant coefficient of variation) models. In Section 10.2.2, Equ. (10.3) and Equ. (10.6) present respectively the joint gamma and log-normal generalized linear models for non-constant variance. The present section describes the discrepancy of regression estimates between the log-normal and the gamma models for non-constant variance, based on real examples (following Das and Lee, 2009). In the present section, these two models are compared with and without structured dispersion, by using three real examples (Das and Lee, 2009).

Example 10.4 Worsted Yarn Experiment:
Myers *et al.* (2002, p. 36) presented the data from a 3^3 factorial experiment for studying the number of cycles to failure of worsted yarn (y) with 3 factors $A - C$. The model matrix and the number of cycles to failure of worsted yarn (y) are given in Myers *et al.* (2002, p. 36). For ready reference, the design matrix along with the responses are reproduced in

TABLE 10.6: Results for mean models with constant variance of worsted yarn data from log-normal and gamma fits. Das, R.N. and Lee, Y.J. (2009). "Log normal versus gamma models for analyzing data from quality-improvement experiments," Quality Engineering, Vol. 21, No. 1, pp. 79-87

		log-normal model				gamma model			
	Covar.	estimate	s.e.	t	P-value	estimate	s.e.	t	P-value
Mean	Const.	6.33	0.04	169.74	<0.01	6.34	0.04	172.95	<0.01
model	A	0.83	0.05	18.06	<0.01	0.84	0.05	18.62	<0.01
	B	−0.63	0.05	−13.82	<0.01	−0.63	0.05	−14.05	<0.01
	C	−0.39	0.05	−8.43	<0.01	−0.38	0.05	−8.45	<0.01
Const.	Const.	−3.28	0.30	−11.13	<0.01	−3.32	0.30	−11.27	<0.01
Var. model									
AIC			325.80					325.06	

Appendix (Table Ap.4) (taken from Myers *et al.*, 2002). Myers *et al.* (2002) have analysed the data using a second-order response surface design treating factors as covariates (with values $-1, 0, 1$). From the residual plot, they noticed that there is an indication of an opening out in the plot, implying possible non-constant of variance. These researchers used Box-Cox (1964) transformation and found that the log transformation is plausible. Thus, they used $\log y$ as the responses. They have also fitted the gamma model (10.1). We fit both of their final models (10.1) and (10.2) and the results are in Table 10.6, along with AICs; results from the two models are almost the same. However, Myers *et al.* (2002) found that the log model is an improvement on the original quadratic fit, but there is still some indication of non-constant variance.

In Figures 10.7(a) and 10.7(b), we plot respectively the absolute residuals of log-normal and gamma fits in Table 10.6 with respect to the fitted values. If the transformation or gamma model with constant variance are satisfactory they should have a flat running means.

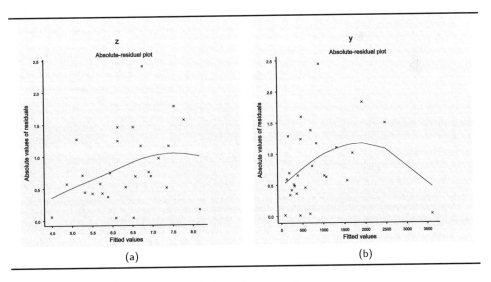

FIGURE 10.7: The absolute residuals plots w.r.to fitted values for const. var. models (Table 10.6) of worsted yarn data for (a) LN and (b) GA fits. Das, R.N. and Lee, Y.J. (2009). "Log normal versus gamma models for analyzing data from quality-improvement experiments," Quality Engineering, Vol. 21, No. 1, pp. 79-87

TABLE 10.7: Results for mean and dispersion models of worsted yarn data from log-normal and gamma fits. Das, R.N. and Lee, Y.J. (2009). "Log normal versus gamma models for analyzing data from quality-improvement experiments," Quality Engineering, Vol. 21, No. 1, pp. 79-87

	Covar.	log-normal model				gamma model			
		estimate	s.e.	t	P-value	estimate	s.e.	t	P-value
Mean	Const.	6.34	0.04	174.87	<0.01	6.35	0.04	178.55	<0.01
model	A	0.85	0.04	19.96	<0.01	0.86	0.04	20.51	<0.01
	B	−0.62	0.04	−16.74	<0.01	−0.62	0.04	−16.90	<0.01
	C	−0.36	0.04	−9.70	<0.01	−0.35	0.04	−9.67	<0.01
Dispers.	Const.	−3.52	0.30	−11.91	<0.01	−3.55	0.30	−12.02	<0.01
model	A	0.78	0.34	2.32	0.03	0.76	0.33	2.27	0.03
AIC		319.50				318.88			

However, the plots show an increasing running means, an indication that the variance is still non-constant, so that the simple log transformation may not be sufficient (Lee *et al.* 2006). Thus, we fit the JGLMs and the results are reported in Table 10.7. Absolute residuals plots for both the models (Table 10.7) in Figures 10.8(a) and 10.8(b) show that the undesirable increasing running means are removed. The decreasing pattern of the right tail of the fitted curves in Figures 10.7(b) and 10.8(b) are due to one small-valued observation at the boundary. We found in Table 10.7 that there is one significant effect A in the dispersion model. Under the AIC rule, we cannot make a choice between the two models in Table 10.7. In consequence, both the models (Table 10.7) give almost same results. The selected models have the smallest AIC value in each class.

This experiment aims to maximize the mean (the number of cycles to failure), while minimizing the variance. Following Tang and Xu (2002) we set the dispersion effects to minimize the variance and then set remaining mean effects to maximize the mean. Under

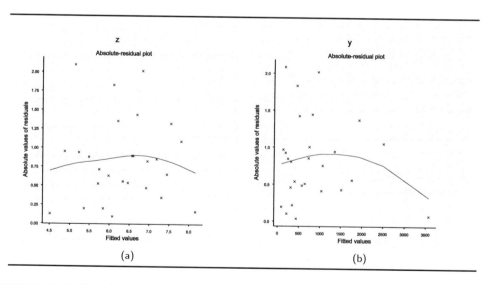

(a)　　　　　　　　　　　　(b)

FIGURE 10.8: The absolute residuals plots w.r.to fitted values for non-const. var. models (Table 10.7) of worsted yarn data for (a) LN and (b) GA fits. Das, R.N. and Lee, Y.J. (2009). "Log normal versus gamma models for analyzing data from quality-improvement experiments," Quality Engineering, Vol. 21, No. 1, pp. 79-87

TABLE 10.8: Results for mean with constant variance models of resistivity data from log-normal and gamma fits. Das, R.N. and Lee, Y.J. (2009). "Log normal versus gamma models for analyzing data from quality-improvement experiments," Quality Engineering, Vol. 21, No. 1, pp. 79-87

		log-normal model				gamma model			
	Covar.	estimate	s.e.	t	P-value	estimate	s.e.	t	P-value
Mean	Const.	5.41	0.03	201.49	<0.01	5.42	0.03	202.51	<0.01
model	A	0.06	0.03	2.29	0.04	0.06	0.03	2.26	0.05
	B	−0.15	0.03	−5.58	<0.01	−0.15	0.03	−5.60	<0.01
	C	0.09	0.03	3.32	0.01	0.09	0.03	3.41	0.01
Const. Var. model	Const.	−4.46	0.41	−10.93	<0.01	−4.47	0.41	−10.95	<0.01
AIC		143.20				143.12			

the two models without structured dispersion in Table 10.6, we select the value A as large as possible and the values of B and C as small as possible to maximize the mean. Under the models with structured dispersion in Table 10.7 we select the value A as small as possible to minimize the variance and the values of B and C as small as possible to maximize the mean. We do not use the value of A in the mean model to adjust the mean because it will affect the variance. By using the effects significant in the mean model, but not significant in the dispersion model, we can maximize the mean without affecting the variance.

Example 10.5 The Resistivity Experiment:

Myers *et al.* (2002, p. 176) presented a data set from an unreplicated factorial experiment with 4 factors $A - D$, each with 2 levels and the experiment was run at a certain stage in a semiconductor manufacturing process for studying the resistivity (y) of the test wafer. Following Myers *et al.* (2002), let us consider the factors as covariates (with values −1 and 1). The model matrix and the response resistivity (y) of the test wafer are given in Myers *et al.* (2002, p. 176). For ready reference, the design matrix along with the responses are reproduced in Appendix (Table Ap.5) (taken from Myers *et al.* 2002). Initially they have analysed the data using the log-transformation of the responses, and found that only three main effects $A - C$ are significant. These researchers have also analysed the data using the gamma GLMs with log link. The results from their final models (10.1) and (10.2), along with the AICs, are almost same as in Table 10.8.

In Figures 10.9(a) and 10.9(b), we plot, respectively, the absolute residuals of log-normal and gamma fits in Table 10.8 against the fitted values. Again both the plots show an increasing running means, an indication that the variance is still increasing with the running means, so that the log transformation and the gamma model with constant coefficient of variation are not satisfactory. Thus, we fit the models with a structured dispersion. Absolute residual plots for both the models ((10.3) and (10.6)) in Figures 10.10(a) and 10.10(b) show that increasing running means are removed by allowing the non-constant variance. We found that there are four significant effects $A - D$ in the dispersion model. Results for the JGLMs ((10.3) and (10.6); both mean and dispersion models) along with the corresponding AICs are given in Table 10.9.

Both the models ((10.3) and (10.6)) give an identical result for the mean, but slightly different results for the dispersion. Resistivity is well known to have a distribution with a heavy right tail, and thus Myers *et al.* (2002, p. 176) stated that a gamma distribution may indeed be appropriate. However, with structured dispersion, the AIC is marginally significant, choosing the log-normal model. From the normal probability plots for mean

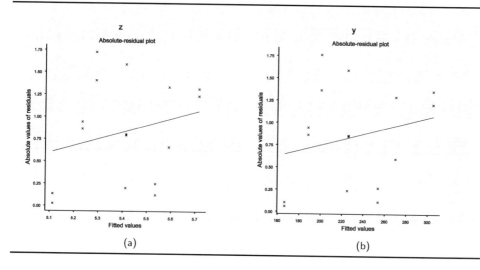

(a)   (b)

FIGURE 10.9: The absolute residuals plots w.r.to fitted values for const. var. models (Table 10.8) of resistivity data for (a) LN and (b) GA fits. Das, R.N. and Lee, Y.J. (2009). "Log normal versus gamma models for analyzing data from quality-improvement experiments," Quality Engineering, Vol. 21, No. 1, pp. 79-87

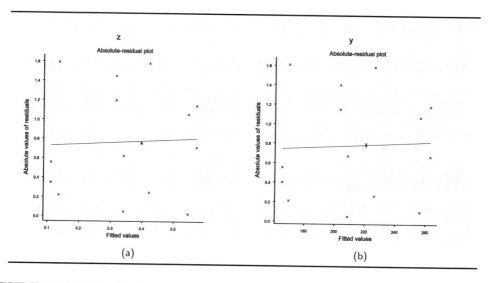

(a)   (b)

FIGURE 10.10: The absolute residuals plots w.r.to fitted values for non-const. var. models (Table 10.9) of resistivity data for (a) LN and (b) GA fits. Das, R.N. and Lee, Y.J. (2009). "Log normal versus gamma models for analyzing data from quality-improvement experiments," Quality Engineering, Vol. 21, No. 1, pp. 79-87

models (Figures 10.11(a) and 10.11(b)), the log-normal model would be slightly better because the gamma model has a gap. Further studies of model choices are of interest when structured dispersion is allowed.

This experiment aims to maximize the mean (resistivity of the test wafer), while minimizing the variance. From the two models without structured dispersion in Table 10.9, we select the values A and C as large as possible and the value of B as small as possible to maximize the mean. For both the models with structured dispersion in Table 10.10, we

TABLE 10.9: Results for mean and dispersion models of resistivity data from log-normal and gamma fits. Das, R.N. and Lee, Y.J. (2009). "Log normal versus gamma models for analyzing data from quality-improvement experiments," Quality Engineering, Vol. 21, No. 1, pp. 79-87

		log-normal model				gamma model			
	Covar.	estimate	s.e.	t	P-value	estimate	s.e.	t	P-value
Mean	Const.	5.35	0.01	912.90	<0.01	5.35	0.01	894.60	<0.01
model	A	0.09	0.01	16.10	<0.01	0.09	0.01	16.00	<0.01
	B	−0.13	0.01	−23.40	<0.01	−0.13	0.01	−23.30	<0.01
	C	0.01	0.00∗	6.10	<0.01	0.01	0.00∗	5.40	<0.01
	A.B	0.01	0.01	2.50	0.05	0.01	0.01	2.50	0.05
Dispers.	Const.	−7.21	0.57	−12.62	<0.01	−7.15	0.58	−12.38	<0.01
model	A	−1.67	0.53	−3.13	0.02	−1.64	0.53	−3.08	0.02
	B	−1.89	0.54	−3.47	0.01	−1.83	0.54	−3.39	0.01
	C	1.91	0.55	3.45	0.01	1.94	0.54	3.57	0.01
	D	−2.09	0.66	−3.19	0.02	−2.04	0.65	−3.12	0.02
AIC		98.30				99.64			

select the values A, B and D as large as possible and the value C as small as possible, to minimize the variance. In this experiment all the mean effects (A, B, C) cannot change the mean without affecting the variance.

Example 10.6 Wiper Switch Experiment:

Wu and Hamada (2000, p. 561) presented an experimental data set for studying the initial voltage drop of four different window wiper switches across multiple contacts. The aim of the experiment is to improve the design of a window wiper switch. An orthogonal design is used in this experiment to study one four-level factor (A) and four two-level factors

%begincenter

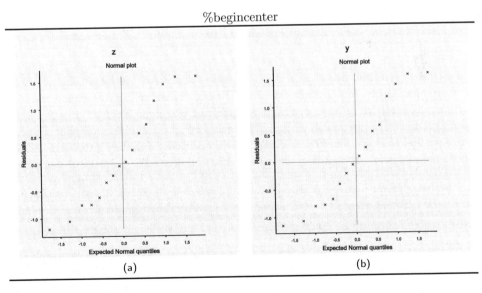

(a) (b)

FIGURE 10.11: The normal probability plots for mean models (Table 10.9) of resistivity data for (a) LN and (b) GA fits. Das, R.N. and Lee, Y.J. (2009). "Log normal versus gamma models for analyzing data from quality-improvement experiments," Quality Engineering, Vol. 21, No. 1, pp. 79-87

TABLE 10.10: Results for mean with constant variance models of voltage drop data for wiper switch experiment from log-normal and gamma fits. Das, R.N. and Lee, Y.J. (2009). "Log normal versus gamma models for analyzing data from quality-improvement experiments," Quality Engineering, Vol. 21, No. 1, pp. 79-87

		log-normal model				gamma model			
	Factor	estimate	s.e.	t	P-value	estimate	s.e.	t	P-value
Mean	Const.	4.18	0.07	62.90	<0.01	4.32	0.06	67.90	<0.01
model	$A2$	0.07	0.07	0.95	0.34	0.04	0.07	0.58	0.56
	$A3$	−0.13	0.07	−1.76	0.08	−0.14	0.07	−2.00	0.05
	$A4$	0.08	0.07	1.06	0.29	0.03	0.07	0.41	0.68
	$B2$	−0.01	0.05	−0.21	0.83	−0.10	0.05	−1.98	0.05
	$D2$	0.13	0.05	2.57	0.01	0.13	0.05	2.73	0.01
	$E2$	0.27	0.05	5.44	<0.01	0.30	0.05	6.14	<0.01
Const.	Const.	−1.60	0.08	−20.02	<0.01	−1.72	0.08	−21.47	<0.01
Var. model									
AIC			3192.00				3174.17		

$(B - E)$. Four different switches are used for each of the eight runs of the orthogonal array. For each switch, the positive response, the initial voltage drop across multiple contacts, is recorded (i.e., first inspection) and then recorded every 20,000 cycles thereafter up to 180,000 cycles, resulting in 10 inspections. For ready reference, the model matrix along with the responses are reproduced in Appendix (Table Ap.6) (taken from Wu and Hamada, 2000, p. 561).

For factors, the constraint that the effects of the first levels are zero is accepted. Therefore, it is taken that the first level of each factor as the reference level by estimating it as zero. Suppose that α_i for $i = 1, 2, 3$ represents the main effect of A. It is taken $\hat{\alpha}_1 = 0$, so that $\hat{\alpha}_2 = \hat{\alpha}_2 - \hat{\alpha}_1$. For example, the estimate of the effect $A2$ means the effect of the difference between the second and the first levels of the main effect A, i.e., $\hat{\alpha}_2 - \hat{\alpha}_1$.

The results for the models (10.1) and (10.2), along with the AICs, are given in Table 10.10. In the first two examples (Examples 10.4 and 10.5), $\hat{\sigma}^2$ is small; in the first example (Example 10.4) $\hat{\sigma}^2$ is 0.038 under the log-normal model and 0.037 under the gamma model and in the second example (Example 10.5) 0.012 under the log-normal model and 0.012 under the gamma model. In such cases analyses from the two models are similar even under the structured dispersion models. Here $\hat{\sigma}^2$ is larger than those in the two previous examples (0.20 under the log-normal model and 0.18 under the gamma model) and both the models give somewhat different results. For the present example, the gamma model is better than the log-normal model under the AIC rule. We also fit the JGLMs with a structured dispersion model. Results for the JGLMs ((10.3) and (10.6)) (both mean and dispersion models) along with the corresponding AICs are given in Table 10.11. In the mean models, the effect $B2$ is significant under the gamma model, but insignificant under the log-normal model, and in the dispersion models, the effect $A2$ is insignificant under the gamma model but significant under the log-normal model. The AIC shows that gamma model gives a better fit.

This experiment aims to maximize the mean (the time to initial voltage drop), while minimizing the variance. For the final gamma model without structured dispersion in Table 10.10, we set the factors (B, D, E) to the levels $(1, 2, 2)$ to maximize the mean. We may set the level of the factor A to be any one among $(1,2,4)$ because the effect is not significant within these levels. But for the final gamma model with structured dispersion we set (A, B, C) to the levels $(4, 2, 1)$ to minimize the variance and set (D, E) to the levels $(2, 2)$ to maximize the mean.

TABLE 10.11: Results for mean and dispersion models of voltage drop data for wiper switch experiment from log-normal and gamma fits. Das, R.N. and Lee, Y.J. (2009). "Log normal versus gamma models for analyzing data from quality-improvement experiments," Quality Engineering, Vol. 21, No. 1, pp. 79-87

		log-normal model				gamma model			
	Factor	estimate	s.e.	t	P-value	estimate	s.e.	t	P-value
Mean	Const.	4.20	0.08	48.67	<0.01	4.35	0.08	54.14	<0.01
model	A2	0.06	0.07	0.82	0.41	0.03	0.07	0.46	0.65
	A3	−0.16	0.07	−2.23	0.03	−0.18	0.07	−2.59	0.01
	A4	0.05	0.07	0.70	0.48	-0.01	0.07	−0.11	0.91
	B2	−0.02	0.05	−0.32	0.75	−0.10	0.05	−2.16	0.03
	D2	0.13	0.05	2.53	0.01	0.13	0.05	2.71	0.01
	E2	0.26	0.05	5.30	<0.01	0.28	0.05	5.93	<0.01
Dispers.	Const.	−0.93	0.21	−4.38	<0.01	−1.13	0.21	−5.29	<0.01
model	A2	−0.48	0.23	−2.08	0.04	−0.42	0.23	−1.81	0.07
	A3	−0.42	0.23	−1.83	0.07	−0.34	0.23	−1.49	0.14
	A4	−0.90	0.23	−3.97	<0.01	−0.84	0.23	−3.68	<0.01
	B2	−1.26	0.16	−7.77	<0.01	−1.12	0.16	−6.92	<0.01
	C2	0.39	0.16	2.40	0.02	0.37	0.16	2.28	0.02
AIC		3122.00				3113.49			

The log-transformation is often recommended for heteroscedastic data. If σ^2 is constant, i.e., ϵ_i is identically distributed with a common $E(\log \epsilon_i)$, parameters from the log-normal and the gamma models have a common interpretation. It is observed that the analysis of data from quality-improvement experiments that the simple log-transformation may not be sufficient, so that a further structured dispersion model is required; this results in different optimal settings. Furthermore, with structured dispersion there is no reason that the two models will give parameterizations with a common interpretation. In all the tables (Tables 10.6 to 10.11) it is found that the standard error of the estimates from these two models are very similar, regardless of the presence of structured dispersions. However, parameter estimates can be sufficiently different as to give different conclusions. In such circumstances the AICs and the model checking plots are useful in selecting a better model. As log-normal and gamma models become popular for the analysis of data from quality improvement experiments, further studies about the model choice are of interest.

Even when the gamma and the log-normal models give different results, the optimal settings from these two models are often the same. However, the optimal settings can be dramatically changed with a structured dispersion. Thus, a proper modeling of a structured dispersion is important for the analysis of data from quality-improvement experiment, giving the optimal setting of the process parameters.

10.4 DISCREPANCY IN FITTING BETWEEN LOG-NORMAL AND GAMMA MODELS

The discrepancy of regression estimates between the log-normal and the gamma model fits (Atkinson, 1982b; Firth, 1988; Wiens, 1999; Dick, 2004; Brynjarsdottir and Stefansson, 2004; Kundu et al., 2005; Das and Lee, 2009; Das and Park, 2012) has been discussed in Sections 10.2 and 10.3. Little is known in literature about the discrepancy in fitting between these two model fits. The present section focuses the discrepancy in fitting between the log-normal and the gamma model fits, for non-constant variance, even though the measure of fitting criteria and regression estimates are almost *identical*, based on a real example (following Das, 2012). In the process, it is shown that some *insignificant* effects may be sometimes very important for fitting a model. An example of silica gel data illustrates this point below.

Example 10.7 Silica gel data: Anand (1997) conducted an orthogonal experiment with four factors each at three levels for improving the yield of silica gel. Silica gel is manufactured from sodium silicate and hydrochloric acid, and a detailed manufacturing process is given in Anand (1997). Four factors are sodium silicate density (A), pH value of sodium silicate and HCL solution (B), setting time (hr.)–solidification time (C), and drying temperature (^0C) in hot-air chamber (D). Thirty kilograms of sodium silicate were processed for each experimental combination, as per layout. Finished material was tested for density and sieved into three groups: (1) crystal size greater than 2 mm (first-grade material); (2) crystal size less than or equal to 2 mm (second-grade material); (3) powder form (scrap material). Responses considered during the experimentation were as follows: (1) quantity of material (Kg) scrapped, expressed as percentage of total quantity processed ($y1$); (2) quantity of material of second-grade, expressed as percentage (y); (3) density of silica gel ($y2$). The experimental layout along with the responses are reproduced in Appendix (Table Ap.7) (taken from Anand, 1997).

In Section 10.2, log-normal and gamma models are described for constant and non-constant variances. In the present section, we analyse the above silica gel data using the joint generalized linear log-normal and gamma models as in Section 10.2.2. Here we analyse the second grade (y) silica gel data, using the JGLMs, and the results for the JGLMs ((10.3) and (10.6)) along with the AICs are given in Table 10.12. The selected models have the smallest AIC value in each class. According to the AIC rule, we cannot make a model choice between the two models in Table 10.12. In consequence, both the gamma and log-normal models ((10.3) and (10.6)) give almost the same results.

For the fitted gamma models in Table 10.12, we plot respectively the histogram of residuals, the absolute residuals plot with respect to the fitted values, the normal probability plots for the mean and the variance models in Figures 10.12(a), 10.12(b), 10.13(a) and 10.13(b). The absolute residuals plot Figure 10.12(b) is a flat diagram with the running means. It indicates that the variance is constant under this fitting. But the remaining Figures 10.12(a), 10.13(a) and 10.13(b) show lack of variable fit for this fitted models. Similarly, for the fitted log-normal models in Table 10.12, we plot respectively the histogram of residuals, the absolute residuals plot with respect to the fitted values, the normal probability plots for

TABLE 10.12: Results for mean and dispersion models ((10.3) and (10.6)) of second grade silica gel data from log-normal and gamma fits. Das, R.N. (2012). "Discrepancy in fitting between log-normal and gamma models: An illustration," Model Assisted Statistics and Applications, Vol 7, No. 1, pp. 23-32

	Covar.	log-normal model				gamma model			
		estimate	s.e.	t	P-value	estimate	s.e.	t	P-value
Mean	Const.	3.61	0.20	17.87	<0.01	3.69	0.20	18.20	<0.01
model	A	−0.23	0.07	−3.17	<0.01	−0.23	0.07	−3.22	<0.01
	B	−0.21	0.05	−4.61	<0.01	−0.24	0.05	−5.08	<0.01
	C	−0.17	0.07	−2.37	0.03	−0.17	0.07	−2.41	0.02
	AC	0.07	0.03	2.13	0.04	0.07	0.03	2.17	0.04
Dispers.	Const.	−0.27	1.19	−0.22	0.82	−0.19	1.13	−0.17	0.87
model	B	−1.65	0.62	−2.67	0.01	−1.68	0.59	−2.88	0.01
AIC			116.90				117.57		

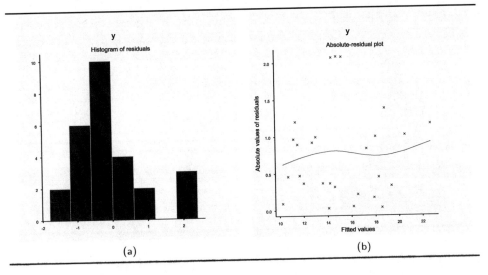

FIGURE 10.12: (a) Histogram plot and (b) the absolute residuals plot w.r.to fitted values for gamma fit (Table 10.12) of second grade silica gel data

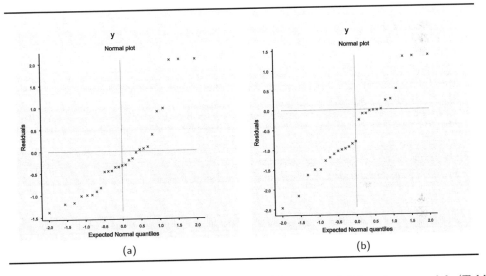

FIGURE 10.13: The normal probability plots for (a) mean and (b) variance models (Table 10.12) of second grade silica gel data for gamma fit

the mean and the variance models in Figures 10.14(a), 10.14(b), 10.15(a) and 10.15(b). But the *exactly* same scenario as in the gamma models (Table 10.12) is observed in case of the log-normal models (Table 10.12). Note that all the effects (Table 10.12) are *significant* in both the fitted models. But there does not exist any significant effect which will be sufficient to explain the lack of fit in both the fitted models given in Table 10.12.

To explain the lack of fit in the models in Table 10.12, according to Hastie *et al.* (2001), we add some *insignificant* effects, and the final models in Table 10.13, overall, are an improvement on the original fits in Table 10.12, and the lack of fit has been removed in the final selected gamma models (Table 10.13). Under the AIC rule, we cannot make a model choice between the two models in Table 10.13. But from graphical analysis, the gamma models fit is much better than the log-normal models fit, which is explained below.

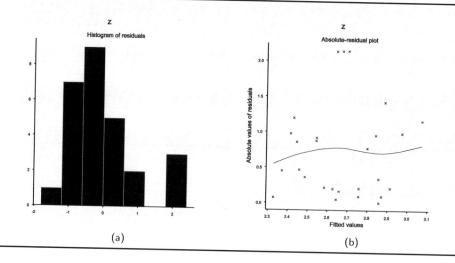

(a)   (b)

FIGURE 10.14: (a) Histogram plot and (b) the absolute residuals plot w.r.to fitted values for log-normal fit (Table 10.12) of second grade silica gel data

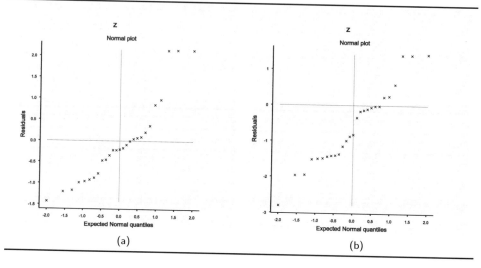

(a)   (b)

FIGURE 10.15: The normal probability plots for (a) mean and (b) variance models of second grade silica gel data for log-normal fit (Table 10.12)

For the fitted log-normal models in Table 10.13, we plot respectively the histogram of residuals, the absolute residuals plot with respect to the fitted values, the normal probability plots for the mean and the variance models in Figures 10.16(a), 10.16(b), 10.17(a) and 10.17(b). Again the same scenario as in the gamma or the log-normal model fits in Table 10.12 is observed in case of the log-normal models fit in Table 10.13.

Similarly, for the fitted gamma models in Table 10.13, we plot respectively the histogram of residuals, the absolute residuals plot with respect to the fitted values, the normal probability plots for the mean and the variance models in Figures 10.18(a), 10.18(b), 10.19(a) and 10.19(b). The absolute residuals plot Figure 10.18(b) presents a flat diagram with the running means. It implies that heteroscedasticity has been removed under the structured

TABLE 10.13: Results for mean and dispersion models ((10.3) and (10.6)) of second-grade silica gel data from log-normal and gamma fits. Das, R.N. (2012). "Discrepancy in fitting between log-normal and gamma models: An illustration," Model Assisted Statistics and Applications, Vol 7, No. 1, pp. 23-32

		log-normal model				gamma model			
	Covar.	estimate	s.e.	t	P-value	estimate	s.e.	t	P-value
Mean	Const.	4.30	0.47	9.13	<0.01	4.78	0.50	9.50	<0.01
model	A	−0.53	0.18	−2.93	0.01	−0.70	0.19	−3.72	<0.01
	B	−0.44	0.15	−2.81	0.01	−0.59	0.17	−3.54	<0.01
	C	−0.19	0.08	−2.48	0.02	−0.21	0.08	−2.59	0.02
	AB	0.10	0.06	1.64	0.12	0.15	0.06	2.40	0.02
	AC	0.08	0.03	2.55	0.02	0.09	0.03	2.83	0.01
Dispers.	Const.	2.65	3.83	0.69	0.50	3.16	3.36	0.94	0.35
model	A	−1.41	1.74	−0.81	0.43	−1.60	1.51	−1.06	0.30
	B	−2.43	1.54	−1.58	0.13	−2.45	1.42	−1.72	0.10
	D	0.06	0.41	0.15	0.88	0.07	0.41	0.18	0.86
	AB	0.33	0.69	0.47	0.64	0.29	0.64	0.46	0.65
AIC		115.30				114.98			

gamma models fit (Table 10.13). Figure 10.18(a) (histogram) does not show any lack of variable fit. Figure 10.19(a) and Figure 10.19(b) do not show any systematic departures, indicating no lack of fit of the fitted gamma models in Table 10.13. So, the final selected models (mean and variance) of the response second grade silica gel are the gamma models as in Table 10.13. Note that the AIC (Table 10.13) shows that the gamma and the log-normal model fits are identical, *but* the graphical analysis shows that the log-normal models fit presents lack of fit, while the lack of fit is disappeared for the fitted gamma models (Table 10.13). Two models in Table 10.13 are almost identical, except AB which is *significant* in the mean model of gamma fit but *insignificant* in the log-normal fit.

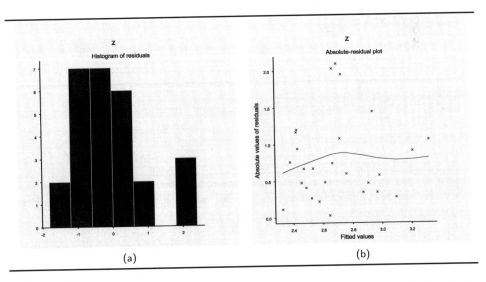

FIGURE 10.16: (a) Histogram plot and (b) the absolute residuals plot w.r.to fitted values for log-normal fit (Table 10.13) of second grade silica gel data

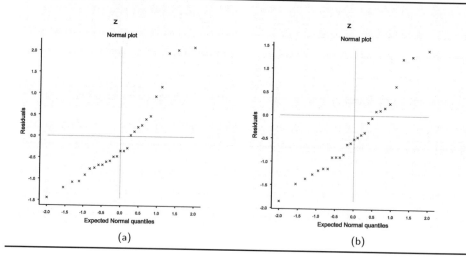

FIGURE 10.17: The normal probability plots for (a) mean and (b) variance models of second grade silica gel data for log-normal fit (Table 10.13)

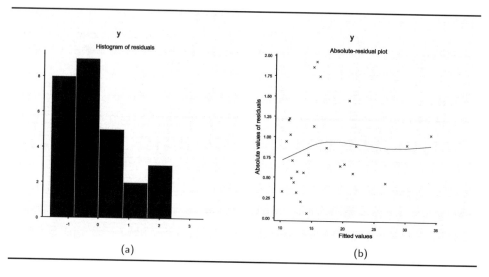

FIGURE 10.18: (a) Histogram plot and (b) the absolute residuals plot w.r.to fitted values for gamma fit (Table 10.13) of second-grade silica gel data

Kolmogorov-Smirnov (K-S) distance (Kundu *et al.*, 2005) measures the closeness or the distance between two distribution functions. The K-S distance between two distribution functions, say $F(y)$ (gamma) and $G(y)$ (log-normal) is defined as $\sup_y |F(y) - G(y)|$, and its value for the response second grade (y) silica gel is 0.0151 (Table 10.14). K-S distance (Table 10.14) shows the difference between the two distributions is *insignificant*. Thus, K-S distance does not support the present graphical analysis. K-S distance is defined for large sample size (Kundu *et al.*, 2005) but it may mislead for small sample size such as $n = 27$.

The present section focuses the discrepancy in fitting between the log-normal and the gamma model fits. It considers mainly five points, namely fitting measure criteria (AICs), nature of estimates, K-S distance, graphical analysis and insignificant effects between the

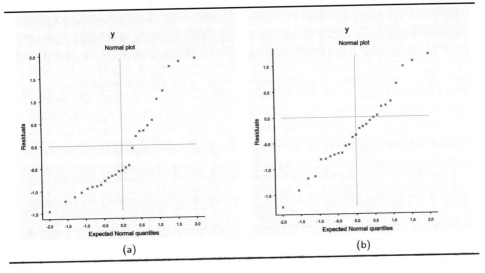

FIGURE 10.19: The normal probability plots for (a) mean and (b) variance models of second-grade silica gel data for gamma fit (Table 10.13)

two models. Moreover, it is pointed that some insignificant effects may be sometimes very important in fitting (Hastie *et al.*, 2001). It presents that even though the measure of fitting criteria and the regression estimates (between the two models) are almost same *but* the fittings may not always be identical. Table 10.12 shows that the estimates are almost identical and significant but both the models show lack of fit. In Table 10.13, numerically most of the estimates are nearly close to each other. Statistically all the estimates of the mean model are significant for the gamma fit, but for the log-normal fit, except *AB*, all other effects are significant. All the estimates for both the dispersion models (Table 10.13) are insignificant. These insignificant effects are included in the model to remove the lack of fit. K-S distance between the log-normal and the gamma distributions for this data set is 0.0151, indicating the distribution difference is *insignificant*. For small sample size ($n = 27$), K-S distance may not always reflect the true situation (Kundu *et al.*, 2005). Even though the estimates are almost same, K-S distance and AICs difference are insignificant, but the fits are *not* identical (Figure 10.16, Figure 10.17 versus Figure 10.18, Figure 10.19), indicating that the appropriate distribution of a given data set may affect in fitting. Practitioners need to check all possible data analysis techniques for selecting a better model.

Based on the log-normal and gamma models the following have been studied by the present author with his seniors. Improvement of resistivity of urea formaldehyde resin has been studied by Das and Lee (2008). An alternative approach of analysing replicated responses in a quality improvement experiment has been proposed by Das and Lee (2010). Resistivity distribution and the discrepancy of classical lifetime distributions in quality

TABLE 10.14: Shape, scale parameters and K-S distance of second-grade silica gel data

Distribution	Parameter	Estimates
Log-normal	Scale (θ)	2.61829604
	Shape (σ)	0.21161545
Gamma	Rate = 1/scale	1.55156740
	Scale (δ)	0.64450950
	Shape (α)	21.7736602
	$\sup_y |F(y) - G(y)|$	0.01510000

engineering have been examined by Das (2011a, 2013a). Das, Dihidar and Verdugo (2011) have studied Indian infant mortality. Das and Dihidar (2013) have studied Indian child mortality. Das and Kim (2012) have examined the ground drinking water quality. Different fields of Medical science such as thyroidology (Das and Mukhopadhyay, 2009; Das 2011b), liver disease (or alcoholic cirrhosis) (Das, 2013b), diabetes, cancer (Das and Sarkar, 2012), relationship of human blood biochemical parameters (Das, 2011c), neonatal low birth weight, etc., have been recently studied by using the log-normal and gamma models. For predicting the optimal level combinations in quality engineering, dual response approach and GLMs/JGLMs approach have been compared by Das (2011d, 2014) and Das and Lee (2010). A few applications of both the models in quality engineering, demography, physical and medical sciences are illustrated in the following sections.

10.5   REPLICATED RESPONSES ANALYSIS IN QUALITY ENGINEERING

Response surface designs are generally used (in quality improvement experiments) for predicting the optimal level combinations of the process parameters. Quality improvement experiments often aim to find the operating condition that achieves the target value for the mean of a process characteristic, and simultaneously minimizes the process variability. To achieve this goal, Taguchi's techniques of analysis based on signal-to-noise ratios and dual response surface approach are commonly used. In the present section, it is described how the generalized linear models approach of modeling the 'mean function' and 'variance function' jointly can be used to achieve the goal. Two examples illustrate the differences among three approaches.

10.5.1   Background of replicated response analyses

Much of response surface methodology (RSM), particularly in the early years, was focused in predicting the operating conditions that resulted in an optimum of the mean response assuming homogeneous variances. During the last two decades, industrial statisticians and practitioners have become aware that they cannot always focus only on the expected value of the response of interest. Instead, the variability of the response also needs to be considered when evidence from real problems suggests that the equal variation assumption may not be practically valid. Industrial practitioners have encountered a need to develop some experimental strategies to achieve the target value of the process characteristic while simultaneously minimizing its variance. Indeed, when the variances are not equal for all variable settings, classical response surface methodology (RSM) can be misleading. The principal aim in an industrial process is to estimate the operating condition that optimizes a response of interest, and simultaneously minimizes the process variability. G. Taguchi has made a significant advances in the use of experimental design and analysis in industry (Box, 1985; Leon *et al.*, 1987). The pioneering work has been credited to Taguchi (1985), who developed a package of tools which were viewed unfavorably by many researchers and practitioners due to lack of statistical foundation (Nair *et al.*, 1992).

Response surface methodology consists of a group of techniques used in the empirical study of the relationship between the response (y) and a number of input variables (x_i's). Consequently, the experimenter is able to find the optimal setting for the input variables that maximizes (or minimizes) the response. A quadratic (second-order) polynomial model, along with least squares fitting, is widely used to study such an empirical relationship. Observations are typically assumed to have an equal variation.

Recently the dual response surface (DRS) approach which was first introduced by Myers and Carter (1973), popularized by Vining and Myers (1990), has received a great deal of attention (Del Castillo and Montgomery, 1993; Lin and Tu, 1995; Del Castillo, 1996; Copeland and Nelson, 1996; Kim and Lin, 1998; Tang and Xu, 2002). Basically, the DRS approach builds empirical models for both the mean and the standard deviation, and then optimizes one of these responses subject to an appropriate constraint on the other's value. The two separate models for the mean and the variance give the analyst a more scientific understanding of the total process, and thus allow them to see what levels of the control factors can give a satisfactory target value of the response as well as the minimum variance.

DRS approach basically assumes normal distribution for the responses. DRS can be extended to GLM family of distributions via joint generalized linear log-normal and gamma models (Section 10.2.2) of the mean and the dispersion (Nelder and Lee, 1991; Lee and Nelder, 1998; Das and Lee, 2009). In this section, we illustrate what we achieve via this extension (following Das and Lee, 2010). For this purpose we briefly review the Taguchi's and DRS approach in Section 10.5.2. Two real examples illustrate the above three approaches in Section 10.5.3.

10.5.2 Taguchi approach and dual-response approach with its extension in GLMs

Taguchi approach: Kacker (1985) presents an excellent review of the basic Taguchi method. Taguchi considered three basic situations 1. "the target is best", where one tries to minimize the mean squared error around a specific target value; 2. "the larger, the better", where one seeks to maximize the characteristic of interest; and 3. "the smaller, the better", where one tries to minimize the characteristic of interest.

Taguchi uses robust parameter design which aims to reduce the performance variation of the process characteristic by choosing the setting of its control variables to make it less sensitive to noise variation. The input variables are divided into two categories: *control variables* and *noise variables*. Control variables are variables whose values remain fixed once they are chosen. Noise variables are variables which are hard to control during the normal process. In conducting an experiment, the settings of the noise factors are controlled to fixed values to find the setting to be less sensitive to the changes of noise variables. Taguchi advocates to use the experiments, which are termed as inner and outer arrays. The inner array represents a design involving the k control variables under the experimenter's control. Each point of the inner array is replicated according to the outer array, which is another experimental design based upon q noise variables. Taguchi relies heavily upon saturated or near saturated orthogonal arrays for both his inner and outer arrays.

Let y_{ij} be the jth response (from the combination of noise variables) at the ith design point (combination of control variables), where $j = 1, 2,\ldots,m$ and $i = 1,2,\ldots,n$. Thus, the m multiple responses at ith design point are $y_{i1}, y_{i2},\ldots,y_{im}$. Note that $\bar{y}_i = \frac{1}{m}\sum_{j=1}^{m} y_{ij}$ and $s_i^2 = \frac{1}{m-1}\sum_{j=1}^{m}(y_{ij} - \bar{y}_i)^2$ are the point estimators of μ_i (mean) and σ_i^2 (variance), respectively, at the ith design point. Taguchi uses the 'signal-to-noise ratios', as his measure of performance. More over 60 different signal-to-noise ratios have been defined by Taguchi (Kacker, 1985; Vining and Myers, 1990), and the three most important SNR_i are 1. $-10\log_{10}[\frac{1}{m}\sum_{j=1}^{m} y_{ij}^2]$, for the smaller is better case; 2. $-10\log_{10}[\frac{1}{m}\sum_{j=1}^{m} 1/y_{ij}^2]$, for the larger is better case; and 3. $-10\log_{10}[\bar{y}_i^2/s_i^2]$, for the target is the best case.

Appropriate SNR_i is chosen to improve the quality of a product. For analysis, $\bar{y}_i$ and SNR_i are used as the responses for the mean and the variance, respectively, and the *active (significant)* control variables are found through analysis of variance (ANOVA).

Dual-response surface methodology: Let x_1, x_2,...,x_k be k control variables. The true response surface is unknown so that it is often assumed be the quadratic of the form

$$y = \beta_0 + \sum_{s=1}^{k} \beta_s x_s + \sum_{s=1}^{k} \beta_{ss} x_s^2 + \sum_{s<j}^{k}\sum^{k} \beta_{sj} x_s x_j + e.$$

Vining and Myers (1990) used the DRS approach. They modeled both the primary responses (w_μ) and the secondary responses (w_σ) jointly,

$$w_\mu = \beta_0 + \sum_{s=1}^{k} \beta_s x_s + \sum_{s=1}^{k} \beta_{ss} x_s^2 + \sum_{s<t}^{k}\sum^{k} \beta_{st} x_s x_t + e_\mu, \tag{10.9}$$

$$w_\sigma = \gamma_0 + \sum_{s=1}^{k} \gamma_s x_s + \sum_{s=1}^{k} \gamma_{ss} x_s^2 + \sum_{s<t}^{k}\sum^{k} \gamma_{st} x_s x_t + e_\sigma. \tag{10.10}$$

Specifically, they optimize one of the fitted response subject to an appropriate constraint on the value of the other fitted response using the Lagrangian multiplier approach. Based on the DRS models, many optimization schemes have been proposed by many researchers (Vining and Myers, 1990; Lin and Tu, 1995; Del Castillo, 1996; Copeland and Nelson, 1996; Kim and Lin, 1998).

As responses w_μ and w_σ, $\bar{y}_i$ and s_i^2 (or s_i, or $\log s_i^2$) are often used; w_μ and w_σ form a dual response system which can be analysed by the techniques developed by Myers and Carter (1973). Let X be the model matrix (or design matrix) (of order $mn \times [1 + \frac{k(k+3)}{2}]$) consisting of the k coordinates of the design points plus the squares and crossproducts of the design point coordinates as defined by the models (10.9) and (10.10). Thus, X represents the design used expanded in terms of the assumed model. Suppose $\mathbf{w}_\mu$ and $\mathbf{w}_\sigma$ denote the vectors of the observed sample means and sample variances, respectively. The usual least squares estimators of $\boldsymbol{\beta}$ and $\boldsymbol{\gamma}$ are

$$\hat{\boldsymbol{\beta}} = (X'X)^{-1}X'\mathbf{w}_\mu \quad \text{and} \quad \hat{\boldsymbol{\gamma}} = (X'X)^{-1}X'\mathbf{w}_\sigma.$$

If the variances of the sample mean values are heterogeneous, weighted least squares are used to estimate $\boldsymbol{\beta}$. Suppose V denotes the variance-covariance matrix of the responses and Myers and Carter (1973) proposed two estimates of V. Suppose that the random errors are independent from design point to design point, so that V is diagonal. The first approach uses the point estimates of the variances as the diagonal elements of V. A second approach uses the predicted values obtained from (10.10) (as $\hat{w}_\sigma = \hat{\gamma}_0 + \sum_{i=1}^{k} \hat{\gamma}_i x_i + \sum_{i=1}^{k} \hat{\gamma}_{ii} x_i^2 + \sum_{i<j}^{k}\sum^{k} \hat{\gamma}_{ij} x_i x_j$) to estimate the diagonal elements of V. Second approach is most frequently used. Thus, using the estimate of V, they proposed to use the weighted least squares estimate

$$\hat{\boldsymbol{\beta}} = (X'V^{-1}X)^{-1}X'V^{-1}\mathbf{w}_\mu.$$

Extension of DRS approach to JGLMs: Because $\bar{y}_i^2$, s_i^2 (or s_i, or $\log s_i^2$) and signal-to-noise ratios are summarizing statistics, their use as response variables often make the analysis inefficient to miss important control variables insignificant (Nelder and Lee, 1991; Das and Lee, 2008). Nelder and Lee (1991) proposes a modelling approach for the analysis of the whole data set y_{ij}. They advocate the use of JGLMs:

$$E(y_{ij}) = \mu_i \quad \text{and} \quad \text{Var}(y_{ij}) = \sigma_i^2 V(\mu_i),$$

where $V(\cdot)$ is the variance function and σ_i^2's are the dispersion parameters. In GLMs the variance consists of two components, $V(\mu_i)$ is the one depending upon the changes of the mean and σ_i^2 is the one independent of mean adjustment. In GLMs the variance function characterizes the distribution of GLM family. For example the distribution is normal if $V(\mu) = 1$, Poisson if $V(\mu) = \mu$, gamma if $V(\mu) = \mu^2$, etc.

Joint models for the mean and the dispersion parameters are

$$\eta_i = g(\mu_i) = \mathbf{x}_i^t \boldsymbol{\beta} \quad \text{and} \quad \xi_i = h(\sigma_i^2) = \mathbf{w}_i^t \boldsymbol{\gamma},$$

where $g(\cdot)$ and $h(\cdot)$ are GLM link functions for the mean and the dispersion, respectively and $\mathbf{x}_i$ and $\mathbf{w}_i$ are vectors for regression models based on the levels of control variables. The estimation methods of $\boldsymbol{\beta}$ and $\boldsymbol{\gamma}$ are given in Section 10.2.2. The DRS models (10.9) and (10.10) with responses $w_\mu = \bar{y}_i$ and $w_\sigma = s_i^2$ can be also fitted by using the JGLMs with $V(\mu) = 1$ (normal distribution),

$$g(\mu_i) = \mu_i = \mathbf{x}_i^t \boldsymbol{\beta} \quad \text{and} \quad h(\sigma_i^2) = \sigma_i^2 = \mathbf{w}_i^t \boldsymbol{\gamma}.$$

For $w_\sigma = \log s_i^2$ we have $h(\sigma_i^2) = \log \sigma_i^2 = \mathbf{w}_i^t \boldsymbol{\gamma}$. With JGLMs, the DRS models can be extended to the non-normal distributions, characterized by a non-constant variance functions $V(\mu_i) \neq 1$. A more discussion on GLM approaches is given in Hamada and Nelder (1997), Lee and Nelder (1998, 2003), Lee et al. (2006), Lesperance and Park (2003), Qu et al. (2000), Das and Lee (2008, 2009, 2010), and Wolfinger and Tobias (1998).

10.5.3   Illustrations with two real examples

Two real data sets with replicated responses are analysed in the present section. One example is for the printing ink data (Box and Draper, 1987), and the other is for a seat-belt experiment data (Wu and Hamada, 2000).

Example 10.8 Printing ink data:

The printing ink data of Box and Draper (1987, Table E7.9, p. 247) has been analysed by Vining and Myers (1990), Lin and Tu (1995), Copeland and Nelson (1996), Kim and Lin (1998). The purpose of the experiment was to determine the effect of speed (x_1), pressure (x_2), and distance (x_3) on the quality (y) of a printing process. The experiment was conducted in a 3^3 factorial design with three replicates at each design point. The data set is reproduced in Appendix (Table Ap.8) (taken from Box and Draper, 1987). The aim of the experiment is to improve the quality of the printing process. Fitted response surfaces (using DRS approach) for $w_\mu = \bar{y}_i$ and $w_\sigma = s_i$ were given in Vining and Myers (1990) under the quadratic models. They analysed the data treating factors as covariates (with values -1, 0, 1).

Table 10.15 shows the analysis from the JGLMs with $V(\mu) = \mu^2$ (gamma response). In the gamma model with $\mu > 0$, it is common to assume the log-link $\log \mu$. To compare with analysis from the DRS we report results from the identity link. The choice of link function does not change the conclusion.

In Figure 10.20, we plot the absolute value of residuals, with respect to the fitted values, for the fitted gamma model (Table 10.15). Figure 10.20 shows a flat running means, indicating that the variance function $V(\mu) = \mu^2$ is appropriate.

Comparison of JGLM analysis with DRS approach for printing ink data:

Using estimation of DRS procedure in Section 10.5.2, Vining and Myers's (1990) predictors for the mean and the variance are as follows:

$$\begin{aligned}
\hat{\mu} = \ & 327.6 + 177.0x_1 + 109.4x_2 + 131.5x_3 + 32.0x_1^2 - 22.4x_2^2 - 29.1x_3^2 \\
& + 66.0x_1x_2 + 75.5x_1x_3 + 43.6x_2x_3,
\end{aligned}$$

TABLE 10.15: Results for mean and dispersion models of printing ink data from gamma fit (using identity link). Das, R.N. and Lee, Y. (2010). "Analysis strategies for multiple responses in quality improvement experiments," International Journal of Quality Engineering and Technology, Vol. 1, No. 4, pp. 395-409

	Covar.	estimate	s.e.	t	P-value
Mean	Const.	317.0	10.32	30.71	<0.01
model	x_1	174.3	11.78	14.80	<0.01
	x_2	115.4	11.87	9.72	<0.01
	x_3	138.5	11.62	11.92	<0.01
	$x_1 x_2$	65.5	13.56	4.83	<0.01
	$x_1 x_3$	78.4	13.28	5.90	<0.01
	$x_2 x_3$	61.2	13.35	4.58	<0.01
	$x_1 x_2 x_3$	70.9	15.19	4.67	<0.01
Dispers.	Const.	-2.49	0.17	-15.04	<0.01
model	x_1	-0.78	0.25	-3.14	<0.01

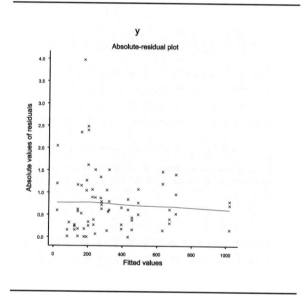

FIGURE 10.20: The absolute residuals plot with respect to fitted values for the gamma fitted models (Table 10.15) of printing ink data. Das, R.N. and Lee, Y. (2010). "Analysis strategies for multiple responses in quality improvement experiments," International Journal of Quality Engineering and Technology, Vol. 1, No. 4, pp. 395-409

and

$$\hat{\sigma}^2 = (34.9 + 11.5x_1 + 15.3x_2 + 29.2x_3 + 4.2x_1^2 - 1.3x_2^2 + 16.8x_3^2 + 7.7x_1x_2$$
$$+ 5.1x_1x_3 + 14.1x_2x_3)^2.$$

The DRS model can be fitted by JGLMs with $V(\mu_i) = 1$. In Table 10.15 we report the gamma fit with $V(\mu_i) = \mu_i^2$, $g(\mu_i) = \mu_i = \mathbf{x}_i^t\boldsymbol{\beta}$ and $h(\sigma_i^2) = \log\sigma_i^2 = \mathbf{w}_i^t\boldsymbol{\gamma}$:

$$\hat{\mu} = 317.0 + 174.3x_1 + 115.4x_2 + 138.5x_3 + 65.5x_1x_2 + 78.4x_1x_3 + 61.2x_2x_3 + 70.9x_1x_2x_3,$$

and
$$\hat{\sigma}^2 = \exp(-2.49 - 0.78x_1),$$
respectively.

In Vining and Myers's mean model x_1, x_2, x_3, x_1^2, x_3^2, x_1x_2, x_1x_3 and x_2x_3 are active. The gamma analysis shows that x_1, x_2, x_3, x_1x_2, x_1x_3, x_2x_3 and $x_1x_2x_3$ are only significant in the mean model. By allowing non-constant variance function $V(\mu) = \mu^2$ the model for the dispersion become much parsimonious. In JGLM, variance can be reduced by using x_1 only, and x_2, x_3 can be used to adjust the mean, but it is very complicated in Vining and Myers's model. From Table 10.15, it is clear that we need to set x_1 at its highest level (1) to reduce the variance, and the quality of the printing process will be improved by setting x_2 and x_3 at their highest levels (1, 1). Then, the predicted mean and variance will be 1021.20 and 0.0380, respectively. Here all the selected optimum levels are 1's. Again all significant second and third order interactions have positive association with the mean of quality of printing process. Thus, the optimal selection (1, 1, 1) improves the mean quality of the printing process, and every significant effect (first-, second- and third-order) has a positive contribution in improving the mean quality of the printing process. Note that the optimal settings suggested by Vining and Myers (1990) are completely different from the present settings as their models are quite different. Vining and Myers (1990) have suggested different optimal settings for different combination of dispersion and mean.

Example 10.9 A seat-belt experiment data:

The present example (seat-belt data) was analyzed by Wu and Hamada (2000) to illustrate the Taguchi approach. This experiment studies the effect of four control factors on the pull strength of truck seat belts following a crimping operation which joins an anchor and cable. The four factors are hydraulic pressure of the crimping machine (A), die flat middle setting (B), length of crimp (C) and anchor lot (D). Each factor has three levels (0, 1 and 2), and a three-level fractional factorial design with 27 runs is considered to study the effects. Each run is replicated three times, the design matrix and the responses are presented in Wu and Hamada (2000, Table 5.2, p. 206), for ready reference it is reproduced in Appendix (Table Ap.9) (taken from Wu and Hamada, 2000).

This experiment studies two quality characteristics: crimp tensile strength (lb), which has a minimum customer specification of 4000 lb, and flash (mm), which is excess metal from the crimping process that cannot exceed 14.00 mm. Strength needs to be maximized to ensure that the seat belt does not come apart in an accident. Therefore, the aim of the experiment is to maximize the strength (as high as possible). On the other hand flash needs to be minimized as it causes cracks in a plastic boot covering this assembly which is put on during a later processing step. Thus, flash needs to be as low as possible. Therefore, the goal of this experiment is to identify the settings that simultaneously maximize the tensile strength and minimize the flash, while both the variances (of responses tensile strength and flash) are to be minimized.

Based on Taguchi's principle, Wu and Hamada (2000, Tables 5.3, 5.6, 5.7, 5.10) presented their analysis using analysis of variance (ANOVA) method. They used ANOVA for mean model, and they considered '$\ln s^2$' for dispersion analysis, but there is no replication, ANOVA method cannot be used to identify significant factors for dispersion. Using informal graphical analysis (half-normal plot) they identified only the factor A as significant for both the dispersions. The present analysis uses all the original observations, rather than the summarizing statistics such as '$\bar{y}_i$' and '$\ln s^2$' in the joint gamma GLMs given as in Section 10.5.2.

To compare the present analysis with Wu and Hamada's model (2000), we analyze the data using identity link and factors under joint gamma GLMs. We take a constraint that the effects of the first levels in the factors are zero (as in Example 10.6). For example the

TABLE 10.16: Results for mean and dispersion models of seat belt tensile-strength data from gamma fit (using identity link and factors). Das, R.N. and Lee, Y. (2010). "Analysis strategies for multiple responses in quality improvement experiments," International Journal of Quality Engineering and Technology, Vol. 1, No. 4, pp. 395-409

	Covar.	estimate	s.e.	t	P-value
Mean	Const.	6087.00	150.70	40.40	<0.01
model	$A2$	776.00	154.90	5.01	<0.01
	$A3$	850.00	179.50	4.74	<0.01
	$C2$	−388.00	230.40	−1.68	0.10
	$C3$	−1255.00	285.80	−4.39	<0.01
	$A2C2$	42.00	337.40	0.12	0.90
	$A2C3$	508.00	310.70	1.64	0.10
	$A3C2$	516.00	253.00	2.04	0.04
	$A3C3$	1272.00	310.80	4.09	<0.01
Dispers.	Const.	−4.17	0.77	−5.44	0.00
model	$A2$	−3.11	0.81	−3.85	<0.01
	$A3$	−1.13	0.94	−1.21	0.23
	$C2$	−1.16	1.14	−1.01	0.31
	$C3$	−0.27	1.00	−0.27	0.78
	$D2$	−1.57	0.83	−1.88	0.06
	$D3$	−0.97	1.02	−0.95	0.34
	$A2C2$	3.52	1.10	3.18	<0.01
	$A2C3$	1.17	1.11	1.05	0.30
	$A3C2$	−2.36	1.21	−1.95	0.05
	$A3C3$	−1.98	1.24	−1.60	0.11
	$C2D2$	2.21	1.26	1.76	0.08
	$C2D3$	2.50	1.42	1.77	0.08
	$C3D2$	2.32	1.39	1.67	0.10
	$C3D3$	3.42	1.34	2.55	0.01

estimate for the effect $A2$ means the effect of difference between the second and the first levels in the main effect A.

The AIC is used to select the final model. In the final model, we retain some insignificant effects in the model in order to respect the marginality rule (Nelder, 1994). Results of the analyses of the tensile strength and flash of seating belt experimental data using the JGLMs are displayed in Table 10.16 and Table 10.17, respectively.

Figures 10.21(a) and 10.21(b) display the absolute residual plots with respect to the fitted values, for the fitted gamma models (Table 10.16 and Table 10.17), of tensile strength and flash, respectively. Both the figures are somewhat flat with the running means, indicating that the gamma models are appropriate.

Comparison of JGLM analysis with Taguchi's approach for seat-belt experiment data

The following are the *main* differences of the present analysis from Wu and Hamada (2000): 1. Wu and Hamada (2000) could not properly analyse the dispersion effects, by using informal half-normal plot, they could only identify the factor A for both (tensile strength and flash) the dispersions. By using all the data, the JGLM analysis shows that the factor A is *only* significant for the dispersion model of flash, but for tensile strength, the factors A, AC and CD are significant. 2. Wu and Hamada (2000) identified factors A, C, D, AB and AC are significant in the mean model of tensile strength, but the JGLM analysis (Table 10.16) shows that factors A, C and AC are only significant. 3. For flash, factors A, C, AC and BC are identified as significant in the mean model by Wu and Hamada (2000), but in the present analysis A, B, C, AC and BC are significant (Table 10.17). 4. Mean and

TABLE 10.17: Results for mean and dispersion models of seat belt flash data from gamma fit (identity link and factors). Das, R.N. and Lee, Y. (2010). "Analysis strategies for multiple responses in quality improvement experiments," International Journal of Quality Engineering and Technology, Vol. 1, No. 4, pp. 395-409

	Covar.	estimate	s.e.	t	P-value
Mean	Const.	12.76	0.11	114.45	<0.01
model	$A2$	1.00	0.18	5.42	<0.01
	$A3$	2.85	0.19	14.65	<0.01
	$B2$	0.30	0.15	2.01	0.05
	$B3$	0.63	0.15	4.10	<0.01
	$C2$	0.06	0.16	0.39	0.70
	$C3$	−0.30	0.15	−1.92	0.06
	$A2C2$	−0.35	0.25	−1.36	0.18
	$A2C3$	−0.29	0.25	−1.14	0.26
	$A3C2$	−1.40	0.26	−5.33	<0.01
	$A3C3$	−1.89	0.26	−7.31	<0.01
	$B2C2$	−0.54	0.21	−2.59	0.01
	$B2C3$	0.01	0.21	0.04	0.97
	$B3C2$	−0.46	0.21	−2.17	0.03
	$B3C3$	−0.57	0.21	−2.73	0.01
Dispers.	Const.	−8.20	0.34	−24.31	<0.01
model	$A2$	1.56	0.47	3.29	<0.01
	$A3$	1.45	0.45	3.21	<0.01

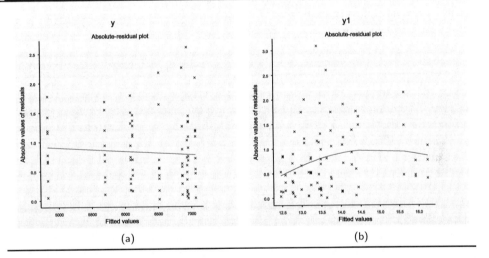

FIGURE 10.21: The absolute residuals plot w.r.to fitted values for gamma fit of (a) tensile strength (Table 10.16) and (b) flash (Table 10.17). Das, R.N. and Lee, Y. (2010). "Analysis strategies for multiple responses in quality improvement experiments," International Journal of Quality Engineering and Technology, Vol. 1, No. 4, pp. 395-409

variance models for tensile strength (using stepwise regression) (Wu and Hamada, 2000, p. 231-232) are

$$\hat{\mu} = 6223.07 + 1116.29A_l - 190.24A_q + 178.69B_l - 589.54C_l + 294.29(AB)_{ql} + 627.94(AC)_{ll} - 191.28D_{01} - 468.42D_{12} - 486.44CD_{l,12},$$ and $\hat{\sigma}^2 = e^{10.89-1.70A_l}$, respectively. Mean and variance models for tensile strength based on JGLM (Table 10.16) are

$$\hat{\mu} = 6087 + 776A2 + 850A3 - 388C2 - 1255C3 + 42A2C2 + 508A2C3 + 516A3C2 + 1272A3C3,$$ and $\hat{\sigma}^2 = e^{(-4.17-3.11A2-1.13A3-1.16C2-0.27C3-1.57D2-0.97D3)} \times$

$e^{(3.52A2C2+1.17A2C3-2.36A3C2-1.98A3C3+2.21C2D2+2.50C2D3+2.32C3D2+3.42C3D3)}$, respectively.

5. Mean and variance models for flash (using stepwise regression) (Wu and Hamada, 2000, p. 232) are $\hat{\mu} = 13.66 + 1.24A_l + 0.18B_l - 0.85C_l + 0.20C_q - 0.94(AC)_{ll} - 0.38(AC)_{ql} - 0.38(BC)_{qq} - 0.30(CD)_{l,12}$, and $\hat{\sigma}^2 = e^{-2.57+1.49A_l}$, respectively. Mean and variance models for flash based on JGLM (Table 10.17) are $\hat{\mu} = 12.76 + 1.00A2 + 2.85A3 + 0.30B2 + 0.63B3 + 0.06C2 - 0.30C3 - 0.35A2C2 - 0.29A2C3 - 1.40A3C2 - 1.89A3C3 - 0.54B2C2 + 0.01B2C3 - 0.46B3C2 - 0.57B3C3$, and $\hat{\sigma}^2 = e^{-8.20+1.56A2+1.45A3}$, respectively. 6. Optimal settings of seat belt experiment according to Taguchi's approach (Wu and Hamada, 2000) are $A = 2$ (highest level), $B = 0$ (lowest level), $C = 1$ (middle level), and $D = 0$ (lowest level). According to JGLMs, optimal settings are $A = 2$ (highest level), $B = 0$ (lowest level), $C = 2$ (highest level), and $D = 0$ (lowest level). 7. At the optimal settings (based on Taguchi's approach), the predicted mean and the variance of tensile strength are 6858.69 and $(127.14)^2$, respectively, and the predicted mean and the variance of flash are 14.37 and $(0.47)^2$, respectively. But for the present optimal settings (using levels -1, 0, 1) based on JGLMs, the predicted mean and variance of tensile strength are 7558 and <0.0001, respectively, and the predicted mean and variance of flash are 12.09 and 0.0056, respectively.

To minimize both the variances simultaneously (with levels -1, 0, 1), we need to set factors A, C and D to the levels (1, 1, -1), and to achieve maximum tensile strength and minimum flash, we have to set factor B at its lowest levels (-1). Under this situation variances of tensile strength and flash will be <0.0001 and 0.0056, respectively, whereas their means will be 7558 and 12.09, respectively, which satisfy the basic requirements of the manufacturing truck seat-belts. Note that this seat-belt experiment desires to obtain a mean tensile strength of 7000 lb with standard deviation of 250 lb, and mean flash of 13.50 mm with standard deviation 0.2 mm. The above optimal settings ensures a very small standard deviations with more than the target mean of tensile strength and smaller than the target mean of flash.

Taguchi's and DRS approaches require a simultaneous modelling of the mean and the dispersion. Appropriate modeling is the principal part in the DRS approach for selecting the optimal level combinations of the process parameters. However, both the approaches cannot exploit all the information in the data because they analyse summarizing statistics. DRS approach always assumes $V(\mu) = 1$ and second-order model for both the mean and the variation. Example 10.8 shows that such assumptions may not always be true. In Example 10.9, we see that wit Taguchis; method one can derive some significant effects for mean but it is quite impossible to detect significant effects for dispersion. By using some informal graphical analysis such as half-normal plot, one may have some idea about the significant factors for dispersion (Wu and Hamada, 2000; Table 5.2, p. 206). With JGLM the use of all the data allows to find significant factors for the dispersion.

In Taguci's and DRS approaches one has to choose summarizing statistics such as ($\bar{y}$ and $\ln s^2$) to use as multiple responses for the analysis. However, appropriate summarizing statistics can be chosen after appropriate models are chosen. Thus, the use of the whole data for the model selection is important to find a final model for the optimum setting. The DRS approach could yield the negative predictive values of the standard deviation, while JGLM always gives positive values. The Taguchi's and DRS approaches can not be applied for non-replicated cases, at each design point for control variables, whereas JGLM can be applied.

10.6   RESISTIVITY OF UREA FORMALDEHYDE RESIN IMPROVEMENT

Gupta and Das (2000) conducted the resin experiment to improve the resistivity of urea formaldehyde through the setting of process parameters. These researchers noticed that the variances of five different response characteristics are non-constant, affected by some factors. In quality-improvement engineering applications, achieving high precision, by minimizing variance is as important as getting the mean at the target. To identify the factors affecting the variance, Gupta and Das (2000) used the analysis of variance method for signal-to-noise ratios. However, such analysis is statistically inefficient to miss important factors as *insignificant*. Das and Lee (2008) proposed to use joint modeling for the mean and the dispersion, which gives completely different analysis for the resin data. In the present section, it is described how to predict the optimal settings of the process parameters to reach the goal, using JGLMs as proposed by Das and Lee (2008).

10.6.1   Resin experiment background

Gupta and Das (2000) analysed the resin data for improving the resistivity of the urea formaldehyde (UF). These researchers have analysed five process characteristics namely resistivity, non-volatile (NV) solid, viscosity, acid value (AV) and petroleum ether tolerance (PET) of UF resin. All the responses (Gupta and Das, 2000, p. 617) related to these five process characteristics of UF resin are positive. The study aim of these researchers is to obtain the resistivity value within the specification by keeping the other responses within their respective specifications. They have transformed the original responses into nominal-the-best type of signal-to-noise ratio (SNR) values. These SNR values were analysed to find the factors having adequate impact on variability of the responses by using the analysis of variance (ANOVA).

SNRs of the Taguchi method (Taguchi and Wu, 1985) have been widely used as responses for the analysis of quality-improvement experiments. However, the use of summarizing statistics such as SNR as response variable is always likely to be a relatively primitive form of analysis. Such a data reduction would be statistically justifiable if it constitutes a sufficiency reduction; however, such a reduction requires distributional assumptions which cannot reasonably be made until after selecting a suitable model, and this requires some model checking. In general, SNR is a complex function of data, which is unlikely to lead to a simple model. Its sampling properties are likely to be complex and difficult to model compared to those of original responses. Mean and variance contributions in SNR are confounded, leading to a difficulty in interpretations. For more discussion on the deficiencies resulting from using SNRs see Box (1988), Welch et al. (1990) and Lee and Nelder (2003).

In the present section, it is described that such a ANOVA method gives inefficient analysis, often resulting in an error that 'the significant factors are classified as insignificant and vice-versa'; the analysis by Gupta and Das (2000) missed many important factors of each process characteristic (Das and Lee, 2008). This is very serious in quality-improvement experiments, because a wrong setting of important factors will results in poor quality of products. Therefore, it is crucial to use efficient statistical method to identify the significant factors of each process characteristic of interest. In statistical literature, models are mainly focused on the mean, so that the modeling of the dispersion has often been neglected. In this section, we describe how to use the joint modeling of mean and dispersion to achieve the goals.

In Section 10.2.1 and Section 10.2.2, log-normal and gamma models are described for constant and non-constant variance, respectively. In the present section, only the log-normal models with constant (10.2) and non-constant (10.6) variance are used to analyse the above five different process characteristics of the urea formaldehyde resin data.

In statistical modeling, it is not wise to use the summarizing statistics such as SNR, sample mean, sample standard deviation etc., as the response variable (Welch et al., 1990). Box (1988) proposed for using linear models with data transformation. If a noise variable appears as a random variable, Myers and Montgomery (1995) advocated the use of the response surface models, which allow to eliminate the variation caused by random noise variables. Another type of variation such as between batch variation can be modeled via random effects (Wolfinger and Tobias, 1998). All these models can be combined into a single general model to analyse the data from quality improvement experiments. Because there is no noise variables and no between batch variation we use the joint GLM for the analysis of resin data.

10.6.2   Urea formaldehyde resin experiment data

Gupta and Das (2000) presented an experimental set of data for studying the resistivity, NV solid, viscosity, AV and PET of UF resin. The aim of the experiment is to improve the resistivity value within the specification by keeping the other responses within their respective specifications. An orthogonal array design is used in this experiment to study one three-level factor A and five two-level factors $(B - F)$. Positive responses of the above process characteristics of UF resin are repeated twice for each trial with 16-experimental runs of the selected design. The model matrix, factors and the responses are given in Gupta and Das (2000, pp. 616-617). For ready reference, the experimental layout and resulting responses are reproduced in Appendix (Table Ap.10 and Table Ap.11) (taken from Gupta and Das, 2000). The experimental factors are amount of NaOH (factor A), reflux time (factor B), solvent distillate percentage (factor C), phthalic anhydride percentage (factor D), water collection time (factor E), solvent distillate collection time (factor F). The authors have prior information from the personnel of the organization that the two factor interaction $D \times E$ and $D \times F$ are believed to improve the resistivity of UF resin. Gupta and Das (2000) used ANOVA method analysis for the responses of SNR formed from the original responses.

10.6.2.1   Resistivity analysis of urea formaldehyde resin data

Gupta and Das (2000) noticed that the factors C and E have a relatively strong impact on the variance of UF resin resistivity, and for other responses they failed to identify any factor for effecting the variance. They also noticed that the factors A and D had a large impact on the mean response of resistivity, whereas the factors B and F had no impact on the mean responses of any process characteristic. We fit the joint log-normal model (10.6) with the factors as recommended by Gupta and Das (2000) in Table 10.18. Analysis shows that none of their recommended factors is significant.

In Table 10.19, we present the final selected models of UF resin resistivity data. For factors (as stated in Example 10.6), similar approach is considered here. For example, the estimate for the effect $A2$ means the effect of difference between the second and the first levels in the main effect A. The selected models have the smallest AIC value in each class.

TABLE 10.18: Results for mean and dispersion models (Gupta and Das, 2000) of UF resin resistivity data from log-normal fit. Das, R.N. and Lee, Y.J. (2008). "Improving resistivity of urea formaldehyde resin through joint modeling of mean and dispersion," Journal of Quality Engineering, Vol. 20(3), pp. 287-295

	Covariate	estimate	s.e.	t	P-value
Mean model	Constant	5.15	0.18	28.77	<0.0000
	$A2$	−0.16	0.21	−0.73	0.4734
	$A3$	−0.27	0.19	−1.37	0.1851
	$D2$	−0.04	0.15	−0.27	0.7898
Dispersion model	Constant	−1.43	0.44	−3.27	0.0036
	$C2$	−0.99	0.64	−1.55	0.1361
	$E2$	0.77	0.64	1.20	0.2435

TABLE 10.19: Results for mean and dispersion models of UF resin resistivity data from log-normal fit. Das, R.N. and Lee, Y.J. (2008). "Improving resistivity of urea formaldehyde resin through joint modeling of mean and dispersion," Journal of Quality Engineering, Vol. 20(3), pp. 287-295

	Covariate	estimate	s.e.	t	P-value
Mean model	Constant	4.48	0.20	21.88	<0.0001
	B2	0.39	0.11	3.45	0.0024
	C2	0.34	0.20	1.65	0.1138
	D2	0.48	0.28	1.73	0.0983
	E2	0.29	0.15	1.97	0.0621
	B2.D2	−0.38	0.16	−2.41	0.0252
	C2.D2	−0.60	0.28	−2.15	0.0433
Dispersion model	Constant	−1.44	0.54	−2.67	0.0143
	C2	−2.79	0.90	−3.11	0.0053
	E2	0.40	0.74	0.55	0.5881
	C2.E2	2.32	1.18	1.96	0.0634

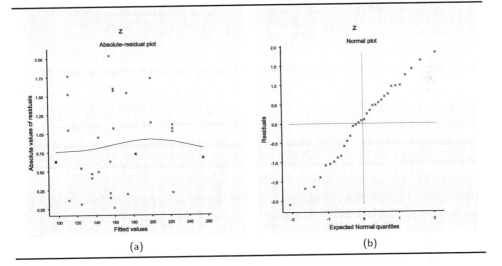

(a)   (b)

FIGURE 10.22: (a) The absolute residuals plot w.r.to fitted values and (b) normal probability plot for mean of resistivity for LN fit (Table 10.19). Das, R.N. and Lee, Y.J. (2008). "Improving resistivity of urea formaldehyde resin through joint modeling of mean and dispersion," Journal of Quality Engineering, Vol. 20(3), pp. 287-295

Following the marginality rule (Nelder, 1994), some insignificant effects are retained in the final selected models. The values of AIC of the final selected model in Table 10.19 is 359.2, while that of Gupta and Das's (2000) is 370.4 (Table 10.18), so that AIC clearly choose the final models in Table 10.19. This is true for the models of other four responses.

In Figure 10.22(a), we plot the absolute value of residuals with respect to the fitted values, and in Figure 10.22(b), we plot the normal probability plot for the mean model in Table 10.19. Figure 10.22(a) has a flat running means which indicates that joint modeling of mean and variance (in Table 10.19) would be satisfactory, and also Figure 10.22(b) does not show any systematic departures, indicating no lack of fit of the final models.

TABLE 10.20: Results for mean and dispersion models of NV solid of UF resin data from log-normal fit. Das, R.N. and Lee, Y.J. (2008). "Improving resistivity of urea formaldehyde resin through joint modeling of mean and dispersion," Journal of Quality Engineering, Vol. 20(3), pp. 287-295

	Covariate	estimate	s.e.	t	P-value
Mean model	Constant	3.99	0.01	603.40	<0.0001
	$C2$	0.06	0.01	6.80	<0.0001
	$D2$	0.03	0.01	3.00	0.0066
	$E2$	0.03	0.01	3.90	0.0007
	$C2.D2$	−0.07	0.01	−5.90	<0.0001
Dispersion model	Constant	−9.16	0.75	−12.27	<0.0001
	$A2$	2.05	0.86	2.39	0.0258
	$A3$	2.00	0.78	2.57	0.0174
	$C2$	−1.13	0.55	−2.04	0.0535
	$F2$	1.04	0.57	1.85	0.7779

From Table 10.19, we need to set the two factors (C, E) in the dispersion model to the levels $(2, 1)$ to reduce the variance and use the two factors (B, D) in the mean model to adjust the mean. We do not use the two factors (C, E) in the mean model to adjust the mean because it will affect the variance. By using factors significant in the mean model, but not significant in the dispersion model, we can adjust the mean without affecting the variance.

10.6.2.2 Non-volatile solid analysis of urea formaldehyde resin data

For the non-volatile (NV) solid of UF resin data, the present final models (using joint log-normal models (10.6)) are presented in Table 10.20. From Table 10.20, we need to set the factors (A, C, F) to the levels $(1, 2, 1)$ to reduce the variance and use the factors (D, E) to adjust the mean. In the present analysis many factors become significant. In Figure 10.23(a) we plot the absolute value of residuals with respect to the fitted values, and in Figure 10.23(b), we plot the normal probability plot for the dispersion, for the fitted log-normal models in Table 10.20. These two plots would indicate no serious lack of fit.

10.6.2.3 Viscosity analysis of urea formaldehyde resin data

In Table 10.21, we present the final models (using joint log-normal models (10.6)) for viscosity of UF resin data. Again it is observed that many factors are significant in the present analysis. From Table 10.21, we need to set the factors (A, C) to the levels $(1, 1)$ to reduce the variance and use the factors (D, F) to adjust the mean. In Figure 10.24(a), we plot the absolute value of residuals with respect to the fitted values, and in Figure 10.24(b), we plot the normal probability plot for mean, for the fitted log-normal models in Table 10.21. These two plots would indicate no serious lack of fit.

10.6.2.4 Acid value analysis of urea formaldehyde resin data

For the acid value of UF resin data, the present final models (using joint log-normal models (10.6)) are displayed in Table 10.22. From Table 10.22, we need to set the factors (A, B, C, D) to the levels $(2, 2, 2, 2)$ to reduce the variance and use the factor E to adjust the mean. In Figure 10.25(a), we plot the absolute value of residuals with respect to the fitted values, and in Figure 10.25(b), we plot the normal probability plot for mean, for the fitted log-normal models in Table 10.22. These two plots would indicate no serious lack of fit.

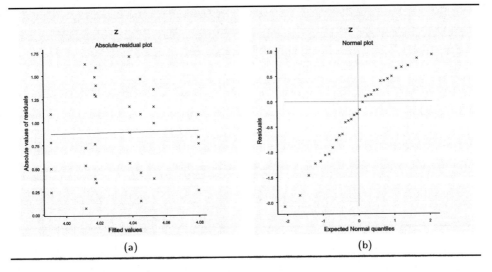

FIGURE 10.23: (a) The absolute residuals plot w.r.to fitted values and (b) normal probability plot for dispersion of NV solid for LN fit (Table 10.20). Das, R.N. and Lee, Y.J. (2008). "Improving resistivity of urea formaldehyde resin through joint modeling of mean and dispersion," Journal of Quality Engineering, Vol. 20(3), pp. 287-295

10.6.2.5   Petroleum ether tolerance value analysis of urea formaldehyde resin data

For the UF resin petroleum ether tolerance value, the present final model (using lognormal model (10.2)) is presented in Table 10.23. From Table 10.23, we use the factor E to adjust the mean. In Figure 10.26(a), we plot the absolute value of residuals with respect to the fitted values, and in Figure 10.26(b), we plot the normal probability plot for mean, for the fitted in Table 10.23. These two plots would indicate no serious lack of fit.

In the current section, it is observed that one should analyse the original responses, rather than the summarizing statistics such as SNR. It is noticed that the analysis of SNR results inefficient analysis to classify significant factors as insignificant. The following are the *main* differences of the present analysis from Gupta and Das (2000): (1.) From the

TABLE 10.21: Results for mean and dispersion models of UF resin viscosity data from log-normal fit. Das, R.N. and Lee, Y.J. (2008). "Improving resistivity of urea formaldehyde resin through joint modeling of mean and dispersion," Journal of Quality Engineering, Vol. 20(3), pp. 287-295

	Covariate	estimate	s.e.	t	P-value
Mean model	Constant	1.90	0.10	18.36	<0.0001
	$C2$	0.55	0.18	3.00	0.0063
	$D2$	0.42	0.14	3.10	0.0051
	$F2$	0.33	0.13	2.60	0.0160
	$C2.D2$	−0.61	0.25	−2.44	0.0228
Dispersion model	Constant	−3.44	0.72	−4.81	<0.0001
	$A2$	1.22	0.80	1.52	0.1421
	$A3$	1.55	0.74	2.10	0.0469
	$C2$	0.85	0.58	1.48	0.1524

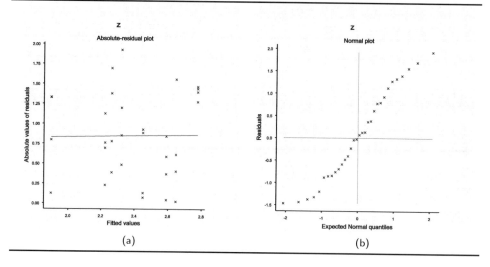

FIGURE 10.24: (a) The absolute residuals plot w.r.to fitted values and (b) normal probability plot for mean of viscosity data for LN fit (Table 10.21). Das, R.N. and Lee, Y.J. (2008). "Improving resistivity of urea formaldehyde resin through joint modeling of mean and dispersion," Journal of Quality Engineering, Vol. 20(3), pp. 287-295

prior information Gupta and Das (2000) considered the two factor interactions $D \times E$ and $D \times F$ as important for improving the resistivity of UF resin. Such prior information should be examined after the data are collected. The present data analysis shows that these two factor interactions $D \times E$ and $D \times F$ are not significant. (2.) Gupta and Das (2000) found that *only* the resistivity has a non-constant variance and the other responses have *constant* variances. But the present analysis identifies that all the response variances are non-constant, *except* the petroleum ether tolerance value (PET) of UF resin data. (3.) For the variance of resistivity, Gupta and Das (2000) found only two significant effects C and E, while in the present analysis, the factors C, E and $C \times E$ are identified. (4.) Gupta and Das (2000) concluded that the two factors (B and F) are insignificant for *all* the responses.

TABLE 10.22: Results for mean and dispersion models of UF resin acid value data from log-normal fit. Das, R.N. and Lee, Y.J. (2008). "Improving resistivity of urea formaldehyde resin through joint modeling of mean and dispersion," Journal of Quality Engineering, Vol. 20(3), pp. 287-295

	Covariate	estimate	s.e.	t	P-value
Mean model	Constant	2.10	0.03	64.27	<0.0001
	A2	0.14	0.03	4.61	0.0002
	A3	−0.01	0.03	−0.34	0.7372
	D2	0.40	0.02	23.82	0.0000
	E2	−0.11	0.03	−3.68	0.0014
Dispersion model	Constant	−3.31	0.86	−3.86	0.0009
	A2	−0.51	0.94	−0.54	0.5948
	A3	1.65	0.72	2.27	0.03385
	B2	−2.03	0.58	−3.51	0.0021
	C2	−1.26	0.60	−2.08	0.0499
	D2	−1.65	0.59	−2.80	0.0107

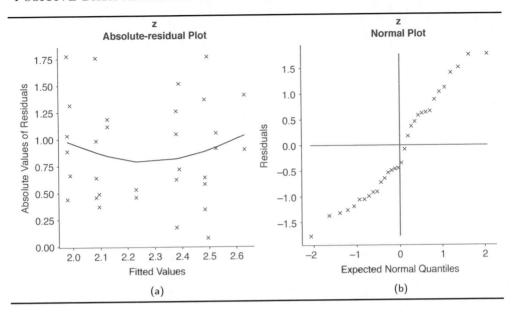

FIGURE 10.25: (a) The absolute residuals plot w.r.to fitted values and (b) normal probability plot for mean of acid value data for LN fit (Table 10.22). Das, R.N. and Lee, Y.J. (2008). "Improving resistivity of urea formaldehyde resin through joint modeling of mean and dispersion," Journal of Quality Engineering, Vol. 20(3), pp. 287-295

TABLE 10.23: Results for mean model of UF resin PET data from log-normal fit. Das, R.N. and Lee, Y.J. (2008). "Improving resistivity of urea formaldehyde resin through joint modeling of mean and dispersion," Journal of Quality Engineering, Vol. 20(3), pp. 287-295

	Covariate	estimate	s.e.	t	P-value
Mean model	Constant	0.77	0.03	26.15	<0.0001
	E2	0.08	0.04	1.99	0.0674
Dispersion model	Constant	−4.27	0.26	−16.55	<0.0001

However, we have found that B is significant in the mean model of resistivity and in the dispersion model of acid value; F is significant in the mean model of viscosity and partially significant in the dispersion model of non-volatile solid. (5.) They found that the factors A and D are significant on the means of all the responses and also the factor C has some significant effect on the mean of the NV solid. It is identified that many additional factors are significant in Tables 10.18–10.23.

From the present analysis, we see that we can identify many significant factors, which could be missed. For quality improvement experiments, identification of important process factors are very important, so that the use of an appropriate statistical method is crucial.

10.7   DETERMINANTS OF INDIAN INFANT AND CHILD MORTALITY

Infant and child mortality is a serious problem in India. In order to better understand the problem previous research has looked at the topic using sophisticated multivariate models assuming that infant mortality is either Gaussian or Log-Gaussian. In the present section,

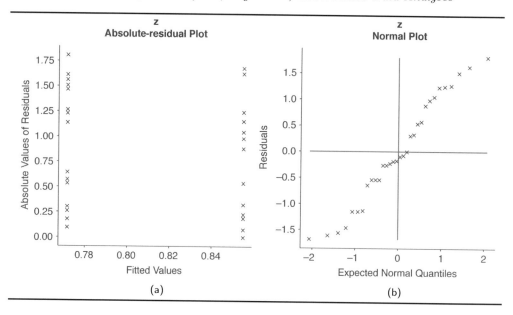

FIGURE 10.26: (a) The absolute residuals plot w.r.to fitted values and (b) normal probability plot for mean of PET data for LN fit (Table 10.23). Das, R.N. and Lee, Y.J. (2008). "Improving resistivity of urea formaldehyde resin through joint modeling of mean and dispersion," Journal of Quality Engineering, Vol. 20(3), pp. 287-295

we argue that infant and child mortality is Log-Gaussian distributed with non-constant variance and that making such an assumption leads to more efficient estimates and a better fit to the data. Using infant and child mortality data from the National Health Survey in Bihar, India we compare two distributions—Log Gaussian and Gamma and find that the Log-Gaussian non-constant variance model does indeed lead to more efficient estimates and a better fit to the data.

10.7.1   Infant and child mortality background

Infant mortality is not only an important factor in population growth; it is also an important measure of economic development. Throughout the world it appears that the infant mortality rate (IMR) has declined since the 1960s. Indeed, between 1960 and 1998 there was a world-wide 55 percent decrease in the IMR (UNICEF, 2001). Variations exited, though. In the Sub-Sahara region of Africa, for example, the IMR declined by 34 percent between 1960 and 1989. Nevertheless, the world-wide decline in the IMR has been good news.

The IMR in India has also been the subject of much research. Though India's IMR has been declining, in 2009, it ranked 143rd in IMR with 55 infant deaths per 1,000 live population (UNICEF, 2001). Such a record is troubling not only because of the loss in human life, but because India is an emerging economic super-power, and it's IMR is clearly not related to its emerging economic status.

Why does India have such a high IMR? What are the factors that contribute to its high IMR? An existing body of research has attempted to answer these and related questions (Jain and Visaria, 1988; Zdachariah and Patel, 1982; Philip, 1985; Puffer, 1985; Sandell *et al.*, 1985; Visaria, 1985; Tilak, 1991; Anilkumar and Ashara, 1993; Khan, 1993; Goyal, 1994; Measham *et al.*, 1999). The problem is complex and scholars have used a wide array of factors to better understand the IMR in India. Yet, there are some important concerns over the assumption that IMR data are Gaussian distributed. That is, past research has estimated IMR using multivariate models that assume a Gaussian distribution. Such an assumption may lead to inefficient and biased estimates, and draw researchers into making important conclusions from dubious results if the distribution is not Gaussian (Myers *et al.*, 2002).

Previous research on India's IMR begins by setting up a model that contains risk factors in three domains: proximate factors, maternal factors, and household/community factors (Jain and Visaria, 1988; Zdachariah and Patel, 1982; Philip, 1985; Puffer, 1985; Sandell *et al.*, 1985; Visaria, 1985; Tilak, 1991; Anilkumar and Ashara, 1993; Khan, 1993; Goyal, 1994; Measham *et al.*, 1999). Proximate factors are those items that involve medical care and non-medical care during the antenatal period, care at birth, and care during the postnatal period. Maternal factors refer to such things as the age, and birth intervals of the mother. And finally, household and community factors refer to such things as sanitation, water supply, and household and community cleanliness. Jain and Visaria (1988) found that significant declines in the IMR are possible without improvement in societal economic development. What the authors found instead was that access to a small number of health and maternal services reduced the IMR: reproductive health services, perinatal care, improved breast feeding, immunization, the treatment of diarrhea, and the introduction to supplementary foods. Other research has reported similar findings (Sandell *et al.*, 1985).

In terms of economics, research has shown that economics is not a major factor in reducing the IMR in India, rather it is those non-economic factors, such as maternal and child health interventions (Zdachariah and Patel, 1982; Philip, 1985; Sandell *et al.*, 1985).

Both sets of studies inform our own study in two ways. First, they assist us in setting up a model for analysis that includes predictors from the three risk-factor domains: proximate, maternal, and household/community. Secondly, it is our sense that previous research may have been in error because they estimated models assuming a Gaussian or Log-Gaussian distribution with constant variance, when, in fact, the distribution may be Gamma distributed and/or with nonconstant variance. For example, Myers *et al.* (2002) found that transforming data for stabilization may not, in fact, stabilize the distribution (see also Das and Lee, 2009 for a discussion). Results from previous research, then, might be inefficient and biased. Estimating these data under the Gaussian assumption could lead to serious errors.

Recently, Das, Dihidar and Verdugo (2011) have examined Indian infant mortality using both the log-normal and gamma models. Das and Dihidar (2013) have examined Indian child mortality using the log-normal distribution. In the present section, we describe Indian infant and child mortality, based on the results obtained by Das, Dihidar and Verdugo (2011) and Das and Dihidar (2013). The present study has two objectives. The first objective is to estimate the impact selected risk factors have on the IMR and child mortality in Bihar, India. The second objective is to evaluate two models based on different assumptions regarding infant mortality: one based on the assumption that infant mortality is Log-Gaussian distributed, and the other assumes that it is Gamma distributed. The present data are from the Indian Survey of the National Family Health Survey-2 (NFHS-2) conducted in 1998-1999. The present study proposes to use the simultaneous modeling of the mean and dispersion in infant and child mortality by using a JGLM.

10.7.2 Infant survival time data, analysis, and interpretation

A. DATA

The National Family Health Survey, Bihar 1998-99: India's first National Family Health Survey (NFHS-1) was conducted in 1992-93. The Ministry of Health and Family Welfare

(MOHFW) subsequently designated the International Institute for Population Sciences (IIPS) in Mumbai, as the agency to initiate a second survey (NFHS-2), which was conducted in 1998-99. An important objective of (NFHS-2) was to provide state-level and national-level information on fertility, family planning, infant and child mortality, reproductive health, child health, nutrition of women and children, and the quality of health and family welfare services.

Another important objective of the NFHS-2 was to examine the above information in the context of socioeconomic and cultural factors. NFHS-2 used three types of questionnaires: the Household questionnaire, the Woman Questionnaire, the Village Questionnaire. The Woman Questionnaire collected information from ever-married women belonging to the age cohort 15-49 who were residents of the sampled household. Female respondents were asked about their background, the details of births of their children during the preceding three years, and whether they practiced contraception. The Child Questionnaire was designed to record details of antenatal care, details of delivery, breastfeeding, and postpartum amenorrhoea, immunization and health care for the two most recent births during the three years preceding the survey. In the present analysis, we have taken a random sample (from NFSH-2 data, India) of 139 ever-married women, aged between 15 and 49 who resided in the state of Bihar, India.

B. VARIABLES

1. Dependent Variable: The dependent variable for the present study is the age at which an infant died. Age is measured in months.

2. Independent Variables: There are three categories of independent variables to be used in the present analysis: proximate, maternal, and household/ community. Table 10.24 presents a description of each set of items and how they are operationalized for the present study.

TABLE 10.24: Operationalization of variables in the IM data analysis. Das, R.N., Dihidar, S. and Verdugo, R. (2011). "Infant mortality in India: Evaluating Log-Gaussian and Gamma," Open Demography Journal, Vol. 4, pp. 34-41

Domain/Variable Name	Operationalization
Proximate	
Tetanus 1. First tetanus shot	1 = At least once, 0 = No
BCG1. Calmette-Gu bacillus (tuberculosis)	1 = Yes, 0 = No
Polio1. First Polio shot	1 = Yes, 0 = No
Polio2. Second Polio shot	1 = Yes, 0 = No
Deliver. Where child was delivered	1 = Hospital or equivalent, 0 = Home
DPT1. First Diphtheria shot	1 = Yes, 0 = No
DPT2. Second Diphtheria shot	1 = Yes, 0 =No
Measles. Measles shot	1 = Yes, 0 = No
Maternal	
Breast (breast feeding)	1 = Yes, 0 = No
Mage (mother's age)	Age in years
Household/Community	
Urban/Rural. Urban or rural residence	1 = urban, 0 = rural
Caste1 (caste is low)	1 = Scheduled, 0 = otherwise
Caste2 (middle caste)	1 = Backward, 0 =otherwise
Religion. Muslim or not	1 = Muslim, 0 = Other
Cfem (gender of infant)	1 = Female, 0 = Male
Dependent Variable	
Cdeath (Age at death of child)	Age in months

TABLE 10.25: Means and Standard Deviations for all Items in the Analysis. Das, R.N., Dihidar, S. and Verdugo, R. (2011). "Infant mortality in India: Evaluating Log-Gaussian and Gamma," Open Demography Journal, Vol. 4, pp. 34-41

Variable Name	Mean/propertion	Standard deviation
Proximate		
Tetanus 1	0.52	0.50
BCG1	0.04	0.19
Polio1	0.10	0.30
Polio2	0.09	0.29
Deliver	0.25	0.43
DPT1	0.05	0.22
DPT2	0.04	0.19
Measles	0.02	0.15
Maternal		
Breast	0.55	0.50
Mage	26.65	6.80
Household/Community		
Urban/Rural		
Caste1	0.36	0.48
Caste2	0.50	0.50
Religion	0.15	3.6
Cfem	0.40	0.49
Dependent Variable		
Cdeath	3.21	5.15

C. FINDINGS

1. Descriptive Statistics: Table 10.25 presents means/ proportions, and standard deviations for all variables to be used in the present models. In terms of Household/ Community items, these data suggest that the vast majority of the sample live in rural areas (96.4%), and are of the lower caste (50.4%). Moreover, it appears that respondents are mostly not Muslim (15%). Finally, note that about 40 percent of the infants that died are female. In terms of Proximate items for infants, most infants do not appear to have received their vaccinations. Only about 4 percent received their first BCG, and 5% their first DPT; 4% their second DPT; and 10% their first polio shot, and 9% their second polio shot. In contrast, about 52 percent have actually been vaccinated for tetanus, and only about two percent of infants received their measles vaccinations.

There are two Maternal items in the present study: breast feeding, and mother's age. A majority of women breast feed their infants (55%), and the average age of mothers in our study is 26.65 years. We should also point out that the average age of death among infants, in months, was 3.21 months.

2. Log-Gaussian and Gamma model analyses

Table 10.26 presents GLM results from the Log-Gaussian Model (LGM) and Gamma Model (GAM) with log-link of infant mortality in Bihar, India.

There are three criteria we apply in conducting the present comparison: We look at the AICs for the best fitting model; we look at the standard errors; and we also examine graphical analysis. First, we look to see if the coefficients appear to be statistically different. Contrary to what we had expected, the models are generally not different in their effects on infant mortality in Bihar, India. The three exceptions are mother age (Mage), place of

TABLE 10.26: Results for mean with constant variance models of infant survival time data from Log-Gaussian & Gamma fits (with all covariates). Das, R.N., Dihidar, S. and Verdugo, R. (2011). "Infant mortality in India: Evaluating Log-Gaussian and Gamma," Open Demography Journal, Vol. 4, pp. 34-41

	Covar.	log-normal model				gamma model			
		estimate	s.e.	t	P-value	estimate	s.e.	t	P-value
Mean	Const.	−1.482	0.81	−1.82	0.07	−1.335	0.83	−1.61	0.11
model	Mage	0.028	0.02	1.76	0.08	0.040	0.02	2.45	0.01
	Rural	0.356	0.57	0.63	0.53	0.282	0.58	0.49	0.62
	Religion	-0.087	0.35	−0.25	0.80	0.107	0.36	0.30	0.76
	Caste1	0.097	0.40	0.24	0.81	0.154	0.41	0.38	0.71
	Caste2	−0.029	0.38	−0.08	0.93	0.029	0.39	0.07	0.94
	Cfem	−0.065	0.22	−0.29	0.77	−0.067	0.23	−0.30	0.76
	Tetanus1	0.131	0.23	0.56	0.57	0.193	0.24	0.81	0.42
	Deliver	0.328	0.29	1.13	0.26	0.607	0.30	2.06	0.04
	BCG1	0.408	1.25	0.33	0.74	1.08	1.36	0.79	0.43
	DPT1	−0.002	1.20	−0.001	0.99	−0.495	1.22	−0.40	0.68
	Polio1	−0.238	1.26	−0.19	0.85	−1.395	0.71	−1.97	0.05
	DPT2	−0.404	1.24	−0.33	0.74	−0.580	1.34	−0.43	0.66
	Polio2	−0.278	1.36	−0.21	0.84	0.625	1.36	0.69	0.49
	Measles	0.066	1.32	0.05	0.96	−0.012	1.36	−0.01	0.99
	Breast	0.974	0.24	4.05	<0.01	1.286	0.24	5.28	<0.01
Const. Var. Model	Const.	0.402	0.13	3.16	<0.01	0.245	0.13	1.93	0.06
AIC		499.100				532.128			

delivery (Deliver) and Polio1, where the Gamma model exhibits significant effects whereas the Log-Gaussian model does not.

How well do each of the models fit the data? Apparently, from the information in Table 10.26, the better fit is the Log-Gaussian model (LGM). The LGM has an AIC of 499.1, while the Gamma Model(GAM) AIC is 532.128.

We also note that the standard errors in the Log-Gaussian model are only slightly smaller than the Gamma model. The implication is that the Log-Gaussian model is only slightly more efficient and bolsters our confidence in using it rather than the Gamma model.

We next examined the model fit based on graphical analysis. In Figure 10.27(a) and Figure 10.27(b), we plot the absolute values of residuals with respect to fitted values for LGM and GAM, respectively. Both the figures show that variances increase with the increase in the means, indicating that variances are non-constant under both the models. The implication is that GLM fits, for both the distributions, are inappropriate. This finding leads us to fit joint GLM for reducing the variance.

Results for fits for the LGM and the GAM are displayed in Table 10.27. In this table we limit the model to those items that were statistically significant.

Table 10.27 shows that joint LGM fit is much better than joint GAM fit based on AIC and standard errors (mean model parameters) as explained above. Indeed, while the GAM fit remains unchanged (532.128 v. 531.897) the LGM fit is reduced by 3.75 percent (480.4 v. 499.1). Figure 10.28(a) and Figure 10.28(b) plot the absolute residuals with respect to fitted values and the normal probability plot for the mean of the joint GAM fit, respectively. Figure 10.28(a) is almost flat with the running means, indicating that the variance is constant under joint GAM fit. Figure 10.28(b) does not show any systematic departures, confirming the final selected gamma model.

Similarly, we plot Figure 10.29(a) (absolute residuals plot) and Figure 10.29(b) (normal probability plot for the mean) for the joint LGM fit. Figure 10.29(a) and Figure 10.29(b)

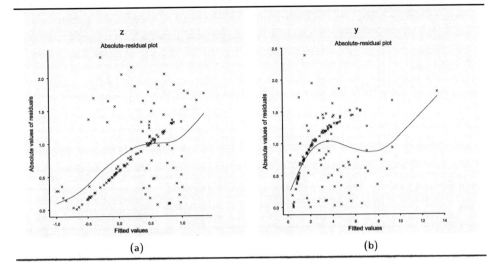

FIGURE 10.27: Absolute residuals plot w.r.to fitted values for constant variance models for (a) LN and (b) GA fits of infant survival time (Table 10.26). Das, R.N., Dihidar, S. and Verdugo, R. (2011). "Infant mortality in India: Evaluating Log-Gaussian and Gamma," Open Demography Journal, Vol. 4, pp. 34-41

show almost the same features as Figure 10.28(a) and Figure 10.28(b). Thus, we conclude that both the fits are satisfactory. In addition, the absolute residuals plot for joint LGM fit is much flatter than the joint GAM fit. Thus, AIC, standard errors for mean model parameters, the number of significant effects, and graphical analysis suggest that the joint LGM fit is much better than the joint GAM fit. The joint LGM fit shows that all the effects are significant (maximum at 6%), whereas in the joint GAM model, the constant and Mother's age fail to meet the 10 percent level of statistical significance (Table 10.27).

Infant mortality in India is an important social and medical problem. In attempting to better understand the issues surrounding this important problem, researchers have focused on factors related to pre-and postnatal factors. The majority of findings from this body of research suggest that medical care trumps economics and social status in reducing infant mortality.

TABLE 10.27: Results for mean & dispersion of infant survival time data from Log-Gaussian & Gamma fit. Das, R.N., Dihidar, S. and Verdugo, R. (2011). "Infant mortality in India: Evaluating Log-Gaussian and Gamma," Open Demography Journal, Vol. 4, pp. 34-41

		log-normal model				gamma model			
	Covar.	estimate	s.e.	t	P-value	estimate	s.e.	t	P-value
Mean	Const.	−1.110	0.35	−3.20	<0.01	−1.257	0.42	−3.03	<0.01
model	Mage	0.025	0.01	1.88	0.06	0.046	0.02	2.96	<0.01
	Deliver	0.513	0.24	2.17	0.03	0.924	0.26	3.52	< 0.01
	Breast	1.025	0.18	5.58	<0.01	1.441	0.20	7.12	<0.01
Dispersion	Const.	−1.681	0.60	−2.79	0.01	−0.960	0.60	−1.61	0.11
Model	Mage	0.035	0.02	1.84	0.06	0.025	0.02	1.29	0.19
	Breast	1.251	0.28	4.42	<0.01	0.598	0.28	2.17	0.03
	Deliver	0.985	0.32	3.05	<0.01	0.600	0.32	1.91	0.06
AIC		480.400				531.897			

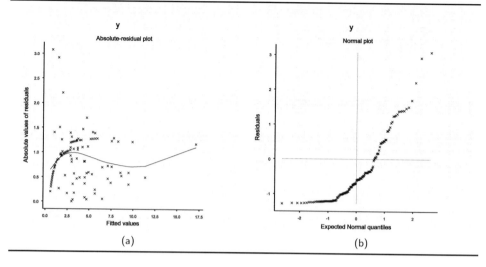

FIGURE 10.28: For GA fit (Table 10.27) (a) absolute residuals plot w.r. to fitted values and (b) normal probability plot for mean of infant survival time. Das, R.N., Dihidar, S. and Verdugo, R. (2011). "Infant mortality in India: Evaluating Log-Gaussian and Gamma," Open Demography Journal, Vol. 4, pp. 34-41

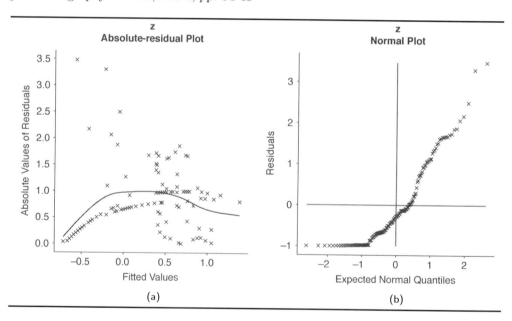

FIGURE 10.29: For LN fit (Table 10.27) (a) absolute residuals plot w.r.to fitted values and (b) normal probability plot for mean of infant survival time. Das, R.N., Dihidar, S. and Verdugo, R. (2011). "Infant mortality in India: Evaluating Log-Gaussian and Gamma," Open Demography Journal, Vol. 4, pp. 34-41

The present study has two purposes. The first is to compare our results to those of previous research. A second purpose is to evaluate the statistical assumption made by previous research regarding the distribution of infant mortality data. Previous research estimated models assuming a Log-Gaussian distribution with constant variance, but infant mortality data appears to be Log-Gaussian or Gamma with non-constant variance. Our concern is that previous research, making the Gaussian assumption and then applying multivariate models, would draw important conclusions from erroneous assumptions. We therefore estimated and compared two models: a Log-Gaussian Model and a Gamma Model. The present results, though not completely conclusive, are revealing.

* The present findings confirm previous research by noting that Proximate and House/Community factors tend to increase the life span of infants.

* Contrary to what we have expected, the models are remarkably similar in their effects on infant mortality.

* There are two exceptions. The intercept and Mage in the joint Log-Gaussian variance model are statistically significant, but not in the joint Gamma variance model. A second exception concerned model fit. The joint Log-Gaussian model exhibited a better fit to the data than the joint Gamma model. The AICs for the joint Log-Gaussian and joint models are: 480.400 vs. 531.897.

10.7.3 Child survival time data, analysis, and interpretation

In section 10.7.2, it is observed that the sample infant survival time data fits Log-Gaussian distribution. In the present section, a random sample of child survival time data is examined (Das and Dihidar, 2013) based on Log-Gaussian and Gamma models. It is found that the Log-Gaussian fit is better than the Gamma fit (based on AIC and graphical analysis), so only the results of Log-Gaussian fit are reported.

A. DATA

In the present analysis, a random sample of 1640 ever-married women, age 15-49 of Bihar State in India is considered from NFSH-2 data. The survival time (or lifetime) of each child in the present study is less or equal to five years. Thus, there is not any censored data. The present study aims to identify the factors which affect on child mortality.

For the present study, all the effective covariates from NFSH-2 data are considered. The used covariates and the levels of the factors (of NFSH-2 data) in the present analysis are the following: (1.) mother's age (in yrs.) (AGE or X), (2.) place of residence (RESD or A) (urban = 1, rural = 2), (3.) religion (RELG or B) (Hindu = 1, Muslim = 2 and other = 3), (4.) caste (CAST or C) (SC = 1, ST= 2 & other = 3), (5.) mother's education (MEDU or D) (below primary = 1, primary to below School Final (SF) = 2, and SF & above = 3), (6.) birth order number (BON or X_1) (eldest = 1 and so on), (7.) age (in months) at death of child (Y), (8.) total children ever born (TCEB or X_2), and (9.) standard of living index (SLI) (E) (low = 1, medium = 2, and high = 3).

B. FINDINGS

We fit the joint Log-Gaussian models (non-constant variance) as in (10.6), with significant, and insignificant factors satisfying marginality rule (Nelder, 1994), and the final results are displayed in Table 10.28. Here we consider survival time (age at death) (y) of children who died on or before reaching age five years as the response, and the other covarites and factors, namely, mother's age (X), place of residence (A), religion (B), caste (C), mother's education (D), birth order number (X_1), total children ever born (X_2), and standard of living index (E) are used as explanatory variables. We select the final model on the basis of the smallest AIC value in each class.

In Figure 10.30(a), we plot the absolute value of residuals with respect to the fitted values, and in Figure 10.30(b), we plot the normal probability plot for the mean model in Table 10.28 of the final selected Log-Gaussian models. Figure 10.30(a) is almost flat with the running means, indicating that joint modeling of mean and variance (Table 10.28) would be

TABLE 10.28: Final results for mean and dispersion models of child survival time data from Log-normal fit

	Covar.	estimate	s.e.	t	P-value
Mean	Const.	−0.08	0.248	−0.32	0.75
model	AGE (X)	0.03	0.005	6.22	<0.01
	RESD ($A2$)	0.25	0.169	1.49	0.13
	CAST ($C2$)	−0.06	0.187	−0.31	0.75
	CAST ($C3$)	−0.26	0.099	−2.61	0.01
	SLI ($E2$)	0.20	0.096	2.08	0.04
	SLI ($E3$)	0.14	0.181	0.79	0.43
Dispers.	Const.	0.71	0.155	4.58	<0.01
model	AGE (X)	0.01	0.004	2.39	0.02
	MEDU ($D2$)	−0.07	0.134	−0.52	0.60
	MEDU ($D3$)	−0.33	0.218	−1.51	0.13

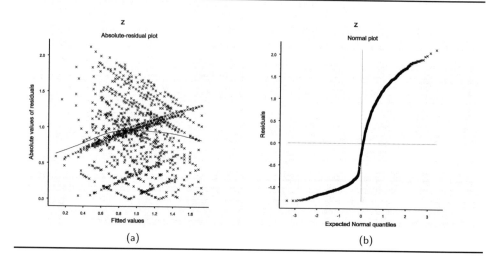

FIGURE 10.30: (a) The absolute residuals plot w.r. to fitted values and (b) normal probability plot for mean model of child survival time (Table 10.28)

satisfactory, and also Figure 10.30(b) does not show any systematic departures, indicating no lack of fit of the final selected Log-Gaussian models in Table 10.28.

Final models (Table 10.28) show many significant factors. In the dispersion model (Table 10.28), mother's age (X) is highly significant and the mother's education level (D) is marginally significant, while in the mean model, mother's age (X), caste (C) and standard of living index (E) are highly significant, and place of residence (A) is marginally significant.

The following are the *main* findings of the present analysis: (1.) It has been identified that the distribution of child survival time sample data is Log-Gaussian (with non-constant variance). (2.) In the mean model, many factors such as mother's age (X), caste (C), and standard of living index (E) are significant. The place of residence (A) is also marginally significant. But in the earlier studies, the significant factors are only the nutritional status and poor living conditions. (3.) Mother's age is an important factor which is positively associated with child survival time. It means that mother's experience increases the child survival time. (4.) Place of residence, a positively associated factor, is also significant in child survival time. Table 10.28 shows that the child survival time is higher in a rural place than an urban place. In Bihar, many people live in a very poor condition mainly in urban areas, so that children are easily affected with many diseases. (5.) Factor caste (*C*) is highly significant in the mean model. In Bihar, it is still observed that people are classified with respect to caste, which is a very sensitive factor in Bihar. (6.) In the mean model, standard of living index is positively significant. This means that higher level of standard of living index increases the child survival time. (7.) It has been detected that the variance of child survival time is non-constant (heteroscedastic). (8.) In the variance model, the covariate mother's age (*X*) is positively significant, and the mother's education level (*D*) is negatively marginally significant (P-value is 0.13). This implies that higher education level of mother and her low age can reduce the variance of child survival time. (9.) Standard error of all the estimates are very small, indicating that estimates are stable (Lee *et al.*, 2006). (10.) Mother's education level should be positively associated with the mean child survival time. Generally, it increases the child survival time. In the *final* fitted mean model, it is *insignificant*, which implies that the mother's education level for this data set is *not* adequate to take care of her child. But in the dispersion model, mother's education level is negatively associated, which implies that higher level of education of mother can reduce the variance of child mortality. Practitioners need to take care for all these factors to reduce the child mortality.

There are two conclusions that can be drawn from the present study. First, in order to reduce infant and child mortality in India, policy and practice should continue to focus on pre-and post natal care. The present findings and those of previous research have continually shown that such practices reduce infant and child mortality.

A second conclusion has to do with the use of statistical models. While further research is called for, we find that a joint Log-Gaussian model is much more effective than either traditional Log-Gaussian (with constant variance) or joint Gamma models because it better fits the data. In short, research should have greater faith in these results than those emanating from the joint Gamma and Log-Gaussian (with constant variance) models.

10.8 AN APPLICATION OF GAMMA MODELS IN HYDROLOGY

In the present section, log-normal and gamma models are used in hydrogeology. As an illustration, groundwater quality characteristics are analysed. Softness, non-alkalinity, content dissolved oxygen, chemical oxygen demand, chloride content and electrical conductivity are all the basic positive characteristics (i.e., values are positive in nature) for good drinking water. The causal factors are identified for these six basic quality characteristics of groundwater sample at Muzaffarpur Town, Bihar, India, using the log-normal and gamma models. Many statistical significant factors are detected and their effects are examined on each groundwater quality characteristic.

10.8.1 Background of drinking groundwater quality characteristics

Water is necessary for most life on Earth. Men can survive for several weeks without food, but for only a few days without water. Most people over the Earth rely on groundwater as the source of drinking water (American Public Health Association, 1995). The attention currently being focused on diffuse pollution has led to an increased recognition of the role of groundwater as a pathway for transporting diffuse pollution to surface water (Grande *et al.*, 1996). Several researches focus that the groundwater is contaminated with many physico-chemical parameters (WHO, 1971; Subbarao *et al.*, 1996; Medha and Patil, 2008; Dhake *et al.*, 2008).

The guidelines of drinking water quality is given by WHO (1971). United States Public Health (USPH) drinking water standards are given by American Public Health Association (APHA, 1995). These are given in many research articles (Mumtazuddin *et al.*, 2009). These are not reproduced here. The basic characteristics of drinking groundwater quality are total alkalinity, total hardness, content dissolved oxygen, chloride content, electrical conductivity and chemical oxygen demand. Dissolved oxygen is an important characteristic for water purity. It is well known that dissolved oxygen depends on temperature and other chemical and biological parameters (Mumtazuddin *et al.*, 2009; Srinivas *et al.*, 2000). Chloride ion concentration bears a direct relationship with mineral contents of the respective water samples. Electrical conductivity is an index of salinity, which greatly affects on taste (Srinivas *et al.*, 2000; Patil and Deore, 2004).

Several researches focus the effects of different factors on quality characteristics of drinking groundwater based on chemical analysis (Grande *et al.*, 1996; Istok and Rautman, 1996). It is quite difficult to examine the effects of many causal factors on a process simultaneously, based on chemical analysis. But it is quite easy to identify the effects through probabilistic models. For positive data analysis, GLM and joint GLM (based on log-normal and gamma) are the most appropriate methods (Firth, 1988; McCullagh and Nelder, 1989; Myers *et al.*, 2002; Das and Lee, 2009).

In the present section, log-normal and gamma models are used in hydrogeology (Das and Kim, 2012). As an illustration, groundwater quality characteristics are analysed. Little is known in literature for detecting casual factors on quality characteristics of drinking groundwater based on statistical analysis. In the present section it is aimed to detect the factors affecting on total alkalinity, total hardness, content dissolved oxygen, chloride content, chemical oxygen demand and electrical conductivity of ground drinking water samples at different sites of Muzaffarpur Town, Bihar, India. Based on these models, the effects of these factors are examined on each characteristic of the sample drinking groundwater. Many interesting facts are focused.

The present analyses (Das and Kim, 2012) identify the following. Chemical oxygen demand (a constant variance response) is negatively associated with electrical conductivity, dissolved oxygen, and is positively associated with chloride content. Total alkalinity is negatively associated with dissolved oxygen, chemical oxygen demand, and is positively associated with electrical conductivity. Total alkalinity is a non-constant variance characteristic, and its variance is affected by pH. Total hardness is negatively associated with dissolved oxygen, temperature, and is positively associated with electrical conductivity, calcium and magnesium. Total hardness is a non-constant variance characteristic, and its variance is affected by pH. Dissolved oxygen is negatively associated with chemical oxygen demand, total dissolved solid, and is positively associated with Magnesium, Chloride content. It is a non-constant variance response, and its variance is negatively associated with total hardness, and is partially positively associated with chemical oxygen demand. Electrical conductivity is negatively associated with total hardness, chemical oxygen demand, and is positively associated with Calcium, temperature, total dissolved solid. It is a non-constant variance response, and its variance is negatively associated with chemical oxygen demand. Finally, chloride content is positively associated with total dissolved solid, dissolved oxygen, total alkalinity, chemical oxygen demand, and is negatively associated with only Calcium. Its variance is partially explained by pH. Some of these facts are well-known and some are little known in literature. For example, variances of the most of these responses are non-constant, and these are little known in literature. Moreover, variances are affected by many factors. In addition, mean model of each response variable has many statistical significant factors. Earlier researches could not identify all these significant factors which are identified in the present analyses.

10.8.2   Description, analyses, and interpretation of groundwater data

The present study is based on the data collected by Mumtazuddin *et al.* (2009), where the authors have studied physico-chemical parameters of bored tube well water samples at different sites of Muzaffarpur Town, India. The parameters such as pH, electrical conductivity (EC), temperature (TEP), total dissolved solids (TDS), dissolved oxygen (DO), total alkalinity (TAK), total hardness (THD), Calcium (Ca), Magnesium (Mg), Chloride content (CLD) and chemical oxygen demand (COD) have been studied. The water samples were selected in the consecutive months of October, November and December 2008 from six bored tube wells in the selected sites of different regions of Muzaffarpur Town. A detailed method of data collection is given in Mumtazuddin *et al.* (2009). For ready reference it is reproduced in Appendix (Table Ap.3).

For analysis of positive data, log-normal and gamma models are generally used (Firth, 1988; McCullagh and Nelder, 1989; Myers *et al.*, 2002; Das and Lee, 2009). The present data set has been examined using both the log-normal and gamma models. It has been found that the gamma model fit is better than the log-normal fit, so only the results of gamma model fit is reported.

It is aimed to detect the significant factors of the basic groundwater quality characteristics such as chemical oxygen demand (COD), total alkalinity (TAK), total hardness (THD), content dissolved oxygen (DO), electrical conductivity (EC) and chloride content (CLD). In the present analyses, each interested characteristic of groundwater quality is considered as the response variable and the remaining others are considered as covariates. All covariates are examined to explain the interested response variable. Only the statistical significant factors are included in the model. Final model is selected based on the smallest AIC value and graphical analysis. Thus, we derive models for each interested basic quality characteristic of groundwater in the following sections. These sections also present interpretations, findings and discussions.

10.8.2.1   Analysis of chemical oxygen demand (COD)

This section presents the analysis of chemical oxygen demand (COD) of the sample data given in Appendix (Table Ap.3). We treat COD as the response, and the remaining other variables as covariates. We start with all other covariates (except COD) as the explanatory variables to predict COD. We include those covariates in the model which explain (i.e., statistically significant) COD. Based on the smallest AIC value in each class and graphical analysis, covariates are finally included in the model. Section 10.2 presents constant and non-constant variance gamma models. Analysis (Table 10.29) shows that the response COD is a constant variance. Thus, we analyse COD data using the gamma GLM, and the results are reported in Table 10.29. The selected model has the smallest AIC value in each class. All the included effects in the model are statistically significant.

TABLE 10.29: Results for mean model of chemical oxygen demand for groundwater data from gamma fit (using log link)

	Covariate	estimate	standard error	t	P-value
Mean	Constant	2.1755	0.0988	22.019	<0.01
model	EC	−0.0006	0.0002	−2.878	0.01
	DO	−0.0667	0.0137	−4.878	<0.01
	CLD	0.0022	0.0005	4.741	<0.01
Constant	Constant	−5.0930	0.3780	−13.480	<0.01
variance					

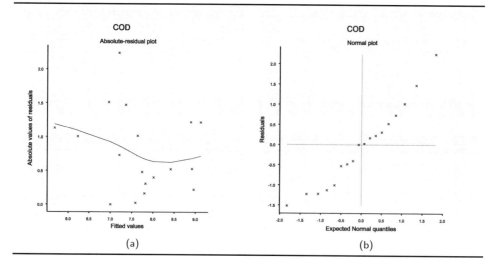

FIGURE 10.31: (a) The absolute residuals plot with respect to fitted values and (b) normal probability plot for mean model (Table 10.29) of COD data

  Figure 10.31(a) and Figure 10.31(b) display respectively the absolute residuals plot with respect to the fitted values and the normal probability plot for the mean model (Table 10.29) of COD data. Figure 10.31(a) is almost flat, except the left tail, with the running means, indicating that variance is constant under the gamma GLM fitting. The left tail of Figure 10.31(a) is little decreasing as the left extreme value is quite large. Figure 10.31(b) does not show any discrepancy in fitting, indicating that there is not any lack of fit. The fitted mean model (with estimated variance $= \hat{\sigma}^2_{CLD} = \exp(-5.093)$) is given below

$$\hat{\mu}_{COD} = \exp(2.1755 - 0.0006EC - 0.0667DO + 0.0022CLD).$$

  It is well-known that variation in COD values may be due to discharge of waste water near the source, and also the COD is maximum in the summer time and minimum in the winter time (Mumtazuddin *et al.*, 2009). Earlier researches do not find any evidence for the variation of COD.

  The present analysis shows that COD (Table 10.29) is negatively associated with electrical conductivity (EC) and dissolved oxygen (DO). This indicates that if EC or DO, or both decreases, COD increases, and vice versa. Again, COD is positively associated with chloride content (CLD), indicating that if CLD increases, COD increases, and vice versa. COD has no direct relation with temperature or any other factors. It is only controlled by EC, DO and CLD. Again from Table 10.33, temperature (TEP) is positively associated with EC. This indicates that if TEP increases, EC increases, consequently, COD decreases. Table 10.33 shows that EC and COD are oppositely associated. It is observed that DO and COD are inversely associated (Table 10.32), and CLD and COD are positively associated (Table 10.34). Table 10.29 shows the relationship of COD with EC, DO and CLD, which is supported by Tables 10.32, 10.33 and 10.34.

10.8.2.2 Analysis of total alkalinity (TAK)

  This section presents the analysis of total alkalinity (TAK). We treat TAK as the response variable, and the remaining other variables as covariates. It is found that TAK is a non-constant variance response. Therefore, we analyse TAK using the gamma joint GLMs

TABLE 10.30: Results for mean and dispersion models of total alkalinity of groundwater data from gamma fit (using log link)

	Covariate	estimate	standard error	t	P-value
Mean	Constant	6.255	0.2205	28.37	<0.01
model	EC	0.001	0.0001	8.79	<0.01
	DO	−0.036	0.0140	−2.54	0.02
	COD	−0.130	0.0212	−6.11	<0.01
Dispersion	Constant	41.600	17.6200	2.36	0.03
model	pH	−6.210	2.3800	−2.62	0.02

as in Section 10.2, and the results are displayed in Table 10.30. The selected models have the smallest AIC value in each class.

Figure 10.32(a) and Figure 10.32(b) present respectively the absolute residuals plot with respect to the fitted values, and the normal probability plot for the mean model (Table 10.30) of TAK data. Figure 10.32(a) is almost flat with the running means, except the right tail, indicating that variance is constant under the gamma JGLM fitting. The right tail of Figure 10.32(a) is little decreasing as the right extreme value is very small. Figure 10.32(b) does not show any lack of fit. The fitted mean and variance models of TAK are respectively,

$$\hat{\mu}_{TAK} = \exp(6.255 + 0.001EC - 0.036DO - 0.130COD),$$

$$\text{and} \qquad \hat{\sigma}^2_{TAK} = \exp(41.60 - 6.21pH).$$

Total alkalinity (TAK) is also a basic quality characteristic of drinking groundwater. Mean model of TAK (Table 10.30) shows that TAK is positively associated with electrical conductivity (EC). This indicates that if EC increases, TAK increases, and vice versa. Again, dissolved oxygen (DO) and chemical oxygen demand (COD) both are negatively associated with TAK. This suggests that if DO or COD, or both increases, TAK decreases, and vice versa. It is well known that oxygen and alkalinity are oppositely associated. This analysis agrees this well known truth. Variance of TAK is negatively associated with pH (Table

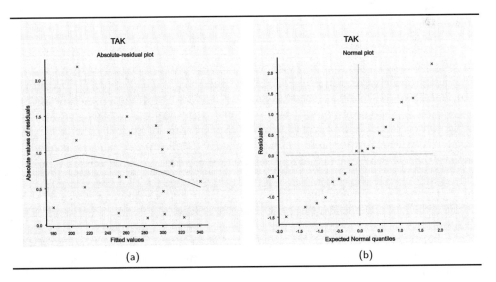

(a)   (b)

FIGURE 10.32: (a) The absolute residuals plot with respect to fitted values and (b) normal probability plot for mean model (Table 10.30) of TAK data

TABLE 10.31: Results for mean and dispersion models of total hardness of groundwater data from gamma fit (using log link)

	Covariate	estimate	standard error	t	P-value
Mean	Constant	11.575	2.017	5.74	<0.01
model	EC	0.010	0.001	6.67	<0.01
	TEP	−0.298	0.055	−5.39	<0.01
	DO	−0.758	0.106	−7.16	<0.01
	Ca	0.945	0.018	53.70	<0.01
	Mg	4.104	0.011	372.17	<0.01
Dispersion	Constant	−46.210	14.867	−3.11	0.01
model	pH	4.540	1.964	2.311	0.04

10.30). This indicates that variance of TAK increases if pH decreases, and vice versa. This analysis suggests that variance of TAK may be minimized with pH, and mean value of TAK can be controlled by DO, COD and EC. Relationship of TAK with its explanatory variables DO, COD, EC and pH can be explained *only* from Table 10.30. Other Tables are unable to focus any such relationship.

10.8.2.3   Analysis of total hardness (THD)

Total hardness (THD) is analysed in the present section. Here THD is considered as the response variable, and all the remaining other variables are treated as covariates. Analysis shows that THD is a non-constant variance response. Thus, we analyse THD data using gamma JGLMs, and the results are reported in Table 10.31. The selected models have the smallest AIC value in each class.

We plot absolute residuals with respect to the fitted values of THD data in Figure 10.33(a), which is almost flat with the running means, except the left tail, indicating that variance is constant under the gamma JGLM fitting. The left tail of Figure 10.33(a) is increasing as the left extreme value is very small. Figure 10.33(b) reveals the normal probability plot for the mean model of THD data, which presents no lack of fit. The fitted mean and variance models of THD are, respectively,

$$\hat{\mu}_{THD} = \exp\left(11.575 + 0.010EC - 0.298TEP - 0.758DO + 0.945Ca + 4.104Mg\right),$$

$$\text{and} \qquad \hat{\sigma}^2_{THD} = \exp(-46.21 + 4.54pH).$$

Mean model of THD (Table 10.31) shows that THD is negatively associated with temperature (TEP) and dissolved oxygen (DO). It indicates that THD reduces if TEP or DO, or both increases, and vice versa. It is well known that THD of water is generally reduced by boiling water. This analysis supports this well known fact. Moreover, this analysis shows that DO can also reduce THD of water. Again THD is positively associated with electrical conductivity (EC), Calcium (Ca) and Magnesium (Mg) (Table 10.31). This indicates THD increases if EC or Ca or Mg, or any two, or all three increases, and vice versa. Table 10.31 shows that Mg and Ca have very strong effect (based on t-value) on THD. Effect of Mg is more than Ca (based on t-value) on THD of water. This finding supports the well known fact of water chemical analysis. Variance of THD is positively associated with pH (Table 10.31). So, variance of THD increases if pH increases, and vice versa. Thus, THD of water is mainly responsible of dissolved Mg, Ca, DO, pH and EC, TEM. Much DO and TEP can reduce THD. Also low EC and small dissolved Ca, Mg can reduce hardness. Small amount of pH reduces the variance of hardness. Thus, a good quality of soft drinking groundwater may be obtained by reducing the amount of Ca, Mg, EC, pH, and increasing DO, TEP.

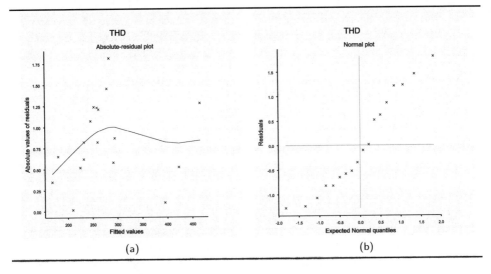

FIGURE 10.33: (a) The absolute residuals plot with respect to fitted values and (b) normal probability plot for mean model (Table 10.31) of THD data

10.8.2.4   Analysis of dissolved oxygen (DO)

Dissolved oxygen (DO) is an important characteristic for pure drinking water. This section presents the analysis of DO. We consider DO as the response variable, and all the remaining other variables as covariates. Analysis shows that DO is a non-constant variance response. So, it is analysed by using the gamma JGLMs, and the results are displayed in Table 10.32. The selected models have the smallest AIC value in each class. It is not necessary that *all* the selected effects in the model are significant (Hastie *et al.*, 2001). COD an *insignificant effect* is included in the dispersion model for an appropriate fitting. All the included effects except COD are significant.

In Figure 10.34(a), we plot the absolute residuals with respect to the fitted values and in Figure 10.34(b), we plot the normal probability plot for the mean model (Table 10.32) of DO data. Figure 10.34(a) shows that variance is constant under the gamma JGLM fitting, as the Figure 10.34(a) is consistent with a flat running means, except the right tail. The right tail of Figure 10.34(a) is little decreasing due to a very small right extreme value. Figure 10.34(b) does not show any discrepancy, indicating that there is not any lack of fit.

TABLE 10.32: Results for mean and dispersion models of dissolved oxygen of groundwater data from gamma fit (using log link)

	Covariate	estimate	standard error	t	P-value
Mean	Constant	3.335	0.245	13.623	<0.01
model	TDS	−0.002	0.001	−3.290	0.01
	Mg	0.008	0.002	4.446	<0.01
	CLD	0.003	0.001	5.945	<0.01
	COD	−0.228	0.031	−7.267	<0.01
Dispersion	Constant	−1.622	4.618	−0.351	0.73
model	THD	−0.036	0.014	−2.555	0.02
	COD	0.923	0.629	1.468	0.16

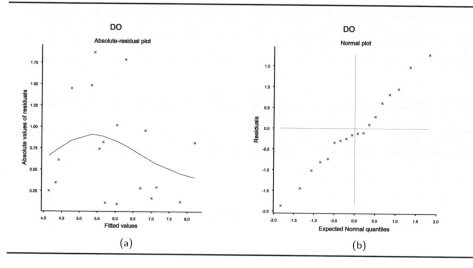

FIGURE 10.34: (a) The absolute residuals plot with respect to fitted values and (b) normal probability plot for mean model (Table 10.32) of DO data

The fitted mean and variance models of DO are respectively,

$$\hat{\mu}_{DO} = \exp(3.335 - 0.002TDS + 0.008Mg + 0.003CLD - 0.228COD),$$

$$\text{and} \quad \hat{\sigma}^2_{DO} = \exp(-1.622 - 0.036THD + 0.923COD).$$

Table 10.32 (mean model of DO) shows that DO is negatively associated with total dissolved solid (TDS) and chemical oxygen demand (COD). This indicates that DO decreases if TDS or COD, or both increases, and vice versa. Also DO is positively associated with Magnesium (Mg) and Chloride content (CLD). This suggests that DO increases if Mg or CLD, or both increases, and vice versa. Variance model of DO (Table 10.32) indicates that DO is negatively associated with total hardness (THD). This implies, if THD increases, variance of DO decreases. It is well-known that if TDS is much in groundwater, DO will be reduced. Contained solids in the groundwater absorbs dissolved oxygen during their dissolving. Consequently, DO will be reduced. For dissolving some organisms in the groundwater, it needs oxygen. This demanded oxygen is obtained from the dissolved oxygen (DO) of the groundwater. Thus, if COD is high, DO will be very low.

10.8.2.5 Analysis of electrical conductivity (EC)

Electrical conductivity (EC) is an index of salinity, which generally affects on taste of the drinking groundwater (Mumtazuddin *et al.*, 2009). We analyse EC, treated as the response variable and the remaining all other variables are considered as covariates. It is identified that EC is a non-constant variance response. Thus, EC is analysed by using the gamma JGLMs and the results are displayed in Table 10.33. The selected models have the smallest AIC value in each class. All the included effects are significant.

Figure 10.35(a) and Figure 10.35(b) display respectively the absolute residuals plot with respect to the fitted values and the normal probability plot for the mean model (Table 10.33) of EC data. Figure 10.35(a) is almost flat with the running means. It indicates that the variance is constant under the gamma JGLM fitting. Figure 10.35(b) does not show any discrepancy in fitting, indicating that there is no lack of fit. The fitted mean and variance

TABLE 10.33: Results for mean and dispersion models of electrical conductivity (EC) of groundwater data from gamma fit (using log link)

	Covariate	estimate	standard error	t	P-value
Mean model	Constant	4.627	0.1749	26.450	<0.01
	TEP	0.029	0.0053	5.409	<0.01
	TDS	0.003	0.0002	12.352	<0.01
	THD	−0.001	0.0002	−3.206	0.01
	Ca	0.008	0.0018	4.400	<0.01
	COD	−0.096	0.0195	−4.893	<0.01
Dispersion model	Constant	0.743	3.3020	0.225	0.82
	COD	−0.866	0.4340	−1.995	0.07

models of EC are respectively,

$$\hat{\mu}_{EC} = e^{(4.627+0.029TEP+0.003TDS-0.001THD+0.008Ca-0.096COD)},$$

and $\quad \hat{\sigma}^2_{EC} = \exp(0.743 - 0.866COD).$

Table 10.33 (mean model of EC) shows that EC is negatively associated with total hardness (THD) and chemical oxygen demand (COD). This satisfies the mean model in Table 10.29 (i.e., the relation between COD & EC) *but* does not satisfy the mean model in Table 10.31 (i.e., the relation between THD & EC). This indicates, EC increases if THD or COD, or both decreases, and vice versa. Also EC is positively associated with temperature (TEP), total dissolved solids (TDS) and Calcium (Ca). It implies, EC increases if TEP or TDS or Ca, or any two, or all three increases. Table 10.33 (variance model of EC) shows that chemical oxygen demand (COD) is negatively associated with the variance of EC. This indicates that variance of EC decreases if COD increases. The same relation between COD & EC holds in the mean model EC (Table 10.33).

10.8.2.6   Analysis of chloride content (CLD)

Chloride content (CLD) is also a basic characteristic of drinking water. This section presents the analysis of chloride content (CLD), treating it as the response variable, and

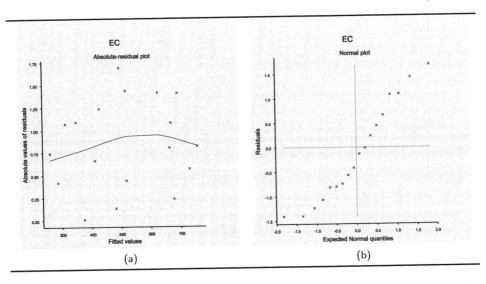

FIGURE 10.35: (a) The absolute residuals plot with respect to fitted values and (b) normal probability plot for mean model (Table 10.33) of EC data

TABLE 10.34: Results for mean and dispersion models of chloride content of groundwater data from gamma fit (using log link)

	Covariate	estimate	standard error	t	P-value
Mean model	Constant	2.412	0.510	4.732	<0.01
	TDS	0.002	0.001	3.216	0.01
	DO	0.078	0.022	3.481	<0.01
	TAK	0.002	0.001	1.838	0.09
	Ca	−0.008	0.003	−2.369	0.03
	COD	0.244	0.044	5.480	<0.01
Dispersion model	Constant	33.360	23.530	1.418	0.18
	pH	−5.080	3.180	−1.598	0.14

the remaining all other variables are considered as covariates. Analysis detects that CLD is a non-constant variance response. Therefore, we analyse the response CLD by using the gamma JGLMs, and the results are reported in Table 10.34. The selected models have the smallest AIC value in each class. pH an *insignificant* effect is included in the dispersion model for better fitting (Hastie *et al.*, 2001). All the included effects except pH are significant.

The absolute residuals plot with respect to the fitted values of CLD data is displayed in Figure 10.36(a). Figure 10.36(a) is almost flat with the running means, except the right tail, an indication that variance is not increasing with the running means. Right tail of Figure 10.36(a) is increasing due to a large right extreme value. Figure 10.36(b) presents the normal probability plot for the mean model of CLD. Figure 10.36(b) does not show any lack of fit, as there is not any discrepancy in the plot. The fitted mean and variance models of CLD are respectively,

$$\hat{\mu}_{CLD} = \exp(2.412 + 0.002TDS + 0.078DO + 0.002TAK - 0.008Ca + 0.244COD),$$

$$\text{and} \qquad \hat{\sigma}^2_{CLD} = \exp(33.36 - 5.08pH).$$

Mean model of CLD (Table 10.34) shows that Calcium (Ca) is negatively associated with CLD. This indicates, CLD increases if Ca decreases, and vice versa. Also CLD is positively

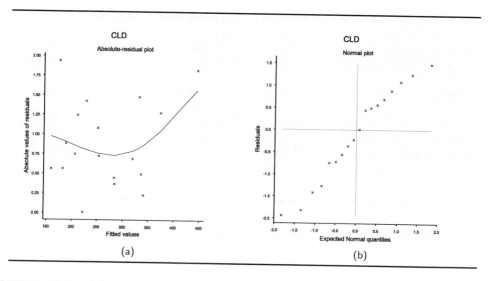

FIGURE 10.36: (a) The absolute residuals plot with respect to fitted values and (b) normal probability plot for mean model (Table 10.34) of CLD data

associated with total dissolved solid (TDS), dissolved oxygen (DO), total alkalinity (TAK) and chemical oxygen demand (COD). This indicates, CLD increases if TDS or DO or TAK or COD, or any two, or any three, or all four increases. Variance of CLD can be reduced by using pH.

For non-constant variance response, joint modeling of both the mean and variance (where both the mean and variance can be modeled simultaneously) is essential (Lee and Nelder, 1998, 2003; Lee et al., 2006). In physical science literature, models are mainly focused on the mean, so that the modeling of the dispersion has often been neglected. This section uses GLMs and JGLMs for modeling the mean and the mean-variance simultaneously in hydrogeology data analyses. Through these models many interesting facts can be explained (Section 10.8.2).

Here we analyse basic quality characteristics of drinking groundwater samples. It detects many statistically significant effects in each characteristic of drinking groundwater samples. Analyses show that all the characteristics are non-constant variance, except the chemical oxygen demand (COD). All the analyses have been done based on the gamma GLMs and JGLMs. It explains the causes of total hardness, total alkalinity, dissolved oxygen, electrical conductivity, chloride content and chemical oxygen demand of groundwater samples. It points many facts that are well established by chemical water analysis. For example, it is well known in physical science that boiling reduces the hardness of water. Dissolved Ca and Mg increase the hardness of water. These analyses supports these facts (Table 10.31). Moreover, it also opens many facts which are very little known in literature. For example, it focuses that the variances of responses (characteristics of drinking groundwater samples) may be non-constant. Variances may be explained by many statistical significant factors. In addition, mean responses (mean models) are also explained by many statistical significant factors. These analyses show that dissolved oxygen (DO) is an important factor for pure drinking water. It reduces the total hardness (THD), total alkalinity (TAK) and chemical oxygen demand (COD). Analyses also show that electrical conductivity (EC) increases both the total hardness and total alkalinity of groundwater samples. pH effects the variance of TAK (Table 10.30), THD (Table 10.31) and CLD (Table 10.34). Included effects in all the models are almost significant except a few. Standard error of all the estimates (Table 10.29 to Table 10.34) are very small, indicating that estimates are stable (Lee et al., 2006). Note that this data set contains only 18 observations along with 1 response and 10 explanatory variables. It is quite difficult to obtain a complete flat diagram for residuals plot only with 18 observations and 10 covariates. Most of the residual plots are almost flat in the present analyses. The relationship of each basic quality characteristic of ground water sample with their explanatory variables is established. Analysts can explain the quality of groundwater sample through these models.

10.9 AN APPLICATION OF LOG-NORMAL AND GAMMA MODELS IN MEDICAL SCIENCE

In the present section, the relationships of human blood biochemical parameters are examined based on the log-normal and gamma models. These studies are performed based on the data collected by Bhattacharyya et al. (2006, 2007) who have studied the body mass index of students of a tertiary level referral and teaching hospital. These authors have studied the family history of the students (i.e., the subjects under study), their lifestyle characteristics, eating habit, and blood biochemical parameters to identify the factors affecting on their body mass index. It has been assumed that the characteristic 'body mass index' is a constant variance response. Here human blood biochemical parameters are examined statistically and many interesting factors are identified.

10.9.1   Background of human blood biochemical parameters

Obesity is an harmful factor for arising many diseases such as diabetes mellitus (DM), hypertension (HT) and coronary heart disease (CHD) etc. (Park, 2005). The risk of mortality from CHD is three to four times more among overweight persons and approximately five times more among obese ones than those with normal weight (Bandolier, 2006). Several studies have been done for detecting significant factors of obesity or body mass index (BMI) (Bhattacharyya *et al.*, 2006, 2007; Katzmarzyk *et al.*, 2004; Obesity, 2004; Poluride, 2002, etc.). To find the significant factors of BMI, many positive characteristics and blood biochemical parameters have been studied (by Bhattacharyya *et al.*, 2006, 2007).

In literature, little is known about the relationship of human blood biochemical parameters. Recently, Das (2011c) has derived the relationship of human blood biochemical parameters based on the data collected by Bhattacharyya *et al.*, (2006, 2007). In the present section, we describe some models on human blood biochemical parameters (following Das, 2011c) such as fasting plasma glucose level (PGL), serum triglyceride (STG), total cholesterol (TC), high-density lipoprotein (HDL), low-density lipoprotein (LDL), and fasting serum insulin (FI), on comparison of the log-normal and gamma models (Sections 10.2), based on the data collected by Bhattacharyya *et al.* (2006, 2007). Many interesting significant factors have been identified for each of the human blood biochemical characteristic which are crucial in any medical treatment process; otherwise wrong selection of factors will increase hazards in human life.

10.9.2   Description, analysis, and interpretations of human blood biochemical parameters

Bhattacharyya *et al.* (2006, 2007) conducted a study on the students of a tertiary level referral and teaching hospital– RG Kar Medical College, Kolkata, West Bengal, India. The subjects of the study were selected by simple random sampling method from the students of RG Kar Medical College, Kolkata, West Bengal, India. The students of this Medical College were taken up in the study; those having extreme dietary habits or any facets influencing the lipid profile were excluded from the study; then 10% of the students were selected by simple random sampling method. Thus, 64 subjects were taken up as a sample in the study after they were informed of the nature of the study and then getting consent from them for inclusion in the study.

The data regarding their identity, sex (SEX) (female = 1, male = 2), age (AGE), height (HGT), weight (WET), obesity (OBS) (under weight = 1, all others = 2), body mass index (BMI), lifestyle (LFT), dietary habits like eating in outside (EOS) (i.e., eating of ready food habit), types of oil consumption (OC) (mustered oil = 1, all others = 2), smoking habit (SMO), family history (FMH) of diabetes mellitus (DMT) (nil = 1, all others = 2), family blood pressure (FBP) (nil = 1, all others = 2), hypertension (HPT) (nil = 1, all others = 2), and coronary heart disease (CHD) (nil = 1, all others = 2), history of past illness, history of any drug intake, and biochemical parameters like fasting plasma glucose level (PGL), serum triglyceride (STG), total cholesterol (TC), high-density lipoprotein (HDL), low-density lipoprotein (LDL), fasting serum insulin (FI), were collected in a predesigned and pretested proforma. The method of collection of this data set and a part of data is given in Bhattacharyya *et al.* (2007). The BMI was calculated by the formula 'weight (in kg)/height^2 (in meter)'. There are two types of characteristics (factors and covariates) in the data set. In the present analysis, we level each factor in two groups, namely low and high, where

TABLE 10.35: Results for mean and dispersion models of fasting serum insulin data from log-normal and gamma fits

| | Covar. | log-normal model | | | | gamma model | | | |
		estimate	s.e.	t	P-value	estimate	s.e.	t	P-value
Mean	Const.	3.54	0.42	8.33	<0.01	3.78	0.42	9.03	<0.01
model	AGE	−0.05	0.02	−2.70	0.01	−0.05	0.02	−2.75	0.01
	OBS2	0.54	0.16	3.41	<0.01	0.46	0.16	2.93	<0.01
	CHD2	−0.22	0.18	−1.19	0.24	−0.25	0.18	−1.39	0.17
	SMO2	−0.17	0.19	−0.89	0.37	−0.28	0.19	−1.47	0.15
Disp.	Const.	2.87	1.55	1.86	0.07	2.53	1.48	1.71	0.09
model	BMI	−0.16	0.06	−2.51	0.01	−0.15	0.06	−2.47	0.02
AIC		455.30				455.97			

high group includes all levels except the low level. In the following sections, all the blood biochemical parameters (fasting plasma glucose level (PGL), serum triglyceride (STG), total cholesterol (TC), high-density lipoprotein (HDL), low-density lipoprotein (LDL) and fasting serum insulin (FI)) are analysed by using the log-normal and gamma models (Section 10.2).

10.9.2.1   Analysis of fasting serum insulin (FI)

In the present section, fasting serum insulin (FI) is analysed, considering FI is the response variable, and the remaining others are considered as the explanatory variables. For factors, we take a constraint that the effects of the first levels in the factors are zero. The selected models have the smallest AIC value in each class.

It is identified that fasting serum insulin is a non-constant variance response, so, FI is analysed using both the log-normal and gamma joint GLMs and the results are displayed in Table 10.35. Two *insignificant* factors 'coronary heart disease (CHD)' and 'smoking habit (SMO)' are included in the mean models for better fitting (following Hastie *et al.*, 2001). These two fitted models (Table 10.35) are almost same according to AIC rule, so there is no model choice.

Figure 10.37(a) reveals the absolute residuals plot with respect to the gamma fitted values (in Table 10.35) which is almost flat with the running means, indicating that variance is constant under the structured gamma models (Table 10.35). Figure 10.37(b) displays the normal probability plot for the gamma fitted mean model (in Table 10.35) which is consistent with the fitting.

Age is negatively and obesity is positively associated with the mean model of FI (Table 10.35). If age increases, FI will decrease. Again, if obesity increases, FI will increase. BMI is negatively associated with the variance of FI (Table 10.35). For the individuals with higher value of BMI, the variance of FI (Table 10.35) will be small.

10.9.2.2   Analysis of total cholesterol (TC)

In the present section, the response total cholesterol (TC) is analysed by using both the joint models (Section 10.2), and the results are reported in Table 10.36. Note that the covariate 'high-density lipoprotein (HDL)' is *significant* in the gamma fitted mean model but it is *alias* in the log-normal fitted mean model. Serum triglycerides (STG) is significant and fasting plasma glucose level (PGL) is *marginally significant* in both the mean models (Table 10.36). Here all the included effects (in both the models) are almost significant. Both the fitted models are completely different, and AIC clearly chooses the gamma model.

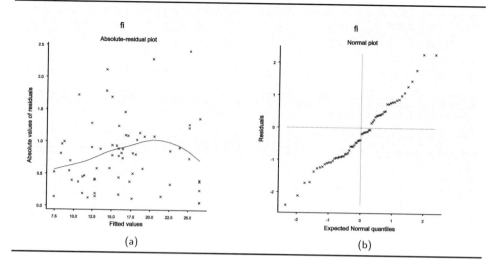

(a)   (b)

FIGURE 10.37: (a) The absolute residuals plot w.r.to fitted values and (b) normal probability plot for GA fitted mean model (Table 10.35) of FI data

In Figures 10.38(a) and 10.38(b), we plot respectively the absolute residuals with respect to the fitted values and the normal probability plot for the gamma fitted mean model (Table 10.36). Both the plots are satisfactory for the final selected gamma fitted models in Table 10.36. Note that the right tail of Figure 10.38(a) is decreasing due to a small value at the right extreme. Similarly, as in Tables 10.3 and 10.4, we use '0.00∗' in the standard error column in Table 10.36 to represent the standard error whose first two decimal places are zeros.

Serum triglycerides (STG), high-density lipoprotein (HDL) and fasting plasma glucose level (PGL) are all positively associated with total cholesterol (TC) (Table 10.36). If STG or HDL or PGL, or any two, or all three increases, TC will increase. Both the HDL and FI are positively and types of oil consumption (OC) is negatively associated with the variance of TC (Table 10.36). If HDL or FI, or both increases, variance of TC will also increase. Variance of TC will increase at the lowest type of oil (mustered oil = 1) consumption. Therefore, TC will be reduced if STG, HDL and PGL will be small. Variance of TC can be reduced by consuming other types of oil, and it will be small if HDL and FI are small.

TABLE 10.36: Results for mean and dispersion models of total cholesterol data from log-normal and gamma fits

	Covar.	log-normal model				gamma model			
		estimate	s.e.	t	P-value	estimate	s.e.	t	P-value
Mean model	Const.	4.56	0.09	47.96	<0.01	4.31	0.13	34.12	<0.01
	HDL	−	−	−	−	0.008	0.00∗	3.05	<0.01
	STG	0.001	0.00∗	8.27	<0.01	0.002	0.00∗	8.80	<0.01
	PGL	0.002	0.00∗	1.86	0.07	0.002	0.00∗	1.67	0.10
Disp. model	Const.	−5.09	0.94	−5.44	<0.01	−4.54	1.04	−4.37	<0.01
	HDL	0.06	0.02	3.10	<0.01	0.04	0.02	1.96	0.05
	FI	0.03	0.01	2.37	0.02	0.03	0.01	2.21	0.03
	OC2	−1.36	0.41	−3.30	<0.01	−1.36	0.43	−3.16	<0.01
AIC		$633.70 + 2 \times 7$				$619.43 + 2 \times 8$			

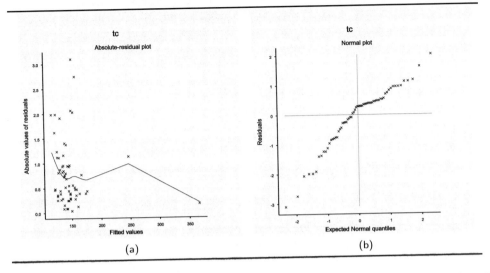

FIGURE 10.38: (a) The absolute residuals plot w.r.to fitted values and (b) normal probability plot for GA fitted mean model (Table 10.36) of TC data

10.9.2.3 Analysis of serum triglycerides (STG)

In the present section, serum triglycerides (STG) is treated as the response variable and the remaining all other factors and covariates are considered as the explanatory variables. It is identified that STG is a constant variance response. Therefore, STG is analysed by using the gamma model (with constant coefficient variance) and the log-normal model (with constant variance) (Section 10.2), and the results are displayed in Table 10.37. Weight (WET) an *insignificant* characteristic is included in both the models for better fit. Estimates from both the fitted models (Table 10.37) are almost identical but AIC chooses the log-normal model.

In Figure 10.39(a) and Figure 10.39(b), we plot respectively the absolute residuals with respect to the fitted values and the normal probability plot for the fitted log-normal mean model (Table 10.37). Middle part of Figure 10.39(a) is almost flat with the running means,

TABLE 10.37: Results for mean model of serum triglyceride data from log-normal and gamma fits

		log-normal model				gamma model			
	Covar.	estimate	s.e.	t	P-value	estimate	s.e.	t	P-value
Mean	Const.	3.28	0.48	6.89	<0.01	3.23	0.49	6.66	<0.01
model	HDL	−0.01	0.01	−2.14	0.04	−0.01	0.01	−2.07	0.04
	TC	0.01	0.00*	5.46	<0.01	0.01	0.00*	5.26	<0.01
	AGE	0.03	0.01	2.53	0.02	0.04	0.01	2.87	0.01
	WET	0.01	0.00*	1.47	0.15	0.01	0.00*	1.42	0.16
	SMO2	−0.19	0.12	−1.61	0.11	−0.22	0.12	−1.87	0.06
	FBP2	−0.20	0.09	−2.11	0.04	−0.19	0.09	−1.97	0.05
Const. Var. model	Const.	−2.17	0.18	−11.80	<0.01	−2.15	0.18	−11.70	<0.01
AIC		655.30				658.94			

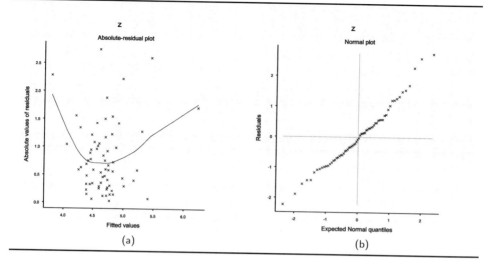

FIGURE 10.39: (a) The absolute residuals plot w.r.to fitted values and (b) normal probability plot for fitted LN mean model (Table 10.37) of STG data

while its left tail is decreasing and right tail is increasing due to two large values at the two extremes. The normal probability plot Figure 10.39(b) does not show any lack of fit.

For the final selected log-normal model (Table 10.37), total cholesterol (TC) and age are positively associated with serum triglycerides (STG). If TC or age, or both increases, STG will increase. Again, high-density lipoprotein (HDL) and family blood pressure (FBP) are negatively associated with STG. If HDL decreases, STG increases. If FBP is at low level, STG will increase.

10.9.2.4   Analysis of low-density lipoprotein (LDL)

In the present section, low-density lipoprotein (LDL) is considered as the response variable, and the remaining other factors and covariates are considered as the explanatory variables. LDL is identified as a non-constant variance response, therefore, it is analysed by using both the joint log-normal and gamma models (Section 10.2), and the results are given in Table 10.38. Estimates in both the models are almost same, but AIC clearly chooses the log-normal models.

In Figure 10.40(a) and Figure 10.40(b), we plot respectively the absolute residuals with respect to the fitted values and the histogram plot of residuals for the fitted log-normal model (Table 10.38). In Figures 10.41(a) and 10.41(b), we plot respectively the normal probability plot for the mean and dispersion for the fitted log-normal model (Table 10.38). Middle part of Figure 10.40(a) is almost flat with the running means, while its left tail is decreasing and right tail is increasing due to two large values at the two extremes. Figures 10.40(b), 10.41(a) and 10.41(b) do not show any lack of fit. So the fitted log-normal model is considered to be an appropriate model for the LDL data.

Note that in Table 10.38, W.P* (first column) denotes the interaction effect between weight and PGL (i.e., WET×PGL). All the factors and covariates included in both the models (Table 10.38) are highly statistically significant. TC, SMO, BMI, FI and WET×PGL are all positively associated with LDL (Table 10.38). If at least any one of these factors (TC, SMO, BMI, FI and WET×PGL) increases, LDL will increase. STG, SEX, FBP, HDL, HGT, OBS, WET, PGL are all negatively associated with LDL. If at least any one of STG, HDL, HGT, WET and PGL decreases, LDL will increase. Note that HDL and LDL are inversely

TABLE 10.38: Results for mean and dispersion models of low-density lipoprotein data from log-normal and gamma fits

		log-normal model				gamma model			
	Covar.	estimate	s.e.	t	P-value	estimate	s.e.	t	P-value
Mean model	Const.	5.09	0.69	7.37	<0.01	5.14	0.70	7.33	<0.01
	TC	0.01	0.00*	19.46	<0.01	0.01	0.00*	18.36	<0.01
	SMO	0.09	0.04	2.36	0.02	0.09	0.04	2.52	0.01
	STG	−0.003	0.00*	−10.59	<0.01	−0.003	0.00*	−10.09	<0.01
	SEX2	−0.08	0.05	−1.74	0.09	−0.09	0.05	−1.85	0.07
	FBP2	−0.14	0.03	−4.72	<0.01	−0.14	0.03	−4.68	<0.01
	HDL	−0.02	0.00*	−11.07	<0.01	−0.02	0.00*	−10.59	<0.01
	BMI	0.04	0.01	3.13	<0.01	0.05	0.01	3.33	<0.01
	HGT	−0.002	0.00*	−4.15	<0.01	−0.002	0.00*	−4.49	<0.01
	CHD2	0.10	0.03	3.26	<0.01	0.10	0.03	3.26	<0.01
	FI	0.01	0.00*	3.10	<0.01	0.01	0.00*	3.52	<0.01
	OBS2	−0.16	0.06	−2.51	0.01	−0.15	0.07	−2.22	0.03
	WET	−0.03	0.01	−3.28	<0.01	−0.04	0.01	−3.38	<0.01
	PGL	−0.03	0.01	−3.01	<0.01	−0.03	0.01	−3.10	<0.01
	W.P*	0.001	0.00*	2.90	0.01	0.001	0.00*	2.99	<0.01
Disp. model	Const.	−16.20	1.73	−9.39	<0.01	−16.65	2.02	−8.25	<0.01
	BMI	0.50	0.07	6.93	<0.01	0.52	0.08	6.24	<0.01
	FI	0.04	0.02	2.31	0.02	0.05	0.02	2.68	0.01
AIC				505.2				513.22	

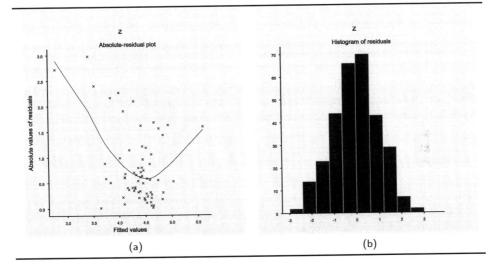

(a)   (b)

FIGURE 10.40: (a) The absolute residuals plot w.r.to fitted values and (b) Histogram plot of residuals for fitted LN model (Table 10.38) of LDL data

related. LDL is inversely related with the factors SEX, FBP and OBS. LDL will be high at the lower level of each of the factors SEX (i.e., female = 1), FBP and OBS. Therefore, LDL will be higher for a female than a male, and it is also higher for an individual with no FBP and under weight. This is also supported by the relation that WET and LDL are inversely related (Table 10.38). Note that, LDL is higher for a shorter individual than a longer.

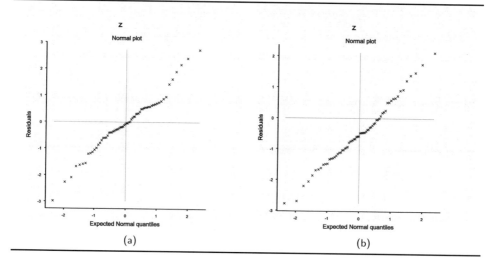

FIGURE 10.41: Normal probability plot for (a) mean and (b) dispersion for fitted LN model (Table 10.38) of LDL data

10.9.2.5   Analysis of high-density lipoprotein (HDL)

In this section, HDL is considered as the response variable and the remaining others are considered as the explanatory variables. It is identified that HDL is a non-constant variance response. Therefore, HDL is analysed by using both the joint log-normal and gamma models (Section 10.2) and the results are presented in Table 10.39. All the included effects are significant. Estimates from both the fitted models are almost identical, and according to the AIC rule, there is no model preference.

In Figure 10.42(a) and 10.42(b), we plot, respectively, the absolute residuals with respect to the fitted values and the normal probability plot for the gamma fitted mean model (Table 10.39) of HDL data. Figure 10.42(a) is completely a flat diagram with the running means, indicating that variance is constant under the structured gamma models. Figure 10.42(b) does not show any systematic departures, indicating no lack of fit of the final selected gamma models (Table 10.39).

Age and sex both are negatively associated with HDL (Table 10.39). It indicates that as age increases, HDL decreases. Normal range of HDL is more than 40 mg./dl. So, it is predicted that HDL will be low at higher age. HDL is higher for a female than a male. BMI is negatively associated with the variance of HDL.

TABLE 10.39: Results for mean and dispersion models of high-density lipoprotein data from log-normal and gamma fits

		log-normal model				gamma model			
	Covar.	estimate	s.e.	t	P-value	estimate	s.e.	t	P-value
Mean	Const.	4.30	0.12	35.77	<0.01	4.30	0.12	35.84	<0.01
model	AGE	−0.02	0.01	−4.46	<0.01	−0.02	0.01	−4.41	<0.01
	SEX2	−0.12	0.04	−2.76	0.01	−0.12	0.04	−2.66	0.01
Disp.	Const.	2.22	1.56	1.42	0.16	2.10	1.56	1.35	0.18
model	BMI	−0.24	0.07	−3.60	<0.01	−0.23	0.07	−3.55	<0.01
AIC		444.90				444.56			

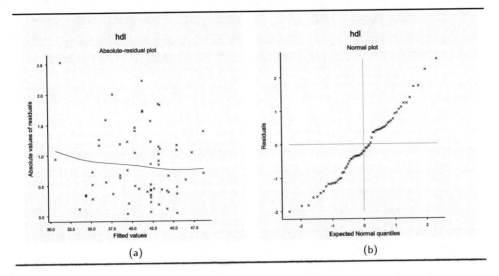

FIGURE 10.42: (a) The absolute residuals plot w.r.to fitted values and (b) normal probability plot for fitted GA mean model (Table 10.39) of HDL

10.9.2.6 Analysis of fasting plasma glucose level (PGL)

In this section, plasma glucose level (PGL) is considered as the response variable and the remaining others are considered as the explanatory variables. PGL is identified as a non-constant variance response. So, PGL is analysed by using both the JGLMs, and the results are reported in Table 10.40. Estimates from both the models are *not* same. Only the covariate 'low-density lipoprotein (LDL)' is significant in both the mean models whereas many factors and covariates are significant in both the dispersion models. Note that the factor 'family blood pressure (FBP)' is *significant* in the fitted gamma dispersion model, but it is *alias* in the fitted log-normal dispersion model (Table 10.40). Here both the fitted models are quite different, and AIC clearly selects the gamma model.

Figure 10.43(a) presents the absolute residuals plot with respect to the fitted values. It is completely a flat diagram with the running means. So, under the structured gamma fitted models, heteroscedasticity has been completely removed. Figure 10.43(b) reveals the normal probability plot for the gamma fitted mean model (Table 10.40), which does not show any lack of fit.

Low-density lipoprotein (LDL) is positively associated with PGL (Table 10.40). If LDL increases, PGL is also increases. Variance of PGL is positively associated with AGE, BMI, SEX, DMT, and it is negatively associated with STG, EOS, FBP. Therefore, variance of PGL increases if AGE or BMI or both increases. Variance of PGL is higher for a male (or a DMT patient) than a female (or a non-DMT patient). Variance of PGL is inversely related with STG, EOS and FBP. Therefore, many factors are highly associated with the variance of PGL.

Relationships of some human blood biochemical parameters are examined here. It is observed that human blood biochemical parameters must have some probabilistic relationships. Based on these relationships, medical practitioners can examine the nature of variability of each blood biochemical parameters given the values of others. Consistency of blood pathological reports can be examined through these relationships. The used data set does not contain many other blood biochemical parameters such as blood urea, uric acid, creatinine etc.

TABLE 10.40: Results for mean and dispersion models of fasting plasma glucose level data from log-normal and gamma fits

		log-normal model				gamma model			
	Covar.	estimate	s.e.	t	P-value	estimate	s.e.	t	P-value
Mean	Const.	4.21	0.02	253.56	<0.01	4.21	0.01	288.82	<0.01
model	LDL	0.001	0.00*	2.17	0.03	0.001	0.00*	2.67	0.01
	TC	0.00*	0.00*	1.02	0.31	0.00*	0.00*	1.61	0.11
	STG	0.00*	0.00*	−0.73	0.47	0.00*	0.00*	−1.17	0.25
Disp.	Const.	−16.06	2.06	−7.79	<0.01	−13.13	1.83	−7.19	<0.01
model	AGE	0.30	0.10	2.93	0.01	0.20	0.09	2.14	0.04
	BMI	0.23	0.04	5.10	<0.01	0.19	0.05	4.15	<0.01
	STG	−0.02	0.01	−4.89	<0.01	−0.02	0.01	−4.31	<0.01
	SEX2	4.36	0.44	9.95	<0.01	3.83	0.45	8.45	<0.01
	EOS	−0.01	0.00*	−5.47	<0.01	−0.01	0.00*	−4.44	<0.01
	DMT2	2.39	0.43	5.54	<0.01	2.24	0.42	5.36	<0.01
	FBP2	−	−	−	−	−1.20	0.40	−2.99	<0.01
AIC		$531.90 + 2 \times 11$				$502.84 + 2 \times 12$			

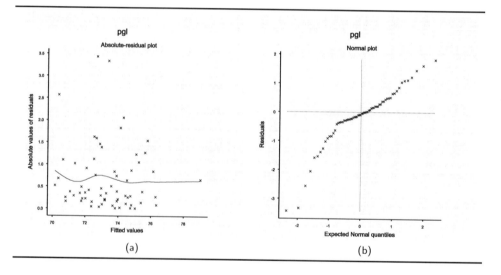

FIGURE 10.43: (a) The absolute residuals plot w.r.to fitted values and (b) normal probability plot for GA fitted mean model (Table 10.40) of PGL

Recently, Das and Mukhopadhyay (2009) have derived the relationship of neonatal weights on thyroid function of mother and her neonate, and maternal urinary iodine excretion. From this relationship, neonatal weight can be predicted given the values of others. Das (2011b) has examined the relationships of thyroid function of mother, and her neonate. Through these relationships, many interesting facts have been explained. Das and Sarkar (2012) have detected the statistically significant impact of personal characteristics and diet on plasma concentrations of retinol and beta-carotene using statistical modeling. Based on probabilistic models, Das (2013b) has shown that wine drinking has individual, interaction and confounding effects on all the components of liver biochemical markers. In the process, the relationship of each biomarker is established with wine drinking and the remaining other biomarkers. Effects of wine drinking on each component of liver biomarkers are identified

(Das, 2013b). Determinants of diabetes mellitus have been identified through statistical modeling (Das, 2014a).

10.10   CONCLUDING REMARKS

Some continuous positive process variables may have non-Normal error distributions, and the class of generalized linear models includes distributions useful for the analysis of data resulting from that process. The simplest examples are perhaps the exponential and the gamma distributions, which are often useful for modeling positive data that have variance with mean relationship, and the variance of the response is non-constant. For heteroscedastic data, the log-transformation is often recommended for stabilizing the variance (Box and Cox, 1964). However, in practice the variance may not always be stabilized under a proper (seems to be suitable) transformation (Myers et al., 2002, Table 2.7, p. 36). It is well known that if the variance is constant, the predicted parameters from the log-normal and the gamma models have a common interpretation (Firth, 1988).

The present chapter focuses the log-normal and gamma models for constant and non-constant variances. Discrepancy of regression estimates and also fittings have been shown in both the models. Taguchi's and dual response approaches are compared with the GLM and JGLM approaches. Dual response approach has been extended to GLM and JGLM. A few applications of both these models are illustrated with real data in the different field of sciences.

Chapter 11

GENERAL CONCLUSIONS AND DISCUSSIONS

The problems considered in this book (described in Chapter 1) have their origin in practice, as in reality the usual assumptions related to the distribution of the response variable and the random errors have to be relaxed and generalized to include some specific pattern in the correlation structure of errors. It is assumed that the response function can be represented by a polynomial of degree one or two. All the described results related to robust rotatability, slope-rotatability, optimality etc., are only for first- and second-order response models. Similar study can be made for higher order polynomials also. Some results on third order models have been communicated which are not included in this book.

This book initiates the concept of robust response surface designs, along with the relevant regression and positive data analysis techniques. It contains mainly three parts namely, response surface designs, regression analysis with correlated errors, and positive data analysis. Randomized block design analysis with correlated errors is included in regression analysis. Robust first- and second-order rotatable, weakly rotatable and optimal designs have been described for normally and non-normally (log-normal and exponential) distributed responses. This book also presents robust second-order slope-rotatability along axial directions, over all directions, with equal maximum directional variance, D-optimal slope and modified slope-rotatability. Regression analysis with correlated errors and its application in block design analysis have been described. Log-normal and gamma models with constant and non-constant variances have been described. Discrepancy of regression estimates and also in fitting between the log-normal and gamma models for constant and non-constant variances have been illustrated.

This is the first book on robust response surface methodology (RRSM) with its relevant and related topics. The research on this area is at infancy. This area is initiated by the present author. Only the present author and with his collaborators have focused some parts of this area. This book contains only the research out put of the present author and with his collaborators. So, a little of this topic is included as an introductory book. Hope that the research on this topic will be enriched in near future, resulting this introductory book will be divided into three or more independent books. At the end of each chapter, a concluding remarks section is included. A general conclusion, some discussions and perspectives of the contents of the book are given below chapter wise.

In Chapter 2, first-order optimality and rotatability have been examined for a normally distributed response with correlated errors under intra-class, inter-class, compound symmetry, tri-diagonal and autocorrelated error structures. The study of first- and second-order rotatability and optimality under intra-class structure is at a certain level, as the results for this error structure are completely identical with the usual (uncorrelated and homoscedastic errors) response surface designs. But for other correlation structures, only a few results are included in this book. Only a few straightforward construction methods of RFORDs and ORFODs are discussed as examples. Of course there may be many more construction methods of RFORDs and ORFODs which will possibly be developed in future. In case of autocorrelated structure, RFORDs have been discussed. Only one general method of construction of RFORDs has been developed for this error structure. It has not been possible to develop a single method of ORFODs under autocorrelated structure. Also for tri-diagonal error structure, the results given in Chapters 2, 3 and 4 are true for lag more than one. But for lag one, it is difficult to derive at least rotatable designs. It is observed (Chapters 2 to 8) that the results of response surface designs for each error structure are completely different from other structures. Therefore, it is interesting to study response surface designs for each error structure independently. In Chapter 2, it is mentioned that there are a lot of other variance-covariance structures, say for instance, the Guttman Circumplex structure, Democratic structure, Quasi-persymmetric structure, Equipredictability structure etc. Similar results of these structures are not included in this book. A few articles on these structures have been communicated. These results will be included in this book in the next revision.

Second-order rotatability and optimality are considered for correlated errors in Chapter 3. Second-order optimality is focused only for the intra-class structure, but for other structures (inter-class, compound symmetry, tri-diagonal and autocorrelated structures) no specific study has been done in this book. Nonspecific general results are stated for optimum robust second-order rotatable designs. For intra-class correlation structure only, specific conditions optimum robust second-order rotatable designs are examined. Second-order rotatability has been examined for intra-class, inter-class, compound symmetry, tri-diagonal and autocorrelated error structures. Only a few construction methods of robust second-order rotatable designs are developed under the above correlation structures. In case of autocorrelated error structure, only one method of construction of RSORDs has been included in this book. There may be many more construction methods of RSORDs which will possibly be developed in future. RSORDs for other correlation structures are currently under investigation which will be communicated in future.

Robust rotatability and optimality have been examined for a normally distributed response in Chapters 2 and 3. In practice, a response distribution may not be normal always. For example, a lifetime response distribution always non-normal. In Chapter 4, first- and second-order designs are examined assuming the response follows log-normal or exponential distribution with correlated errors. First-order D-optimal and second-order rotatable designs are examined for log-normal and exponential lifetime distributions with correlated errors under intra-class, inter-class, compound symmetry and tri-diagonal error structures. Under autocorrelated errors, combined error variance-covariance structure for the log-normal or exponential distribution model, will not reduce to an autocorrelated structure. It is difficult to study response surface designs for Gamma, Weibull and extreme value distributions with correlated errors. Some recent works on response surface designs with correlated errors for these distributions have been communicated which are not included in this book.

In Chapters 2, 3 and 4, it is observed that D-optimal or even rotatable designs do not exist for some correlated error structures. To examine the degree of robust rotatability, the concept of weakly robust first- and second-order rotatability has been introduced in Chapter 5. Two measures of first-order rotatability have been introduced based on the dispersion matrix of regression coefficients and moment matrix. Also two measures of second-order rotatabilty have been introduced based on moment matrix. The concept of cost function has been introduced in conducting an experiment. Based on the variance and the cost function, one can compare between a robust rotatable design and a weakly robust rotatable design. Some rules have been suggested to compare any two designs (separately for first- and second-order designs) based on variance and cost function. Some examples illustrate all the concepts of weakly robust rotatability and comparison rules. There may be many more measures which can be extended in future research. With the help of these measures, one can examine the *robust rotatability* of a first- or second-order design with respect to any error variance-covariance structure.

The problem of estimation of slopes occurs frequently in many practical situations. For example, if differences at points close together are involved, estimation of the local slopes (the rate of change) of the response is of interest. Therefore, robust second-order slope and modified slope rotatable designs along axial directions are considered in Chapter 6. Robust second-order slope-rotatability and modified slope-rotatability conditions along axial directions are derived for a general correlated error structure. These conditions are simplified for intra-class error structure, and are compared with the usual (uncorrelated and homoscedastic errors) modified second-order slope-rotatability conditions. There is no method of construction of second-order slope-rotatable or modified slope-rotatable designs for any correlation structure. It is very hard even for the intra-class structure. Therefore, modified slope-rotatable designs are introduced. Some robust second-order slope-rotatable designs over all directions, *D*-optimal slope, and with equal maximum directional variance slope are examined with respect to modified robust second-order slope-rotatability. It is observed that robust second-order slope-rotatable designs over all directions, *D*-optimal slope, and with equal maximum directional variance slope are *not* modified robust second-order slope-rotatable designs. Second-order slope-rotatable designs along axial directions and modified second-order slope-rotatable designs with correlated errors are unknown in literature. Construction of robust second-order slope-rotatable and modified slope-rotatable designs along axial directions under different error structures is currently under investigation. We propose to discuss our findings in subsequent articles.

A few concepts of optimal (*A*-, *D*- and with equal maximum directional variance) robust slope-rotatable designs have been considered in Chapter 7. Slope-rotatable designs over all directions are known as *A*-optimal slope. The necessary and sufficient conditions for slope-rotatable designs over all directions, with equal maximum directional variance and *D*-optimal slope have been derived for a general correlated error structure. These designs have been examined for intra-class, inter-class, compound symmetry, tri-diagonal and autocorrelated error structures. A class of each design (*A*-, *D*- and with equal maximum directional variance) has been constructed for each correlation structure. The other concepts of optimality (for example, *E*-, D_s- etc.) have not been considered in this book. These are currently under investigation.

In Chapter 6, it is observed that there is no method of construction of robust second-order slope-rotatable designs along axial directions. Again in Chapter 7, it is found that robust second-order rotatable designs are optimal slope-rotatable designs. But for some error structures, it is very hard to obtain a rotatable design. Therefore, some general measures of robust second-order slope-rotatability along axial directions, over all directions and with equal maximum directional variance have been discussed in Chapter 8. These measures are illustrated with few examples. Based on these measures, one can examine the *robust slope-rotatability* of any second-order design with respect to any error variance-covariance structure. These measures can be used to compare the degree of robust slope-rotatability of the designs with the same number of factors. Note that the degree of robust second-order slope-rotatability with equal maximum directional variance and *D*-optimal slope (for $k > 2$) can be examined based on the measures of robust second-order rotatability as given in Chapter 5. Many advantages of these measures are listed in Chapter 8.

Robust first- and second-order rotatable (or *D*-optimal) designs constructed under the autocorrelated and tri-diagonal error structures are *not* invariant under some permutations (i.e., the order of the runs of the experiment) of the design points with respect to robust rotatability. This is also true for inter-class and compound symmetry structures. It is also true for the optimal robust second-order slope-rotatable designs (RSOSRDs) under all other correlation structures except the intra-class structure. The order of the runs of the experiment is very important in case of any optimal RSOSRDs. The practitioners need to maintain the order of the runs of the experiment. Accordingly, the experimental level combinations of the optimal RSOSRDs are to be arranged for conducting the experiment. It is not possible to conduct the experiment (for any optimal RSOSRD) at a time (one after another) for the similar level combinations (say central points) to reduce the cost of the experiment. In general, the number of design points of an optimal RSOSRD is more than a usual SORD. Consequently, the cost of the experiment of an optimal RSOSRD is greater than a usual SORD. Any design, which is an optimal RSOSRD under a particular structure, may not be an optimal RSOSRD for other correlation structure. An optimal RSOSRD depends on the pattern of the correlation structure but independent of the values of the correlation parameters involved in it. Note that the optimal RSOSRDs are always SORDs (for $\rho = 0$) but the converse is not always true.

Response surface designs are mostly used in quality improvement experiments and in many field of sciences. Observations may be correlated, and the practitioners only need to know the pattern of the correlation structure for using these optimal or rotatable designs. Even the distribution of the response is normal or non-normal (log-normal or exponential), the practitioners may use the same designs if the correlation structure is identified. Evidence from the experimental data shows that the experimental errors may be correlated, and they may have different patterns of correlation structures. Response surface designs are widely used in different fields such as quality engineering, chemical process improvement, industrial and engineering chemistry, lifetime improvement, biological experiments etc. Thus, it is reasonable to consider that the experimental responses may be correlated in many cases. Response surface practitioners need to know the pattern of the correlation structure depending on the experimental conditions, responses (replicated, non-replicated), etc. before conducting the experiment. Accordingly, appropriate optimal or rotatable designs may be used in every field to get better information about the unknown process parameters. Sometimes it is observed that log-normal or exponential distribution is more appropriate for the response variable. In that case, 'logarithm of the response' may be considered as the response.

In Chapters 2, 3 and 4 we have considered some regression models with correlated errors for normally or non-normally distributed response variable. In Chapter 9, we have considered the regression analysis with correlated errors under the compound symmetry and compound autocorrelated structures, assuming the response follows Normal distribution. It is observed that intra-class, inter-class and autocorrelated structures are some particular cases of compound autocorrelated structure. Therefore, similar technique can be used for these error structures. The best linear unbiased estimates of all the regression coefficients have been derived *except* the intercept. Hypothesis testing and index of fit have been introduced. Regression analyses with correlated errors have been used in the analysis of a randomized block design, as an application. Several interesting results have been derived. There may be many more applications. Some papers have been communicated for the analysis of a general block design, balanced incomplete block design, row-column design etc. These results are not included in this book. Regression analysis for non-normal distributions (as in Chapter 4) is not included in this book. Some articles on this topic have been communicated. There are a lot of problems regarding regression analysis and its applications (as introduced in Chapter 9) which are currently under investigations. These results will be edited in the book in the next revision.

In Chapter 10, log-normal and gamma models are examined for constant and non-constant variances. The discrepancy of regression estimates and also fittings between these two models have been examined with real data analysis. Taguchi's and dual response approaches are compared with the GLM and JGLM approach. Dual response approach has been extended to GLM and JGLM. A few applications of both these models are illustrated with real data in the different field of sciences.

This book contains only 10 chapters. Only a few basic concepts of response surface methodology are included in this book. There are many more concepts of RSM which are still untouched with correlated errors. It is also mentioned that each chapter will be extended in future. Also the regression analysis with correlated errors have been introduced in this book. It will be extended more in future. Applications of regression analysis included in this book are really interesting. Through this correlated regression analysis, correlated block designs and row-column designs can be studied. A few articles on correlated block designs and row-column designs have been communicated. A few applications of log-normal and gamma models are included in this book. The author has produced many papers in medical science using the log-normal and gamma models. Many interesting results are focused on different diseases. It is possible to write a book on medical science based on the results obtained by the present author. Currently, the log-normal and gamma models are used by the present author in the different field of sciences. It is expected that the contents of this book will be extended in future, and this book will be divided into many independent books.

APPENDIX

TABLE Ap.1: Experimental Layout and Responses for the Roman Catapult Data

Trail No.	Run order	x_1	x_2	x_3	Rep.1	Rep.2	Rep.3
1	20	−1	−1	−1	39	34	42
2	9	−1	−1	1	80	71	91
3	11	−1	1	−1	52	44	45
4	14	−1	1	1	97	68	60
5	15	1	−1	−1	60	53	68
6	10	1	−1	1	113	104	127
7	13	1	1	−1	78	64	65
8	1	1	1	1	130	79	75
9	7	−1.682	0	0	59	51	60
10	4	1.682	0	0	115	102	117
11	19	0	−1.682	0	50	43	57
12	18	0	1.682	0	88	49	43
13	12	0	0	−1.682	54	50	60
14	8	0	0	1.682	122	109	119
15	6	0	0	0	87	78	89
16	5	0	0	0	86	79	85
17	17	0	0	0	88	81	87
18	3	0	0	0	89	82	87
19	2	0	0	0	86	79	88
20	16	0	0	0	88	79	90

Source: Data from Luner, J. J. (1994). Achieving continuous improvement with the dual response approach: A demonstration of the Roman catapult, *Quality Engineering*, 6, pp. 691-705.

TABLE Ap.2: Design Matrix and Lifetime Data of Threading Machine Experiment

Trail No.	A	B	C	D	E	F	Rep.1	Rep.2	Rep.3
1	-1	-1	-1	-1	-1	L	137	24	58
2	-1	-1	-1	1	1	R	89	41	26
3	-1	-1	1	-1	1	L	56	34	199
4	-1	-1	1	1	-1	L	545	105	106
4	-1	-1	1	1	-1	L	122	132	168
4	-1	-1	1	1	-1	R	74	66	49
5	-1	1	-1	-1	1	L	428	157	288
6	-1	1	-1	1	-1	R	352	68	97
7	-1	1	1	-1	-1	L	320	750	988
8	-1	1	1	1	1	R	632	73	529
9	1	-1	-1	-1	1	R	44	41	7
10	1	-1	-1	1	-1	R	2	3	4
11	1	-1	1	-1	-1	R	112	64	59
12	1	-1	1	1	1	R	59	45	9
13	1	1	-1	-1	-1	R	40	7	32
14	1	1	-1	1	1	R	21	42	120
15	1	1	1	-1	1	R	14	19	28
15	1	1	1	-1	1	R	168	34	58
16	1	1	1	1	-1	R	70	57	18

Source: Data from Watkins, D., Bergman, A. and Horton, R. (1994). Optimization of tool life on the shop floor using design of experiments, *Quality Engineering*, 6, pp. 609–620.

TABLE Ap.3: Physico-chemical Characteristics of Groundwater Samples

Site	Month of 2008	pH	EC	TEP	TDS	DO	TAK	THD	Ca	Mg	CLD	COD
1	Oct.	7.51	545	24.9	469	6.7	255.2	263	75.9	45.4	275.6	8.2
1	Nov.	7.42	529	22.8	480	7.3	224.7	280	74.3	49.9	280.1	7.6
1	Dec.	7.53	533	21.5	495	6.5	248.7	252	69.4	44.3	293.4	7.9
2	Oct.	7.65	476	25.7	410	6.4	322.7	168	51.2	28.3	221.6	6.2
2	Nov.	7.45	441	22.2	380	7.4	288.9	210	53.4	38.0	191.2	5.8
2	Dec.	7.31	515	22.0	495	7.9	313.6	180	58.7	29.4	185.2	5.3
3	Oct.	7.34	741	25.4	645	6.7	345.8	465	80.2	93.5	348.1	8.2
3	Nov.	7.70	695	22.4	615	7.0	290.3	422	79.4	83.2	355.6	8.0
3	Dec.	7.54	640	19.8	590	7.8	320.0	395	88.3	74.5	360.1	7.5
4	Oct.	7.28	372	25.9	320	3.6	280.7	232	72.2	38.8	146.3	7.6
4	Nov.	7.12	360	21.9	380	4.9	246.2	260	69.6	46.2	158.4	8.4
4	Dec.	7.65	395	20.1	415	5.1	252.7	245	71.9	42.0	161.7	7.0
5	Oct.	8.23	675	24.8	570	4.0	283.6	290	91.6	48.2	320.2	9.1
5	Nov.	7.73	716	22.5	612	4.1	324.2	286	83.5	47.9	342.7	8.4
5	Dec.	7.2	680	18.9	620	4.2	310.7	295	87.7	49.4	325.6	8.6
6	Oct.	7.17	282	24.8	320	4.5	156.8	232	49.8	44.2	221.7	8.2
6	Nov.	7.59	255	22.7	310	5.6	179.2	260	57.4	49.2	222.6	8.7
6	Dec.	7.34	298	20.3	370	6.2	272.7	225	54.7	41.3	230.7	7.8

Source: Data from Mumtazuddin, S., Azad, A. K., Kumar, M. and Gautam, A. K. (2009). Determination of Physico-chemical parameters in some ground water samples at Muzaffarpur Town, *Asian J. Chemical and Environ. Research*, 2, pp. 18-20.

TABLE Ap.4: The Worsted Yarn Data

Run Number	Length, x_1	Amplitude, x_2	Load, x_3	Cycles to Failure, y
1	−1	−1	−1	674
2	0	−1	−1	1414
3	1	−1	−1	3636
4	−1	0	−1	338
5	0	0	−1	1022
6	1	0	−1	1368
7	−1	1	−1	170
8	0	1	−1	442
9	1	1	−1	1140
10	−1	−1	0	370
11	0	−1	0	1198
12	1	−1	0	3184
13	−1	0	0	266
14	0	0	0	620
15	1	0	0	1070
16	−1	1	0	118
17	0	1	0	332
18	1	1	0	884
19	−1	−1	1	292
20	0	−1	1	634
21	1	−1	1	2000
22	−1	0	1	210
23	0	0	1	438
24	1	0	1	566
25	−1	1	1	90
26	0	1	1	220
27	1	1	1	360

Source: Data from Myers, R. H., Montgomery, D. C., and Vining, G. G. (2002). *Generalized Linear Models with Applications in Engineering and the Sciences*, John Wiley and Sons, New York (p. 36).

TABLE Ap.5: Resistivity Data

Run Number	x_1	x_2	x_3	x_4	Resistivity (y)
1	-1	-1	-1	-1	193.4
2	1	-1	-1	-1	247.6
3	-1	1	-1	-1	168.2
4	1	1	-1	-1	205.0
5	-1	-1	1	-1	303.4
6	1	-1	1	-1	339.9
7	-1	1	1	-1	226.3
8	1	1	1	-1	208.3
9	-1	-1	-1	1	220.0
10	1	-1	-1	1	256.4
11	-1	1	-1	1	165.7
12	1	1	-1	1	203.5
13	-1	-1	1	1	285.0
14	1	-1	1	1	268.0
15	-1	1	1	1	169.1
16	1	1	1	1	208.5

Source: Data from Myers, R. H., Montgomery, D. C., and Vining, G. G. (2002). *Generalized Linear Models with Applications in Engineering and the Sciences*, John Wiley and Sons, New York (p. 176).

TABLE Ap.6: Design Matrix and Voltage Drop Data, Wiper Switch Experiment

Factor					Inspection								
0	−	−	−	−	24	37	40	65	72	77	90	101	117	128
					22	36	47	64	71	86	99	118	127	136
					17	34	40	52	66	79	91	98	115	119
					24	30	38	46	57	71	73	91	98	104
0	+	+	+	+	45	60	79	90	113	124	141	153	176	188
					51	68	84	104	122	136	148	166	191	197
					42	58	70	82	103	119	128	143	160	175
					41	56	56	70	81	89	98	108	113	128
1	−	−	+	+	28	40	56	69	87	86	110	121	132	146
					46	50	81	95	114	130	145	161	185	202
					45	54	79	90	111	132	143	168	185	202
					37	58	81	99	123	143	166	191	202	231
1	+	+	−	−	54	51	64	66	78	84	90	93	106	109
					47	45	50	53	58	57	61	55	61	66
					47	54	63	68	70	77	88	86	91	102
					53	55	66	68	91	90	98	104	118	120
2	−	+	−	+	18	35	48	56	65	81	89	98	117	124
					20	37	52	53	67	75	85	95	112	122
					32	54	76	98	119	143	158	181	205	231
					28	39	54	73	89	98	117	127	138	157
2	+	−	+	−	44	50	48	46	55	63	65	71	68	76
					43	44	55	56	58	62	66	66	72	72
					40	46	45	49	55	62	61	61	64	66
					55	67	73	75	91	88	102	111	115	119
3	−	+	+	−	39	47	58	72	84	104	109	129	143	154
					29	42	55	67	82	91	104	117	130	136
					36	45	56	80	93	101	121	138	154	170
					31	40	60	72	82	98	103	117	130	146
3	+	−	−	+	61	67	69	86	86	88	95	103	107	118
					68	75	82	90	95	109	107	118	120	133
					60	72	85	84	87	98	99	111	113	125
					65	68	69	75	79	84	95	96	101	100

Source: Data from Wu, C.F.J. and Hamada, M. (2000). *Experiments and Planning Analysis, and Parameter Optimization*, John Wiley and Sons, New York (p. 561).

TABLE Ap.7: The Experimental Layout and Responses of Silica Gel Expt. (Anand, 1997)

Trail No.	A	B	C	D	Second Grade	Scrap	Density
1	1	1	1	1	14.0	3.6	7.10
2	1	1	2	2	13.5	3.8	7.25
3	1	1	3	3	18.3	4.5	6.90
4	1	2	1	2	17.4	4.2	6.41
5	1	2	2	3	16.3	3.9	6.53
6	1	2	3	1	13.9	2.8	6.20
7	1	3	1	3	14.1	2.7	6.90
8	1	3	2	1	12.0	2.9	6.98
9	1	3	3	2	12.0	3.0	6.51
10	2	1	1	2	17.0	4.3	6.83
11	2	1	2	3	10.5	4.2	6.98
12	2	1	3	1	15.3	4.0	6.41
13	2	2	1	3	21.0	4.9	6.23
14	2	2	2	1	20.5	4.7	6.47
15	2	2	3	2	19.8	4.5	6.01
16	2	3	1	1	12.3	3.2	7.27
17	2	3	2	2	11.3	2.1	7.32
18	2	3	3	3	10.5	2.3	7.12
19	3	1	1	3	15.2	3.1	7.00
20	3	1	2	1	12.3	2.9	7.00
21	3	1	3	2	12.0	2.7	6.98
22	3	2	1	1	11.1	1.9	6.27
23	3	2	2	2	12.8	2.4	6.32
24	3	2	3	3	13.4	2.2	5.90
25	3	3	1	2	10.3	1.8	7.22
26	3	3	2	3	10.4	2.2	7.28
27	3	3	3	1	11.7	2.5	7.12

Source: Data from Anand, K. N. (1997). Improving the yield of silica gel in a chemical plant, *Quality Engineering*, **9(3)**, pp. 355-361.

TABLE Ap.8: The Printing Process Study Data

Trail No.	x_1	x_2	x_3	Rep. 1	Rep. 2	Rep. 3
1	−1	−1	−1	34	10	28
2	0	−1	−1	115	116	130
3	1	−1	−1	192	186	263
4	−1	0	−1	82	88	88
5	0	0	−1	44	178	188
6	1	0	−1	322	350	350
7	−1	1	−1	141	110	86
8	0	1	−1	259	251	259
9	1	1	−1	290	280	245
10	−1	−1	0	81	81	81
11	0	−1	0	90	122	93
12	1	−1	0	319	376	376
13	−1	0	0	180	180	154
14	0	0	0	372	372	372
15	1	0	0	541	568	396
16	−1	1	0	288	192	312
17	0	1	0	432	336	513
18	1	1	0	713	725	754
19	−1	−1	1	364	99	199
20	0	−1	1	232	221	266
21	1	−1	1	408	415	443
22	−1	0	1	182	233	182
23	0	0	1	507	515	434
24	1	0	1	846	535	640
25	−1	1	1	236	126	168
26	0	1	1	660	440	403
27	1	1	1	878	991	1161

Source: Data from Box, G.E.P. and Draper, N.R. (1987). *Empirical Model-Building and Response Surfaces*, John Wiley and Sons, New York (p. 247).

TABLE Ap.9: The Seat-Belt Experiment Data

					Tensile strength			Flash		
Run	A	B	C	D	Rep. 1	Rep. 2	Rep. 3	Rep. 1	Rep. 2	Rep. 3
1	0	0	0	0	5164	6615	5959	12.89	12.70	12.74
2	0	0	1	1	5356	6117	5224	12.83	12.73	13.07
3	0	0	2	2	3070	3773	4257	12.37	12.47	12.44
4	0	1	0	1	5547	6566	6320	13.29	12.86	12.70
5	0	1	1	2	4754	4401	5436	12.64	12.50	12.61
6	0	1	2	0	5524	4050	4526	12.76	12.72	12.94
7	0	2	0	2	5684	6251	6214	13.17	13.33	13.98
8	0	2	1	0	5737	6271	5843	13.02	13.11	12.67
9	0	2	2	1	5744	4797	5416	12.37	12.67	12.54
10	1	0	0	1	6843	6895	6957	13.28	13.65	13.58
11	1	0	1	2	6538	6328	4784	12.62	14.07	13.38
12	1	0	2	0	6152	5819	5963	13.19	12.94	13.15
13	1	1	0	2	6854	6804	6907	14.65	14.98	14.40
14	1	1	1	0	6799	6703	6792	13.00	13.35	12.87
15	1	1	2	1	6513	6503	6568	13.13	13.40	13.80
16	1	2	0	0	6473	6974	6712	13.55	14.10	14.41
17	1	2	1	1	6832	7034	5057	14.86	13.27	13.64
18	1	2	2	2	4968	5684	5761	13.00	13.58	13.45
19	2	0	0	2	7148	6920	6220	16.70	15.85	14.90
20	2	0	1	0	6905	7068	7156	14.70	13.97	13.66
21	2	0	2	1	6933	7194	6667	13.51	13.64	13.92
22	2	1	0	0	7227	7170	7015	15.54	16.16	16.14
23	2	1	1	1	7014	7040	7200	13.97	14.09	14.52
24	2	1	2	2	6215	6260	6488	14.35	13.56	13.00
25	2	2	0	1	7145	6868	6964	15.70	16.45	15.85
26	2	2	1	2	7161	7263	6937	15.21	13.77	14.34
27	2	2	2	0	7060	7050	6950	13.51	13.42	13.07

Source: Data from Wu, C.F.J. and Hamada, M. (2000). *Experiments and Planning Analysis, and Parameter Optimization*, John Wiley and Sons, New York (p. 206).

TABLE Ap.10: Experimental Layout Conducted by Gupta and Das (2000) for Improving Resistivity of UF Resin

Trail No.	A	B	C	D	E	F
1	1	1	1	1	1	1
2	1	1	1	2	1	1
3	2	1	1	1	2	2
4	2	1	1	2	2	2
5	2	2	2	2	1	1
6	2	2	2	1	1	1
7	1	2	2	2	2	2
8	1	2	2	1	2	2
9	3	1	2	2	1	2
10	3	1	2	1	1	2
11	3	1	2	2	2	1
12	3	1	2	1	2	1
13	3	2	1	1	1	2
14	3	2	1	2	1	2
15	3	2	1	1	2	1
16	3	2	1	2	2	1

Source: Data from Gupta, A. and Das, A. K. (2000). Improving resistivity of UF resin through setting of process parameters, *Quality Engineering*, **12(4)**, pp. 611-618.

TABLE Ap.11: Five Responses (Each with Two Replicates) Conducted by Gupta and Das (2000) for Improving Resistivity of UF Resin

Trail No.	NV- 1	-Solid 2	Visco- 1	-sity 2	Acid- 1	-Value 2	PE- 1	-T 2	Resist- 1	-ivity 2
1	53.77	53.89	5.5	7.5	7.41	8.72	1.9	2.1	60	135
2	56.30	55.00	9.0	9.5	10.55	12.05	2.14	2.31	220	160
3	57.60	56.90	7.5	13.0	9.87	7.06	2.21	2.44	85	180
4	57.50	53.30	17.0	12.5	11.86	11.76	2.63	2.6	330	110
5	55.30	54.67	5.0	21.5	14.08	13.78	2.57	1.81	95	130
6	58.40	56.60	12.0	18.0	9.42	9.20	2.12	1.83	190	175
7	56.20	57.00	11.0	14.5	11.12	10.82	2.29	2.06	145	200
8	59.65	59.00	21.5	11.5	7.01	7.48	2.21	2.33	300	210
9	57.60	54.50	9.5	13.0	11.60	12.77	2.3	1.8	110	100
10	59.60	57.00	36.0	7.0	6.94	6.41	2.24	2.66	125	130
11	56.00	58.20	15.0	12.0	9.55	9.75	2.71	2.06	300	170
12	58.46	58.40	7.0	12.5	10.90	6.54	2.30	2.21	65	160
13	56.90	53.50	7.0	8.5	7.49	10.56	2.5	2.01	170	90
14	56.50	59.00	14.0	25.0	12.44	11.01	2.49	2.1	70	250
15	57.50	55.00	11.0	7.0	8.31	8.50	2.65	2.83	380	80
16	57.79	55.50	6.5	21.0	11.23	10.92	2.35	1.94	105	200

Source: Data from Gupta, A. and Das, A. K. (2000). Improving resistivity of UF resin through setting of process parameters, *Quality Engineering*, **12(4)**, pp. 611-618.

Bibliography

Aalen, O.O. (1988). Heterogeneity in survival analysis, *Statistics in Medicine*, **7**, pp. 1121–1137.

Aalen, O.O. (1989). A linear regression model for the analysis of lifetimes, *Statistics in Medicine*, **8**, pp. 907–925.

Aalen, O.O. (1992). Modeling heterogeneity in survival analysis by the compound poisson distribution, *Annals of Applied Probability*, **2**, pp. 951–972.

Aalen, O.O. (1994). Effects of frailty in survival analysis, *Statistical Methods in Medical Research*, **3**, pp. 227–243.

Adhikary, B. and Panda, R.N. (1992). Quality-quantity type second order rotatable designs, *Sankhya*, **B54**, pp. 25–29.

Adhikary, B. and Sinha, B.K. (1976). On group divisible rotatable designs, *Cal. Statist. Assoc. Bull.*, **25**, pp. 79–93.

Aitkin, M. and Clayton, D.G. (1980). The fitting of exponential, Weibull and extreme value distributions to complex censored survival data using GLIM, *Appl. Statist.*, **29**, pp. 156-163.

Akaike, H. (1977). On entropy maximisation principle, in *Applications of Statistics*, P.R. Krishnaiah, (Eds.), North Holland, Amsterdam, pp. 27–41.

Anand, K.N. (1997). Improving the yield of silica gel in a chemical plant, *Quality Engineering*, **9(3)**, pp. 355–361.

Anderson, T.W. (1969). *Statistical Inference for covariance matrices with linear structure*. In P.R. Krishniah (Eds.), Multivariate Analysis II, Academic Press, New York, pp. 55–69.

Anderson, T.W. (1970). Estimation of covariance matrices which are linear combinations or whose inverses are linear combinations of given matrices. In R.C. Bose, I.M. Chakravarti, P.C. Mahalanobis, C.R. Rao and K.J.C. Smith (Eds.) *Essays in Probability and Statistics*, Chapel Hill, N.C.: University of North Carolina Press, pp. 1–12.

Anderson, T.W. (1984). *An Introduction to Multivariate Statistical Analysis*; Second Edition, John Wiley and Sons, New York.

Anilkumar, K. and Ashara, A. (1993). Mortality change in India since independence. Paper presented at the XVII IASP Annual Conference, Annamalai University, Annamalainagar, Tamilnadu, December 16- 19, 1993.

Ash, A. and Hedayat, A. (1978). An introduction to design optimality with an overview of the literature, *Commun. Statist. Theo. Metho.*, **A7**, pp. 1295–1325.

Atkinson, A.C. (1969). A test for discriminating between models, *Biometrika*, **56**, pp. 337–347.

Atkinson, A.C. (1970a). The design of experiments to estimate the slope of a response surface, *Biometrika*, **57**, pp. 319–328.

Atkinson, A.C. (1970b). A method for discriminating between models (with discussion), *J. R. Statist. Soc. B*, 32, pp. 323–353.

Atkinson, A.C. (1972). Planning experiments to detect inadequate regression models, *Biometrika*, **59**, pp. 275–293.

Atkinson, A.C. (1982a). Developments in the design of experiments, *International Statistical Review*, **50**, pp. 161–177.

Atkinson, A.C. (1982b). Regression Diagnostics, Transformations and constructed variables, *J. R. Statist. Soc.*, **B44**, pp. 1–36.

Bandolier (2006). Being unfit: as life threatening as other risk factors. Bandolier–*Evidence based Health J* 14-10-2006. http://www.jr 2.ox.ac.uk/bandolier/booth/ hliving/unfit.html (on access date 14-10-2006).

Bandemer, H. (1980). Problems in foundation and use of optimal experimental design in regression models, *Mathematische Operationsforschung and Statistik Series Optimization*, **11**, pp. 89–113.

Berenblut, I.I. and Webb, G.I. (1974). Experimental design in the presence of autocorrelated error, *Biometrika*, **61**, 427–437.

Bhattacharyya, S., Mukhopadhyay, M., Bhattacharyya, I., Lahiri, S.K., Mitra, P.K. Dutta, U.K. (2006). A study on body mass index, serum insulin level and lipid profile in normoglycemic young adults *Ind. J. Med. Biochem*, **10(2)**, pp. 33–37.

Bhattacharyya, S., Mukhopadhyay, M., Bhattacharyya, I., Lahiri, S.K., Mitra, P.K. Dutta, U.K. (2007). A study on body mass index (BMI) and some biochemical parameters of the medicos with family history of diabetes mellitus, hypertension and coronary heart disease, *J. of the Indian Med. Assoc.*, 107(7), pp. 370–379.

Bickel, P.J. and Herzberg, A.M. (1979). Robustness of design against autocorrelation in time I: asymptotic theory, optimality for location and linear regression, *Ann. Statist.*, **7**, pp. 77–95.

Bickel, P.J., Herzberg, A.M. and Schilling, M. (1981). Robustness of design against autocorrelation in time II: optimality, theoretical and numerical results for the first order-autoregressive process, *J. Am. Statist. Assoc.*, **76**, pp. 870–877.

Bischoff, W. (1992). On exact D-optimal designs for regression models with correlated observations, *Ann. Inst. Statist. Math.*, **44**, pp. 229–238.

Bischoff, W. (1995a). Determinant Formulas with Applications to Designing when the observations are correlated, *Ann. Inst. Statist. Math.*, **47**, pp. 385–399.

Bischoff, W. (1995b). Lower bounds for the efficiency of designs with respect to the D-criterion when the observations are correlated, *Statistics*, **27**, pp. 27–44.

Bischoff, W. (1996). On maximin designs for correlated observations, *Statistics and Probability Letters*, **26**, pp. 357–363.

Bose, R.C. and Carter, R.L. (1959). Complex representation in the construction of rotatable designs, *Ann. Math. Statist.*, **30**, pp. 771–780.

Bose, R.C. and Draper, N.R. (1959). Second order rotatable designs in three dimensions, *Ann. Math. Statist.*, **30**, pp. 1097–1112.

Box, G.E.P. (1952). Multifactor designs of first order, *Biometrika*, **39**, pp. 49–57.

Box, G.E.P. (1954). The exploration and exploitation of response surfaces: some general considerations and examples, *Biometrices*, **10**, pp. 16–60.

Box, G.E.P. (1963). The effect of errors in the factor levels and experimental design, *Technometrics*, **5**, pp. 247–262.

Box, G.E.P. (1976). Science and Statistics, *J. Amer. Statist. Assoc.*, **71**, pp. 791–799.

Box, G.E.P. (1985). Discussion of off-line quality control parameter design, and the Taguchi method by R.N. Kackar, *J. Quality Technology*, **17**, pp. 189-190.

Box, G.E.P. (1988). Signal-to-Noise Ratios, Performance Criteria, and Transformations (with discussion), *Technometrics*, **30**, pp. 1–40.

Box, G.E.P. and Behnken, D.W. (1960a). Some new three level designs for the study of quantitative variables, *Technometrics*, **3**, pp. 576–596.

Box G.E.P. and Behnken, D.W. (1960b). Simplex-sum designs: A class of second order rotatable designs derivable from those of first order, *Ann. Math. Statist.*, **31**, pp. 838–864.

Box, G.E.P. and Cox, D.R. (1964). An analysis of transformations, *J. R. Statist. Soc.*, **B26**, pp. 211–252.

Box, G.E.P. and Draper, N.R. (1959). A basis for the selection of a response surface design, *J. Am. Statist. Assoc.*, **54**, pp. 622–654.

Box, G.E.P. and Draper, N.R. (1963). The choice of a second order rotatable design, *Biometrica*, **50**, pp. 335–352.

Box, G.E.P. and Draper, N.R. (1975). Robust designs, *Biometrika*, **62**, pp. 347–352.

Box, G.E.P. and Draper, N.R. (1987). *Empirical Model-Building and Response Surfaces*, John Wiley and Sons, New York.

Box, G.E.P. and Draper, R.N. (2007). *Response Surfaces, Mixtures, and Ridge Analyses*, 2nd Eds. John Wiley and Sons, New York.

Box, G.E.P. and Hunter, J.S. (1957). Multifactor experimental designs for exploring response surfaces, *Ann. Math. Statist.*, **28**, pp. 195–241.

Box, G.E.P. and Hunter, J.S. (1961a). The 2^{k-p} fractional factorial designs, Part I, *Technometrics*, **3**, pp. 311–351.

Box, G.E.P. and Hunter, J.S. (1961b). The 2^{k-p} fractional factorial designs, Part II, *Technometrics*, **3**, pp. 449–458.

Box, G.E.P. and Wilson, K.B. (1951). On the experimental attainment of optimum conditions, *J. Roy. Statist. Soc.*, **B13**, pp. 1–45.

Brynjarsdottir, J. and Stefansson, G. (2004). Analysis of cod catch data from Icelandic ground fish surveys using generalized linear models, *Fisheries Research*, **70**, pp. 195–208.

Chatterjee, S. and Hadi, A. (1988). *Sensitivity Analysis in Linear Regression*, John Wiley and Sons, New York.

Chatterjee, S. and Price, B. (2000). *Regression Analysis by Examples*, 3rd ed., John Wiley and Sons, New York.

Chen, W.W. (1980). On the tests of separate families of hypotheses with small sample size, *J. Statist. Comp. Simul.*, **2**, pp. 183–187.

Chernoff, H. (1953). Locally optimal designs for estimating parameters, *Ann. Math. Statist.*, **24**, pp. 586–602.

Copeland, K. and Nelson, P. (1996). Dual response optimization via direct function minimization, *J. Quality Technology*, **28**, pp. 331-336.

Condra, L.W. (1993). *Reliability improvement of experiments*, Marcel Dakker, New York.

Cook, R.D. and Nachtsheim, C.J. (1980). A comparison of algorithms for constructing exact D-optimal designs, *Technometrics*, **22**, pp. 315–324.

Cox, D.R. (1961). Tests of separate families of hypotheses, *Proceedings of the Fourth Berkely symposium in Mathematical Statistics and Probality*, Berkeley, University of California Press, pp. 105–123.

Cox, D.R. (1962). Further results on tests of separate families of hypotheses, *J. R. Statist. Soc.*, **B24**, 106–124.

Cox, D.R. and Reid, N. (1987). Parameter orthogonality and approximate conditional inference, *J. R. Statist. Soc.*, **B49**, pp. 1–39.

Cutler, D.R. (1993). Efficient block designs for comparing test treatments to a control when the errors are correlated, *J. Statist. Planning Inference*, **37**, pp. 393–412.

Das, M.N. (1958). On reinforced incomplete block designs, *J. Indian Soc. Agric. Statist.*, **10**, pp. 73–77.

Das, M.N. (1961). Construction of rotatable designs from factorial designs, *J. Indian Soc. Agric. Statist.*, **13**, pp. 169–194.

Das, M.N. (1963). On the construction of second order rotatable designs through balanced incomplete block designs with blocks of unequal size, *Cal. Statist. Assoc. Bull.*, **12**, pp. 31–46.

Das, M.N. and Dey, A.K. (1967). Group-divisible rotatable designs, *Ann. Inst. Math. Statist.*, **19**, pp. 331–347.

Das, M.N. and Narasimham, V.L. (1962). Construction of rotatable designs through balanced incomplete block designs, *Ann. Math. Statist.*, **33**, pp. 1421–1439.

Das, R.N. (1997). Robust second order rotatable designs: Part-I, *Cal. Statist. Assoc. Bull.*, **47**, pp. 199–214.

Das, R.N. (1999). Robust second order rotatable designs: Part-II, *Cal. Statist. Assoc. Bull.*, **49**, pp. 65–78.

Das, R.N. (2003a). Robust second order rotatable designs: Part-III, *J. Indian Soc. Agric. Statist.*, **56**, pp. 117–130.

Das, R.N. (2003b). Slope rotatability with correlated errors, *Cal. Statist. Assoc. Bull.*, **54**, pp. 57–70.

Das, R.N. (2004). Construction and analysis of robust second order rotatable designs, *J. Statist. Theo. and Appli.*, **3**, pp. 325–343.

Das, R.N. (2009). Response surface methodology in improving mean life time, *ProbStat Forum*, **2**, pp. 8–21.

Das, R.N. (2010). Regression analysis for correlated data, *J. Quality Technology and Quality Management*, **7(3)**, pp. 263–277.

Das, R.N. (2011a). Resistivity Distribution: Gamma or Else Other?, *Journal of Quality*, **18(1)**, pp. 49–60.

Das, R.N. (2011b). The role of iodine in the thyroid status of mothers and their neonates, *Thyroid Science*, **6(2)**, pp. 1–15.

Das, R.N. (2011c). Modeling of biochemical parameters, *Model Assisted Statistics and Appli.*, **6(1)**, pp. 1–12.

Das, R.N. (2011d). Dual response surface methodology: Applicable always?, *ProbStat Forum*, **4**, pp. 98–103.

Das, R.N. (2012). Discrepancy in fitting between log-normal and gamma models: An illustration, *Model Assisted Statistics and Appli.*, **7(1)**, pp. 23–32.

Das, R.N. (2013a). Discrepancy in classical lifetime model classes: Some illustrations, *Journal of Quality*, **20(5)**, pp. 521–532.

Das, R.N. (2013b). Relationships of liver biochemical parameters and effects of wine drinking, *Model Assisted Statistics and Appli.*, **8(2)**, pp. 163–175.

Das, R.N. and Dihidar, S. (2013). Determinants of causal factors on child mortality, *J. Indian Statist. Assoc.*, To appear in **51(2)**, 2013, (in Press).

Das, R.N., Dihidar, S. and Verdugo, R. (2011). Infant mortality in India: Evaluating Log-Gaussian and Gamma, *Open Demography Journal*, **4**, pp. 34–41.

Das, R.N. and Huda, S. (2011). On D-optimal robust designs for exponential lifetime distribution, *J. Statist. Theo. and Appli.*, **10(2)**, pp. 198–208.

Das, R.N. and Kim, J. (2012). GLM and joint GML techniques in hydrogeology: An illustration, *Int. J. Hydrology Science and Technology*, **2(2)**, pp. 185–201.

Das, R.N. and Lee, Y.J. (2008). Improving resistivity of urea formaldehyde resin through joint modeling of mean and dispersion, *Quality Engineering*, **20(3)**, pp. 287–295.

Das, R.N. and Lee, Y.J. (2009). Log normal versus gamma models for analyzing data from quality-improvement experiments, *Quality Engineering*, **21(1)**, pp. 79–87.

Das, R.N. and Lee, Y. (2010). Analysis strategies for multiple responses in quality improvement experiments, *Int. J. Quality Engineering and Technology*, **1(4)**, pp. 395–409.

Das, R.N. and Lin, D.K.J. (2011). On D-optimal robust first order designs for lifetime improvement experiments, *J. Statist. Planning Inference*, **141(12)**, pp. 3753–3759.

Das, R.N. and Mukhopadhya, B.B. (2009). Affects of thyroid function and maternal urinary iodine on neonatal weights, *J. Neonatal Nursing*, **15**, pp. 204–211.

Das, R.N., Pal, P. and Park, S.H. (2014). Modified robust second order slope-rotatable designs, *Commun. Statist. Theo. Method*, **43**.

Das, R.N. and Park, J.S. (2012). Discrepancy in regression estimates between Log-normal and Gamma: Some case studies, *J. Applied Statistics*, **39(1)**, pp. 97–111.

Das, R.N. and Park, J.S. (2014). A reinforced randomized block design with correlated errors, *Commun. Statist. Theo. Method.* **43(1)**, pp. 191–209.

Das, R.N. and Park, S.H. (2006). Slope rotatability over all directions with correlated errors, *Appl. Stochastic Models Bus. Ind.*, **22**, pp. 445–457.

Das, R.N. and Park, S.H. (2007). A measure of robust rotatability for second order response surface designs, *J. Korean Statist. Soc.*, **36(4)**, pp. 557–578.

Das, R.N. and Park, S.H. (2008a). On efficient robust rotatable designs with autocorrelated errors, *J. Korean Statist. Soc.*, **37(2)**, pp. 97–106.

Das, R.N. and Park, S.H. (2008b). On exact D-optimal robust designs with tri-diagonal errors, *J. Advanc. and Applica. Statist.*, **9 (1)**, pp. 37–51.

Das, R.N. and Park, S.H. (2008c). Analysis and multiple comparison of treatments of an extended randomized block design with correlated errors, *J. Statist. Theo. and Appli.*, **7**, pp. 245-262.

Das, R.N. and Park, S.H. (2009). A Measure of robust slope rotatability for second order response surface designs, *J. Applied Statistics*, **36(7)**, pp. 755–767.

Das, R.N. and Park, S.H. (2010). On D-optimal robust first order designs, *J. Statist. Theo. and Appli.*, **9(2)**, pp. 217–232.

Das, R.N., Park, S.H., and Aggarwal, M.L. (2010a). Robust second order slope-rotatable designs with maximum directional variance, *Commun. Statist. Theo. Metho.*, **39(5)**, pp. 803–814.

Das, R.N , Park, S.H., and Aggarwal, M.L. (2010b). On D-optimal robust second order slope-rotatable designs, *J. Statist. Planning Inference*, **140(5)**, pp. 1269–1279.

Das, R.N. and Sarkar, P.K. (2012). Lifestyle characteristics and dietary impact on plasma concentrations of beta-carotene and retinol, *BioDiscovery*, **3(3)**, pp. 1–12.

Davies, O.L. (1954). *Design and Analysis of Industrial Experiments*, Hafner Press, New York.

DeBaun, R.M. (1959). Response surface designs for three factors at three levels, *Technometrics*, **1**, pp. 1–8.

Del Castillo, E. (1996). Multiresponse process optimization via constrained confidence regions, *J. Quality Technology*, **28**, pp. 61–70.

Del Castillo, E. and Montgomery, D.C. (1993). A nonlinear programming solution to the dual response problem, *J. Quality Technology*, **25**, pp. 199–204.

Dey, A. and Nigam, A.K. (1968). Group divisible rotatable designs– some further considerations, *Ann. Inst. Statist. Math.*, **20**, pp. 477–485.

Dick, E.J. (2004). Beyond log-normal versus gamma: discrimination among error distributions for generalized linear models, *Fisheries Research*, **70**, pp. 351–366.

Dhake, R.B., Phalak, R.P. and Waghulde, G.P. (2008). Ground water quality assessment at Bhusawal town: a case study, *Asian J. Chem. Environ. Res.*, **1(1)**, pp. 54-58.

Draper, N.R. (1960a). Second order rotatable designs in four or more dimensions, *Ann. Math. Statist.*, **31**, pp. 23–33.

Draper, N.R. (1960b). Third order rotatable designs in three dimensions, *Ann. Math. Statist.*, **31**, pp. 865–874.

Draper, N.R. (1960c). Third order rotatable designs in four dimensions, *Ann. Math. Statist.*, **31**, pp. 875–877.

Draper, N.R. (1961). Third order rotatable designs in three dimensions: some specific designs, *Ann. Math. Statist.*, **32**, pp. 910–913.

Draper, N.R. (1962). Third order rotatable designs in three factors: analysis, *Technometrics*, **4**, pp. 219–234.

Draper, N.R. and Guttman, I. (1988). An index of rotatability, *Technometrics*, **30**, pp. 105–111.

Draper, N.R. and Herzberg, A.M. (1971). On lack of fit, *Technometrics*, **13**, pp. 231–241.

Draper, N.R. and Herzberg, A. M. (1973). Some designs for extrapolation outside a sphere, *J. R. Statist. Soc.*, **B35**, pp. 268–276.

Draper, N.R. and Herzberg, A.M. (1979). An investigation of first-order and second-order designs for extrapolation outside a hypersphere, *Canad. J. Statist.*, **7**, pp. 97–101.

Draper, N.R. and Lawrence, W.E. (1965). Designs which minimize model inadequacies: cuboidal regions of interest, *Biometrika*, **52**, pp. 111–118.

Draper, N.R. and Lin, D.K.J. (1996). *Response surface designs*, In: Ghosh, S., Rao, C.R.(Eds.), Handbook of Statistics, **13**, North-Holland, NewYork, pp. 323–375 (Chapter11).

Draper, N.R. and Pukelsheim, F. (1990). Another look at rotatability, *Technometrics*, **32**, pp. 195–202.

Draper, N.R. and Smith, H. (1998). *Applied Regression Analysis*, John Wiley and Sons, New York.

Edwards, C.H. (1973). *Advanced Calculus of Several Variables*, Academic Press, New York.

Egrioglu, E. and Gunay, S. (2010). Bayesian model selection in ARFIMA models, *Expert Syst. Appl.*, **12**, pp. 8359-8364.

Elfving, G. (1952). Optimum allocation in linear regression theory, *Ann. Math. Statist.*, **23**, pp. 255–262.

Elfving, G. (1955). Geometric allocation theory, *Skand Aktuarietidskr*, **37**, pp. 170–190.

Elfving, G. (1959). *Design of Linear Experiments*, Cramer Festschrift Volume, John Wiley and Sons, New York.

Federer, W.T. (1956). Augmented (or hoonuiaku) designs, *Hawaiian Planters' Record*, **55**, pp. 191–208.

Federer, W.T. (1961). Augmented designs with one-way elimination of heterogeneity, *Biometrics*, **17**, pp. 447–473.

Finney, D.J. (1960). *An Introduction to the Theory of Experimental Design*, The University of Chicago Press, Chicago.

Firth, D. (1988). Multiplicative errors: log-normal or gamma?, *J. R. Statist. Soc.*, **B50**, pp. 266–268.

Galil, Z. and Kiefer, J. (1977). Comparison of design for quadratic regression on cubes, *J. Statist. Planning Inference*, **1**, pp. 121–132.

Galil, Z. and Kiefer, J. (1980). Time and space-saving computer methods, related to Mitchell's DETMAX, for finding D-optimum designs, *Technometrics*, **22**, pp. 301–313.

Gardiner, D.A., Grandage, A.H.E. and Hader, R.J. (1959). Third order rotatable designs for exploring response surfaces, *Ann. Math. Statist.*, **30**, pp. 1082–1096.

Gennings, C. Chinchilli, V.M. and Carter, W.H. Jr. (1989). Response surface analysis with correlated data: a non-linear model approach, *J. Am. Stat. Assoc.*, **84**, pp. 805–809.

Geramita, A.V. and Seberry, J. (1979). *Orthogonal Designs*, Marcel Dekker, Inc., New York and Basel.

Gill, P.S. and Shukla G.K. (1985). Experimental designs and their efficiencies for spatially correlated observations in two dimensions, *Commun. Statist. Theo. Metho.*, **14**, pp. 2181–2197.

Goon, A.M., Gupta, M.K. and Dasgupta, B. (2002). *Fundamental of Statistics*, Volume 2, India: The World Press Private Limited, Kolkata.

Goyal, R.P. (1994). Mortality in India: Trend and prospects, *Demography India*, **23(1/2)**, pp. 103–116.

Grande, J.A., Gonzalez, A.B., Beltran, R. and Sanchez-Rodas, D. (1996). Application of factor analysis to the study of contamination in the aquifer system of Ayamonte-Huelva (Spain), *Groundwater*, **34(1)**, pp. 155-161.

Gupta, A. and Das, A. K. (2000). Improving resistivity of UF resin through setting of process parameters, *Quality Engineering*, **12(4)**, pp. 611–618.

Hacker, R.S. and Hatemi-J, A. (2008). Optimal lag-length choice in stable and unstable VAR models under situations of homoscedasticity and ARCH, *J. Applied Statistics*, **35**, pp. 601-615.

Hader, R.J. and Park, S.H. (1978). Slope-rotatable central composite designs, *Technometrics*, **20**, pp. 413–418.

Hamada, M. and Nelder, J.A. (1997). Generalized linear models for quality-improvement experiments, *J. Quality Technology*, **35**, pp. 2-12.

Hannan E.J. and Quinn, B.G. (1979). The determination of the order of an autoregression, *J. R. Stat. Soc.*, **B41**, pp. 190-195.

Hartley, H.O. (1959). Smallest composite design for quadratic response surfaces, *Biometrics*, **15**, pp. 611–624.

Hastie, T., Tibshirani, R. and Friedman, J. (2001). *The Elements of Statistical Learning*, Springer-Verlag, New York.

Hedayat, A. and Wallis, W.D. (1978). Hadamard matrices and applications, *Ann. Statist.*, **6**, pp. 1184–1192.

Herzberg, A.M. (1964). Two third order rotatable designs in four dimensions, *Ann. Math. Statist.*, **35**, pp. 445–446.

Herzberg, A.M. (1966). Cylindrically rotatable designs, *Ann. Math. Statist.*, **37**, pp. 242–247.

Herzberg, A.M. (1967). A method for the construction of second order rotatable designs in k dimensions, *Ann. Math. Statist.*, **38**, pp. 177–180.

Herzberg, A.M. and Andrews, D.F. (1976). Some considerations in the optimal design of experiments in non-optimal situations, *J. R. Statist. Soc.*, **B38**, pp. 284–289.

Hill, W.J. and Hunter, W.G. (1966). A review of response surface methodology: A literature review, *Technometrics*, **8**, pp. 571–590.

Huda, S. (2006). Design of experiments for estimating differences between responses and slopes of the response, In Khuri, A.I. (Eds.), *Response Surface Methodology and Related Topics*, World Scientific Publishing Co. Pt. Ltd., pp. 427–446.

Huda, S. and Benkherouf, L. (2010). Rotatability is a sufficient condition for A- and D-rotatability, *Commun. Statist.–Simul. Comp.* **39**, pp. 1174–1182.

Huda, S., Benkherouf, L. and Alqallaf, F. (2007). On A- and D- Rotatability of two dimensional third-order designs, *Ali Feitschrift*, Muncie, Indiana: Ball State University, pp. 71–77.

Huda, S., Benkherouf, L. and Alqallaf, F. (2008). Necessary and sufficient conditions for D- rotatability of second-order designs, *Aligarh J. Statistics*, **28**, pp. 47–53.

Istok, J.D. and Rautman, C.A. (1996). Probabilistic assessment of ground-water contamination: 2. Results of case study, *Groundwater*, **34(6)**, pp. 1050-1064.

Jain, A.K., Visaria, P. (1988). Infant mortality in India: An overview. In Jain A.K. and Visaria, P. (Eds.) Infant Mortality in India: Differentials and Determinants, Sage Publications, India, New Delhi.

Jang, D.H. and Park, S.H. (1993). A measure and a graphical method for evaluating slope rotatability in response surface designs, *Commun. Statist. Theo. Metho.*, **22**, pp. 1849–1863.

Johnson, M.E. and Nachtsheim, C.J. (1983). Some guidelines for constructing exact D-optimal designs on convex design space, *Technometrics*, **25**, pp. 271–277.

Johnston, J. (1984). *Econometric Methods*, McGraw-Hill Book Co., Singapore.

Kackar, R.N. (1985). Off-line quality control, parameter design, and the Taguchi method (with discussion), *J. Quality Technology*, **17**, pp. 176-209.

Kao, J.H.K. (1959). A graphical estimation of mixed Weibull parameters in life testing of electron tubes, *Technometrics*, **1**, pp. 389–407.

Karlin, S. and Studden, W.J. (1966). Optimal experimental designs, *Ann. Math. Statist.*, **37**, pp. 783–815.

Katzmarzyk P.T., Srinivasan S.R., Chen, W., Malina, R.M., Bouchard, C., Berenson, G.S. (2004). Body mass index, waist circumference, and clustering of cardiovascular disease risk factors in a biracial sample of children and adolescents, *Pediatrics*, **114**, pp. 198–205.

Kenward, M.G. and Roger, J.H. (1997). Small sample inference for fixed effects from restricted maximum likelihood. *Biometrics*, **53**, pp. 983–997.

Khan, M.E. (1993). Cultural determinants of infant mortality in India, *J. Fam. Welf.*, **39(2)**, pp. 3–13.

Khuri, A.I. (1988). A measure of rotatability for response surface designs, *Technometrics*, **30**, pp. 95–104.

Khuri, A.I., and Cornell, J.A. (1996). *Responses Surfaces*, Marcel Dekker, New York.

Khuri, A.I. (2006). *Response Surface Methodology and Related Topics*, (Ed. Khuri, A.I.), World Scientific Publishing Co. Pt. Ltd.

Kiefer, J. (1958). On the non-randomized optimality and the randomized optimality of symmetrical designs, *Ann. Math. Statist.*, **29**, pp. 675–699.

Kiefer, J. (1959). Optimum experimental designs, *J. R. Statist. Soc.*, **B21**, pp. 272–319.

Kiefer, J. (1960). Optimum experimental designs V, with applications to systematic and rotatable designs, Fourth Berkeley Symposium, Math. Statist. and Probability **1**, pp. 381–405.

Kiefer, J. (1961). Optimum designs in regression problems, II, *Ann. Math. Statist.*, **32**, pp. 298–312.

Kiefer, J. (1962a). Two more criteria equivalent to D-optimality of designs, *Ann. Math. Statist.*, **33**, pp. 792–796.

Kiefer, J. (1962b). An extremum results, *Canad. J. Math.*, **14**, pp. 597–601.

Kiefer, J. (1974). General equivalence theory for optimum designs (approximate theory), *Ann. Statist.*, **2**, pp. 849–463.

Kiefer, J. (1975). Optimal design: variation in structure and performance under change of criterion, *Biometrika*, **62**, pp. 277–288.

Kiefer, J. and Wolfowitz, J. (1959). Optimum designs in regression problems, *Ann. Math. Statist.*, **30**, pp. 271–294.

Kiefer, J. and Wolfowitz, J. (1960). The equivalence of two extremum problems, *Canad. J. Math.*, **12**, pp. 363–366.

Kiefer, J. and Wynn, H.P. (1981). Optimum balanced block and latin square designs for correlated observations, *Ann. Statist*, **9**, pp. 737–757.

Kiefer, J. and Wynn, H.P. (1983). *Autocorrelation-robust design of experiments Scientific Inference and Data Analysis and Robustness*, Academic, New York.

Kiefer, J. and Wynn, H.P. (1984). Optimum and minimax exact treatment designs for one-dimensional autoregressive error processes, *Ann. Statist.*, **12**, pp. 431–449.

Kim, J., Das, R.N., Sengupta, A., and Paul, J. (2009). Regression Analysis for Correlated Data under Compound Symmetry Structure, *J. Statistical Theo. and Appli.*, **8(3)**, pp. 269–282.

Kim, K. and Lin, D.K.J. (1998). Dual response surface optimization: A fuzzy modeling approach, *J. Quality Technology*, **30**, pp. 1–10.

Kmenta, J. (1971). *Elements of Econometrics*, Macmillan, New York.

Kotz, S. and Adams, Jhon W. (1964). Distribution of sum of identically distributed exponentially correlated gamma-variables, *Ann. Math. Statist.*, **35**, pp. 277–289.

Kundu, D., Gupta, R.D. and Manglick, A. (2005). Discriminating between the log-normal and the generalized exponential distributions, *J. Statist. Planning Inference*, **127**, pp. 213–227.

Kunert, J. (1985). Optimal repeated measurements designs for correlated observations and analysis by weighted least squares, *Biometrika*, **72**, pp. 375–389.

Kunert, J. (1988). Considerations on optimal design for correlations in the plane. In Dodge, Y., Fedorov, V.V. and Wynn, H.P. (Eds.), *Optimal Design and Analysis of Experiments*, pp. 115–122, North-Holland, Amsterdam.

Lawless, J.F. (1982). *Statistical Models and Methods for Lifetime Data*, John Wiley and Sons, New York.

Lee, Y. and Nelder, J. A. (1998). Generalized linear models for the analysis of quality improvement experiments, *Canad. J. Statist.*, **26**, pp. 95–105.

Lee, Y. and Nelder, J.A. (2003). Robust Design via Generalized Linear Models, *J. Quality Technology*, **35**, pp. 2–12.

Lee, Y., Nelder, J. A. and Pawitan, Y. (2006). *Generalized Linear Models with Random Effects (Unified Analysis via H-likelihood)*, Chapman & Hall, London.

Leon, R.V., Shoemaker, A.C. and Kacker, R.N. (1987). Performance measures independent of adjustment an explanation and extension of Taguchi's signal-to-noise ratios, *Technometrics*, **29 (3)**, pp. 253-265.

Lesperance, M.L. and Park, S. (2003). GLMs for the analysis of robust designs with dynamic characteristics, *J. Quality Technology*, **35**, pp. 253–263.

Lieblein, J. and Zelen, M. (1956). Statistical investigation of the fatigue life of deep groove ball bearings, *J. Res. Nat. Bur. Stand.*, **57**, pp. 273–316.

Lin, D.K.J. and Tu, W. (1995). Dual response surface optimization, *J. Quality Technology*, **27**, pp. 34–39.

Lindsey, J.K. (1993). *Models for Repeated Measurements*, Oxford Clarendon Press.

Luner, J.J. (1994). Achieving continuous improvement with the dual response approach: A demonstration of the Roman catapult, *Quality Engineering*, **6**, pp. 691–705.

Martin, R.J. (1982). Some aspects of experimental design and analysis when errors are correlated, *Biometrika*, **69**, pp. 597–612.

Martin, R.J. (1986). On the design of experiments under spatial correlation, *Biometrika*, **73**, pp. 247–277.

Martin, R.J. (1996). Spatial experimental design. In Hand book of Statistics (S. Ghosh and C.R. Rao, Eds.), **13**, 477–514, North-Holland, Amsterdam.

Martin, R.J. and Eccleston, J.A. (1993). Incomplete block designs with spatial layouts when observations are dependent, *J. Statist. Planning Inference*, **35**, pp. 77–92.

McCullagh, P. and Nelder, J.A. (1989). *Generalized Linear Models*, Chapman & Hall, London.

Mead, R. (1994). *The Design of Experiments*, Cambridge University Press, London.

Measham, A.R., Jamison, D.T., Wang, J., Rao, K.D. and Singh A. (1999). The performance of India and Indian states in reducing infant mortality and fertility, 1975-1990, *Econ. Pol. Wkly.*, **34(22)**, pp. 1359–1367.

Medha, K.A. and Patil, C.L. (2008). Determination of heavy metals and physicochemical parameters in some groundwater samples in Dombivli region (MS), *J. Chemtracks.*, **10(1-2)**, pp. 345-350.

Meeker, W.O. and Escobar, L.A. (1998). *Statistical Methods for Reliability Data*, John Wiley and Sons, New York.

Mood, A.M. (1946). On Hotelling's weighing problem, *Ann. Math. Statist.*, **17**, pp. 432–446.

Mood, R. and Pike, D.J. (1975). A review of response surface methodology from a biometric viewpoint, *Biometrics*, **31**, pp. 803–822.

Morgan, J.P. (1990). Some series constructions for two-dimensional neighbor designs, *J. Statist. Planning Inference*, **24**, pp. 37–54.

Morgan, J.P. and Uddin, N. (1991). Two-dimensional designs for correlated errors, *Ann. Statist.*, **19**, pp. 2160–2182.

Morgan, J.P. and Uddin, N. (1998). A class of neighbor balanced block designs and their efficiencies for spatially correlated errors, *Statistics*, **30**, pp. 1–14.

Morrison, D.F. (1976). *Multivariate Statistical Method*, McGraw Hill, New Delhi.

Mukherjee, B.N. (1981). A simple approach to testing of hypotheses regarding a class of covariance structures, Proceeding of the Indian Statistical Institute Golden Jubilee International Conference On Statistics: Applications and New Directions, Kolkata, 16 Dec.- 19 Dec., 1981, pp. 442–465.

Mukherjee, R. and Huda, S. (1985). Minimax second and third order designs to estimate the slope of a response surface, *Biometrika*, **72**, pp. 173–178.

Mukhopadhyay, A.C. (1969). Operability region and optimum rotatable designs, *Sankhya*, **B31**, pp. 75–84.

Mukhopadhyay, A.C., Bagchi, S.B. and Das, R.N. (2002). Improvement of quality of a system using regression designs, *Cal. Statist. Assoc. Bull.*, **53**, pp. 225–233.

Mumtazuddin, S., Azad, A.K., Kumar, M. and Gautam, A.K. (2009). Determination of Physico-chemical parameters in some ground water samples at Muzaffarpur Town, *Asian J. Chemical and Environ. Research*, **2**, pp. 18–20.

Murty, V.N. and Studden, W.J. (1972). Optimal designs for estimating the slope of a polynomial regression, *J. Am. Statist. Assoc.*, **67**, pp. 869–873.

Myers, R.H. and Carter, W.H. (1973). Response surface techniques for dual response systems, *Technometrics*, **15**, pp. 301–317.

Myers, R.H., Khuri, A.I. and Carter, Jr., W.H. (1989). Response surface methodology: 1966-1988, *Technometrics*, **31**, pp. 137–157.

Myers, R.H. and Lahoda, S.J. (1975). A generalization of the response surface mean square error criterion with a specific application to the slope, *Technometrics*, **17**, pp. 481–486.

Myers, R.H. and Montgomery, D.C. (1995). *Response Surface Methodology*, John Wiley, New York.

Myers, R.H., Montgomery, D.C., and Vining, G.G. (2002). *Generalized Linear Models with Applications in Engineering and the Sciences*, John Wiley and Sons, New York.

Myers, R.H., Montgomery, D.C., Vining, G.G., Borrow, C.M. and Kowalski, S.M. (2004). Response Surface Methodology: A retrospective and literature survey, *J. Quality Technology*, **36**, pp. 53–77.

Nair, V.N., Abraham, B., Mackay, J., Box, G., Kacker, R., Lorenzen, T., Lucas, J., Myers, R.H., Vining, G., Nelder, J.A., Phadke, M., Sackes, J., Welch, W., Shoemaker, A., Tsui, K., Taguchi, S., and Wu, C.F.J. (1992). Taguchi's parameter design: A panel discussion, *Technometrics*, **34**, pp. 127–161.

Nelder, J.A. (1965a). The analysis of randomized experiments with orthogonal block structure. I. Block structure and the null analysis of variance, *Proc. R. Soc. Lond. Ser.*, **A283**, pp. 147–162.

Nelder, J.A. (1965b). The analysis of randomized experiments with orthogonal block structure. II. Treatment structure and the general analysis of variance, *Proc. R. Soc. Lond. Ser.*, **A283**, 163–178.

Nelder, J.A. (1994). The statistics of linear models: back to basics, *Statistics and Computing*, **4**, pp. 221–234.

Nelder, J.A. and Lee, Y. (1991). Generalized linear models for the analysis of Taguchi-type experiments, *Applied Stochastic Models and Data Analysis*, **7**, pp. 107–120.

Nelson, W.B. (1972). Graphical analysis of accelerated life test data with the inverse power law model, *IEEE Trans. Reliab.*, **R21**, pp. 2–11.

Nigam, A.K. and Das, M.N. (1966). On a method of construction of rotatable designs with smaller number of points controlling the number of levels, *Cal. Statist. Assoc. Bull.*, **15**, pp. 153–174.

Obesity (2004). Dial for Health India Ltd: P-1. http://www.dialforhealth.net/diet/obesity.asp (on access date, 16-08-2006).

O' Hagan, A. (1975). A general structure for Bayes inference about variance and covariances. Department of Statistics, University of Warwick, Conventry, Warwickshire, England, Mimeographed Report.

Ott, L. and Mendenhall, W. (1972). Designs for estimating the slope of a second order linear model, *Technometrics*, **14**, pp. 341–353.

Palta, M. (2003). *Quantitative Methods in Population Health: Extensions of Ordinary Regression*, John Wiley and Sons, New York.

Panda, R.N. and Das, R.N. (1994). First order rotatable designs with correlated errors, *Cal. Statist. Assoc. Bull.*, **44**, pp. 83–101.

Park, C. (1999). On the estimation in regression models with multiplicative errors, *J. Korean Data Infor. Science Soc.*, **10**, pp. 193–198.

Park, K. (2005). *Preventive and Social Medicine*, 18th Eds., Jabalpur; Banarsidas Bhanot.

Park, S.H. (1987). A Class of Multifactor Designs for Estimating the Slope of Response Surfaces, *Technometrics*, **29**, pp. 449-453.

Park, S.H. (1996). *Robust Design and Analysis for Quality Engineering*, Chapman and Hall, London.

Park, S.H. (2006). Concepts of slope-rotatability for second order response surface designs. In Response Surface Methodology and Related Topics, (Eds. Khuri, A.I.), World Scientific Publishing Co. Pt. Ltd., pp. 409–426.

Park, S.H., Jung S.H. and Das, R.N. (2009). Slope rotatability of second order response surface regression models with correlated error, *J. Quality Technology and Quality Management*, **6(4)**, pp. 471–493.

Park, S.H. and Kim, J.H. (1992). A measure of slope-rotatability for second order response surface designs, *J. Applied Statistics*, **19**, pp. 391-404.

Park, S.H. and Kwon, H.T (1998). Slope-rotatable designs with equal maximum directional variance for second order response surface models, *Commun. Statist. Theo. Metho*, **27**, pp. 2837–2851.

Park, S.H., Lim, J.H. and Baba, Y. (1993). A measure of rotatability for second order response surface designs, *Ann. Inst. Statist. Math.*, **45**, pp. 655–664.

Patil, C.L. and Deore, H.O. (2004). Physico-chemical analysis of water in Dhule region (MS), *Environ. Media Karad Poll. Res.*, **23(2)**, pp. 395–397.

Pawitan, Y. (2001). *In All Likelihood: Statistical Modelling and Inference Using Likelihood*, Oxford Science Publications.

Pearce, S.C. (1960). Supplemented balance, *Biometrika*, **47**, pp. 263–271.

Peto, R. and Lee, P. (1973). Weibull distributions for continuous carcinogenesis experiments, *Biometrics*, **29**, pp. 457–470.

Philip, E. (1985). Why infant mortality is low in Kerala. *Indian J. Ped.*, **52(48)**, pp. 439–43.

Pike, M.C. (1966). A method of analysis of a certain class of experiments in carcinogenesis, *Biometrics*, **22**, pp. 142–161.

Placket, R.L. and Burman, J.P. (1946). The design of optimum multifactorial experiments, *Biometrika*, **33**, pp. 305–325.

Poluride, G. (2002). Impact of obesity on glucose and lipid profiles in adolescents at different age groups in relation to adulthood, *BMC Fam Pract.*, **3**, pp. 18–21.

Press, S.J. (1972). *Applied Multivariate Analysis*, Holt. Rinehart and Winston, New York.

Puffer, R.R. (1985). *Mortality in Infancy and Childhood in India*, USAID Report, India, New Delhi.

Pukelsheim, F. (1993). *Optimal Design of Experiments*, John Wiley and Sons, New York.

Qu, Y., Tan, M. and Rybicki, L. (2000). A unified approach to estimating association measures via a joint generalized linear model for paired binary data, *Commun. Statist. Theo. Metho.*, **29**, pp. 143–156.

Raghavarao, D. (1963). Construction of second order rotatable designs though incomplete block designs, *J. Indian Statist. Assoc.*, **1**, pp. 221–225.

Rao, C.R. (1971). Unified theory of linear estimation, *Sankhya*, **A33**, pp. 371–394.

Rao, C.R. (1973). *Linear Statistical Inference and its Applications*, 2nd Eds., John Wiley and Sons, New York.

Rees, D.H. (1967). Some designs for use in serology, *Biometrics*, **23**, pp. 779–791.

Rogers, G.S. and Young, D.L. (1977). Explicit maximum likelihood estimators for certain patterned covariance matrices, *Commun. Statist. Theo. Metho.*, **6**, pp. 121–133.

Roy, J. (1954). On some tests of significance in samples from bipolar normal distributions, *Sankhya*, **A14**, pp. 203–210.

Roy, S.N. (1953). On a heuristic method of test construction and its use in multivariate analysis, *Ann. Math. Stat.*, **24**, 220–238.

Roy, S.N. and Bose, R.C. (1953). Simultaneous confidence interval estimation, *Ann. Math. Stat.*, **24**, pp. 513–536.

Rubin, D. and Szatrowski, T. (1982). Finding maximum likelihood estimates of patterned covariance matrices by the (EM) algorithm, *Biometrika*, **69(3)**, pp. 657–660.

Sacks, J. and Ylvisaker, D. (1966). Design for regression problems with correlated errors I, *Ann. Math. Statist.*, **37**, pp. 66–89.

Sacks, J. and Ylvisaker, D. (1968). Design for regression problems with correlated errors II, *Ann. Math. Statist.*, **39**, pp. 49–69.

Sacks, J. and Ylvisaker, D. (1970). Design for regression problems with correlated errors III, *Ann. Math. Statist.*, **41**, pp. 2057–2074.

Sakamoto, Y., Ishiguro, M. and Kitagawa, G. (1986). *Akaike Information Criterion Statistics*, Scientific Publisher, Japan, Tokyo.

Sandell, J., Upadhya A.K. and Mehrotra, S.K. (1985). A study of infant mortality rate in selected groups of population in district Gorakhpur, *Indian J. Public Health*, **29(1)**, pp. 37–42.

Satpati, S.K., Parsad, R. and Gupta, V.K. (2007). Efficient block designs for dependent observations-A computer-aided search, *Commun. Statist. Theo. Metho.*, **36**, pp. 1187–1223.

Scheffe, H. (1953). A method of judging all contrasts in the analysis of variance, *Biometrika*, **40**, pp. 87–104.

Scheffe, H. (1959). *The Analysis of Variance*, John Wiley and Sons, New York.

Seber, G.A.F. (1984). *Multivariate Observations*, John Wiley and Sons, New York.

Silvey, S.D. (1978). Optimal design measures with singular information matrices, *Biometrika*, **65**, pp. 553–562.

Silvey, S.D. (1980). *Optimal Design*, Chapman and Hall, London.

Shwarz, G. (1978). Estimating the dimension of a model, *Ann. Stat.*, **6**, pp. 461-464.

Smith, K. (1918). On the standard deviations of adjusted and interpolated values of an observed polynomial function and its constants and the guidance they give towards a proper choice of the distribution of observations, *Biometrika*, **12**, pp. 1–85.

Srinivas, C.H., Ravi Shankar, P., Venkateshwar, C., Satyanarayan Rao, M.S. and Reddy, R. R. (2000). Studies of ground water quality of Hyderabad, *Poll. Res.*, **19(2)**, pp. 285-289.

St. John, R.C. and Draper, N.R. (1975). D-optimality for regression designs: A review, *Technometrics*, **17**, pp. 15–23.

Subbarao, C., Subbarao, N.V. and Chandu, S.N. (1996). Characterization of groundwater contamination using factor analysis, *Environ. Geol.*, **28(4)**, pp. 175-180.

Szatrowski, T. (1978). Explicit solutions, one iteration convergence and averaging in the multivariate normal estimation problem for patterned means and covariances, *Ann. Inst. Stat. Math.*, **30**, pp. 81–88.

Taguchi, G. (1985). *Introduction to quality engineering: designing quality into products and processes*, Tech. Report, Asian Productivity Organization, Tokyo.

Taguchi, G. (1986). *Introduction to Quality Engineering*, Asian Productivity Organization, UNIPUB, White Plains, NY.

Taguchi, G. (1987). *System of Experimental Design: Engineering Methods to Optimize Quality and Minimize Cost*, UNIPUB/Kraus International, White Plains, NY.

Taguchi, G. and Wu, Y. (1985). *Introduction to Off-line Quality Control*, Central Japan Quality Control Association, Nagoya, Japan.

Tang, L.C. and Xu, K. (2002). A unified approach for dual response surface optimization, *J. Quality Technology*, **34**, pp. 437–447.

Tilak, J.B.G. (1991). Socioeconomic correlates of infant mortality in India, Washington, DC: The World Bank, (Population, Health, and Nutrition Division, Population and Human Resources Dept.).

Tukey, J.W. (1952). Allowances of various types of errors rates, unpublished invited address, Blacksburg meeting of the Institute of Mathematical Statistics, March, 1952.

Uddin, N. (1997). Row-column designs for comparing two treatments in the presence of correlated errors, *J. Statist. Research*, **31**, pp. 75–81.

Uddin, N. (2000). Optimal cylindrical block designs for correlated observations, *J. Statist. Planning Inference*, **84**, pp. 323–336.

Uddin, N. and Morgan, J.P. (1997a). Universally optimal designs with block size $p \times 2$ and correlated observations, *Ann. Statist.*, **25**, pp. 1189–1207.

Uddin, N. and Morgan, J.P. (1997b). Efficient block designs for setting with spatially correlated errors, *Biometrika*, **84**, pp. 443–454.

Vataw, D.F. (1948). Testing compound symmetry in a normal multivariate population , *Ann. Math. Stat.*, **19**, pp. 447–473.

Victorbabu, B.Re. (2005). Modified slope rotatable central composite designs, *J. Korean Statist. Soc.*, **34**, pp. 153–160.

Victorbabu, B.Re. (2006). Modified second order slope rotatable designs using balanced incomplete block designs, *J. Korean Statist. Soc.*, **35**, pp. 179–192.

Victorbabu, B.Re. (2009). Modified second-order slope rotatable designs with equi-spaced levels, *J. Korean Statist. Soc.*, **38**, pp. 59–63.

Victorbabu, B.Re. and Vasundharadevi, V. (2005). Modified second order response surface designs using balanced incomplete block designs, *Sri Lankan Journal of Applied Statistics*, **6**, pp. 1–11.

Vining, G.G. and Myers, R.H. (1990). Combining Taguchi and response surface philosophies: A dual response surface, *J. Quality Technology*, **22**, pp. 38–45.

Visaria L. (1985). Infant mortality in India: Level, trends, and determinants, *Econ. Pol. Wkly.*, **20(32)**, pp. 352–450.

Vuchkov, I.N. and Boyadjieva, L.N.(1983). The robustness of experimental designs against errors in the factor levels. *J. Statist. Comput. Simul.*, **17**, pp. 31–41.

Wald, A. (1943). On the efficient design of statistical investigations, *Ann. Math. Statist.*, **14**, pp. 134–140.

Wallis, W.D., Street, A.P. and Wallis, J.S.(1972). *Combinatorics, Room squares, Sum free sets, Hadamard matrices*, Springer-Verlag, New York.

Wang, Z. and Bessler, D.A. (2009). Finite sample performance of the model selection approach in co-integration analysis, *J. Stat. Comput. Simul.*, **79**, pp. 349-360.

Watkins, D., Bergman, A. and Horton, R. (1994). Optimization of tool life on the shop floor using design of experiments, *Quality Engineering*, **6**, pp. 609-620.

Welch, W.J., Yu, T.K., Kang, S.M. and Sacks, J. (1990). Computer Experiments for Quality Control by Parameter Designs, *J. Quality Technology*, **22**, pp. 15–22.

WHO (1971). The guidelines for drinking water quality recommendations, international standards for drinking water, World Health Organization.

Wiens, B.L. (1999). When Log-normal and Gamma Models give different results: A case study, *Amer. Statist.*, **53**, pp. 89–93.

Wolfinger, R.D. and Tobias, R.D. (1998). Joint estimation of location, dispersion, and random effects in robust design, *Technometrics*, **40**, pp. 62–71.

Wu, C.F.J. and Hamada, M. (2000). *Experiments and Planning Analysis, and Parameter Optimization*, John Wiley and Sons, New York.

Yates, F. (1965). A fresh look at the basic principles of the design and analysis of experiments. In: L.M. LeCam and J. Neyman (Eds.), *Proc. 5th Berkeley Symp. Math. Statist. and Probability, Vol. 4*, University of California Press, Berkeley, pp. 777–790.

Ying, L.H., Pukelshiem, F., and Draper, N.R. (1995a). Slope rotatbility over all directions designs, *J. Applied Statistics*, **22**, pp. 331–341.

Ying, L.H., Pukelshiem, F., and Draper, N.R. (1995b). Slope rotatbility over all directions designs for $k \geq 4$, *J. Applied Statistics*, **22**, pp. 343–354.

Zdachariah, K. and Patel, S. (1982). Trends and determinants of infant and child mortality in Kerala. Washington, DC: The World Bank. (Population and Human Resources Division. Paper No. 82–2).

Index

normed moment matrix, 92, 97, 115, 119–121, 123
null analysis, 185
null analysis of variance, 185
null hypothesis, 188
null treatments, 185

odd moments, 72, 126, 131
operating conditions, 12
optimal design point, 51
optimal experimental designs, 6
optimal factor, 4
optimality, 1
optimality of regression designs, 5
optimum response, 78
ORFOD, 16, 18, 19, 272
ORSORD, 51
orthogonal array, 4, 225, 234
orthogonal blocking, 90
orthogonal designs, 5
orthogonal structure, 10, 22, 24, 29
outer array, 225
outliers in the resulting data, 7
outward-opening, 207
over all directions, 3, 11, 109, 124, 125, 143
overall design, 58
overhead cost, 101

parabola, 51
partial derivatives, 110
per unit of cost efficiency, 94
performance measure, 12
performance variable, 12
performance variation, 225
perinatal care, 241
permutation invariant, 131
permutation of design points, 90, 91, 94, 102
physico-chemical parameters, 251
polar transformation, 5
positive definite, 2, 18, 51, 157
power transformation, 208
precision matrix, 126
predicted additive errors, 201
predicted mean, 229, 232
predicted value, 4
prediction variance, 4
principal diagonal, 30
principles of experimental design, 183
process parameters, 77
proportional hazards, 79
proximate factors, 241

quantitative factors, 1
quantitative measure, 5
Quasi-persymmetric structure, 40, 272

random effects, 79, 234
random ellipsoid, 162, 170
random interval, 171
randomization, 183
randomized block, 158
randomized block design, 12
reciprocal misspecification, 199
region of experimentation, 4
regression coefficients, 162
regression designs, 2, 6
regression parameters, 3
regular pattern, 20, 52
reinforced, 183
reinforced incomplete block design, 183
reinforced randomized block design, 12
reinforced RBD, 183, 184
reliability, 78
reliability theory, 78
repeated measures, 7, 88
replicated measures, 207
resistant to errors, 8
resistivity, 3, 12, 213, 232
response function, 3
response surface, 2
response surface designs, 1
response surface function, 51
response surface methodology, 1, 4, 124, 156, 271
response variable, 1
restricted MLE, 188
RFORD, 9, 16, 18, 22, 35, 82, 94, 271
right-tailed distribution, 7, 142
robust balanced design, 131
robust first-order rotatability, 15
robust first-order rotatable design, 94
robust optimality, 46
robust parameter design, 225
robust rotatability, 90
robust rotatable designs, 11
robust second-order slope rotatable design, 112
robust second-order slope-rotatability, 11
robust slope-rotatability, 109, 110, 146
robust slope-rotatable, 11
robustness, 7
robustness to correlation parameter, 7
robustness to errors, 7